BIOINOCULANTS FOR INTEGRATED PLANT GROWTH

BIOINOCULANTS FOR INTEGRATED PLANT GROWTH

Dr H C Lakshman
Professor
P.G. Department of Botany
Karnatak University
Dharwad-580 003 India

MD PUBLICATIONS PVT LTD
NEW DELHI
www.mdppl.com

"Views represented in the book are personal views of the authors and have no association with the organisations with which the authors are working with."

Published by :

MD Publications Pvt Ltd
"MD House", 11, Darya Ganj
New Delhi-110 002
Phone : +91-11-41563325 (Customer Service)
+91-11-41562846 (Editorial)
E-mail (orders): order@mdppl.com
Website : www.mdppl.com

ISBN : 978-81-7533-253-9

Published and Printed by Mr. Pranav Gupta on behalf of **MD Publications Pvt Ltd** at Times Press, New Delhi.

PREFACE

In recent years the use of microbial inoculants, as a source of biofertilizers has become a hope for most of countries, as far as economical and environmental view points are concerned.

Here, biological origin should be referred to as Bioinoculants or microbial inoculants. They are involved microbiological process synthesizing complex compounds and their further release into the outer medium, to the close vicinity of plant roots, which are again taken up by plants. In developing countries like India, it can solve the problem of high cost of fertilizers and help in saving the economy of the country.

Phosphorus is the key element for plant growth, it involved in various processes such as cell division, and photosynthesis biological oxidation. Phosphorus application at recommended rates for many crops is needed for sustainable plant growth. Therefore, the AM fungi significantly affect the physiological process of their hosts, especially in the root absorption. The beneficial effect of mycorrhiza is of special importance for those plants having coarse and poorly branched root system, since the external hyphae can extend as much as 8cm away from the roots absorbing nutrients from much larger soil volume then the absorption zone surrounding a non-mycorrhizal root. This is of particular importance for the absorption of nutrients of low mobility in soil solution such as P, Zn and Cu. Majority of plant species depend upon mycorrhizal associates for adequate nutrient uptake; those plants lacking Arbuscular mycorrhizal fungi can be stunted with low growth. Arbuscular mycorrhiza fungi play a significant role in adverse soil conditions. Afforestation of these lands by suitable tree species inoculated with AMF will release the pressure on reserved forests for meeting the increasing demand for fuel, fodder and timber.

Nitrogen is another important major element for plant growth and development. The phenomenon of fixation of atmospheric nitrogen by biological means is known as 'biological nitrogen fixation'. Microorganisms which establish symbiotic relationships with different parts of plant and may develop (or may not) special structures as the site of nitrogen fixation. Root nodules developed by *Rhizobium* in leguminous plants and are the site of nitrogen. Cynobacteria such as *Lyngbya, Oscillotoria, Plectonoma,* and *Nostoc* where, N_2 - fixation process takes place in internally organized cells having reduced conditions. *Azotobactor chroococcum* and *bacillus megaterium* significantly increase the crop yield about 10-20 per cent. In rhizosphere, bacteria secrete growth substances and antibiotic secondary metabolites which contribute to seed germination and growth of seedlings.

Phosphate solubilizing bacteria (*Bacillus magaterium* and *Pseudomonas striata*) and their effect on *sorghum* with *A. brasilense* and *Pseudomonas striata.* They found a significant increase in root nitrogenase activity, dry matter and seed yields as compared to control.

This book comprises articles on AM fungi, *Rhizobium, Azotobactor, Azospirillum,* phosphate solubilizing bacteria, bioremediation and microbial biotechnology. I profusely thank all the contributors.

In preparation and compiling of this book, I recollect my research guide Prof. S. Raghevendra (Rtd.), who made me to learn a,b,c of research. Prof. M. Krishnappa, Applied Botany Kuvempu University, Shankaragatta and my colleague Dr. T.C. Taranath, who always inspired me, while I was writing this book. I cannot forget the co-operation and support of my better half Smt. Tulasa. B. Netrekar, my son Mr. L. Suharsh and daughter Miss L Suprita. I wish to record my gratitude to my research students Shweta J Sabannavar, Mahesh B Byatanal, Sandeepkumar K, Romana M Mirdhe and Pushpa K Kavatagi, who have helped me one or the other way in completing this book. I am very grateful to Mr Pranav Gupta of MD Publications Pvt Ltd, New Delhi, for publishing this book with patience, care and interest.

This book is edited to serve as a reference copy for Botanists, Plant scientists, Research scholars of agriculture, horticulture and forestry.

Dr HC Lakshman

CONTENTS

LIST OF CONTRIBUTORS

C Maya
Department of Botany Bangalore University
Jannanabharathi Campus
Bangalore-560 056, India

PK Kavatagi
Post Graduate Department of Studies in Botany (Microbiology Lab.)
Karnatak University
Dharwad- 580 003, India

S Agarwal
Plant Biochemistry and Mycorrhizal Laboratory
School of Studies in Botany
Jiwaji University
Gwalior – 474011 (M.P.)

SJ Sabannavar
Post Graduate Department and Research in Botany
(Microbiology Laboratory),
Karnataka University
Dharwad -580 003, India

MB Byatanal
Post Graduate Department and Research in Botany
(Microbiology Laboratory),
Karnataka University
Dharwad -580 003, India

M Jayashankara
Post Graduate Department of Microbiology
Mangalore University
Field March Kariyappa College Campus
Madikeri – 571201, India

K Muthuchelian
Centre for Biodiversity and Forest Studies
School of Energy, Environment and Natural Resources
Madurai Kamaraj University
Madurai – 625 021, India

RM Mirdhe
Post Graduate Department and Research in Botany
(Microbiology Laboratory),
Karnataka University
Dharwad -580 003, India

HC Lakshman
Post Graduate Department Studies in Botany
(Microbiology Lab)
Karnatak University
Dharwad-580 003, India

RF Inchal
Department of Botany S.K. and H.S.K.
Science Institute Vidyanagar
Hubli- 580031

M Deiveekasundaram
Department of Microbiology, Faculty of Agriculture
Annamalai University
Annamalai nagar - 608 002, Tamilnadu

RH Ratageri
P. G. Department of Studies in Botany (Microbiology Lab)
Government Science College
Chitradurga -577 501, India

PA Hosamani
Department of Botany
Bangurnagar Degree College
Dandeli-581 325, India

VS Bheemreddy
Department of Botany
JSS College
Vidyagiri
Dharwad – 580 004, India

N Mathur
Department of Biotechnology
Mahila P. G. Mahavidhyalaya
Jodhpur- 342 001

RK Jambotkar
J.S.S's Shri Manjunatheshwar P.U. Science College
Vidyagiri
Dharwad-580005, India

D Vyas
Lab of Microbial Technology & Plant Pathology
Department of Botany
Dr. H.S. Gour University
Sagar (M.P.)-470 003

SG Hiremath
Post Graduate Department and Research in Botany
Microbiology Laboratory, Karnataka University
Dharwad -580 003, India

TC Taranath
Post Graduate Department and Research in Botany
Environment Laboratory, Karnataka University
Dharwad -580 003, India

Sandeepkumar K
Post Graduate Department and Research in Botany
Microbiology Laboratory, Karnataka University
Dharwad -580 003, India

Geeta B Patil
Post Graduate Department and Research in Botany
Microbiology Laborator, Karnataka University
Dharwad -580 003, India

Girish
Department of Microbiology
Basaveshwar Science College
Bagalkot – 587 101, India.

MG Nadagouda
Department of Botany
A.S.M. College for Women, Gandhinagar,
Bellary-583 101, India

AK Venu
Department of Microbiology
Mangalore University, P. G. Centre
F. M. Kariyappa College
Madekeri, India

SB Gadi
Department of Botany
JSS College, Vidyagiri
Dharwad – 580 004, India

Avita K Marihal
Department of Agricultural Microbiology
University of Agricultural Sciences
Dharwad-580 005, India

GROWTH RESPONSE AND IMPROVEMENT OF PLANT GROWTH, YIELD AND BIOMASS PRODUCTION

1

AM Fungi
A Promising Bio Fertilizer for the Healthy Seedlings Stock of Forest Nursery

[1]*C Maya*, [2]*H C Lakshman*, [2]*Shweta J Sabannavar, and* [2]*Mahesh B Byatanal*

INTRODUCTION

Mycorrhizae are symbiotic associations of non pathogenic soil fungi with living cell of plant roots. Mycorrhizae form the primary absorptive organs of most vascular plants and have been described as "metabolic harmony" between two organisms (Anderson, 1988). This arbuscular mycorrhizal continued fungi geographically ubiquitous and occur over a broad ecological range. They are commonly associated with plants in agriculture, horticulture pastures and tropical forests. About 85% of vascular plants establish mutualistic relationship with AM fungi may have access to thousands of hosts.

1. Department of Botany Bangalore University, Jannanabharathi Campus, Bangalore-560 056, India.

2. Post Graduate Deptt. and Research in Botany (Microbiology Laboratory) Karnataka University, Dharwad – 580 003 dr.hclakshman@gamil.com

While in tropical rainforests usually have arbuscular mycorrhizae, the reasons for this contrast are worth considering. The tree flora in tropical rainforests is extremely diversified. It is common to find 100 different species of trees per hectare, and the total number of woody species may approach 400 ha^{-1} . As many as 208 species of trees have been recorded in 1000 m^2 and individual specimens of the same tree species may be 800 m apart. This diversity may not be reflected in the arbuscular mycorrhizal fungi associated with the roots of the trees (Douds and Milner, 1999).

Arbuscular mycorrhizae are as we mentioned in the introduction, by far the oldest established and commonest form of mycorrhiza. They have been detected in some of the oldest land plants *Rhynia* and *Asteroxylon*, they occur scattered throughout the subsequent fossil record, and are present in the roots of the great majority in living plants. The only woody family among the entirely non-mycorrhizal plants is the Proteaceae, which is characterized by finely branched roots and extensive root hairs.

Many of the rest are weeds, vigorous pioneer herbaceous annuals with highly opportunistic lifestyles. They germinate quickly and grow rapidly in disturbed soils, and in some cases flower and seed set in a phenomenally short period. This means that they have no time to sit around and wait for the local arbuscular mycorrhizal fungi to find and colonize their roots, indeed in some of the habitats occupied by weeds there may be little or no AMF inoculum so the plants have evolved systems of root hairs that enable them to get along without mycobionts (Raven, 1986). The endomycorrhizal relationship is extremely old, and it is sharply surprising that some plants may now, or after hundreds of millions years, be evolving alternate strategies. That this should happen mainly among the herbaceous annuals is even more understandable when we consider that they are the most recently evolved group of plants.

Fungus root or mycorrhizal association was known since (Frank, 1885). These symbiotic fungi, which transform plant roots into unique morphological entities called mycorrhizae undoubtedly, constitute one of nature's most ubiquitous, widespread, persistent and interesting examples of parasitism. Most plants of ornamental, agricultural, medicinal, horticultural and floricultural importance are mycorrhizal. Allen (1991) defined that mycorrhiza is a mutualistic symbiosis, between plant and fungus localized in a root in which energy moves primarily from plant to fungus and inorganic resources move from fungus to plant. Plants release photosynthetic carbon, which is utilized by the arbuscular mycorrhizae and are geographically ubiquitous and occur in varied soils. These fungi belong to zygomycotina and are commonly associated with cereals, pulses, fibre crops, horticultural, ornamental, medicinal, aromatic, trees, weeds, rhizoids and roots of bryophytes/pteridophytes etc. Arbuscular mycorrhizal fungal species are reported (Anonymous, 1997). A number of plants belonging to approximately 200 families are colonized by the arbuscular mycorrhizal fungi. All the arbuscular mycorrhizal fungi are morphologically and physiologically not the same. Although, these fungi have extremely wide host range, they are not specific in their effect on host plant species. Their effect can vary on cultivars within a single plant species and also on plants in different ecosystems, soil types and similar soils differing in their physico-chemical constituents.

The fundamental problem which the world faces today is the rapidly increasing pressure of population on the limited resources of the land. To meet the ever-increasing demand of expanding, population, agriculture production has been raised through the abundant use of inorganic fertilizers, adopting multicropping system and liberal application of chemical pesticides (fungicides, bactericides, etc.). Though the use of chemicals has increased the yield dramatically it has also resulted in rapid deterioration of land and water resources apart from wastage of scarce resources. This has adversely affected the biological balance and lead to the presence of toxic residues in food, soil, and water in addition to imposing economic constrains to the developing countries. India is among the top in fertilizer consuming countries in the world and its use is increasing rapidly. Use of agrochemicals results in an increase of the nitrates in water, accumulation of heavy metals, cadmium, and phosphates. Nutrient enrichment, eutrophication, and deterioration of surface water quality due to transportation of nutrients applied through fertilizers via leaching and/or run off and sediment erosion are other problems.

It is becoming increasing clear that the present agricultural production system has been responsible for degradation of earth's environment. For sustainable development, a clear priority has to be given to productive agriculture, essential to efficiently manage the agricultural inputs for sustaining high crop productivity on a long term basis with minimum damage to ecological and socioeconomic environment. To reduce cost of agrochemicals and harm rendered by biofertilizers are used. These systems minimise the use purchased inputs by substitution of farm-generated inputs like green manures or exploit biological systems such as symbiotic N_2 fixation to increase soil fertility and potentially through management of (EM and AMF) mycorrhizal fungi to improve the efficiency of P use. The mycorrhizal association appears to have evolved as survival mechanism for both the fungi and the higher plants, allowing each to survive in the existing environments of low temperatures, soil fertility, periodic drought, diseases, extreme temperatures and other natural stresses. Mycorrhiza appears to be the resist line of biological defense against stress.

CATEGORIZATION OF PLANTS INTO DEPENDENCY CLASSES

Mycorrhizal dependency of plants has been mostly related to soil fertility. According to Gerdemann (1974), mycorrhizal dependency is the degree to which a plant relies on the mycorrhizal condition to produce its maximum growth or yield at a given level of soil fertility. According to Menge, Johnson, and Platt (1978) defined mycorrhizal dependency as the dry mass of a mycorrhizal plant expressed as a percentage of dry mass of non-mycorrhizal plants at a given level of soil fertility. According to Menge (1984) defined mycorrhizal dependency as the dry mass of a mycorrhizal plant expressed as a percentage of dry mass of non-mycorrhizal plants at a given level of soil fertility. Plenchette, Fortin, and Furlan (1983) expressed their apprehension that this method of calculation may lead to exteremely high percentages. According to them, the highest level of mycorrhizal dependency should not exceed 100% and the lowest should be 0%. A new method was thus proposed by them to calculate mycorrhizal dependency:

$$\text{RFMD} = \frac{\text{Dry mass of mycorrhizal plant dry matter of non-mycorrhizal plant}}{\text{Dry mass of mycorrhizal plant}}$$

RFMD is relative field mycorrhizal dependency. It was proposed that RFMD have a superscript that represents the level (in mg/g) of the available phosphorus of the soil as the mycorrhizal dependency is linked with a given level of soil fertility. Since then, the method given by Plenchette, Fortin, and Furlan (1983) has been followed by various authors to determine mycorrhizal dependency.

Studies conducted at the University of Agadir, Faculty of Science, Department of Biology, Agadir, Morocco, on in-vitro micropropogated plantlets of *Argania spinosa*, an important agro-forestry tree in semiarid and arid zones, showed that mycorrhizal inoculation increased total shoot length, stem girth and biomass of plants when cultivated in controlled condition with a phosphorus fertilization of 31 mg phosphorus per kg added to the substrate already containing 13.5 mg/per kg available phosphorus as per Olsen's method. The positive effect increased with time and after 6 months, the

above ground dry matter plants. The relative mycorrhizal dependency index (RMDI) calculated on the total dry matter, the RMDI was 81–83% placing the organ tree among the plants most ratio was markedly (40–50%) reduced by inoculation, demonstrating the higher efficiency of a mycorrhizal root system (Nouiam and Chaussod, 1994).

Plants were grouped into three dependency classes: (a) highly dependent with 41% and above increased in growth (b) intermediate with 10–40% increase in growth, and (c) non-dependent with less than 10% increase in growth over uninoculated plants (Cruz, Zarate, Aggangan *et al.*, 1992).

The dry weight expressed as low phosphorus supply. In studies conducted at the Citrus Research and Education Centre, University of Florida, Lake Alfred, Florida, mycorrhizal dependency of the five citrus genotypes was calculated as follows:

$$\text{Mycorrhizal dependency} = \frac{\text{Dry weight of mycorrizal plants}}{\text{Dry weight of non-mycorrhizal plants}}$$

Habte and Manjunath (1991) suggested determining mycorrhizal dependency of plants, at critical soil solution phosphorus concentrations at which maximum growth increase is achieved by mycorrhizal plants over non-mycorrhizal plants. In studies conducted at the Department of Agronomy and Soil Science, University of Hawaii, Honolulu, Hawaii, *Brassica nigra* and selected species of *Leucacena* and *Sesbania* were grown in a greenhouse in the oxisol with different concentrations of established soil solution phosphorus in the presence or absence of *Glomus aggregatum*. The extent to which the plant species depended on the fungus for dry matter production diminished with increased concentrations of soil solution phosphorus, but the magnitude of this decrease varied from species to species. Five distinct mycorrhizal categories were proposed based on the differences observed ranging from non-dependent to very highly dependent species. The critical soil solution phosphorus concentrations that were useful for separating host species into distinct AM dependent groups were 0.02 and 0.2 mg/litre. Species differing in their mycorrhial dependency differed with respect to the soil solution phosphorus concentration required to express maximum AM infection, and the pattern of immobile nutrient accumulation (Habte and Manjunath 1991).

Research programme related to the role of endo-mycorrhizae in nurseries of hard wood tree species has been started at Banthara station at Lucknow involving eight nitrogen fixing tree species: *Acacia nilotica, A. nilotica,* var. *cupressiformis, A . auriculiformis, Cassia siamea, Leucaena leucocephala, Parkinsonia aculata, Prosopis juliflora* and *Tamarindus indica* and certian non-leguminous tree species like *Casuarina glauca, C. obessa* and *C. equisetifolia* (Dehn and Schuepp, 1989). Mycorrhizal association and dominant species have been identified and isolated to see the effect of these isolates on growth of nursery seedlings. Initial results suggested that these fungi were not only beneficial to growth of the seedlings in nurseries but promote their survival and growth in field.

There is a wide variation among and within different species of AM fungi in their ability for promoting growth of tree seedlings (Mohan *et al.,* 1995). There are a number of studies for preliminary screening of different VAM fungi for different tree species (Paul and Peter, 1993; Feldmann and Lieberei, 1992). A preliminary screening trial using 18 different AM fungi obtained from different parts of the world against 4 cultivars of *Leucaena,* showed *Gigaspora margarita, Glomus fasciculatum* and *G. monosporum* to the promising for *Leucaena* (Paul and Peter, 1993). In a subsequent study, seven AM fungi, three from the preliminary trial and other four local isolates were screened for symbiotic response with four Hawaiian giant cultivars (K8, K28, K67 and K72).

HARDWOODS

Four hardwood trees were inoculated with six AMF species (*Glomus mosseae, G. fasciulatum, G. etunicatum, G. macrocarpum, G. epigaeum,* and *Gigaspore margarita* grown for 16 weeks in a soil with low available phosphorus. Mycorrhizal dependency values were greatest for *Fraxinus pennsylvania,* followed by *Liquidambar styraciflua, Leriodendron tulipifera,* and *platanus occidentales,* and were significantly influenced by the fungal species. Inoculation with *G. macrocarpum* resulted in the highest mycorrhizal dependency values (Pope, Chaney, and Rhodes 1983).

RMD (Relative Mycorrhizal Dependency) and phosphorus concentrations in shoot tissues. *Alzelia africana, Landolphia heudelottii* and *Saba senegalensis* did not form symbiotic associations and the growth of *A. africana* decreased following mycorrhizal inoculation *Adansonia digitata, Aphania senegalensis, Anacardium occidentale, Balanites aegyptiaca,* and *Sclerocarya birrea* were well colonized with AM fungi but did not significantly increase biomass production. *Zizyphus mauritiana* was found to be very highly dependent (RMD 77.5%); *Tamarindus indica* was highly dependent (50%-75% RMD); and *Dialium guineensis, Parkia biglobosa* and *Cordyla pinnata* were moderately dependent (25%-50% RMD). Phosphorus absorption probably contributed to this dependency more than the absorption of potassium, *G. aggregatum* was most effective AM fungus (Ba, Plenchette, Danthu, *et. al.*, 2000).

With *Gigaspora margarita, Glomus etunicatum, Glomus macrocarpum* plants highly dependent to AMF were those that exhibited the best growth when compared to uninoculated plants (41% and above increase in growth). These were *Acacia mangium, A. auriculiformis, Paraserianthes falcataria, Albizia lebeckoides, Samanea saman corn* (*Zea maize*), Cacao (*Theobroma*

Table 1.1 : Showing the effect of minimum level of soil phosphorus for maximum mycorrhizal effects in plants

Plant species	Tested fungus	Minimum soil phosphorus level for maximum mycorrhizal effect	Stated dependency class	Reference
Acacia nilotica	mixed local VAM inoculum	20 ppm P	-	Sharma, Gaur, Bhatia et.al. (1995)
Albizia ferruginea	*Glomus fasciculatum* *G. aggregatum*	0.02 mg P/litre	High	Habte and Musoko (1994)
Capsicum annum	*Glomus fasciculatum*	0.03 ppm P	-	Waterer and Coltman (1987)
Cassia siamea	*Glomus fasciculatum* *G. aggregatum*	0.02 mg P/litre	-	Habte (1995)
Citrus limonia	*Glomus etunicatum*	0-50 ppm P	-	Antunes and Cardoso (1987)
Dendrocalamus strictus	*Unknown VAM sp.*	4 ppm P	-	Bhattacharya, Misra, Saha etal. (1999)
Enterolobium cyclocarpum	*Glomus fasciculatum* *G. aggregatum*	0.02 mg P / litre	High	Habte and Musoko (1994)
Eucalyptus tereticornis	Mixed local VAM inoculum	10 ppm P	-	Sharma, Gaur, Bhatia et al. (1999)
Eucalyptus urophylla	*Pisolithus tinctorius* *Scleroderma cepa*	8 mg P / kg		Aggangan, Dell, Malajczuk et al. 1996
Leucaena leucocephala	*Glomus fasciculatum*	0.21 mg P / ml		Habte and Manjunath (1987)
	G. aggregatum	0.02-0.2 mg P / ml	V. high	
Artocarpus heterophyllus	*G. mosseae*	6 mg P/kg	V. high	
Largerstemia lanceolatus	mixed local AMF	4 mg P / kg	high	
Pterocarpus marsupium	*G. mosseae*	0.5 mg P / kg	V. Low	

cacao), and langka (*Artocarpus heterophylla*) were not dependent on mycorrhizas as these showed no growth increases with mycorrhiza inoculation (Lakshman, 1996).

Some tree genera such as *Alnus, Eucalyptus, Casunna, Cupressus, Juniperus, Tilia* and *Arbutus* form ecto and endomycorrhizae another type if mycorrhizae, singly or together depending on soil condition and tree age. Harley (1989) suggested the name 'mycorrhizosphere' for the soil surrounding and influenced by mycorrhizae.

SOURCES OF VAM INOCULUM

The inoculation of experimental plants can be carried out with colonized soil, or colonized roots, or spores, or a mixture of all three. The first two kinds of inocula have some profound disadvantages. Soil may contain more than one mycorrhizal fungus, and it may also contain pathogenic organisms. Root inoculum can be used if it has been grown under more or less aseptic conditions and if the original inoculum used to infect the roots was of a single, named species. The features diagnostic of individual species are present only in the spores, which develop primarily on extrametrical hyphae, so these are perhaps the best inoculum for laboratory experiments.

Since the fungi cannot be grown in axenic culture, the only source of spores is the soil near the roots of infected plants. Agricultural soil has been found to contain varying numbers of AM fungal spores. Sutton and Barron (1972) recorded up to 70 ±17 spores g^{-1} dry soil under maize and up to 86 ±11 under strawberry, all in the uppermost 24 cm of soil. Bowen (1980), estimated that the upper 10 cm of soil in an undisturbed Acer-dominated hardwood forest in Michigan contained nearly 7000000 sporocarps ha^{-1} of AM species.

Although inoculum for experiments could be derived from naturally colonized soil, this kind of inoculum will usually contain several different AM fungi, and is also present at a fairly low level. It was discovered that individual AM fungi could be propogated in a dual culture system, fungus plus host plant, in what is called pot culture. Soil in which onions inoculated with *Gigaspora calospora* had been grown for 19 months contained 325 spores g^{-1} dry soil (Read, 1991).

Gerdemann devised the first useful technique for extracting spores from soil. A soil sample was suspended in four times its volume of water, heavier particles were allowed to settle for a few seconds, and then the liquid was decanted through a sieve with 1mm mesh. Whatever passed through this sieve was then poured through another sieve with 0.25 mm (250 um) mesh. Material retained by this sieve was washed and transferred to a petri dish and the spores picked out by hand under a dissecting microscope. This technique was slightly refined by Gerdemann who used the following series of sieves: 1mm, 710um, 420um, 250um, 149um, 105um, 74um and 44um. They found that most of the desired spores fill in the 420–149 um range, and they used this fraction for their study. Since it is recommended that from 50 to 500 spores are required for inoculation of a single pot, and a well-designed experiment might need scores of pots, an investigator was condemned to spend an excessive amount of time simply to collect enough spores for each experiment. The initial sieving could, however, be speeded up by nesting the set of sieves in a vertical series.

Ohms (1956), reported that Gerdemann's technique yielded only about 100 spores per operator per day, and tried to make spore recovery less laborious by centrifuging the fraction of soil on a sucrose gradient. He sieved the soil initially to obtain the fraction retained by a 177 um mesh, and then suspended this material in water. Lighter material was decanted after the spores and heavier soil particles had settled, or had been centrifuged for 3 min at 3100 r.p.m. (presumably in a clinical centrifuge). A suspension of the spores and the remaining debris was added to a 50 ml centrifuge tube containing a sucrose gradient established by placing 10 ml of 50 % sucrose at the bottom, 15 ml of 25% sucrose above it, and 10 ml of water at the top. The upper layers were added very carefully with a hypodermic syringe equipped with a curved extension that allowed the liquid to be placed directly on the side wall of the vertically held tube. After centrifuging for 5 min at 3100 r.p.m., the spores were largely to be found in the middle layer, while the soil particles had settled to the bottom layer. The spores in the middle layer were pipetted into a beaker, and the sucrose solution diluted with water to decrease the specific gravity of the solution and allow the spores to sink to the bottom. The solution was decanted and the spores were washed several times with water. They could then be readily hand-sorted. This method yielded over 1500 spores per man-day. One disadvantage of this technique was the osmotic pressure exerted by the sugar solution, which could apparently damage young spores.

To avoid this problem, Mosse and Jones (1975), separated the spores by differential sedimentation on gelatin columns. Gelatin was dissolved in distilled water to make solutions of 5%, 15% and 20% (w/v), which had specific gravities of 1.02, 1.04 and 1.05, respectively, at 32°C. Liquid 20% gelatin at 30° C was poured into glass tubes 30 cm long and with an internal diameter of 37 mm to a depth of about 5 cm and allowed to solidify at 4° C. A 5 cm layer of 15% gelatin was then added and allowed to solidify, then a similar layer of the 5% solution, and finally a 10 cm layer of water. Columns were stored at 4°C until used. For spore separation, the columns were placed upright in a water bath at 32°C and allowed to liquify. Then the fraction of a 50 g soil sample that passed through a 250 um sieve but was retained on a 100 um sieve was washed into the water layer at the top of a column. After 30 min of sedimentation, the columns were cooled to 4°C again. Spores were extracted from the columns as follows: The water layer was discarded. The column was then held under a hot water tap until the gelatin became free from the wall of the glass tube, and could be slid out of the tube. The core could now be cut into segments. It was found that vesicular arbuscular mycorrhizal spores tended to sediment near the 5–15% interface. A 1 cm segment of the column (0.5 cm on either side of the interface) was cut out, gently melted, then strained through a piece of finely woven material supported on a 100 um sieve. This gave good recovery of four different kinds of arbuscular mycorrhizal spores. Mosse and Jones claimed that their gelatin columns had the following advantages over the sucrose solutions used by Ohms (1956), (1) high viscosity and relatively high specific gravity; (2) low osmotic pressure; (3) solidification below 30°C, which makes it easy to set up columns, to maintain sharp boundaries between layers, even in very wide columns, and to segregate portions of the resolidified columns for recovery of the material they contain.

Sutton and Barron (1972), extracted spores from soil by a flotation of spores on water, the hydrophobic nature of their surface layer and their tendency to adhere to glass surfaces. A 10 g subsample of soil was placed in a 50 ml cylinder which was then filled with water within about 2 cm of the brim, and shaken for 10–15 s, the suspension was allowed to settle for about 2 min, while the vesicular arbuscular mycorrhizal fungal spores accumulated in the scum floating at the meniscus. Most of the aqueous layer was then decanted into a separatory funnel, carrying most of the spores with it. Further water was added to the cylinder and the shaking, sedimentation and decanting processes repeated twice more. The accumulated liquid in the separatory funnel was then allowed to settle for 2–3 min. this allowed the spores to collect around the margin of the meniscus in the funnel. The water was then drained at 75–100 ml min^{-1} from the funnel into a second 250 ml separatory funnel, where it was allowed to settle once more before being drained at a similar rate. The scum and spores adhering to the insides of the two funnels were then washed onto a filter paper. Preliminary experimentation showed that this technique 94–98% of the spores in any given soil sample, unfortunately including spores that are dead.

In an effort to simplify and speed up the process of collecting the thousands of spores needed for experimental purposes. Furlan and Fortin (1975) introduced a flotation-bubbling collection technique. The soil was first sieved to concentrate the spores, then material was added to a column containing a 50% aqueous solution of glycerol, which was found to have a higher specific gravity (1.13) than the spores without exerting any deleterious osmotic effects on them. Compressed air was injected at the bottom of the column through a fritted disc, so that the small bubbles would both agitate the suspension, helping to separate the spores from soil particles, and propel the spores toward the surface, where they could be recovered. Furlan and Fortin (1975) found that the 74–500 μm soil fraction was most appropriate for this technique. The apparatus consisted of a glass tube of 6 mm internal diameter and 55 cm long, held firmly in the vertical position, and with a fritted disc with pore size 4–5.5 um fused in position 5 cm above the bottom. Compressed air (monitored by a pressure gauge) was supplied to the bottom of the column at approximately 83 kPa. A litre of 50% glycerol solution was added to the tube, the air pressure raised to 103.4 kPa, and 10 g of soil fraction added. After the bubbling had continued for 1-2 min the air was turned off, the foam allowed to subside, soil particles adhering to the wall of the tube above the liquid were carefully washed down and the suspension was allowed to settle. After 30 min the supernatant was drawn off, the tip of the aspiration tube being held just below the surface of the liquid to pick up all the spores. All supernatant was aspirated to within 1 cm of the deposit at the bottom. The supernatant was now sieved and the spores were washed thoroughly in tap water and stored in Ringer's solution (adjusted to a pH of 7.4 with NaOH, filtered and autoclaved). This technique was found to give 94.5% recovery of spores. If virtually total recovery was required, a second treatment would increase the yield to 99.5%.

Smith and Skipper, (1979) used discontinuous sucrose gradients to recover large number of spores from massive soil samples. They decanted and wet-sieved 18 kg of soil with cold water and found that most spores were present in the 425–250 um fraction. Spores were separated from most of the remaining debris using discontinuous 30% (w/v) aqueous sucrose gradients. The sieved material was layered on 600 ml water over 200 ml sucrose in a 11 beaker. After settling, the spores and debris that collected at the interface were removed by vacuum aspiration, rinsed in cold water and centrifuged for 1–5 min at 1600 × g in a clinical centrifuge on a second gradient (15 ml water over 20 ml sucrose in a 50 ml tube), the duration of centrifugation being determined by the kind and amount of debris present. Fresh pot culture material needed 5 min of centrifugation to ensure complete separation of the finest root fragments. The spores were then aspirated from the gradient interface, rinsed and stored in cold water. In this manner, up to 10,000 spores seed^{-1} could be recovered by a single operator.

APPLICATION OF VAM INOCULUM

This section examines the problems related to the actual application of the inoculum, first in the laboratory, then in the greenhouse and finally in the field.

Laboratory Experiments

For complete precision in the establishment of aseptic dual cultures (host + fungus) the methods described are appropriate, and the inoculum usually takes the form of Spores. (1) *The paper substrate method*: A seed is germinated between a paper strip and a slide standing in a tube with its lower end immersed in a nutrient solution. Spores are then placed on the young roots, (2) The *Fahraeus slide* method: A seed is germinated between a slide and a coverslip held parallel to, but at a distance of 800 um from, the slide. The lower part of the slide is again immersed in nutrient solution in a tube, and spores are added to the young roots.

It is possible to infect a root with a single spore, and Hepper (1994) showed that inocula of 3 spores seed^{-1} and 250 spores seed^{-1} produced similar levels of infection in mature tomato plants. However, we would not suggest using very low spore numbers, since other workers have found that the level of inoculum used can affect the rate of infection. Higher levels of inoculum seem to shorten the lag phase in mycorrhiza development or hasten the achievement of the extensive infection desired, the plateau phase. When spores are used as inoculum, the number applied usually ranges from 50 to 500 per plant.

Greenhouse Experiments

Crops that are grown in the greenhouse, or that are usually transplanted to field situation, can be infected before planting out without involving significant extra investment of time in an already labour-intensive process. This is appropriate for all container-grown seedlings, and has particular relevance to many horticultural crops. Since relatively small volumes of soil are involved, it should not be difficult to incorporate a reasonable concentration of inoculum in this soil during its preparation. The chances for early and widespread infection are presumably increased as the number of spores or other infective propagules increases, up to a point. The speed of infection can presumably be increased by ensuring that the propagules are in close contact with the roots. Inoculum can be mixed with the soil, or placed below or to one side of the seed or the roots of the seedling, or even pipetted onto the roots. But at least one worker has suspended the spores in a viscous, non-toxic medium containing carboxymethylcellulose, and then dipped the roots of seedlings in this medium before planting.

Development of AMF fungi

Nicolson and Gerdemann (1968), Mukerji and Kapoor (1987) emphasized that AM though designate as endotropic, is composed of two phase mycelium system:- (i) an internal mycelium within the cortex of the mycorrhizal root, and (ii) an external mycelium in soil which varies considerably in extent of different hosts. There is no organized fungal growth on the root surface. The AM association does not cause alterations in the root morphology.

The vesicular-arbuscular mycorrhizae are the most complex group of mycorrhizae which forms interradical structures (i) intracellular hyphae forming coils, often found in the outer layers of cortical parenchyma; (ii) the intercellular hyphae; (iii) the intracellular hyphae with numerous ramifications i.e., the arbuscules; (iv) the inter or intracellular hypertrophied hyphae i.e., the vesicles.

The mycelial system surrounding the roots is dimorphic (Mosse 1973 and Nicolson, 1967): (i) with coarse thickwalled irregular non-septate hyphae, and (ii) smaller, thin walled ephemeral lateral branches. The thickwallled hyphae penetrate the host root and cause internal infection. At the entry point, the penetrating hyphae form appressoria in the host plants (Gianinazzi-Pearson *et al.*, 1991). The penetrating, infecting hyphae spread inter and intracellularly in the host root cortex. Characteristic H and Y connections are seen in the cortex. The highly branched structures called arbuscules are usually formed in the inner cortex. These are ephemeral structures formed by the repeated dichotomousely branching and complete their development in 4 or 5 days (Brundrett *et al.*, 1985). Arbuscules are the key sites for nutrient exchange and remain active only for 4 to 15 days (Carling and Brown, 1982; Cox and Tinker, 1976).

Many but not all endomycorrhizal fungi which forms arbuscules later also form terminal or intercalary vesicles in the root cortex. These are expanded, thin-walled structures which are not delimited by a septum but often contain a large quantity of lipids. They may be spherical, oval or lobed and may become thick walled and resemble resting spores. They serve as the endophytic storage organs and are rich in lipids.

Arbuscular Mycorrhizal Fungi in Nutrient Uptake Phosphorus

The relative benefit that a plant obtains may vary with different genera and with same species according to their relative mycorrhizal dependency (Plenchette *et al.*, 1983) have devised a scheme for calculating a relative filled mycorrhizal dependency index that can be useful in the importance of mycorrhizae to particular host plant.

EM enhances uptake of P, N, K, Zn etc. The fungal hyphae absorb PO_4 ions from the soil solution. They promote dissolution adjacent particles of mineral phosphate and transport to higher plants (Baltruschat and Dehne, 1988). The fungal sheath is the storage organ of PO_4 to the host in the absence of any external source. The increased levels of mineral nutrient supply and enhance the rate of C_5 - fixation (Grenville *et al.*, 1985) and vigour.

Vesicular arbuscular mycorrhizal also possess an active mechanism of transferring PO_4 to the plant throughout their relationship. Polyphosphate granules are seen in the fungal cytoplasm and in the vacuoles of inter and intracellular hyphae. The branches of arbuscules have been found to have acid and alkaline phosphatases (Gianinazzi *et al.*, 1979).

Carbon

Mycorrhizal fungi are entirely dependent on their host for carbon which they absorb with considerable avidity from direct photosynthetic products. Harley (1970) estimated carbon demand of these fungi to be 10% of potential wood production.

Harley (1969) and Harley and Smith (1983) have reported a decrease in dry weight and size of the young seedling due to competition between the partners of limited supply of carbon. But as more foliage appears, competition decreases and the deleterious effect of the fungus is replaced by the benefits it brings to the host nutrition.

Various studies (Finlay and Soderstrom, 1992; Peterson and Farquhar, 1994), indicate that as much as atleast 15% of the carbon assimilated by the host plant may be exported to fungal partner. This carbon flow should be regarded as an investment rather than as a loss.

The carbon flow via the mycorrhizal network to the soil allowed a more active saprophytic microflora to thrive in close conjugation with the mycelium, thereby increasing the rate of decomposition of organic matter (Entry *et al.*, 1991).

Nitrogen

Improvement of nitrogen status of the host by mycorrhizal infection one of the ecologically significant consequence (Abnziandah and Read, 1986a; Aubziandah *et al.*, 1986b). It is of particular importance in Scandinavian conifer ecosystem (Linder, 1989), where primary production appears to be restricted by nitrogen availability Nitrogen is not assimilated by most mycorrhizae but ammonium is easily assimilated by EM and transferred to plant partner as glutamine/glutamic acid. Certain amino acids such as arginine and citrulline accumulate in plant roots under N deficient conditions. This indicates a blockage of the "Omithine cycle" of amino acid synthesis, which is alleviated by mycorrhizal colonization. Fungi capable of reducing with the aid of nitrate reductase will also utilise NO_3 as source of nitrogen (Finlay *et al.*, 1989).

The ability of EM fungi to use organic nitrogen source is of great importance and further information is now required about the spatial and temporal dynamics of different organic nitrogen compounds in forest ecosystems. Additional of mycorrhizal fungi to utilize organic nitrogen and how these are related to different species at different successional stages of forest development or under different regimes of nitrogen fertilization.

Synergism between VAM and nitrogen-fixing bacteria has been reviewed (Barea and Azcon Aguilar, 1984; Hayman, 1986). VAM hyphae can translocate and assimilate ammonium and nitrogen nutrition of N fixing plants (Ames *et al.*, 1983; Barea *et al.*, 1987).

Nutrients other than Nitrogen and Phosphorus

Comparatively little work has been done with nutrients other than P and N. Finlay and Soderstrom (Finlay and Solderstrom, 1992) have shown that Rubidium (^{86}Rb) and to a lesser extent Calcium (^{45}Ca) has been translocated in the mycelium.

(1) The role of volatile substances of host roots and growth regulating substances of the fungus in recognition and colonisation under natural conditions.

(2) The regulatory and functional processes of the symbiosis like the actual transfer of nutrients between the two symbionts.

(3) Correlation of physio-biochemical parameters with drought resistance due to mycorrhizal infection.

(4) Employment of phosphorus tolerance strains for inoculating high yeilding varieties of field and horticultural crops demanding heavy doses of fertilizers.

Forestry

(1) The selection, propagation, manipulation and management of superior strains of EM fungi can contribute towards the success of reforestation

Table 1.2 : Effect of *Glomus mosseae* (Nicol. & Gerd) on growth, colonization and biomass production in six timber tree species.

Duration	Plant height (cm)	No. of nodes	Shoot diam (mm)	Shoot		No. of leaves	Leaf length (cm)	Leaf breadth (cm)	Leaf		Total root length (cm)	Root diam. (mm)		Root	Total dry wt. Shoot (g)	Root/ shoot ratio (g)	%p shoot	% Col onization	No. spores/ 100g. soil
				Fresh wt (g)	Dry wt				Fresh wt (g)	Dry wt		Fresh wt (g)	Dry wt						
										A. heterophyllus									
NM	8.51	3.53	2.04	1.0	0.72	3.41	4.62	2.50	0.89	0.58	11.31	0.58	0.69	0.30	1.30	0.23	0.05	-	-
3M	17.52	7.32	2.30	5.2	2.80	3.82	9.70	4.55	1.30	0.92	10.51	1.90	0.88	0.52	3.72	0.14	0.10	59.2	513
NM	10.50	4.85	2.71	1.9	0.83	3.41	4.61	2.54	0.93	0.64	11.32	0.62	0.71	0.32	1.47	0.21	0.05	-	-
6M	21.10	11.33	3.48	8.5	3.30	5.53	11.20	4.62	2.20	1.20	23.62	2.20	0.90	0.56	4.50	0.12	0.12	72.4	529
NM	12.45	5.90	3.70	2.6	1.30	5.72	4.70	2.74	1.21	0.72	12.50	0.64	0.71	0.32	2.02	0.15	0 10	-	-
9M	24.60	13.81	3.95	10.2	4.20	8.62	14.53	5.32	4.71	2.10	24.51	2.32	1.40	0.70	6.30	0.11	0.24	75.3	633
NM	15.21	7.95	3.73	3.3	1.50	5.91	4.92	2.94	1.10	0.75	13.23	0.69	0.72	0.34	2.25	0.15	0.11	-	-
12M	31.54	14.83	4.12	11.1	4.60	13.42	16.22	6.21	7.30	2.70	34.21	2.33	1.61	0.74	7.30	0.28	0.28	83.4	842
										E. bracteata									
NM	7.75	4.48	1.29	0.94	0.45	3.31	6.51	3.82	0.52	0.31	10.50	0.47	0.48	0.20	0.76	0.26	0.05	-	-
3M	19.22	12.20	3.32	4.20	1.82	12.20	9.21	4.62	0.93	0.64	17.80	0.53	0.76	0.28	2.46	0.11	0.14	72.5	497
NM	9.71	4.87	2.20	1.0	0.56	3.41	7.82	4.30	0.52	0.33	10.40	0.47	0.48	0.20	0.89	0.22	0.10	-	-
6M	23.40	19.31	4.35	6.7	2.11	14.30	9.42	4.52	1.21	0.74	18.61	0.59	0.81	0.36	2.84	0.13	0.14	82.2	529
NM	14.01	5.10	2.51	1.7	0.81	3.84	8.30	4.40	0.77	0.34	11.50	0.53	0.59	0.22	1.15	0.19	0.07	-	-
9M	30.21	22.52	3.52	8.6	3.51	18.30	12.10	5.20	2.31	1.20	20.21	0.64	0.92	0.43	4.35	0.10	0.28	87.4	647
NM	17.32	6.30	2.85	3.8	1.50	3.92	8.30	4.40	1.30	0.78	13.31	0.54	0.63	0.23	2.28	0.10	0.12	-	-
12M	36.42	24.30	4.63	13.10	5.21	22.21	14.20	5.30	3.40	1.60	24.11	0.65	0.93	0.44	6.80	0.06	0.34	87.4	783
										L. Lanceolata									
NM	9.13	2.32	1.20	0.80	0.36	3.51	3.10	1.60	0.35	0.18	8.21	0.84	0.24	0.10	0.54	0.18	0.05	-	-
3M	15.25	7.31	3.53	3.30	1.51	4.91	7.82	4.01	0.72	0.31	14.10	1.83	0.56	0.13	1.81	0.07	0.10	49.20	561
NM	7.70	3.34	1.92	0.81	0.37	3.51	4.12	2.81	0.42	0.19	8.41	1.21	0.35	0.12	0.56	0.21	0.11	-	-
6M	19.01	9.18	3.70	3.50	1.71	7.10	8.10	4.53	2.00	0.90	15.72	2.12	0.61	0.19	2.60	0.07	0.13	51.51	743
NM	9.32	5.32	2.42	0.89	0.42	3.59	4.20	1.82	0.45	0.23	9.82	1.20	0.30	0.17	0.65	0.18	0.10	-	-
9M	21.90	12.72	4.27	7.00	2.20	8.32	10.21	4.40	2.41	1.30	16.31	2.30	0.63	0.27	3.50	0.08	0.14	54.12	904
NM	11.40	5.30	2.42	9.70	4.80	4.62	4.21	1.82	0.68	0.27	10.61	1.42	0.48	0.22	5.07	0.04	0.12	-	-
12M	26.32	13.57	4.30	7.30	2.81	8.65	11.20	5.62	3.30	1.50	21.50	2.30	0.71	0.32	4.30	0.07	0.18	56.41	941

Duration	Plant height (cm)	No. of nodes	Shoot diam (mm)	Shoot		No. of leaves	Leaf length (cm)	Leaf breadth (cm)	Leaf		Total root length (cm)	Root diam. (mm)		Root	Total dry wt. Shoot (g)	Root/ shoot ratio (g)	%p shoot	% Col onization	No. spores/ 100g. soil
				Fresh wt (g)	Dry wt				Fresh wt (g)	Dry wt		Fresh wt (g)	Dry wt						
										P. marsupium									
NM	5.50	2.60	0.81	0.68	0.30	1.45	7.92	3.44	0.37	0.16	7.12	0.48	0.15	0.09	0.46	0.19	0.10	-	-
3M	12.40	4.80	1.62	2.10	0.97	3.75	5.52	4.15	0.68	0.30	12.31	1.32	0.36	0.14	1.27	0.11	0.10	36.12	868
NM	7.80	3.62	0.81	0.75	0.32	2.45	7.92	3.45	0.37	0.16	7.42	0.53	0.24	0.10	0.48	0.21	0.10	-	-
6M	13.51	4.95	1.73	3.30	1.50	4.25	6.52	5.21	0.84	0.51	14.52	1.60	0.39	0.18	2.01	0.09	0.14	39.19	871
NM	9.50	3.74	1.11	0.75	0.32	3.12	8.42	3.82	0.61	0.19	9.30	0.69	0.26	0.11	0.51	0.21	0.11	-	-
9M	15.32	6.25	2.49	4.00	1.90	5.37	8.12	5.32	1.30	0.72	14.19	1.90	0.45	0.22	2.62	0.08	0.14	45.22	1014
NM	10.15	3.75	1.22	0.76	0.33	3.14	8.42	3.81	0.43	0.20	10.60	0.78	0.36	0.14	0.58	0.24	0.12	-	-
12M	19.42	8.32	2.62	4.50	2.20	8.32	11.41	6.52	2.00	0.91	11.30	2.10	0.70	0.31	3.11	0.10	0.17	48.32	987
										T. tomentosa									
NM	8.13	3.52	2.65	1.50	0.76	3.81	4.21	2.32	0.80	0.50	9.41	1.49	0.31	0.14	1.26	0.11	0.10	-	-
3M	20.42	13.60	2.43	5.80	2.30	14.35	8.42	3.91	3.20	1.40	15.80	1.70	0.69	0.30	3.70	0.08	0.14	63.2	713
NM	11.52	4.72	2.72	2.60	1.20	3.85	4.25	2.32	0.81	0.52	10.22	0.52	0.31	0.14	1.72	0.08	0.10	-	-
6M	24.21	15.32	4.28	9.30	4.10	16.70	9.25	4.21	5.80	2.30	19.32	1.91	0.70	0.31	6.40	0.05	0.19	75.3	816
NM	14.50	5.21	2.72	3.30	1.50	4.35	5.21	2.10	2.10	1.00	10.22	0.57	0.35	0.16	2.50	0.06	0.12	-	-
9M	29.32	16.38	4.31	11.30	5.40	12.30	13.30	5.22	7.40	3.50	21.50	2.20	0.82	0.36	8.90	0.04	0.22	81.5	843
NM	15.72	5.21	2.82	3.60	1.80	4.71	5.25	2.15	2.50	1.30	11.31	0.69	0.40	0.18	3.10	0.05	0.14	-	-
12M	32.41	19.32	4.32	13.50	6.10	21.42	15.31	5.41	9.40	4.30	29.15	2.25	0.91	0.44	10.40	0.04	0.41	91.4	821
										L. Lanceolata									
NM	8.91	3.30	2.70	1.40	0.81	4.06	5.51	2.40	0.94	0.57	11.30	0.44	0.39	0.17	1.38	0.12	0.10	-	-
3M	18.51	8.41	3.18	5.70	2.60	6.15	12.10	3.15	3.00	1.30	18.42	1.52	0.54	0.23	3.90	0.06	0.12	73.4	635
NM	12.30	5.83	2.73	1.60	0.84	4.12	6.80	2.40	0.95	0.62	14.20	0.47	0.39	0.17	1.46	0.12	0.10	-	-
6M	24.10	13.90	3.21	6.80	3.20	11.65	13.11	6.21	3.30	1.50	22.61	1.82	0.59	0.28	4.70	0.06	0.16	84.5	784
NM	16.91	9.73	2.83	1.80	0.87	4.17	7.11	3.32	0.96	0.64	16.42	0.49	0.37	0.16	1.51	0.10	0.12	-	-
9M	31.12	22.01	3.84	9.40	4.10	17.32	14.50	6.72	4.20	2.10	26.81	2.01	0.76	0.32	6.20	0.05	0.29	84.5	943
NM	17.32	11.70	2.93	3.00	1.40	4.62	7.12	3.30	1.00	0.74	16.82	5.15	0.35	0.19	2.14	0.09	0.14	-	-
12M	36.41	21.40	4.17	14.30	6.70	19.85	17.30	8.70	7.60	3.40	32.31	2.32	0.87	0.43	10.10	0.04	0.37	85.50	875

* Mean values of 4 replicates. NM = non-mycorrhizal, 3M = 3 months mycorrhizal plants, 6M = 6 months mycorrhizal plants, 9M = 9 months mycorrhizal plants, 12M = 12 months mycorrhizal plant

programmes, especially in the reclamation of drastically disturbed lands.

(2) Even a slight improvement of mycorrhizal efficiency in tropical soils where P deficiency is one of the bottlenecks of tropical forestry has great practical importance.

(3) The study of impact of mycorrhizal root residues on the ecology of VAM fungi in soil deserves attention.

(4) The inoculation seed for direct seedling operations could become very important alternative to planting seedlings, especially remote sites in very rough terrains.

AM FUNGI IN SYNTHESIS OF PROTEINS OF UNKNOWN FUNCTIONS

These physiological changes in roots have a direct bearing on changes in rhizosphere microflora e.g. Arginine in root extract is known to reduce chlamydospore production in *Thelaviopsis basicola* (Dehne *et al.*, 1978). Similarly higher chitinolytic activity is implicated in reduced fungal infection. Deposition of lignin in the cell walls of endodermis and stele is responsible for restriction of endophyte to the cortex (Miller *et al.*, 1985). Accumulation of phenols in host cells during fungal colonisation increase the resistance against fungal penetration and acts as physical barrier (Krishna *et al.*, 1982 Morandi, 1996). Increased concentration of soluble carbohydrate and reducing sugars in mycorrhizal plants is indirectly linked to decreased pathogen activity (Amijee *et al.*, 1993; Bansal and Mukerji, 1994; Schenck, 1981). AM fungal spores of *Glomus mosseae* also contain certain hydrolytic enzymes (pectinase, cellulase, hemicellulase) that are involved in fungal penetration in host root (Garcia Garrido *et al.*, 1992). Higher ATPase activity is associated with energisation of cells and is involved in transport of phosphorus (Gianinazzi-Pearson *et al.*, 1991).

Endomycorrhizal (AM) symbiosis causes a modification of free sterols which could influence the development of biotrophic foliar pathogen feeding via haustoria (Bhandari and Mukerji, 1993; Dugassa *et al.*, 1996; Losel, 1991). The nutrient transfer between hosts as mediated by sterols. The higher amount of sterols could stimulate the pathogen development e.g. some members of Pythiaceae are highly dependent on external sterols (Hendrin, 1970).

The VAM fungus after establishing inside a fine root, alters photosynthate leakage from root (Graham *et al.*, 1981). The influence of root exudates on microbial population in vicinity of roots is well known (Bansal and Mukerji, 1996; Curl and Truelove, 1983). Quantitative decrease in number of fungal colonies in the rhizosphere of *Leucaena leucocephala* is positively correlated with quantitative decrease in exudation of amino acids and sugars from mycorrhizal roots (Bansal and Mukerji, 1994). There is also possibility of exudation of antifungal substances from mycorrhizal roots. In case of nematodes also AM fungi alter the host root exudates and change their chemotactic attraction (Pinochet *et al.*, 1995). AM fungi and rhizobacteria are also suggested to interact via root exudation (Prikryl and Vancura, 1980).

AM linseed plants show a higher CO_2 assimilation, besides direct damage, a biographic pathogen causes indirect damages through changes in assimilated allocation. It induces a sink in infected tissue which competes for assimilates in the sink, i.e. active developing plant parts such as shoot apex. This damage may be correlated with pathogen activity, which can be measured in terms of sporulation rates. Although the sporulation of the mildew fungus is estimated in AM linseed plants, these plants contained even more sucrose at the shoot apex, than non-AM ones. Consequently the AM plants suffered less from pathogen attack in terms of fresh weight production. This is evidence that positive correlation between disease interrity and its damages, enabling the host plants to maintain their productivity, this AM effect inducing tolerance which is based on a stronger sink activity of shoot apex (Duggass *et al.*, 1996; Trotta *et al.*, 1996). This effect could be multiple and the increased concentration of the growth promoting, phytohormones such as cytokinins, auxins and gibberellins are one of the significant effects of AM/AM association, these phytohormones increase photosynthetic activity (Drunge and Schonbeck, 1992) and translation in the plant sinks (Miyamoto *et al.*, 1993).

DIRECT MECHANISMS

There are various examples in literature which show direct stimulatory or antagonistic relations between VA mycorrhizal fungi and other microorganisms (Tilak *et al.*, 1987). Involvement of various water soluble, volatile or diffusible substances from rhizosphere bacteria in stimulation of AM fungal hyphal growth and formation of secondary vesicles is suggested (Azcon, 1987).

Direct spatial association of AM fungi *(G. fasciculatum, G. intraradices, G. mosseae* and *Gigaspora gilmori)* with some free living di-nitrogen fixers (*Azospirillum, Pseudomonas)* is reported (Tilak *et al.*, 1987). *Azotobacter brasilense* has also been isolated from root cortex with AM fungi besides rhizosphere (Tilak and SubbaRao, 1987). This supports direct interaction between AM fungus and *Azospirillum* within the plant.

Involvement of growth hormones viz. auxins, gibberellins, cytokinins in interaction of bacteria with VAM fungi is also known (Azcon *et al.*, 1978). *Azotobacter vinelandii* and *A. beijerinckii* are known to produce phyto hormones like auxins, gibberellins and cytokinins in culture. Their effects on AM fungi can be stimulated by exogenous addition combination of gibberellic acid, kinetin and IAA (Azcon *et al.*, 1978). The growth promoting substances might also act directly on AM fungi or these might produce morphological changes in plants which in turn influence mycorrhizal symbiosis (Paulitz and Linderman, 1991).

Direct parasitism (Mycoparasitism) of VAM fungi may occur (Bhattactharjee *et al.*, 1982; Sylvia and Schenck, 1983), where spores of AM fungi are found to be parasitized by variuos other fungi including chytridiaceous fungi, *Fusarium* sp., *Penicillium* sp., *Trichoderma* sp., *Humicola* sp., etc.

PROSPECTUS

Since as Trappe so felicitiously put it, 'Most woody plants require mycorrhizae to survive, and most herbaceous plants need them to thrive', we envisage increasing exploitation of both AM fungi. Man, after all requires food to survive, and lumber to thrive.

The world's endomycorrhizal rainforests are rapidly being destroyed by extraction of timber, slash and burn agriculture, and other processes. Since these forests do not regenerate easily or quickly, it is very likely that in many areas they will be replaced by plantations of exotic conifers which often grow much faster than indigenous trees. Such plantations may also be initiated on many other currently treeless areas (Crew *et al.*, 1978; Tiwari, 2001). And existing forests need to be re-established after harvesting. In every case, biotechnological investments in arbuscular ectomycorrhizal fungi will be well repaid by the ease and speed with which the trees become established, and by the grains in growth of the maturing forests.

Tree nurseries and field crops are often heavily fertilized. This effectively discourages many mycorrhizal fungi, which are successfully weaned forest nurserymen and farmers away from the chemicals that have served them so well. In the long run, increasing energy costs may provide the required impetus.

Horticulturists commonly use their own mixture of sand, peat and wood shavings as a propagating medium for cuttings, often adding lime and fertilizer. Needless to say, the mix does not contain AM inoculum. The addition of AM fungi would confer all the usual benefits upon these plants when they were outplanted, and would have the added advantage of reducing the amount of fertilizer that must be added to the mixture. Lakshman (1996) have suggested that AM fungi could be added to any substrate used in the greenhouse to produce stock for transplanting to the field. At present, tomatoes raised in greenhouses have their root systems plunged in fertilizer before outplanting. This step may eventually be replaced by plunging in AM inoculum.

Starter cultures of AM fungi have been available in the past from universities and from industry, but the difficulties associated with maintaining the purity of the inoculum have been major obstacles to their commercialization. Although pure inoculum of a single species is often used in laboratory experiments, (Mukerji, Chamola and Singh 2000), this might not be not indeed be desirable, for field applications. As we mentioned in previous section, a special mixture of selected AM fungi was recently placed on the market by Biofertec Inc. in Canada. The producers claimed that this mixed inoculum can make most species of garden plants, fruit trees and experience of this product, and we can neither confirm nor refute the claims made for it. It is certainly possible that a mixture of several fungi might constitute a more generally applicable inoculum by increasing the chance that at least one of the potential mycobionts would be well suited to the particular host-soil-climate combination (Peterson Farquhar, 1994). We have learned that Biofert has been temporarily taken off the market because of quality control problems.

This treatment would have damaged the hyphae, but not the spores or infected roots. Hayman suggested that direct drilling and minimum tillage might reduce disruption of the mycelial network and favour established populations of AM fungi. It seems to us that this knowledge can be turned to good account in another way. When we need to introduce highly efficient strains, through cultivation should actually help them to become establishing in cutting down competition from indigenous organisms.

Another way of doing this, or of eliminating known pathogens, is to fumigate the soil with methyl bromide, indeed in California the law requires that this be done when plants destined for sale are to be produced. This treatment is extremely effective in eradicating soil microorganisms, and once the soil has been detoxified for the recommended time, introduced mycorrhizal fungi can proliferate without competition. (Lakshman *et al.,* 2006). Soil fumigation does, however require equipment, time and money which may not always be necessary.

For many years, substances with specific toxicity toward fungi have been deliberately introduced to the plant or to the soil in order to control pathogens. Most early fungicides were toxic to a wide range of fungi, and also to many non-target organisms, not excluding the very plants they were designed to protect Dodd (2000). Members of the second generation of fungicides are far less phototoxic, and are also systemic. They enter the plant but remain apoplastic, moving only upward in the transpiration stream. Third generation fungicides are not only systemic, but symplastic, moving upward in the transpiration stream and downward in the phloem. The early fungicides posed a threat to mycorrhizal fungi only when deliberately applied to the seed or the soil, or when washed down into the soil. The most recent fungicides will reach the interior of the roots no matter where the application is made. Obviously, there is concern that these newer fungicides, in addition to controlling the target pathogen, may damage or displace the desirable mycorrhizal fungi present in or near the roots. Fortunately, some of the most recently introduced fungicides are highly selective in their action, and have negligible toxicity to many non-target organisms (Carbon, 1989). In fact, it has recently been demonstrated that foliar applications of a selectively anti-oomycete fungicide, Aliette, actually stimulate the colonization of the roots of onion by a species of *Glomus.* The possible impact on indigenous or introduced mycorrhizal fungi of the wide range of biocides currently used in agriculture and forestry has not usually been considered in the past, but we believe it should be a subject for ongoing research, and a concern in the choice of strategies for the control of pests and pathogens.

The possible role of mycorrhizae in the management of marginally productive lands, such as arid and semi-arid range lands where animals graze, has been considered (Fitter, 1991). Overgrazing is deleterious to the mycorrhizal system. Perhaps because the animals selectively eat forage plants that are mycorrhizal annuals, leaving behind unpalatable, non-mycorrhizal perennials. Non-disruptive management practices, such as the introduction of efficient mycorrhizal fungi, removal of non-mycorrhizal perennials and planting of mycorrhizal perennials, might result in higher productivity from such marginal lands.

Surveying the AM fungi of citrus orchards in California and Florida, Verma and Scheup (1995) found that plants of different ages tended to have different AM symbionts. They concluded that the age of the plant plays a role in determining which AM fungus will be the dominant mycorrhizal partner. It seems quite likely that under natural conditions, over the long term, a succession of AM fungi will associate with a single perennial plant (Bagyaraj, 2006). Since the source of inoculum for the later stages of this succession will in all probability be the community of AM fungi in the surrounding soil programs aiming at introducing efficient AM fungi to perennials should concentrate on those best adapted to the earliest stages of growth, those which will bring about quick and extensive infection, and increase the probability of the young plant's survival.

The spores of AM will germinate well in the absence of host roots (Daniel and Trappe, 1980) a responsive frost root is

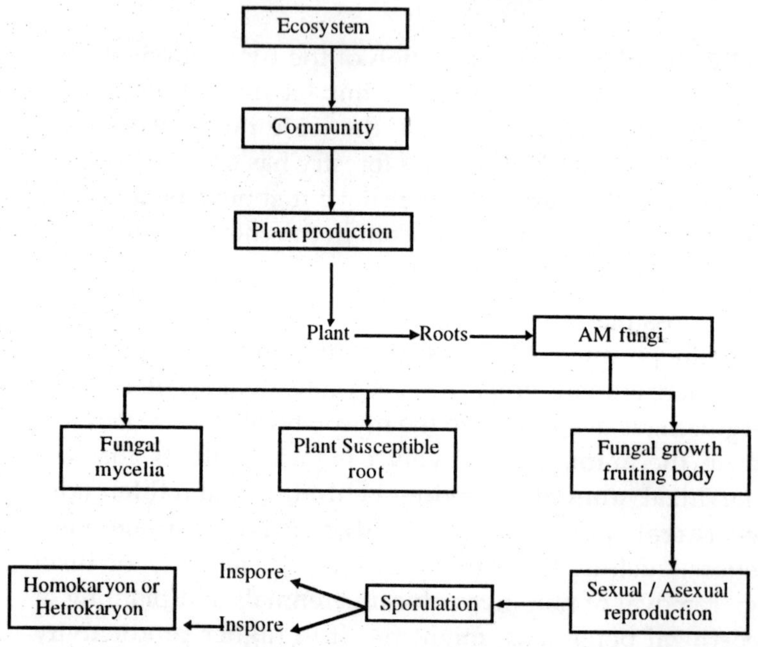

not encountered, the spore will withdraw the cytoplasm from the germtube and may germinate and may reduce the vulnerability of the fungus to grazing by mycophagous soil microfauna (Koske and Gemma, 1997) and ensure the successful colonization of the plant by the fungus.

The spores of AM germinate and if the germtube or hyphae encounters a receptive root or root hair it forms an appresorium (Hall, 1977) the hyphal branch from the appresorium penetrates the epidermis in the zone of differentiation and elongation of feeder roots. The root tips are not normally infected.

The hyphae grow between the cortical cells and never enter the meristemetic cells and the endodermal cells. Specialized branches enter individual cortical cells and form finely arborescent structures called "arbuscules" which are the sites of exchange between the fungus and the plant individual arbuscule function for 4–15 days. The progress of AM

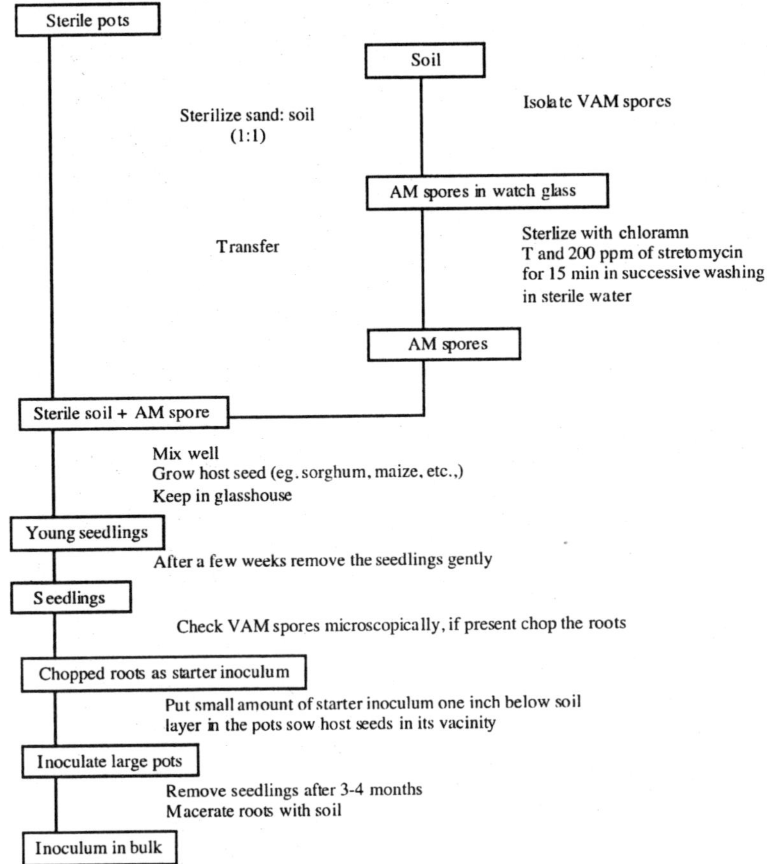

colonization in a single root system commonly follows a sigmoid curve. The mathematical modeling of progressive measurement of infection has highlighted the need for more detailed experimental study of biology of individual infection.

The development of fungus inside the root constitute intramatrical phase and the proliferation outside in the soil is extramatrical phase. The latter phase enables the plant to exploit greater volumes of soil available to nonmycorrhizal plant. The fungal efficiency depends upon the extent of extramatrical hyphae which in turn is based on the availability of the substrate in root.

The role of volatile substances emitted by host roofs and gibbereilin and cytokinin like substances secreted by the spore of the fungus *(Glomus mosseae)* (Azocon, 1987) in recognition and colonization under natural conditions deserves further investigation.

Limitations and prospects for further research

(*i*) The final effect of AM fungi as biofertiliser/ biocontrol agent can be seen in as series of events. This includes spore germination, growth of mycelium through soil, stimulation and attachment of infective hyphae to root, penetration of root, secondary spread inside the root, production of vesicles, arbuscules and spores. Most of the studies to date have given more emphasis on the root colonisation as a fungal parameter measure. Since AM fungi are obligate symbionts and cannot be grown independently of plant, this makes it very difficult to study the exact mechanisms and stage of interaction involved.

(*ii*) As AM fungi cannot be cultured axenically the effect can only be observed on plant root. In order to work out the mechanism involved in interactions and to harness their potential as viable biocontrol/ biofertilizer, workable technique for commercial production of arbuscular mycorrhizal fungi needs to be designed,

(*iii*) The mere presence of mycorrhizal fungi on roots of a plant does not ensure a positive effect on a disease or suppression of a pathogen, rather the effects are very specific, both on host and companion microorganisms. Therefore, the mycorrhizal fungi on roots of a plant need to be very specifically managed for each AM microbe combination. Standardisation of a particular AM pathogen or AM fungus-biofertilizer partner combination for plant is also very important.

(*iv*) There is a parent difficulty in determining the biologically significant differences in population; density of microbes in the rhizosphere and statistical differences in many cases are not obvious.

(*v*) The influence of AM fungi on other rhizosphere microorganisms has not been demonstrated under field conditions despite the promising results in green house studies. Little effort has been made to use natural systems in evaluating the AM fungi and their interactions with other microorganisms. For this reason the results from studies to date are very difficult to correlate with field situation.

CONCLUSION

To improve the understanding of the interactions between the heavy metals, mycorrhizal fungi, and roots in contaminated soils, parameters like the tolerance of the mycorrhizal fungi and host plant, the nutritional status of both organisms, soil properties of the metals and their specific behaviour should be considered. The extent to which mycorrhizal fungi can alleviate metal toxicity to plants under field conditions remains to be established.

The protective effect of ectomycorrhizal and ericoid mycorrhizal fungi against metal toxicity in plants is quite clear and various mechanisms have been demonstrated, the main one being fungal structures acting like barriers mechanisms have been demonstrated to metal up take in plants. More work remains to be done on AM fungi to understand the mechanism involved in metal immobilization in the hyphae inside and/or outside the roots. The compartment system mentioned already is a promising approach for such studies.

There is relatively large amounts of extraradical hyphae can be produced and separated from the roots, which should allow the study of metal adsorption and uptake by AM hyphae, transfer to the roots, which should allow the study of metal adsorption and uptake by AM hyphae, transfer to the roots, and translocation to the shoots. The functioning of different fungi could be compared in such systems. Other experimental devices including normal or transformed roots might also be useful to study metal tolerance of AM fungi. The results with ectomycorrhizal fungi show metal specific tolerance mechanisms and competition between metals for uptake. This should also be investigated for all kind of mycorrhizae, and especially for the possible use of mycorrhizal fungi in phytoremediation.

There is no doubt that genetic engineering aimed at producing plants, which are more tolerant and resistant to heavy metals may play an important role in the future. But such attempts must not overlook the possible changes in the susceptibility of the plants to mycorrhizal infection. A metal-resistance plant breed should still be susceptible to mycorrhizal symbiosis, and when colonized by a metal-resistant mycorrhizal fungus, genetically modified or derived mycorrhizal fungus, genetically modified or derived from the natural resources, should be of great value for the rehabilitation of metal contaminated soils. The management of soil microorganisms including mycorrhizal fungi is a prerequisite for the success of future restoration programmes.

REFERENCES

Abuziandah, R.A., Finlay, R.D. and Read, D. J. (1986b). The role of proteins in the nitrogen nutrition of ectomycorrhizal plants. II. Utilization of protiens by mycorrhizal plants of "*Pinus contorta*". *New Phytol*. 103: 495-506.

Abuziandah, R.A. and Read, D.J. (1986a). The role of proteins in the nitrogen nutrition of ectomycorrhizal plants. III. Protein utilization by *Betula, Picea* and *Pinusin* mycorrhizal association with *Hebeloma crustuliniforme*. *New Phytol.* 103:507-514.

Allen, M. F. (1991). The Ecology of Mycorrhizae Cambridge Univ. Press. New York, pp 184

Ames, R.N., Reid, C.P.P., Porter, L.K., Cabardella, C. (1983). Hyphal uptake and transport of nitrogen from two $^{1\&}$N - labelled sources by *Glomus mosseaea* vesicular - arbuscular mycorrhizal fungus. *New Phytol.* 95: 381-396.

Anderson, A. J. (1988). Mycorrhizae - host specificity and recognition. *Phytopathol.* 78: 375-378.

Anonymous (1997). Classification of Glomales. Updated on 19 Dec. 1997. IMVAM's home page.

Azcon, R. (1987). Germination of hyphae growth of *Glomus mosseae in vitro,* Effect of rhizosphere bacteria and cell free culture media, *Soil Biol. Biochem.* 19:417-419.

Azcon, R., Azcon Aguilar, C. and Barea, J.M. (1978). Effects of plant hormones present in bacterial culture on formation and responses to VA endomycorrhiza, *New Phytol.* 80:359-364.

Azcon-Aguilar, C. and Barea, J.M. (1996). Arbuscular mycorrhizas of biological control of soil-borne plant pathogen, on overview of mechanics involved, *Mycorrhiza.* 6:457-464.

Azon, R. (1987). Germination and hyphal growth of *Glomus mosseae* in vitro: Effect of rhizosphere bacteria and cell free culture media. *Cell. Biol. Bichemist.* 19: 417-419.

Ba A. M., Plenchette C. Danthu P, Duponnois, R. and Guisson, T. (2000). Functional compatibility of two arbuscular mycorrhizae with thirteen fruit trees in Senegal. *Agroforestry Sytems.* 50(2): 95-105.

Bagyaraj, D.J. (2006). Arbuscular mycorrhizal fungi in sustainable agriculture. In: *Techniques in mycorrhizae.* M.J. Bukhari and B.F. Rodrigues eds. Department of Botany. Govt. College. Quepem. Goa. Pp. 1-8.

Baltruschat, H. and Dehne, H.W. (1988). The occurence of vesicular arbuscular mycorrhiza in agro ecosystems. I. Influence of nitrogen fertilization and green manure in continuous monoculture and in crop rotation of the inoculum potential of winter wheat. *Plant and Soil.* 107:279-284.

Bansal, M. and Mukerji, K.G. (1994). Positive correlation between root exudation and VAM induced changes in rhizosphere mycoflora, *Mycorrhiza.* 5:39-44.

Bansal, M. and Mukerji, K.G. (1996). Root exudates in rhizosphere biology, in : *Concept in Applied Microbiology and Biotechnology,* K.G. Mukerji and V.P. Singh, eds., Aditya Book, New Delhi, India, pp. 98-120.

Barea, J.M. and Azcon Aguilar, C. (1982). Production of plant growth regulating substances by the vesicular - arbuscular mycorrhizal fungus, *Glornus mosseae. Appl. Environ. Microbiol.* 43: 810-813.

Barea, J.M. and Azcon Aguilar, C. (1984). Mycorrhizas and their significance in noduiating nitrogen - fixing plants. *Adv. Agron.* 36: 1-54.

Barea, J.M., Azcon Aguilar, C. and Azcon, R. (1987). Vesicular - arbuscular mycorrhiza improve both symbiotic N_2 fixation and N, uptake from soil as assessed with a ^{lb}N technique under field conditions. *New Phytol.* 106: 717-725.

Bhandari, N.N. and Mukerji, K. G. (1993). *The Haustorium,* Research Studies Press Ltd, U.K. and John Willey & Sons Inc., New York, U.S.A. pp. 308.

Bhattacharjee, M., Mukerji, K.G., Tewari, J.P. and Skorapad, W.P. (1982). Structure and hyperparasitism of a new species of *Gigaspora, Trans. Bri. Mycol. Soc.* 78:184-188.

Bowen, G.D. (1980). Mycorrhizal roles in tropical plants and ecosystems. In: *Tropical mycorrhizal Research* (Eds. P.Mikola), Oxford University Press, Oxford. Pp. 165-190.

Brundrett, M., Melville, L. and Peterson, R. L. (1994). Practical methods in mycorrhiza Research, *Mycologue waterloo, Canada*. pp. 1-161.

Brundrett, M.C., Piche, Y. and Peterson, R.L. (1985). A developmental study of the early stages in Vesicular-arbuscular mycorrhiza formation. *Can. J. Bot.* 63: 194-206.

Carbon, M. (1989). Potential use of mycorrhiza in control of soil boon diseases. *Can. J. Plant. Pathol.* 11: 177-179.

Carling, D.E. and Brown, M.F. (1982). Anatomy and physiology of Vesicular-arbuscular and non-mycorrhizal roots. *New Phytol.* 72: 1108-1114.

Cox, G. and Tinker, P.B. (1976). Translocation and transfer of nutrients in Vesicular-arbuscular mycorrhizas. The arbuscule and phoshorus transfer. A quantitative ultra structural study. *New Phytol.* 77: 371-378.

Crews, C.L., Johnson, E.R. and Jainer, J.N. (1978). Benefits of growth and development of three woody ornamentals. *Hot.Sci.*13:429-430.

Cruz R. E., de la, Zarate J. T., Aggangan N. S. and Lorilla E. B. (1992). Differential mycorrhizal dependency of some agricultural horticultural and forestry crops to inoculation with arbuscular mycorrhizal fungi. In Proceedings of the International symposium on Management Mycorrhizas in Agriculture, *Horticulture and Forestry*, pp. 54 edited by D. Jasper.

Curl, E. A. and Truelove, B. (1983). The *Rhizosphere, Springer Verlag,* Heidelberg.

Daniel, B.A. and Trappe, J.M. (1980). Factors affecting spore germination of the Vesicular-arbuscular mycorrhizal fungus, *Glomus epigaeus. Mycologia.* 72:457-471.

Dehn, B. and Schuepp, H. (1989). Influence of VA- mycorrhizae on the uptake and distribution of heavy metals in plants. *Agri.EcoSy. Environ.* 29: 79-83.

Dodd, J.C. (2000). The role of arbuscular mycorrhizal fungi in agro and natural ecosystem. *Outlook on agriculture.* 29 (1): 55-62.

Douds, J.R.D. and Miller, P.D. (1999). Biodiversity of arbuscular mycorrhizal fungi agroecosystems. *Agri.Ecos.Environ.* 74: 77-93.

Drunge, U. and Schonbeck, F. (1992). Effect of vesicular-arbuscular mycorrhizal infection on transpiration, photosynthesis and growth of flora *Linum usitatissimum* L., in relation to cytokinin levels. *J.Plant. Physiol.* 141: 40-48.

Dugassa, G.D., Alten, N. and Schoenbeck, F. (1996). Effect of arbuscular mycorrhiza AM, on health of *Linum usitatissimum* L. infected by fungal pathogens, *Plant Soil.* 185: 171 -182.

Entry, J. A., Rose, C. L. and Cromack, Jr. K. (1991). Litters decomposition and nutrient release in ectomycorrhizal mat soils, of a Doublas fir ecosystem. *Soil Boil Biochem.* 23:285-290.

Feldmann, F. and Lieberei, R. (1992). Vescicular arbuscular mycorrhiza in rubber tree monocultures. *Mycorrhiza News.* 5(4): 1-6.

Finlay, R. D. and Soderstrom, B. (1992). Mycorrhiza and carbon flow to the soil. In mycorrhizal functioning, Ed. M. F. Allen Chapman and Hall, London, 134-160.

Finlay, R.D., EK, H., Odham, G. and Soderstrom, B. (1989). Uptake, translocation and assimilation of nitrogen from ^{15}N - labeled ammonium and nitrate sources by intact ectomycorrhizal systems of *Fagus sylvatica* infected with *Paxillus involutus*. *New Phytol.* 113: 47-55.

Fitter, A.H. (1991). Costs and benefits of mycorrhizas; In : Implications and fumctioning under natural conditions. *Experientia.* 47: 350-355.

Frank, A. B. (1885). Uber die auf Wurzel - symboise beruhende Ernahrung gewisser Baume duroh unterirdische Pilze. Ber. Deutsch, *Bot. Ges.* 3: 128-145.

Furlan, V. and Fortin, J.A. (1975). A flotation-bubbling system for collecting Endogonaceae spores from sieved soil. *Naturaliste can.* 102: 663-667.

Garcia-Garrido, J.M., Garicia Romera, I. and Ocampo, J.A. (1992). Factors stimulate ^{32}P uptake and plasmalemma ATPase activity in vesicular arbuscular mycorrhizal fungus *Gigaspora margarita, New Phytol.* 118:289-294.

Gianinazzi, S. V., Gianinazzi - Pearson V. and Dexheimer, J. (1979). Enzymatic studies on the metabolism of vesicular - arbuscular mycorrhiza. Ultrastructural localization of acid and alkaline phosphatase in onion roots infected by *Glomus mosseae* (Nicol and Gerd). *New Phytol.* 82: 127-132.

Gianinazzi-Pearson, V., Smith, S.R., Gianinazzi, S. and Smith, F.A. (1991). Enzymatic studies on metabolism of vesicular-arbuscular mycorrhizas, V, Is H^+ ATPase a component of ATP hydrolysing enzyme activities in host fungus interface? *New Phytol.* 117:67-75.

Gerdemann, J.W., and Trappe, J.M. 1974., Endogonaceae in the pacific Northwest. Mycol. Mem., 5 : 1-76.

Graham, J.H., Leonard, R. T. and Menge, J. A. (1981). Membrane mediated decreases in root exudation responsible for phoshorus inhibition of vesicular-arbuscular mycorrhizal formation. *Plant Physiol.* 18: 548-552.

Grenville, D. J., Piche, Y. and Peterson, R. L. (1985). Sclerotia as viable sources of mycelia for the establishment of ectomycorrhizae. *Can. J. Microbiol.* 31: 1085-1088.

Habte and Manjunath (1991). Categories of Vesicular-arbuscular mycorrhizal dependency of host species. *Mycorrhiza.* 1(1):3-12.

Hall, I. R. (1977). Species and mycorrhizal infections of New Zealant Endogonaceae. *Trans Br. Mycol. Soc.* 68: 341-356.

Harley, J. L. (1969). The Biology of mycorrhiza, 2nd ed. Leonard Hill, London, p. 334.

Harley, J. L. (1989). The Significance of mycorrhiza. *Mycol. Res.* 92:129-139.

Harley, J. L. and Smith, S. E. (1983). Mycorrhizal symbiosis. Academic Press, New York, p. 483.

Harley, J.L. (1970). Mycorrhiza and nutrient uptake in forest trees. In: *Physiology of tree crops.* L.C. Lukwill and C.V. Cutting eds. Academic press, London, pp. 1-24.

Hayman, D. S. (1986). Mycorrhizae of Nitrogen-fixing legumes MIRCEN Journal, 2: 121-145.

Hendrin, J.W. (1970). Sterols in growth and reproduction of fungi, *Ann. Rev. Phytopathol.* 8:111-130.

Hepper, C.M. (1994). Spore germination of AM fungus Acaulospora laevis by different soil pH and other features. *Trans.Br.Mocl.Soc.* 83: 154-156.

Koske, R. and Gemma, J.N. (1997). Mycorrhizae and succession in plantings of beachgrass in sand dunes. *Am. J. Bot.* 84: 118-130.

Koske, R. E. (1981). Multiple germination by spores of *Gigaspora gigantea* Trans. *Br. Mycol. Soc.* 76: 328-330.

Krishna, K.R., Balakrishna, A.N. and Bagyaraj, D.J. (1982). Interactions between a vesicular-arbuscular mycorrhizal fungus and *Streptomyces cinnamomeous* and their affects on finger millet, *New Phytol.* 92:401-405.

Lakshman, H.C. (1996). VA mycorrhizal studies in some important timber tree species. Ph.D, thesis, Karnatak University Dharwad, India. Pp. 249.

Lakshman, H.C., Bheemreddy, V.S. and Santhosh, G. H. (2006). Microbial inoculants for *Capsicum annum* in copper polluted soil. *J.Micro.World.* 8 (2). 184-186.

Linder, S. (1989). Nutritional control of forest Yield. In: *The Marcus Wallenberg Foundation Symposia Proceedings*, 6:62-87.

Losel, D.M. (1991). Synthesis and functioning of membrane lipids in fungi and unfeated plants, *Pesti. Sci.* 32:353-362.

Menge J. A., Johnson E. L. V. and Platt R. G. (1978). Mycorrhizal dependency of two Hawiian endemic tree species: koa and mamane. *Journla of Plant Nutrition.* 16(7): 1339-1356.

Menge, J.A. (1984). Inoculum production. In: VA mycorrhiza, Eds. C.L. Powell and D.J. Bagyaraj. CRC press Boca Raton, FL. pp.187-203.

Miller, J.C., Rajapakse, S. and Garbaer, R.K. (1985). Vesicular arbuscular mycorrhizae in vegetable crops, *Hort. Scie.* 21:974-984.

Miyomoto, K., Veda, J. and Kamisaka, S. (1993). Gibberellins enhanced sugar accumulation in growing subhooks of eliolated *Pisum sativa* seedlings. Effects to gibberellic acid, indol acetic acid and cyclohexamide in invertage activity, sugar accumulation and growth. *Physiol. Plant.* 88: 301-306.

Mohan, V., Neelam Verma and Singh, Y.P. (1995). Distribution of VAM fungi in nurseries and plantations of Neem *Azadirachta indica* in arid zone of Rajastan. *Indian Forest.* 121: 1069-1076.

Morandi, D. (1996). Occurrence of phytoalexins and phenolic compounds in endomycorrhizal interaction and their potential role in biological control. *Plant Soil.* 185:241-251.

Mosse, B. (1973). Advances in the study of Vesicular-arbuscular mycorrhiza. *Ann. Rev. phytopathol.* 11: 171-196.

Mosse, B. and Jones, G.W. (1975). Separation of endogone spores from organic soil debris by deferential sedimentation on gelatin column. *Trans. Br. Mycol. Soc.* 51: 5-604-608.

Mukerji, K.G. and Kapoor, A. (1987). Vesicular mycorrhiza patterns of Modern Research. *Fronties Appl. Microbiol.* 2: 189-200.

Mukerji, K.G., Chamola, B.P. and Jagjit Singh. (2000). *Mycorrhizal Symbiosis*. Kluwer academic Newyork. Pp. 338

Nicolson, T.H. (1967). Vesicular-arbuscular mycorrhiza—A universal plant symbiosis. *Sci. progress.* 6: 511-581.

Nicolson, T.H. and Gerdemann, J.W. (1968). Mycorrhizal Endogone species. *Mycologia.* 60: 313-325.

Nouaim, R. and Chaussod, R. (1994). Mycorrhizal dependency of micropropogated argan tree (*Argania spinosa*): Growth and biomass production. *Agroforestry Systems.* 27(1) : 53-65.

Ohms, R.E. (1956). A Flotation method for collecting spores of a phycomycetous mycorrhizal parasitic soil. *Phytopathology.* 47: 751-752.

Paul, W. and Peter, M. (1993). Examination of VAM communities in forest ecosystems using both spore isolation and direct observation methods. *Mycorrhiza News.* 5(1): 4-7.

Paulitz, T.C. and Linderman, R.G. (1991). Mycorrhizal interaction with soil organisms. In: *Hand book of Applied Mycology.* Vol. I. Soil and Plants. D.K. Arora, B. Rai, K.G. Mukerji and G.R. Knudsen. Eds, Marcel Dekker Inc. Newyork. Pp. 77-129.

Peterson, R. L. and Farquhar, M. L. (1994). Mycorrhizas – integrated development between roots and fungi Mycologia. 86: 311-326.

Pinochet, J., Calvet, C, Camprubi, A. and Fernandez, C. (1995). Interaction between the root leision nematode *Pratylenchus vulnus* and the mycorrhizal association of *Glomus intraradices* and Santa Lucia 64 cherry stock, *Plant Soil.* 170:323-329.

Plenchette, C., Fortin, J. A. and Furlan, V. (1983). Growth response of several plant species to mycorrhiza in a soil of moderate P fertility. I: mycorrhiza dependency under field conditons. *Plant and Soil.* 70(2):191-209.

Pope, P. E., Chaney, W. R. and Rhodes, J. D. (1983). The mycorrhizal dependency of four hardwood tree species. *Canadian Journal of Botany,* 61:412-417.

Prikryl, Z. and Vancura, V. (1980). Root exudation as dependent on growth, concentration gradient of exudates and the presence of the bacteria. *Plant Soil*. 57: 69-83.

Raven, J.A. (1986). Evaluation of plant life forms. In: On the economy of plant form and function. Ed. T. Givnish. Cambridge Uni. Press. Pp. 421-492.

Read, C.P.P. and Bowen, G.D. (1979). The soil root interface. J. L. Harley and R.S Russell, eds. Academic Press, London, pp. 211-219.

Read, D. (1991). Mycorrhizas in ecosystems. *Experientia*. 47: 376-391.

Schenck, N.C. (1981). Can mycorrhizae control root diseases, *Plant Diseases*. 65:232-234.

Smith, G.W. and Skipper, H.D. (1979). Comparison of methods to extract spores of Vesicular-arbuscular mycorrhizal fungi. *Soil. Sci. Soc. Am*. 43: 722-725.

Sutton, J.C. and Barron, G.L. (1972). Population dynamics of Endogone spores in soil. *Can. Bot*. 50: 1909-1914.

Sylvia, D.M. and Schenck, N.C. (1983). Soil fungicides for controlling chytridiaceous mycoparasits *of Gigaspora margarita* and *Glomus fasciculatum, App. Environ. Microbiol*. 45: 1306-1309.

Tilak, KV.B.R. and Subba Rao, N.S. (1987). Association of *Azospirillium brasilense* with pearl millet *Pennisetum americanum* (L) Lecke, *Biol. Ferti. Soils*. 4: 97-102.

Tiwari, S.C. (2001). Vesicular-arbuscular association of tree species in humid tropical forest of Arunachal Pradesh. *Ecol. Env. And Cons*. 7(1): 21-23.

Trotta, A., Varese, G.C., Gnavi, E., Fusconi, A., Sampo, S. and Berta, G. (1996). Interactions between the soil borne root pathogen phytophthora nicotianae var. parasitica and the arbuscular mycorrhizal fungus *Glomus mosseae* in Tomato plants. *Plant Soil*. 185: 199-209.

Verma, A. and Scheup, H. (1995). Mycorrhizae: Biofertilizers for the future (Eds. A, Adholeya Sujan Singh) TERI, New Delhi. Pp. 322-328.

2

Screening of Potential AM Fungi for the Improvement of Costus Speciosus Smith. A Rare Medicinal Plant

Pushpa K Kavatagi and H C Lakshman

INTRODUCTION

Plants have fed the world and cured ills from immemorial time. A vast knowledge of medicinal plants must therefore have been accumulated, especially in India, where a large number of plants of medicinal value are found.

From ancient times until recently, peoples healed themselves with traditional herbal medicines which in many cases, by trial and error, proved efficaceous. In the recent past one could notice a global trend for the revival of interest in the traditional system of medicine.

Medicinal plants are potential renewable natural resources. Therefore, the conservation and sustainable utilization of medicinal plants must necessarily involve a long term, integrated, scientifically oriented action programme.

Post Graduate Department and Research in Botany (Microbiology Lab.)
Karnatak University Dharwad- 580 003, India
Email.id: dr.hclakshman@gmail.com

This should involve the pertinent aspects of protection, preservation, maintenance, exploitation, conservation and sustainable utilization.

The improvement phase of medicinal plant starts from to time the domestication of plant is undertaken. It is certainly necessary to follow certain models like collection and evaluation, colonal selection hybridization, mutation and other methods. Fertilizer consumption increased tremendously to maximum in yield per unit are by cultivation of hybrid crops after green revolution. One such perhaps may be the use of AM fungi for growth improvement (Smith and Read, 1997). Mycorrhizal plants are therefore adapted to scope with nutrient deficient situations or prevent pathogenesis by other organisms. Mycorrhizal fungi will not prove effective with all plants when the plants have little or mycorrhizal dependency Kumar and Jamalluddine (2007). Mohankumar and Mahadevan (1984) have reported the absence of AM fungal association in medicinal plants and attributed this to the presence of various secondary metabolites in the host plants. However there are many reports of medicinal plants with secondary substance harboring AM fungal associations in their root system (Abbott and Robson, 1982; Khade *et al.*, 1982; Lakshman and Raghavendra, 1992; Prasad Reddy, 1998). Kumar and Murugesh (2002) have tested the dependency of three AM fungal species *Glomus mosseae, Glomus fasciculatum* and *Glomus monosporum* ten medicinal plants and showed clearly that mycorrhizal inoculation is beneficial in improvement of their growth (Basu and Srivastava 1998) Biradar and Reddy (2007), have tested more than ten species of AM fungi and observed that AM fungal association enhanced the growth of medicinal compounds. Therefore, of there is a need for research in improving the quality and quality of drugs produced from native medicinal plants in relatively shorter period and a lower expense by using selective AM fungi. This paper reports the inoculation of difference between AM fungal species to understand the efficacy of *Costus speciosus* Smith. Pot experiments were conducted using unsterilized soil.

MATERIALS AND METHODS

In the present study *Costus specious* Smith. Plant was selected. This is an important medicinal plant. Its leaves juice used for skin disease and tuber juice is best medicine for diabetic patients. Leaf juice and tuber juice proportionally mixed and taken along with milk to reduce hypertension. Stem tuber cuttings measuring 6 cm were planted in earthen pots (20 × 30) filled with 4 kg of unsterilised soil. The experimental soil has following physiochemical characteristics viz E.C: 0.04 (E.mohos/cm) organic matter: 0.19%, total nitrogen: 0.05%, available Phosphorous: 5.3 (ppm) and available Potassium: 110.6 (ppm). Before planting the stem cutting 10 g of mixed inoculum of all the four different AMF inoculum was spread 1 cm uniformly in each earthen pots except control or non- inoculated one. Maize-root based cultures of two AM species; *Glomus faciculatum, Glomus mosseae*. John son grass-root based culture of other two AM species; *Gigaspora margarita* and *Sclerocystis dussii* were developed with single of the species isolated from rhizosphere soil of grass hosts in fallow alluvial soil of the university form. The mother cultures were maintained in steam-sterilized sand soil mixture in PVC (Poly vinyl chloride) pots in isolation under transparent plastic sheet cover in a green house. Infected mother culture root were maintained in green house, watering once in two days. Plants were harvested at 60, 120, 180, 240 and 300 days. The roots were cleared by dipping them in water for several times till the adhering soil particles were recovered. Both inoculated and uninoculated root samples cut into 1 cm bits and fixed in slandered FAA (Formalin acetoalcohol) and processed further for the assessment of AMF colonization. Root pieces were mixed in 10% KOH and autoclaved for 30 minutes and washed in distilled water and stained in 0.05% try pan blue and mounted on clean slide with in lacto phenol according to (Phillips and Hayman, 1970). For every harvest plant parts were removed and oven dried for 72 hr at 70°C. Roots and shoot systems were then weighed. All 60 days the following observation were recorded plants height, number of tuber, number of leaves, leaf length, length of tuber, diameter of tuber. The total chlorophyll content in leaves was estimated according (Earanna *et al.*, 2001) and phosphorus content in shoot estimated according to (Jackson, 1973).

RESULTS AND DISCUSSION

It is clearly evident from the data that root system of *Costus speciosus* Smith grown in both control and different species inoculated were infected in unsterilized soil invariably found to harbor AMF association. Figures 2.3–2.6 shows the macerated root section of the host plants. This clearly brought enhanced plant height, stem diameter, leaf length, number of leaves and total chlorophyll content of leaves (Table 2.1). Similarly, underground tuber with root system were found in significantly increased, number of tuber, length, number, dry weight and P content of tuber in AM fungi inoculated plants (Table 2.2). Overall result indicates that *Sclerocystis dussii* was the most efficient AM fungus which influences in increasing plant height, leaf length, leaf number, chlorophyll content, and bellow ground on enlarged tube diameter its length and phosphorous content of tuber was higher and compare to uninoculated plants however the variation among AM fungal species inoculated on *Costus specious* Smith exhibited variedly. Figure 2.2 showing the effect of *Glomus fasciculatum*, *Glomus mosseae*, *Gigaspora margarita*, *Sclerocystis dussii* on *Costus speciosus* at 60 days . Plants height was increased with the inoculation of *Sclerocystis dussii* it followed by *Glomus faciculatum* and *Gigaspora margarita*. Where as chlorophyll content increased in plants inoculated with *Sclerocystis dussii* followed by *Gigaspora margarita* and *Glomus mosseae* respectively. The increased leaf length, leaves number was recorded in plants inoculated with *Sclerocystis dussii* followed by *Glomus fasciculatum* and *Gigaspora margarita.* However the underground parts such as stem tuber length, tuber diameter, tuber dry weight and P content was influenced by *Sclerocystis dussii* which is followed by *Glomus mosseae.* Figure 2.1, showing the effect AM fungi on dry weight and phosphorous content in tuber of *Costus speciosus* Smith. The present

Table 2.1 :Table Showing the Effect of AM Fungi on Growth Characters in *Costus speciosus* Smith for 300 Days.

Treatment	Plants Height (cm)	Stem Diameter (cm)	Number of Leaves/Plant	Leaf Length (cm)	Total Chlorophyll (mg/gm tissue)
Control	23.2e	1.96c	6.2b	3.4b	0.30d
Glomus fasciculatum	65.4b	3.5d	27.1de	6.6c	0.44c
Glomus mosseae	52.1bc	3.4b	20.8bc	6.6d	0.55e
Gigaspora margarita	67.1d	3.6bc	25.2d	5.3de	0.57cd
Sclerocystis dussii	79.2a	3.8a	30.3a	7.6a	0.59a

Mean values followed by the letter within a column do not differ significantly at P=0.05 by DMRT.

Table 2.2 : Effect of AM Fungi on Tuber Yield Content in *Costus speciosus* Smith for 300 Days.

Treatment	Number of Tuber/Plants	Length of Tuber/Plants (cm)	Diameter of Tuber/Plants (cm)	Tuber dry weight (mg/plant)	Phosphorus content/tuber (%)
Control	11.1b	9.4c	2.2b	10.2c	0.142d
Glomus fasciculatum	9.2b	10.2d	2.2d	15.2e	0.384e
Glomus mosseae	10.5c	12.4e	2.2e	21.4de	0.424b
Gigaspora margarita	11.3bc	10.8d	2.4dc	18.5b	0.450de
Sclerocystis dussii	11.4a	12.2a	2.6a	24.2a	0.610a

Mean values followed by the some letter within a column do not differ significantly at P=0.05 by DMRT.

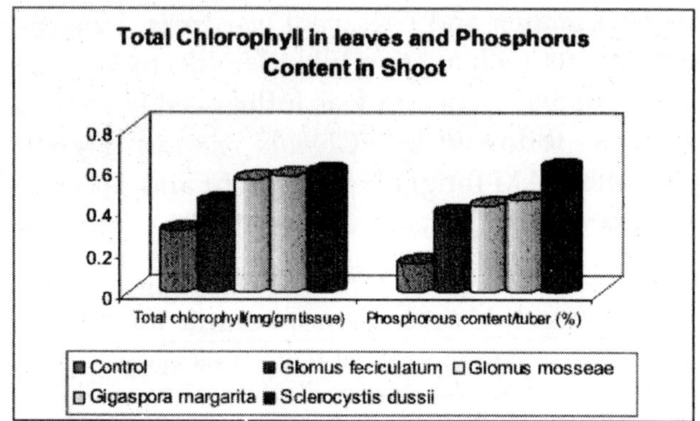

Fig 2.1 :Showing the Effect of AM Fungi on *Costus speciosus* Smith Showing Total Chlorophyll in Leaves and Phosphorus Content in Shoot for 300 Days.

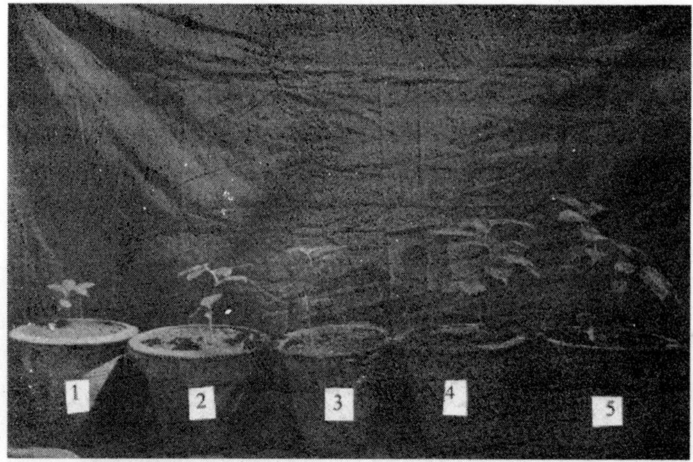

Fig 2.2 Showing the Effect of Glomus fasciculatum, Glomus mosseae, Gigaspora margarita, *Sclerocystis dussii on Costus speciosus* at 60 Days.

1. Control without mycorrhizal inoculation
2. Inoculated with *Glomus fasciculatum* mixed inoculum
3. Inoculated with *Glomus mosseae* mixed inoculum
4. Inoculated with *Gigaspora margarita* mixed inoculum
5. Inoculated with *Sclerocystis dussii* mixed inoculum

Fig 2.3 : Showing Macerated Section of the Host Plant Maize (Zea mays L) was Significantly Colonized with *Glomus fasciculatum* (Thaxter) Gerd & Trappe. (No Scale). C.cy=Central cylinder, V=Vesicles.

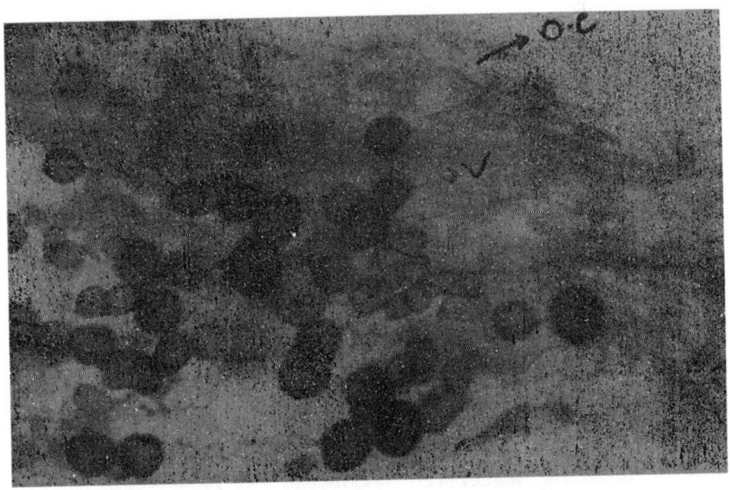

Fig 2.4 : Showing Macerated Section of the Host Plant Maize (Zea mays L) was Significantly Colonized with *Glomus mosseae* (Nicol & Gerd) Gerd & Trappe. (No Scale). o.c=Outer cortex, V=Vesicles.

Fig 2.5 : Showing Macerated Section of the Host Plant Sorghum halpensis Brot. was Significantly Colonized with *Gigaspora margarita* Becker&Hall. (No Scale). hy=Hypha, V=Vesicles.

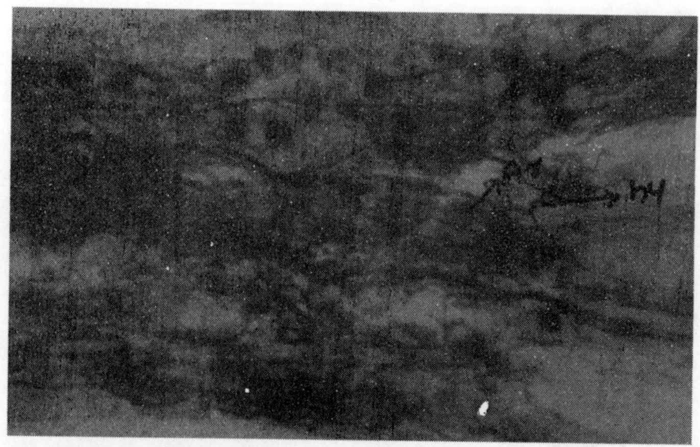

Fig 2.6 : Showing Macerated Section of the Host Plant Sorghum Halpensis Brot. was Significantly Colonized with *Sclerocystis dussii* (Pat)V.Hohn. (No Scale). hy=Hypha, Ar= Arbuscules.

investigation on the effect of four AM fungal species inoculation brought growth response when grown in unsterile soil. These results confirm the earlier reports of (Roa *et al.*, 1989; Lakshman and Jayashankar, 2004; Lakshman and Inchal, 2005; Lakshman, Biradar and Narayana Reddy, 2007) and have shown that medicinal plant contain several secondary metabolites to response to AMF colonization when inoculated artificially. A pronounced growth response was followed by the inoculation of *Sclerocystis dussii* in all parameters.

CONCLUSION

Costus speciosus Smith is a rare medicinal plant, its parts are used in diabetes. Improvement of this medicinal plant may be achieved with the inoculation of AM fungi. This work suggests that those plants, which are easily propagated from stem cutting /vegetative propagation needs inoculation of indigenous AM fungi. It is concluded that *Sclerocystis dussii* was the most efficient AM fungi to bring significant increase in plant height, chlorophyll content, increase in tubers and phosphorus content in shoots of *Costus speciosus* compare to *Glomus faciculatum*, *Gigaspora margarita* and *Glomus mosseae*.

REFERENCES

Addott, L.K. and Robson A.D. (1982). The effect of root density, inoculum placement and infectivity of inoculum on the formation of vesicular mycorrhizals. *New Phytologist*. 97:285-299.

Basu, M. and Srivasatava, N.K. (1998). Root endophyts in medicinal plants: there population and effect. *Mycorrhiza news*. 1(3): 2-3.

Biradar, R. and Reddy, C.N. (2007). Response of five medicinal plants to VAM inoculation in unsterile soil. *Mycorrhiza news*. 19(2): 25-29.

Earnna, N., Mallikarjunaya, R.B., Bhagyraj, D.J. and Suresh, C.K. (2001). Response of coleus aromaticus to *Glomus fasciculatum* and other beneficial soil microflora. *J.species. Aromatic crops*. 10(2):141-143.

Lakshman, H.C. and Jayashankar N. (2004). Screening of AM fungi associated with *Coriandrum Sativum* L. In Dharwad Distict and effect of introduced strain. *Nature Env.And Pollution Technology*. 3(4): pp 505-509.

Lakshman, H.C., Jayashankar. M. and Kolkar, K. P. (2004). Effect of Glomus fasciculatum (AMF) and municipal sewage water on growth of *Bauhinia racemosa Ind. J.Env and Ecoplan*. 8(2):309- 314.

Lakshman, H.C. and Inchal, R.F. (2006). Arbuscular mycorrhizal fungal association in some rare/endangered endemic medicinal plants of Hubli-Dharwad region in Karnataka. *Nature Environment and Pollution Technology*. 5(4):631-634.

Jackson, M.L. (1973). Soil chemical analysis prantisal of India limited. New Delhi. pp 248 .

Kumar, G.S. and Murugesh, S. (2002). Studies on the VAM Fungi improve growth of some medicinal plants. *Advance in Plants Science*. 15(1):43-46.

Kumar, V.M. and Mahadevan, A. (1984). The secondary substance inhibits mycorrhizal association. *Current Science*. 53:377-378.

Kumar, A.S. and Jamalluddian. (2007). Arbuscular mycorrhiza fungi associated with *Vitex nedundo* a important medicinal plant. *Mycorrhiza news*. 19(3):18-20.

Khade, S.W., Bukhara. M.J., Jaiswal.V., Gaonkar. U.C. and Rodrigues, B.F. (2002). Arbuscular mycorrhizal status of medicinal plants. *J.Eco.Taxo.Bot*. 26(3):571-578.

Lakshman, H.C. (2006). Dose a vesicular mycorrhizal fungi is need to inoculate plants. *Current Research*. 21:21-23.

Lakshman, H.C. and Raghavendra, S. (1992). Occurrence of VAM in importance medicinal plants. *Current research*. Pp.21-23.

Phillips, J. M. and Hayman, D. S. (1970). Improved procedure for clearing roots and staining parasitic and vesicular arbuscular mycorrhizal fungi for rapid assessment of infection. *Translation of the British Mycologist Society*. 55:158-161.

Prasad, V. and Reddy, C.N. (1998). VA-myceotrophy in some important ornamental and medicinal plants of Gulbarga, *India. Bulletin of pure and Applied Science*. 17B(1):23-29.

Rao, Y.S.G., Suresh, C.K., Suresh, N.S., Mallikarjunaiah, R. R., Bagyaraj, D.J. (1989).Vesicular arbuscular mycorrhizae in medicinal plants. *Indian Phytopathology*. 92(3):17-21.

Smith, S.E. and Read, D.J. (1997). Mycorrhizal symbiosis academic press California, pp. 498.

3

Arbuscular Mycorrhizal Fungi - Essential Partner in Sustainable Plant Growth

Shuchi Agarwal, K K Koul, Rafiq Lone

INTRODUCTION

The word "mycorrhiza" was coined by Frank in 1885 for the fungus specifically those of the roots (Greek, Myco – fungus, rhiza – roots). These fungi are described as the universal symbionts of plants, since these forms show an association practically with every taxonomic group; the plants showing association far outweigh to those without association. These fungi are obligatory biotrophic organisms that live symbiotically with the roots of most plants (Kirk *et al.*, 2001).

Arbuscular mycorrhiza, or AM (formerly known as vesicular arbuscular mycorrhizas) are mycorrhiza whose hyphae enter into the root cell walls producing structures that are either balloon-like (vesicles) or dichotomously branching invaginations (arbuscules). The fungal partners are members

Plant Biochemistry and Mycorrhizal Laboratory, School of Studies in Botany, Jiwaji University, Gwalior - 474011 (M.P)
Corresponding Author: Prof. K K Koul, SOS in Botany, Jiwaji University, Gwalior- 474011 (M.P) kirteekumar@gmail.com

of the Glomeromycota (a newly designated phylum whose members were earlier placed in the Zygomycota).

The fungal hyphae do not penetrate the protoplasm but invaginate the cell membrane. These arbuscular structures greatly increase the contact surface area of the hyphae and the cell continuum so as to facilitate the increased transfer of nutrients between the two.

Estimates show that about 95 percent of plant families from all phyla of land plants are hosts to AM fungi (Trappe, 1987; Remy *et al.*, 1994) except for some angiospermic families *viz.* Proteaceae (Nicholson, 1967; Brundrett *et al.*,1996), Cruciferae, Zygophyllaceae (Verma, 1998), Dipterocaceae, Betulaceae, Myrtaceae, Fagaceae (Nicholson, 1967) and Cactaceae, Chenopodiaceae, Cyperaceae, Amaranthaceae and Juncaceae (Hirrel *et al.*, 1978).

The establishment of a functional symbiosis between AM fungi and host plants involves a sequence of recognition events leading to the morphological and physiological integration of the two symbionts. The developmental switches in the fungi are triggered by host signals which induce changes in gene expression and a process leading to unequivocal recognition between the two partners of the symbiosis (Smith and Read, 1997). Lack of host specificity is even more characteristic of this symbiosis than other types known. Wide spread distribution both in terms of habitats and host species, evolution of symbiosis due to terrestrialization, host protection and growth promotiveness obligate nature and non specificity for host, positive interaction with other rhizosphere microbes and several other characteristics of AM fungi have obviously forced plant biologists to work out their practical aspects. With progress of isolation techniques, mass production methods and inoculation, AM fungi have been regarded as a boon for agriculture, forestry and restoration of disturbed niches.

Mycorrhiza is the mutualistic symbiosis (non pathogenic

association) between soil borne fungi with the roots of vascular plants (Sieverding, 1991). The partners in this association are members of fungi (basidiomycetes, ascomycetes, zygomycetes) and most vascular plants (Harley and Smith, 1983; Brundrett, 1991). Fitter and Moyersoen (1996) have provided a succinct definition as "a sustainable nonpathogenic biotropic interaction between a fungus and a root". In the mycorhizal literature, the term symbiosis is used to describe the highly interdependent mutualistic relationships where the host plant receives mineral nutrients while the fungus obtains photosynthetically derived carbon atoms (Harley and Smith, 1983). Mycorrhizal associations involve three-way interactions between host plants, mutualistic fungi and soil factors. Available fossil and molecular evidences have been supportive of that the symbiosis is of ancient origin, exemplifying co-evolution of plants and fungi for adaptations and establishments of land flora around 450 million years back (Simon *et al.*, 1993; Remy *et al.*, 1994).

Mycorrhiza is presently classified into seven types. Firstly scientists described five types, those of Arbuscular, Ericoid, Arbutoid, Ectomycorrhiza and Ectendomycorrhiza, whereas these were then described as seven types with the seggregation of Monotropoid and orchid types from the earlier five (Peterson *et al.*, 2004).

TAXONOMY

Presently the taxonomic status of AMF stands raised to a separate phylum of fungi Glomeromycota (Schubler *et al.*, 2001). The Glomeromycota is divided into four orders, eight families and ten genera. The genera which include most of the described species are *Glomus, Acaulospora, Scutellospora,* and *Gigaspora,* (Khade, 2008). More than 150 species are described within the phylum Glomeromycota on the basis of their spore development and morphology. However, the recent molecular analyses indicate that the definite number of AM taxa may be much higher (Daniell *et al.*, 2001; Vandenkoornhuyse *et al.*, 2002).

CULTURE CONDITIONS

AMF being biotrophic obligate symbionts (Koske and Polson, 1984), therefore, mostly either grow poorly or do not grow at all under *in vitro* conditions. Addition of spores or mycelia as inoculums rarely results in the root colonization under non-sterile condition (Danielson and Visser, 1989; Horton and Bruns, 2001).

Three major methods have been commonly used for inoculum production of arbuscular mycorrhizal fungi.

Pot culture- AMF are cultivated by sand based pot cultures (Menge, 1984). However, it is a labour-intensive process and not economical for mass production of inoculum.

Solution culture- Solution culture techniques such as nutrient film (Elms and Mosse, 1984; Mosse and Thompson, 1984) and aeroponics (Sylvia and Hubbell, 1986) are mainly adapted for the production of clean, soilless inoculum of AMF. Nevertheless, these methods have proven successful only in cultivating *Glomus* species (Hung and Sylvia, 1988; Sylvia and Jarstfer, 1994).

Root organ culture- This process is also known as Tissue culture techniques. In this technique contaminant-free AMF inoculum are produced. Ri-plasmid transformed-root cultures seem to offer the most efficient method for growing colonized roots (Mugnier and Mosse, 1987). An excellent piece of successful experimentation has been done on the root organ cultures of *Glomus intraradices* (Becard and Piche, 1989, 1990; Chabot *et al.*, 1992 a, b) and *Gigaspora margarits* (Chabot *et al.*, 1992 b).

ARBUSCULAR MYCORRHIZAL FUNGI AS BIOFERTILIZERS

The AM fungal mycelium appears to be more resistant than the root itself to abiotic stresses such as drought (Panwar and Vyas, 2002) toxic element and soil acidity (Sylvia & Williams, 1992). A plant with mycorrhiza remains in close contact with the soil for a longer period of time than a plant without mycorrhiza. The lifespan of the external mycelium appears to decline rapidly three to four weeks after the first infection of the plant by the fungus (Schubert *et al.*, 1987). Amongst the principal function of mycorrhiza ascertained is to increase the volume of soil explored for nutrient uptake (Panwar and Vyas, 2002) and thereby to enhance the efficiency of nutrient absorption. It has been shown that mycorrhizal plants can assimilate more carbon and accumulate several times more phosphate from the soil or solution than non-mycorrhizal plants (Mosse 1973; Mosse *et al.*, 1981; Smith and Dowd, 1981; Hayman, 1986; Schweiger, 2007). Mycorrhizal plants also accumulate P, K, Ca (Khan *et al.*, 2000), Cu and Mn in the leaf in higher concentrations than non-mycorrhizal plants (Ross and Nopamombodi *et al.*, 1987).

Biofertilizers are commonly called as microbial inoculants which are capable of mobilizing important nutritional elements in the soil from non usable to usable form by the crop plants through their biological processes. For the last two decades or so biofertilizers are used extensively as an eco-friendly approach to minimize the use of chemical fertilizers, improve soil fertility status and for sustainable enhancement of crop production by their biological activity in the rhizosphere. Extensive research has and is being carried out on the use of VAM fungi as biofertilizers for supplementing nitrogen and phosphorus fertilizers. Considerable improvement in the growth of medicinal plants, herbs and agricultural crops and forest plants has been the capacity of AM fungi to act as biofertilizers, bioregulators and bioprotectors (Mulongoy *et al.*, 1992; Lovato *et al.*, 1994; Linderman *et al.*, 1992) and has repeatedly been demonstrated. Thus, they play a key role in sustainable conservation of tropical gene pool and diversity (Panwar and Vyas, 2002). AM fungi are well-known to bring about biochemical changes in plants by increasing various enzymatic activities (Panwar and Vyas, 2002), accumulation of growth hormones such as Indole acetic acid (IAA), cytokinins and gibbberellin like substances, nodulation in legumes, transplant shock (Mosse, Stribley and Le Tacon, 1981; Bagyaraj, 1990; Bagyaraj and Verma, 1995), and plant establishment in disturbed and hostile soils (Gianinazzi *et al.*, 1991). Acid phosphatase and nitrate reductase are important enzymes of phosphorus and nitrogen metabolism, respectively. Peroxidase (PRO) and polyphenol oxidase (PPO) are important components of the defence mechanism of plants against pathogens (Schenck *et al.*, 1977; Schonbeck, 1980; Schenck, 1981; Schonbeck and Dehne, 1981; Dehne, 1982; Bagyaraj, 1984; Graham, 2001; Sylvia and Williams, 1992; Reddy *et al.*, 2006). Phenols and phenolic compounds (Baltruschat and Schonbeck, 1972; Ling-Lee *et al.*, 1977; Dehne *et al.*, 1978; Krishna and Bagyaraj, 1983; Morandi *et al.*, 1984) are also important in plant disease resistance. The beneficial effects of AMF fungi on plant health improvement are well established (Prasad 2000, 2002, Parkash *et al.*, 2005; Javot *et al.*, 2007), therefore, can play major role in the improved cultivation of plants.

The needs for indigenous AM species isolation, screening and identification have been emphasized for these purposes. It has been a common experience that although chemical fertilizers have doubled the agricultural productivity however, the intensity of mycorrhizal infection and also spore production have decreased. Therefore, keeping in view the co-evolution of AM and plants, the level of synthetic fertilizers have to be minimized to increase AM in the soil by inoculum so as to pursue sustainable agricultural practices of the shrinking available land for the purpose.

Although a few studies have been conducted on medicinal plants, the results are encouraging as the growth of seedlings and productivity was found to enhance. In countries like India the use of AM fungi in the forestry, cultivation of medicinal plants and herbs and traditional agriculture appears to be more important due to large scale deficiencies in facilities like irrigation fertilizer and plant protection. The practical use of AM fungi is further enhanced as these are effective in overcoming the stress conditions like drought, disease and deficiencies of nutrients.

PHYSIOLOGY AND BIOCHEMISTRY OF AM FUNGI

Major benefits, associated with AM symbiosis are the enhanced phosphorous (P) status of mycorrhizal plants. Thus P is the most important 'currency' exchange between the host and symbionts in the symbiosis. After absorbing P from the soil in the dissolved form as a solution, the fungi first incorporate it into the cytosolic pool, and the excess P thus accumulated is transferred to the vacuoles. The vacuolar P pool probably plays a central role in the P supply to the plant. The main forms of inorganic P in fungal vacuoles are orthophosphate and polyphosphate, however, organic P molecules may also be present. Long distance translocation of P from the site of uptake in the external mycelium to the site of transfer to the plant is probably achieved via transfer of vacuolar components. This transport is supposed to be mediated either by protoplasmic streaming or through the motile tubular vacuole-like system. The site of release of P into the interfacial apoplast and then to the plant is most probably the fungal arbuscules. The biochemical and biophysical processes involved in P metabolism and transfer between cellular compartments in the symbiosis are currently not well understood, though it can be suggested, without reservations that the enhancement is due to the uptake of P from the soil by the AM fungus and then its subsequent translocation and transfer to the host plant (Smith and Smith, 1989). The phosphate flux has been estimated at 13 nmol m^{-2} s^{-1} across the arbuscule in the mycorrhiza; although this rate can increase if exogenous phosphate is supplied (Cox and Tinker, 1976; Smith *et al.*, 1994). A high affinity phosphate transporter has been cloned from *Glomus versiforme*; however, the gene transcripts corresponding to the transporter are only detected in the extraradical mycelium and not in the fungal structures within the root (Harrison and van Buuren, 1995). This suggests that this transporter must be responsible for the initial uptake from the soil matrix but the mechanisms for the transfer of P to the host plant have yet to be identified (Harrison and Van Buuren, 1995).

The arbuscule is thought to be the major site for nutrient exchange between the two symbiotic partners. Therefore, mechanisms should be present to facilitate the transport of these compounds. Membrane H^+-ATPases are considered the major transport proteins controlling ionic and molecular transport processes at the cellular level in plants and fungi. Studies have revealed changes in distribution and activation of particular H^+-ATPases in the peripheral, peri-arbuscular and arbuscular membranes of functional AM interfaces (Marx *et al.*, 1982; Gianinazzi-Pearson *et al.*, 1991; Bago *et al.*, 1997). Utilizing a PCR cloning approach based on the use of highly degenerate primers, five partial genomic clones encoding P-type ATPases (GmHA1-5), were isolated from the AM fungus, *G. mosseae* (Ferrol *et al.*, 2000). These clones represent putative isoforms, whose function remains to be analysed, but expression analysis of GmHA1 and GmHA2 demonstrated that the former was expressed in the intraradical hyphae and the latter in the extraradical hyphae (Ferrol *et al.*, 2000).

BRASSICAS ARE NOT ASSOCIATED WITH AM FUNGI

Brassicas are not associated with arbuscular mycorrhizal (AM) fungi because *Brassicas* are known to release some inhibitory compounds. They enhance the growth of *Brassica* crops, particularly *B. napus annua* (canola) and *B. juncea* (Indian mustard), reducing the incidence of root diseases to a greater extent than non-*Brassica* break-crops (Angus *et al.*, 1991). This effect has been termed 'biofumigation' and is thought to result from biocidal compounds being released when glucosinolates (GSLs) in *Brassica* tissues are hydrolysed to form compounds including isothiocyanates (ITCs). However, the impact of biofumigation on beneficial organisms such as rhizobia and arbuscular mycorrhizal (AM) fungi has not been investigated. However, *Brassicas* may also negatively affect AM colonization in subsequent crops, and are generally considered to be non-hosts for AM fungi and also while a field is planted with a *Brassica* no new AM inoculum is produced and the viability of existing inoculum declines. In addition, the release of ITCs by brassicas may directly affect AM fungi through reducing the viability of AM fungal inoculum or affecting growth of the fungi (Schreiner and Koide, 1993; Lakshman *et al.*, 2001)

IMPACT ON PLANT PROTECTION AND MICROBIAL INTERACTIONS

AM fungi are recognized as high potential agents in plant protection and pest management (Quarles, 1999; Sharma *et al.*, 1996; St. Arnaud *et al.*, 1995). In several cases direct biocontrol potential has been demonstrated, especially for plant diseases caused by *Phytophtora, Rhizoctonia*, and *Fusarium* pathogens (Abdel-Aziz *et al.*, 1997; St. Arnaud *et al.*, 1997; Vigo *et al.*, 2000). Several studies have confirmed synergism between AMF and biocontrol agents such as *Burkholderia cepacia* Palleroni & Holmes (Ravnskov *et al.*, 2002), *Pseudomonas fluorescens* Migula (Edwards *et al.*, 1998), *Trichoderma harzianum* Rifai (Datnoff *et al.*, 1995), and *Verticillium chlamydosporium* Kamyschko ex Barron and Onions (Rao *et al.*, 1997). These interactions suggest that AM might affect plant and soil microbial activity by stimulating the production of root exudates, phytoalexins, and phenolic compounds

(Morandi, 1996; Norman *et al.*, 2000). A small increase of activity of plant defence genes, especially for the production of chitinases, glucanases, flavonoid biosysthesis, and phytoalexins, has been observed during mycorrhizal growth, however, these mycorrhizal defence induction mechanisms remain transitory (Harrison, 1997).

MATERIALS AND METHODS IN AMF

Collection of samples

Soil and root plant samples are collected at the same time. Soils are sampled at the depth of about 15 cm by digging beside the plant root. Soil is collected from the field. All of the samples are stored in polyethylene bags at 2–5°C until the spores are extracted and counted.

Examination of vesicles and arbuscules in the root tissue

The soil is washed from the plant roots, the feeder roots are removed from each plant. They are cleared and stained to determine the percent of root colonization using the Phillips and Hayman (1970) method. The roots are cleared in 10% KOH, boiled for about 15 minutes and then stained in lactoglycerol trypan blue stain. Roots are randomly cut into 1-cm lengths (100 pieces) and placed on slides. A microscope (x100) is used to estimate the percentage of root colonization (Nicolson, 1975) by counting the arbuscules or surface mycelia or clusters of echinulate vesicles. (Miller *et al.*, 1985)

Per cent root colonization is calculated using following relationship.

$$\text{Percent colonization} = \frac{\text{Total number of colonized root pieces}}{\text{Total number of root pieces examined}} \times 100$$

EXTRACTION AND COUNTING OF AM SPORE PROPAGULES FROM THE SOILS

The AM spores are extracted from the soil samples by wet sieving and decanting as per method described by

Gerdemann (1963). Soil samples of approximately 25 gm are suspended in one litre of water which is gently stirred. After heavier particles are allowed to settle for a few minutes and the liquid is decanted through a sieve (250 microns) to remove the large particles of organic matter and allow the desired spores to pass through. The suspension is then passed through a series of sieves of (240 microns, 120 microns, 60 microns and 30 microns). The spores and small amounts of debris thus remaining on each sieve are transferred from the beakers separately and the spores are collected under the stereoscopic microscope. Spores from each sample are counted based upon 100 grams of soil (Gaur and Adholeya, 1994), propagule numbers for every sample recorded and spores of different types are collected and transferred on the glass slide containing polyvinyl lacto glycerol (PVLG) (Omar *et al.*, 1979) with or without Melzer reagent (Morton, 1988).

Identification and Taxonomic characterization of Arbuscular mycorrhizal fungi

The AM fungi are identified using manual provided by Schenck and Perez (1990). Identifications are done according to their spore morphology and wall characters (Walker and Treppe, 1993; Wu, 1993; Schenck and Perez, 1990).

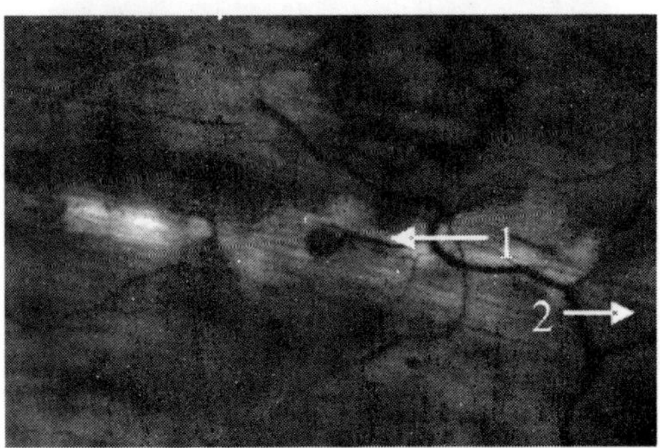

Fig. 3.1: AMF fungal hyhae (2) and vesicles (1) inside the root

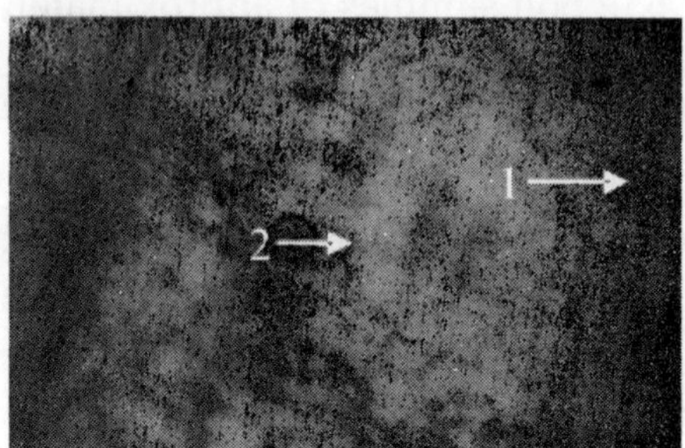

Fig 3.2: AMF fungal hyphe (1) and spore (2) inside the root

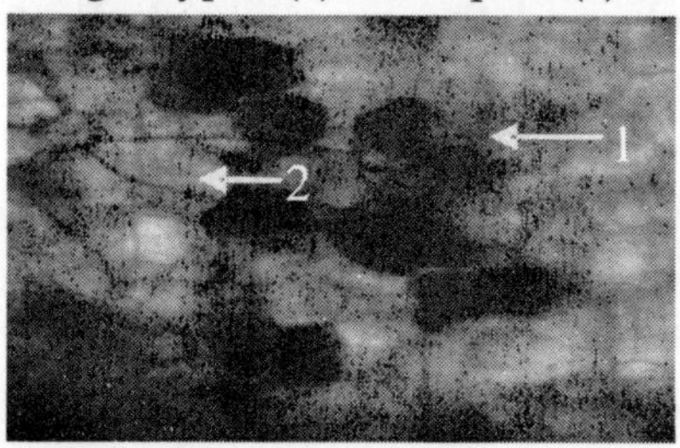

Fig 3.3:AMF Fungal vesicles (2) and arbuscules (1) inside the root

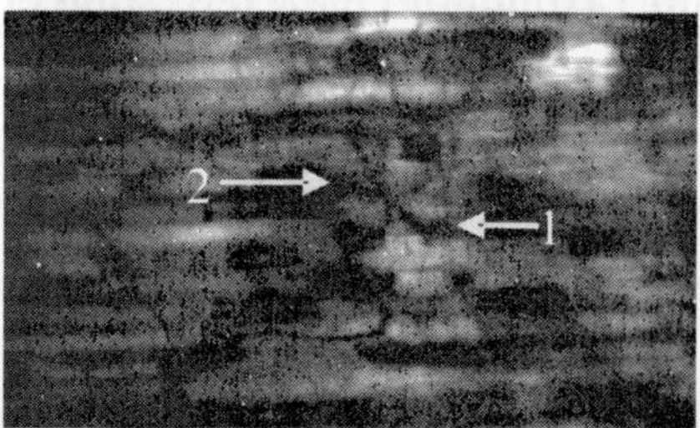

Fig. 3.4: AMF Fungal hyphae (2), auxillary cells (1) inside root

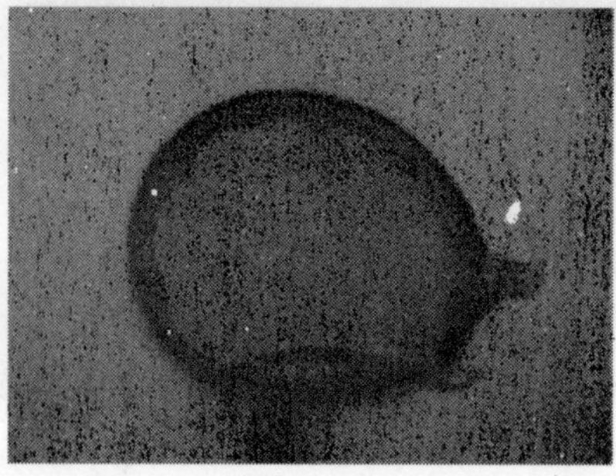

Fig. 3.5: AMF Fungal spore; *Glomus geosporum*

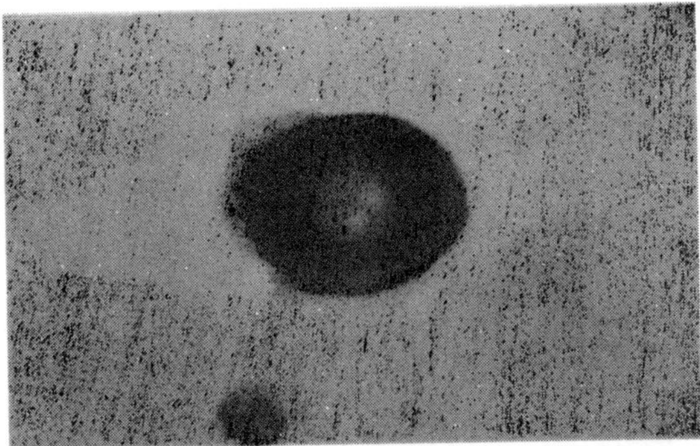

Fig 3.6: AMF Fungal spore; *Glomus Mosseae*

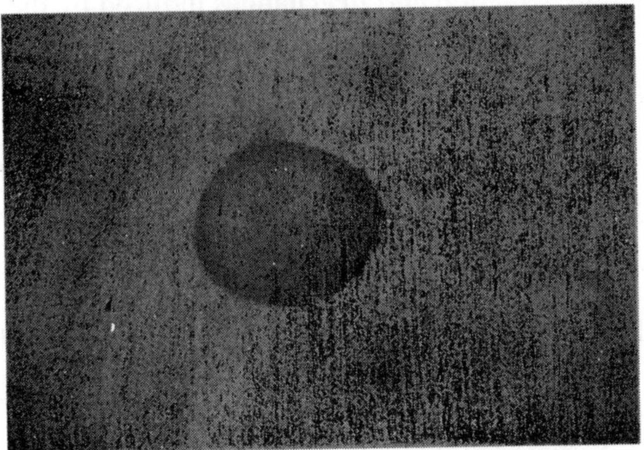

Fig. 3.7: AM Fungal spore *Glomus intraradices*

CONCLUSION

There are more than seven type of mycorrhizae are know among which arbuscular mycorrhizal fungi (AMF). Estimates about 95 percent of plant families from all phyla of land plants. This fungi considered to be the most promising biofertilizer. How one can know its taxonomy its cultural charcterstics and physiological micro-organisms interaction. This aid in minimizing the use of chemical fertilizers, improve the soil fertility status and sustatainable enhancement of crop production by their biological activity in the rhizosphere.

REFERENCES

Abdel-Aziz, R. A., Radwan, S. M. A., Abdel-Kader, M. M. and Barakat, M.I.A. (1997). Biocontrol of faba bean root-rot using VA mycorrhizae and its effect on biological nitrogen fixation. *Egypt. J. Microbiol.* 31:273–286.

Angus, J. F., Gardner, P. A., Kirkegaard, J. A. and Howe, G. N. (1991). Productivity and break-crop effect of winter growing oilseeds. *Australian Journal of Experimental Agriculture.* 31:669–77.

Bago, B., Vierheilig, H., Piche, Y. and Azcon-Aguilar, C. (1996). Nitrate depletion and pH changes induced by the extraradical mycelium of the arbuscular mycorrhizal fungus *Glomus intraradices* grown in monoxenic culture. *New Phytologist.* 133:273–280.

Bagyaraj, D.J. (1995). Mycorrhizal association in crop plants and their association in agriculture. In: Nair, M.C., Balakrishnan, S., (eds), Beneficial fungi and their utilization, Scientific publishers, Jodhpur, India, 61-71.

Bagyaraj, D. J. and Manjunath, A. (1990). Mycorrhizal symbiosis and plant growth. Mycorrhiza Net Work Asia, University of Agriculture Sciences, Bangalore,

Bagyaraj, D. J. (1984). Biological interactions with mycorrhizal fungi. In: *VA mycorrhiza* Powell, C. L. and Bagyaraj, D. J. (eds.) CRC Press, New York, pp: 131–153.

Baltruschat, H. and Schonbeck, F. (1972). The influence of endotrophic Mycorrhiza on the injestation of tobacco by *Thielaniofsis basicola. Phytophathol.* 284:171–188.

Becard, G. and Piche, Y. (1989). Fungal growth stimulation by CO_2 and root exudates in vesicular-arbuscular mycorrhizal symbiosis. *Appl Environ Microbiol.* 55: 2320–2325.

Becard, G. and Piche, Y. (1990). Physiological factors determining vesicular-arbuscular mycorrhizal formation in host and non-host Ri T-DNA transformed roots. *Can J Bot.* 68:1260–1264.

Burndett, M. (1991). Mycorrhizas in natural ecosystems. *Advances in Ecological Research.* 21:171–313.

Brundrett, M., Bougher, N., Dell, B., Grove, T. and Malajczuk, N. (1996). *Working with mycorrhiza in forestry and agriculture*, ACIAR.

Chabot, S., Becard, G. and Piche, Y. (1992a). Life cycle of *Glomus intraradix* in root organ culture. *Mycologia.* 84:315–321.

Chabot, S., Bel-Rhid, R., Chênevert, R. and Piché, Y. (1992b). Hyphal growth promotion *in vitro* of the VA mycorrhizal fungus *Gigaspora margarita* Becker & Hall, by the activity of structurally specific flavonoid compounds under CO_2 enriched conditions. *New Phytol.* 122: 461–467.

Cox, G. and Tinker, P.B. (1976). Translocation and transfer of nutrients in vesicular arbuscular mycorrhizas. I. The arbuscule and phosphorus transfer: a quantitative ultrastructural study. *New Phytol.* 77:371–378.

Danielson, R. M. and Visser, S. (1989). Host response to inoculation and behavior of introduced and indigenous ectomycorrhizal fungi of jack pine grown on oil-sands tailing. *Can. J. Forest Res.*, 19:1412–1421.

Daniell, T. J., Husband, R., Fitter, A. H. and Young, J. P. W. (2001). Molecular diversity of arbuscular mycorrhizal fungi colonising arable crops. *FEMS Microbiol. Ecol.* 36:203–209.

Datnoff, L. E., Nemec, S. and Pernezny, K. (1995). Biological control of fusarium crown and root rot of tomato in Florida using *Trichoderma harzianum* and *Glomus intraradices. Biol. Control.* 5:427–431.

Dehne, H. W. and Schonbeck, F. (1975). The influence of the endotrophic mycorrhiza on the fusarial wilt of tomato; Z. *Pflanzenkr. Pflanzenschutz.* 82:630–639.

Dehne, H. W., Schonbeck, F. and Baltruschat, (1978). The influence of endotrophic mycorrhiza on plant diseases 3. Chitinase-activity and the omithine-cycle; Z. *Pflanzenkr. Pflanzenschutz.* 85:666–678.

Dehne, H. W. (1982). Interaction between vesicular-arbuscular mycorrhizal fungi and plant pathogens, *Phytopathology.* 72:1115–1119.

Edwards, S. G., Young, J. P. W. and Fitter, A. H. (1998). Interactions between *Pseudomonas fluorescens* biocontrol agents and *Glomus mosseae*, an arbuscular mycorrhizal fungus, within the rhizosphere. *FEMS Microbiol. Lett*. 166:297–303.

Elmes, R. P. and Mosse, B. (1984). Vesicular-arbuscular endomycorrhizal inoculum production. II. Experiments with maize *(Zea mays)* and other hosts in nutrient flow culture. *Canad. J. Bot.* 62:1531–1536.

Ferrol, N., Miguel-Barea, J. and Azcon-Aguilar, C. (2000). The plasma membrane H^+-ATPase gene family in the arbuscular mycorrhizal fungus *Glomus mosseae*. *Current Genetics*. 37:112–118.

Fitter, A.H. and Moyersoen, B. (1996). Evolutionary trends in root-microbe symbiosis. *Philos. Trans. R. Soc. Lond. B. Boil. Sci.* 351:1367–1375.

Frank, A.B. (1885). Uber die auf Wurzelsymbiose beruhende Ehrnahrung gewisser Baum durch unterirdische Pilze. Berichte der Deutschen Botanischen Gesselschaft, 3:128-145. (Cited).

Gaur A. and Adholeya A. (1994). Estimation of VAM spores in the soil – A modified method. *Mycorrhiza News*. 6(1):10-11.

Gianinazzi-Pearson, V., Smith, S. E., Gianinazzi, S. and Smith, F.A. (1991). Enzymatic studies on the metabolism of vesicular-arbuscular mycorrhizas. V. Is Hc-ATPase a component of ATP-hydrolysing enzyme activities in plant-fungus interfaces? *New Phytol*. 117: 61–74.

Gerdemann, J.W., Nicolson, T.H. (1963). Spores of mycorrhizal *Endogone* species extracted from soil by wet sieving and decanting. *Trans. Br. Mycol. Soc*. 46: 235–244.

Giovannetti, M. and Mosse, B. (1980). An evaluation of techniques for measuring VAM infection in roots. *New Phytologist*. 84: 489–500.

Grace, C. and Stribley, D. P. (1991). A safer procedure for routine staining of vesicular arbuscular mycorrhizal fungi. *Mycological Research*. 95: 1160-2.

Graham, J. H., and Abbott, L. (1998). Functional diversity of mycorrhizal fungi in the wheat rhizosphere. In 'Proceedings of the ICOM II Conference'. Pp. 72-3. (Uppsala, Sweden.)

Harley, J.L. and Smith, S.E. (1983). *Mycorrhizal symbiosis* (1st ed.). Academic Press, London.

Harrison, M.J. Van Buuren, M.L. (1995). A phosphate transporter from the mycorrhizal fungus *Glomus versiforme*. Nature. 378: 626–629.

Harrison, M. J. (1997). The arbuscular mycorrhizal symbiosis: an underground association. *Trends in Plant Science*. 2: 54–60.

Hayman, D. S. (1986). Mycorrhizae of nitrogen-fixation legumes; *MIRCEN J. Appl. Microbiol Biotechnol.* 2: 121–145.

Hirrel, M.C., Mehravaran, H., Gerdemann, J.W. (1978). Vesicular-arbuscular mycorrhizae in the Chenopodiaceae and Cruciferae: do they occur? *Can J Bot.* 56:2813–2817.

Horton, T. R., Bruns, T. D. (2001). The molecular revolution in ectomycorrhizal ecology: peeking into the black-box. *Mol. Ecol.* 10: 1855–1871.

Hung, L. L. and Sylvia., D. M. (1988). Production of vesicular-arbuscular mycorrhizal fungus inoculum in aeroponic culture. *Appl. Environm. Microbiol.* 54: 353–357.

Javot, H., Pumplin, N., Harrison, M. J. (2007). Phosphate in the arbuscular mycorrhizal symbiosis: transport properties and regulatory roles. *Plant Cell and Environment*. V. 30, p.310–322.

Khade, S.W. (2008). Morpho-taxonomy of synonyms: Glomus species (Glompromycota). *Anale de Biologia*. 30:55–59.

Khan, A.G., Kuek, C., Chaudary, T. M., Khoo, C. S. Haeys, W. J., (2000). Role of plants, mycorrhizae and phytochelators in heavy metal contaminated land remediation. *Chemosphere*. 41, 197–207.

Kirk, P.M., Cannon, P.F., David, J.C., Stalpers, J. (2001). Ainsworth and Bisby's Dictionary of the Fungi. 9th ed. CAB International, Wallingford, UK.

Kirkegaard, J. A., and Sarwar, M. (1998). Biofumigation potential of brassicas I. Variation in glucosinolate profiles of diverse field-grown brassicas. *Plant and Soil*. 201: 71–89.

Koske, R.E., Polson, W.R., (1984). Are VA-mycorrhizae required for sand dune stabilization, *Bioscience*. 34: 420–424.

Krishna, K. R., Bagyaraj, D. J. (1983). Interaction between *Glomus fasciculatum* and *Sclerotium rolfsii* in peanut; *Can. J. Bot.* 61: 2349–2351.

Lovato, P., Schiiepp, H., Trouvelot, A. and Gianinazzi, S. (1994). In *Mycorrhiza: Structure, Function and Molecular Biology and Biotechnology* (eds. Varma, A. and Hock, B.), Springer-Verlag, Heidelberg, pp. 443–457.

Linderman, R. G. (1992). In *Mycorrhiza and Plant Health* (eds. Pfleger, F. L. and Linderman, R. G.), APS Press, St. Paul, M. N., pp. 1–26.

Ling-Lee, M., Chilvers, G. A., Ashford, A. E. (1977). A histochemical study of phenolic materials in mycorrhizal and uninfected roots of *Eucalyptus fastigata* Deane and Maiden, *New Phytol.* 78: 313–328.

Lakshman, H.C., Mulla, F.I., Inchal, R.F. and Srinivasalu,Y. (2001). Prevalence of arbuscular mycoprrhizal fungal colonization in some disputed plants. *Mycorrhiza News*. 13(3): 16–22.

Marx, C., Dexheimer, J., Gianinazzi-Pearson, V., Gianinazzi, S. (1982). Enzymatic studies on the metabolism of vesicular arbuscular mycorrhiza. IV. Ultracytoenzymological evidence ATPase for active transfer processes in the host arbuscular interface. *New Phytologist*. 90: 37–43.

Menge, J. A. (1984). Inoculum production. *In*: VA mycorrhiza., Powell, C. L. and Bagyaraj, D. J. Eds. CRC Press, Inc., Boca Raton, Florida. Pp. 187–203.

Morandi, D. (1996). Occurrence of phytoalexins and phenolic compounds in endomycorrhizal interactions and their potential role in biological control. *Plant Soil*. 185: 241–251.

Morton, J. B. (1988). Taxonomy of VA mycorrhizal fungi: Classification, nomenclature, and identification. *Mycotaxon*. 32: 267–324.

Mosse, B. (1973). Advances in the study of vesicular-arbuscular mycorrhiza; *Annu. Rev. Phytopathol.* 11: 171–196.

Mosse, B., Stribley, D.P., Tacon, F.Le. Ecology of mycorrhizae and mycorrhizal fungi, in: Alexander, M. (ed.), Advances in microbial ecology, vol. 5, Plenum Press, New York and London, 1981, p. 250.

Mosse, B., Warner, A. & Clarke, C. A. (1981). Plant growth responses to *Vesicular-Arbuscular Mycorrhiza*. XIII. Spread and introduced VA endophyte in the field and residual growth effect in the second year. *New Phytologist*. 90: 521–528.

Mosse, B. and Thompson, J. P. (1984). Vesicular-arbuscular endomycorrhizal inoculum production. I. Exploratory experiments with beans (*Phaseolus vulgaris*) in nutrient flow culture. *Canad. J. Bot*. 62: 1523–1530.

Miller, J.C. Rajapakse, S. and Garbaer, R.K. (1985). Vesicullar arbuscular mycorrhizae in vegetable crops. Hort.Sci. 21:974–984.

Mugnier, J., Mosse, B. (1987). Vesicular-arbuscular mycorrhizal infection in transformed root-inducing T-DNA roots grown axenically. *Phytopathology*. 77: 1045–1050.

Mulongoy, K., Gianinazzi, S., Roger, P. A. and Dommergues, Y., (eds) in *Microbial Technology: Economic and Social Aspects* (Da Silva, E., Sasson, A. and Rotledge, C.) (1992), Cambridge Univ. Press, Cambridge, pp. 59–69.

Norman, J. R. and Hooker, J. E. (2000). Sporulation of *Phytophthora fragariae* shows greater stimulation by exudates of non-mycorrhizal than by mycorrhizal strawberry roots. *Mycol. Res*. 104: 1069–1073.

Nicolson, T.H. (1967). Vesicular-arbuscular mycorrhiza - a universal plant symbiosis. *Science Progress,* oxford 55: 561–581.

Nicolson, T.H. (1975). In endomycorrhiza, (Eds). F.E. Sanders, B.Mosse and P.B. Tinker. Academic press, London pp. 25–44.

Nopamornbodi, O., Thamsurakul, S., Charoensook, P. (1987). Selection of efficient VA mycorrhizal species for increasing soybean mungbean and peanut growth. Annual Report, 1987. Soil Sci. Div. Bangkok. Pp. 16.

Omar, M.B., Bolland, L., Heather, W.A. (1979). P.V.A. (polyvinyl alcohol). A permanent mounting medium for fungi. *Bull. Brit. Mycol. Soc.* 13, 31-32.

Ortas, I. (2003). Effect of selected mycorrhizal inoculation on phosphorus sustainability in sterile and no-sterile soils in the Harran Plain in South Anatolia. *J. Plant Nutr.* 26(1): 1–17.

Panwar, J., Vyas, A. (2002). AM fungi: A biological approach towards conservation of endangered plants in Thar desert, India. *Current Science.* 82 (5): 576–578.

Phillips, J. M. and Hayman, D. S. (1970). Improved procedures for clearing roots and staining parasitic and vesicular-arbuscular mycorrhizal fungi for rapid assessment of infection; *Trans. Br. Mycol. Soc.* 55: 158–161.

Peterson, R.L., Massicotte, H.B. and Melville L.H. (2004). *Mycorrhizas: anatomy and cell biology.* National Research Council Research Press.

Pfleger, F.L., Linderman, R.G. (1994). Mycorrhizae and Plant Health. American Phytopathological Society Press, St. Paul, MN. Pp 344.

Prakesh, V., Aggarwal. A, Sharma. S., Sharma, D. (2005). Effect of endophytic mycorrhizae and fungal bioagent on the development and growth of *Eucalyptus saligna Sm.* Seedlings. *Bulletin National Institute of Ecology,* 15: 127–131.

Prasad, K. (2000). Growth responses of *Acacia nilotica* (L). Del. inoculated with Rhizobium and *Glomus fasciculatum* VAM fungi. *Journal of Tropical Forestry.* 16 (1): 22–27.

Prasad, K. (2002). Effect of arbuscular mycorrhiza on biomass yield, uptake and translocation of nitrogen, phosphorous and potassium in *Azadirachta indica* L. In: *Frontiers in Microbial Biotechnology and Plant Pathology.* (Manoharachary C., Purohit D.K., Reddy S.R, Singaracharya M.A. and Girisham S., eds.). *Scientific Publishers Jodhpur,* pp. 187–191.

Quarles, W. (1999). Plant disease biocontrol and VAM fungi. *IPM Pract.* 21:1–9.

Reddy, N., Raghavender, C.R., Sreevani, A. (2006). Approach for enhancing Mycorrhiza mediated disease resistance of tomato damping-off. *Indian Phytopath.* 59(3): 299–304.

Rao, M. S., Kerry, B. R., Gowen, S. R., Bourne, J. M. and Parvatha, R. P. (1997). Management of *Meloidogyne incognita* in tomato nurseries by integration of *Glomus deserticola* with *Verticillium chlamydosporium*. *Pflanzenkrank. Pflanzenschutz.* 104: 419–422.

Ravnskov, S., Larsen, J., and Jakobsen, I. (2002). Phosphorus uptake of an arbuscular mycorrhizal fungus is not affected by the biocontrol bacterium *Burkholderia cepacia. Soil Biol. Biochem.* 34:1875–1881.

Remy, W., Taylor, T.N., Hass, H., Kerp, H. (1994). Four hundred million year old vesicular-arbuscular mycorrhizae. *Proc. National Academy of Sciences.* 91: 11841–11843.

Ross, J.P., Harper, J.A. (1970). Effect of *Endogane mycorrhiza* on soyabean yields. *Phytopathology*. 60: 1552–1556.

Schenck, N. C. (1981). Can mycorrhiza control root disease. *Plant Dis.* 65: 230–234.

Schenck, N.C. (1987). Vesicular-arbuscular mycorrhizal fungi and the Control of Fungal Root Disease. In: "Innovative Approches to Plant Disease Control". Chet I. ed. John Wiley and Sons, New York. (179–191).

Schenck, N. C., Ridings, W. B., Cornell, J. A. (1977). Interaction of two vesicular-arbuscular mycorrhizal fungi and *Phytophthora parasitica* of two citrus root stocks, in *Proceedings of the third North American Conference on Mycorrhizae* (Oregon: Corvallis), p. 9.

Schenck, N. C. and Perez Y. (eds.). (1990). Manual for the identification of VA mycorrhizal fungi. Synergistic Publications, Gainesville.

Schonbeck, F. (1980). Endomycorrhiza: Ecology, function and phytopathological aspects; *Forum Microbiol.* 3: 90–96.

Schonbeck, F. and Dehne, H. W. (1981). Mycorrhiza and plant health; *Gesunde Pflanz.* 33: 186–190.

Schreiner, R. P. and Koide, R. T. (1993). Mustards, mustard oils and mycorrhizas. *New Phytologist*. 123: 107–13.

Schüβler, A., Gehrig, H., Schwarzott D., Walker C. (2001). Analysis of partial Glomales SSU rRNA gene sequences: implications for primer design and phylogeny. *Mycol. Res.* 105: 5–15.

Schüßler A, Schwarzott A.D., Walker, C. (2001). A new fungal phylum, the Glomeromycota: phylogeny and evolution. *Mycological Research*. 105: 1413–1421.

Schubert, A., Marzachi, C., Mazzitelli, M., Cravero, M. C., Bonfortepasolo P. (1987). Development of total and viable extraradical mycelium in the vesicular-arbuscular mycorrhizal fungus *Glomus calrum* Nicol. & Schenck. *New phytol.*107: 183–190.

Schweiger, P., F, Robson, A. D., Barrow, N. J., Abbot, L. K. (2007). Arbuscular mycorrhizal fungi from three genera induce two phase plant growth response on a high P-fixing soil., *Plant and Soil*. 292 (1–2): 181–192.

Sieverding, E. (1991). Vesicular-Arbuscular Mycorrhiza management in Tropical Agrosystems. *Deutche Gesellschaft für Technische Zusammenarbeit*, GTZ No. 224. Eschborn, p. 371.

Sharma, S., and Dohroo, N. P. (1996). Vesicular-arbuscular mycorrhizae in plant health and disease management. *Int. J. Trop. Plant Dis.* 14: 147–155.

Sieverding, E. (1991). *Vesicular-Arbuscular Mycorrhiza Management in Tropical Agrosystems*. GTZ, Eschborn, Germany.

Simon, L., Bousquet, J., Lévesque, R.C., Lalonde, M. (1993). Origin and diversification of endomycorrhizal fungi and coincidence with vascular land plants. *Nature*. 363: 67–69.

Smith, W.H., Dowd, M. L. (1981). Biomass production of mycorrhiza in Florida. *Forestry*. 79: 508–511.

Smith, S. E., Read, D. J. (1997). Mycorrhizal symbiosis. New York: Academic Press.

Smith, S., Gianinazzi-Pearson, E.V., Koide, R., and Cairney, J.W.G., (1994). Nutrient transport in mycorrhizas: structure, physiology and consequences for efficiency of the symbiosis. *Plant soil*. 159: 103–133.

Smith, F.A., Smith, S. E. (1989). Membrane transport at the biotrophic interface: an overview. *Australian Journal of Plant Physiology*. 16: 33–43.

St. Arnaud, M., Hamel, C., Caron, M., and Fortin, J. A. (1995). Endomycorhizes VA et sensibilite des plantes aux maladies: Synthese de la litterature et mecanismes d interaction potentielsin: La Symbiose Mycorhizienne. Etat des Connaissances.(J. A. Fortin, Y. Piche, C. Charest, eds.). Editions Orbis, Frelisghburg, Quebec. P. 51–87.

St. Arnaud, M., Hamel, C., Vimard, B., Caron, M., and Fortin, J. A. (1997). Inhibition of *Fusarium oxysporum f. sp. dianthi* in the non-VAM species *Dianthus caryophyllus* by co-culture with *Tagetes patula* companion plants colonized by *Glomus intraradices*. *Can. J. Bot.* 75, 998–1005.

Sylvia, D. M., and Hubbell, D. H. (1986). Growth and sporulation of vesicular-arbuscular mycorrhizal fungi in aeroponic and membrane systems. *Symbiosis*. 1: 259–267.

Sylvia, D.M., and Williams, S. F. (1992). Vesicular-arbuscular mycorrhizae and environmental stress. In Mycorrhizae and Sustainable Agriculture. Bethlenfalvay, G.J., and. Linderman, R.G., eds. ASA Special Publication 54. Madison, WI. Pp. 101–124

Sylvia, D.M., Jarstfer, A. G. (1994). Production of inoculum and inoculation with arbuscular mycorrhizal fungi. *In*: Management of mycorrhizas in agriculture, horticulture and forestry. Eds., Robson, A. D., Abbott, L. K., Malajczuk, N., Kluwer Academic Publishers, Dordrecht, Netherlands. Pp. 231-238.

Trappe, J.M. (1987). Phylogenetic and ecologic aspects of mycotrophy in the angiosperms from an evolutionary standpoint. In Ecophysiology of VA Mycorrhizal Plants., Safir G.R. (ed.), CRC Press, Florida.

Vandenkoornhuyse, P., Husband, R., Daniell, T. J., Watson, I. J., Duck, J. M., Fitter, A. H., Young, J. P. W. (2002). Arbuscular mycorrhizal community composition associated with two plant species in a grassland ecosystem. *Mol. Ecol.* 11: 1555–1564.

Verma N.K. (1998). Effect of VA mycorrhiza on the growth and P uptake in *Eupatorium adenophorum* Spring. (Asteraceae) grown in soil amended with soluble phosphate. *J. Natl. Bot.Soc.* 52: 41–45.

Vigo, C., Norman, J. R., Hooker, J. E. (2000). Biocontrol of the pathogen *Phytophthora parasitica* by arbuscular mycorrhizal fungi is a consequence of effects on infection loci. *Plant Path.* 49: 509–514.

Walker, C., Trappe, J. M. (1993). Name and epithets in the Glomales and endogonales. *Mycological Research.* 97: 339–344.

Wu, C.G. 1993. Glomales of Taiwan: III. A comparative study of spore ontogeny in *Sclerocystis* (Glomaceae, Glomales). *Mycotaxon*. 47: 25–39.

4

Utilization of AM Fungi on Growth and Nutrition in *Phyllanthus Emblica* L. (Var. NA-6)

Mahesh B Byatanal and H C Lakshman

INTRODUCTION

In recent times an enormous amount of research has been carried out on the potential role of mycorrhizae in the revegetation of disturbed landscapes (Reeves *et al.*, 1979; Allen and Allen, 1980; Zak *et al.*, 1982; Jasper *et al.*, 1987; Stathl *et al.*, 1988; Mahadevan *et al.*, 1989). Tropical forests are an abode of medicinal plants. They are in great demand due to the increased acceptance of traditional medicines, because of their property of less or no side effects. In fact, large scale collection of medicinal plants from forests has accelerated the depletion of tropical forests, thus making them endangered and threatened group of plant species. In order to maintain a sustained supply of raw materials to the drug industry, these plants are encouraged to be cultivated outside the forest

---

Post Graduate Department Studies in Botany (Microbiology Laboratory), Karnatak University Dharwad -580 003, India

ecosystem in recent years (Crew *et al.*, 1978; Plenchette *et al.*, 1983; Aldrich-Wolfe *et al.*, 2007).

Mycorrhiza helps seedling survival by increasing rooting intensity and surface area of root. Such an increase will have little impact on highly mobile nutrients such as NO_3-N, but will have a large impact on uptake of poorly mobile nutrients and water. For example, with AM the inflow of phosphorus into mycorrhizal plants is sometimes 4-5 time that in non-mycorrhizal plants (Sanders and Tinker, 1973). However, many soil factors as well as nutrient deficiency can lead to low rooting intensity such as high acidic or low soil temperature, high aluminium, root disease and high salinity. Bowen (1980, 1991) suggested that rather than restricting our thinking only to mycorrhiza function in low nutrient soils, mycorrhizas should be considered as an alternate mode for ion uptake in a range of soil conditions deleterious to root growth.

The early development of most tree seedlings is dependent on an adequate mycorrhizal association. Seedlings which receive adequate light, soil and water resources will develop mycorrhizae if viable propagules are present in the rhizosphere. Juvenile legumes are dependent on a tripartite relationship with both mycorrhizal fungi and *Rhizobium*. Nursery cultural practice should be designed to foster a healthy rhizosphere environment for tree seedlings.

Tree seedling response to AM fungal inoculation in the nursery is highly dramatic. Inoculation of some tree genera (e.g., *Citrus, Liquidambar, Juglans*) have resulted in 2–5 times increase in root and shoot biomass in the nursery. Other nursery benefits include increased water use efficiency of seedlings, improved nutrition, and reduced incidence and damage from soil-born pathogens and pollutants. Their response is highly dependent on host and symbiont genetics. Matching the appropriate symbionts with host seed source in an area which merits further testing.

INOCULUM PRODUCTION OF AM FUNGI

AM fungi cannot be grown in pure culture and consequently the techniques used for inoculum production are different from those generally employed for other biotechnological interesting microbes (Gianinazzi and Gianinazzi-Pearson, 1981). AM inoculum has to be produced on living roots which is usually considered a major disadvantage. However, this process can have the advantage of minimizing the risk of maintaining fungi that have lost their symbiotic properties. On the other hand, producing AM fungi on living plants is problematic because of the precaution that has to be taken in order to obtain pure inoculum. It is possible to produce AM on excised roots or whole plants under axenic conditions using disinfected spores as inoculum (Hepper, 1984), but for the moment this procedure is not viable for mass production.

METHOD OF AM FUNGAL INOCULATION FOR TREE SEEDLINGS AT GREENHOUSE LEVEL

For many greenhouse studies inoculum of any type can be mixed thoroughly with the soil. In other cases, it may be desirable to add inoculum next to the roots when plants are transplanted. For more precise studies, it is often necessary to assume that a single plant or root becomes colonized from single or relatively few spores. To accomplish this, it is necessary to confine a spore close to the root or assure that roots grow near spores. Several techniques are effective. Spores may be pipette out directly onto roots when transplanting seedlings. However, spores applied in this manner become dislodged from the roots easily. Alternatively, spores can be pipette onto a small piece of filter paper and the paper wrapped around rootlets to keep spores in place. The filter paper will decompose within 2 to 3 weeks.

The buried slide techniques may be useful in following mycorrhizal infection from various inocula or under different soil conditions. Powell (1976), coated slides with agar, placed spores or root fragments on one end of the slide and onion seed on the other and buried the slides. Onion roots grew down the slide and became infected.

Phyllanthus emblica L. (var. NA-6) is one such medicinal plant with numerous medicinal properties. It is a branched tree grows up to a height of 10–20 m. In the tropical region, it is supposed to be an evergreen tree but behaves as deciduous tree due to complete defoliation of leaves. Every part of *Phyllanthus emblica* L. (var. NA-6) has medicinal use and is used for treating chronic dysentery, bronchitis, diabetes, fever, diarrhoea, jaundice, dyspepsia, cough, toothache, fever, anaemia, epilepsy, sores, pimples, fistula, goner, gravel, cholera, sterility in women, anthrax. (Pathak, 2003).

The current day emphasis is on sustainable agriculture, which uses less of chemical inputs like fertilizers and pesticides having adverse effect on soil health, fertility and environment. Thus, use of microbial inoculants play an important role in sustainable agriculture (Bagyaraj, 2006). Arbuscular mycorrhizal fungi are known to improve the nutritional status, growth and development of plants, protect plants against root pathogens and offer resistance to drought and salinity (Jeffries, 1987). Though these fungi are not host specific, recent studies have clearly brought out host preference in arbuscular mycorrhizal fungi. Host preference has been reported in many forest tree species like *Casuarina equisetifolia* (Vasanthakrishna *et al.*, 1995); *Tectona grandis* (Rajan *et al.*, 2000); *Garcinia indica* (Lakshmipathy *et al.*, 2003) and a few medicinal plant species like *Phyllanthus amarus* and *Withania somnifera* (Earanna, 2001), and *Coleus forskohlii* (Gracy *et al.*, 2005). Hence, in the present investigation, it is envisaged the need for selecting efficient arbuscular mycorrhizal fungi for inoculating *Phyllanthus emblica* L. for its cultivation.

MATERIALS AND METHODS

Experimental Soil

The physical and chemical soil characteristics used for pot experiment were estimated according to Jackson (1973). Per cent of organic matter is determined according to Piper (1950). Electric conductivity is measured using Bridge meter and pH by 1:2.5 (w/v) soil to water ratio. The experimental soil was sandy loam with pH 8.1, Electric conductivity 320 uS/cm, Moisture 4.86%, Total organic Carbon 1.71, Nitrogen 0.08%, Phosphorus 4.52%, Potassium 7.94%, Magnesium 0.121%, Calcium 0.472%, Copper 0.03ppm, Zinc 3.86ppm, Manganese 0.97ppm, Iron 8.24ppm. The soil was steam sterilized for 1 hour on two consecutive days.

Arbuscular mycorrhizal inoculum production

Arbuscular mycorrhizal fungal spores were isolated from the rhizosphere soil of species of *Phyllanthus emblica* L. (var. NA-6) collected from National Bureau of Plant Genetic Resources. Arbuscular mycorrhizal fungal spores were isolated following the wet sieving and decanting method (Gerdemann *et al.*, 1963). The spores were identified by using manual written by Schenck *et al.* (1990). Species of *Acaulospora, Gigaspora, Glomus, Scutellospora* and *Sclerocystis* were isolated from 100 gram rhizosphere soil samples. Four species *Sclerocystis dussi, Acaulospora laevis, Gigaspora margarita* and *Glomus fasciculatum* were dominant in this soil and were therefore selected for use in this experiment. Isolated spores were multiplied using onion as host plants, for three months in culture pots measuring 15 cm height, 30 cm diameter pots were filled with autoclaved soil (121°C for 1h at 15 psi) from the same site. After three months onion was cut at ground level, the roots were chopped into 1 cm pieces and mixed with the soil from rhizosphere of host plants. This soil based inoculum was used for inoculation.

Soil preparation and seed sowing

For raising plants, soil was collected from garden and stored in a greenhouse. 10 kg of sterilized soil was amended in experimental pots. Arbuscular mycorrhizal inoculum was added at the rate of 10 g and seeds of *Phyllanthus emblica* L. (var. NA-6) were scarified in 5% H_2SO_4 for 10 min, washed with sterilized distilled water. Seedlings were raised in polythene bags containing soil inoculated with isolates of different AM fungi. Seedlings were selected for uniformity and one seedling per pot was planted.

Establishment of test plants and greenhouse experiments

The studies were conducted under greenhouse conditions with temperature ranges from 28–31°C. Arbuscular mycorrhizal fungal treatments were given on a layer below the germinated seedlings to the pot soil. Control (without mycorrhiza inoculation), 10g inoculum/10kg soil was given to each pot inoculated with *Sclerocystis dussi, Acaulospora laevis, Gigaspora margarita* and *Glomus fasciculatum*. The experiment was completely randomized with three replication per treatment of the NA-6 Variety. All the pots were maintained under greenhouse condition. To maintain moisture, pots were watered on every alternate day. 20 ml of Hogland solution was added to all the pots once in fifteen days.

Growth parameters

Three seedlings per treatment were harvested with rhizosphere soil at 120 days interval. Roots were washed thoroughly with tap water to remove adhering soil particles. Shoot and root fresh and dry weight, per cent root colonization, spore number and phosphorus content in shoots were recorded. Shoot and root dry weight were obtained after drying for 72h in a hot air oven at 80°C.

Mycorrhizal status

Roots were processed using Phillips *et al.*, 1970, technique to study per cent colonization (Figure 4.2). Arbuscular mycorrhizal fungal spores were isolated by wet sieving and decanting method (Gerdeman *et al.*, 1963).

Phosphorus content

The phosphorus content in the shoots was determined by the vanado-molybdate phosphoric acid yellow color method outlined by Jackson, 1973.

Statistical analyses

The data were statistically analyzed using Analysis of Variance (ANOVA) and the means were separated by Duncan's Multiple Range Test (DMRT) using SPSS 7.5 (Snedecor *et al.*, 1980).

RESULTS AND DISCUSSION

In general, mycorrhizal inoculation resulted in a significant increase in plant height, biomass and phosphorus content of *Phyllanthus emblica* L. (var. NA-6). In *Phyllanthus emblica* L. plant height was highest in plants inoculated with *Glomus fasciculatum* and differing significantly from other treatments and the least being in the uninoculated control plants (Table 4.1 and Figure 4.2). Highest plant biomass was observed in plants inoculated with *Glomus fasciculatum* followed by *Sclerocystis dussi, Acaulospora laevis* and *Gigaspora margarita*. Uninoculated control plants had significantly the least plant biomass compared to all other inoculated treatments (Table 4.1 and Fig. 4.1).

In *Phyllanthus emblica* L. (var. NA-6) all the four inoculation treatments showed higher shoot phosphorus content and differed significantly among each other. The highest shoot phosphorus content was observed in plants inoculated with *Glomus fasciculatum* and the least P content was observed in uninoculated plants (Table 2 and Fig. 4.1).

In this study highest mycorrhizal percent colonization was observed in plants treated with *Glomus fasciculatum* (Table 1 and Fig.1), followed by other treatments. Highest number of mycorrhizal spores in the root zone soil was also observed in *Glomus fasciculatum* treated plants. Least number of spores occurred in the root zone of uninoculated plants (Table 2).

Phyllanthus emblica L. (var. NA-6) plants varied in their response to inoculation with different mycorrhizal fungi. Most of the AM fungi resulted in significant increase in plant height (Figure 4.2), plant biomass and phosphorus content of *Phyllanthus emblica* L. (var. NA-6). Host preference among AM fungi has been reported by earlier workers (McGraw *et al.*, 1981; Gracy *et al.*, 2005), hence, the need for inoculating different mycotrophic plants has been stressed (Jeffries *et al.*, 1987; Bagyaraj, 2007). Improved plant height, because of AM fungal inoculation has been reported in other medicinal plants like Coleus (Boby *et al.*, 2003; Lakshman *et al.*, 2002) and Kalmegh (Tharun *et al.*, 2006). The plant biomass is an important parameter for selecting a fungus for its symbiotic efficiency. Plant biomass (shoot + root) was enhanced due to

Table 4.1 : Effect of soil inoculation with AM fungi on shoot and root length, fresh weight of shoot and root, dry weight of shoot and root of *Phyllanthus emblica* L. (NA-6 variety) for 120 days.

	SL(cm)	FWS(gm)	DWS(gm)	RL(cm)	FWR(gm)	DWR(gm)
CN	16.25e	6.28e	2.09e	41.52e	10.25e	2.83e
SD	27.56b	18.52b	6.72b	58.21b	25.38b	7.25b
AL	25.82c	14.24c	4.39c	53.82c	20.49c	5.96c
GM	21.52d	11.62d	3.89d	50.29d	15.38d	4.10d
GF	36.58a	27.82a	10.25a	62.19a	30.58a	9.76a

SL-Shoot Length; FWS-Fresh Weight of Shoot; DWS-Dry Weight of Shoot; RL-Root Length; FWR-Fresh Weight of Root; DWR-Dry Weight of Root; CN-Control; SD-*Sclerocystis dussii*; AL-*Acaulospora laevis*; GM-*Gigaspora margarita*; GF-*Glomus fasciculatum*; Means sharing a letter in columns are not significantly different according to Duncan's test $P<0.05$.

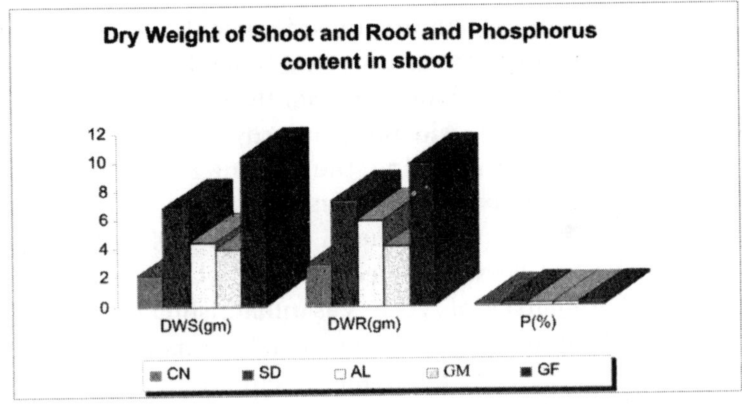

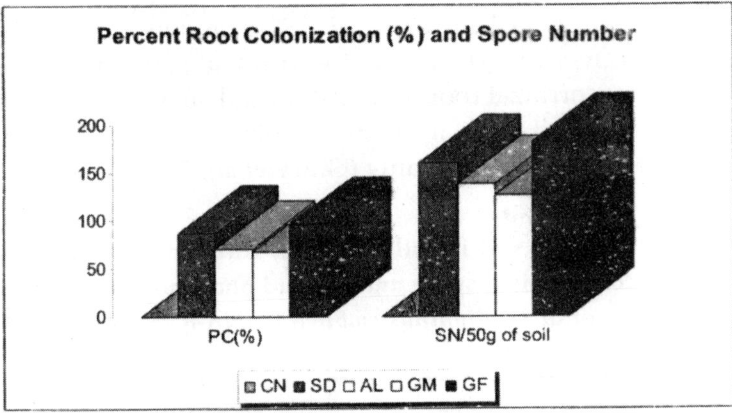

Fig. 4.1. : Effect of AM fungi on dry weight of shoot, dry weight root, phosphorus uptake in shoot, per cent colonization and spore number of *Phyllanthus emblica* L. (var. NA-6). CN-Controley; SD-*Sclerocystis dussi*; AL-*Acaulospora laevis*; GM-*Gigaspora margarita*; GF-*Glomus fasciculatum*.

Glomus fasciculatum inoculation compared with uninoculated plants. Increased plant biomass because of inoculation with AM fungi has been reported in aromatic plants like palmarosa (Gupta *et al.*, 1991), eucalyptus(Oliveria *et al.*, 1995), bergamot mint (Kotharis *et al.*, 1999), *Tinospora cardifolia* (Lakshman, *et al.*, 2004) and sweet basil (Copetta *et al.*, 2007).

Increased plant growth due to AMF inoculation is mainly through improved uptake of diffusion limited nutrients such as P (Krishna *et al.*, 1991; Lambert *et al.*, 1979; Gracy *et al.*, 2005; Lakshman *et al.*, 2004). AM fungi improving plant biomass were also good in increasing the P content of the host, significantly highest being in plants inoculated with *Glomus fasciculatum*. Selected efficient fungi enhancing plant biomass and P uptake has been reported in other plants by several workers (Ulfath Jaiba *et al.*, 2006; Vasanthakrishna *et al.*, 1995). Such higher P content in AMF inoculated plants is attributed to higher influx of P into the plant system through AM fungi which explores the soil volume beyond P depletion zone (Bagyaraj *et al.*, 1995; Hattingh *et al.*, 1973; Sanders *et al.*, 1971). The enhancement in growth and nutritional status was also related to mycorrhizal root colonization and spore numbers in the root zone soil. This upholds the observations made by earlier workers on other plants (Gracy *et al.*, 2005; Tharun *et al.*, 2006).

Table 4.2 : Effect of soil inoculation with AM fungi on per cent colonization, spore number and phosphorus uptake in shoot of *Phyllanthus emblica* L. (NA-6 variety) for 120 days.

	PC(%)	SN/50g of soil	P(%)
CN	0.00e	0.00e	0.07e
SD	86.58b	160.22b	0.18b
AL	70.58c	138.65c	0.15c
GM	68.00d	126.21d	0.11d
GF	95.62a	182.00a	0.21a

PC-Per cent root colonization; SN-Spore number; P- Phosphorus uptake in shoots; CN-Control; SD-*Sclerocystis dussii*; AL-*Acaulospora laevis*; GM-*Gigaspora margarita*; GF-*Glomus fasciculatum*; Means sharing a letter in columns are not significantly different according to Duncan's test $P<0.05$.

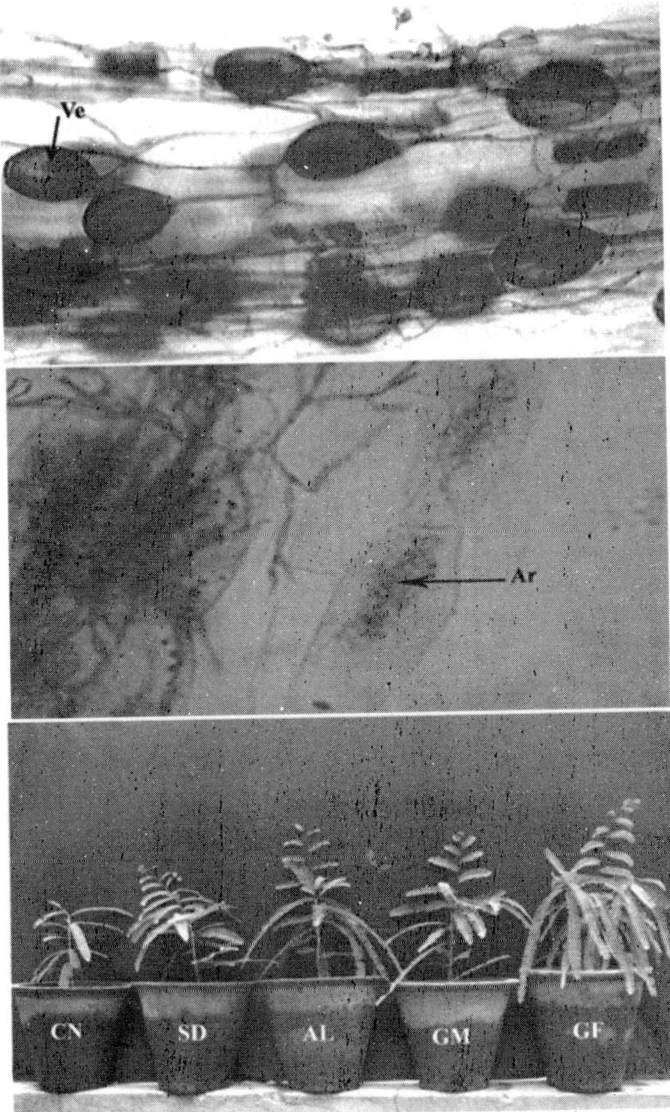

Fig. 4.2. : Showing vesicles (Ve), arbuscules (Ar) and treatment of different arbuscular fungi (SD-Sclerocystis dussii, AL-Acaulospora leavis, GM-Gigaspora margarita and GF-Glomus fasciculatum) and control (CN).

CONCLUSION

Phyllanthus emblica L. (var. NA-6) is an important medicinal tree species, its improvement by inoculation of AM fungi (*Glomus fasciculatum*) at greenhouse level, as not carried out so far. Therefore the present study brought out some significant results at greenhouse level with providing AM (*Glomus fasciculatum*) fungal inoculum brought a significant results in shoot and root length, fresh weight of shoot and root, dry weight of shoot and root, percent root colonization, spore number/50 g of soil, phosphorus uptake in shoot. Therefore it can be concluded that efficient AM (*Glomus fasciculatum*) fungi is more suitable for improving seedlings of *Phyllanthus emblica* L. (var. NA-6) at greenhouse level.

ACKNOWLEDGEMENT

Authors are indebted to U.G.C, New Delhi for awarding UGC research fellowship in Science and financial support.

REFERENCES

Aldrich-Wolfe, L. (2007). Distinct mycorrhizal communities on new and established hosts in a transitional tropical plant community. *Ecology*.**83** (3): 559–566.

Allen, E.B. and Allen, M.F. (1980). Natural re-establishment of vesicular-arbuscular mycorrhizae following strip mining reclamation in Wyoming. *J.App.Ecol.,* 17- 139–147.

Bagyaraj, D.J. (2007). Arbuscular mycorrhizal fungi and their role in Horticulture. *In: Recent trends in horticultural biotechnology for sustainable development and productivity*. Ed., Keshavchandran, R. Scientific publishers, New Delhi. Pp. 53–58.

Bagyaraj, D.J., Varma, A. (1995). Interactions between arbuscular mycorrhizal fungi and plants: their importance in sustainable agriculture in arid and semiarid tropics. *Advances in Microbial Ecology,* 14: 119–142.

Bagyraj, D.J. (2006). Arbuscular mycorrhizal fungi in sustainable agriculture. *In: Techniques in mycorrhizae*. Eds. M.J., Bukhari and B. P. Rodrigues. Dept of Botany Goa University, Goa. Pp. 1–8.

Boby, V.U and Bagyaraj, D.J. (2003). Biological control of root-rot of *Coleus forskholii* Briq. using microbial inoculants. *World Journal of Microbiology and Biotechnology*, 19: 175–180.

Bowen, G.D. (1980). Mycorrhizal roles in tropical plants and ecosystems. In: *Tropical Mycorrhizal Research,* (Ed. P. Mikola), Oxford University Press, Oxford, pp. 165–190.

Bowen, G.D. (1991). Soil temperature, root growth and plant function. In: "*Plant Roots the Hidden Hay*" (Eds. Y.Waisel, A. Eshel and Kafkafi, U), Marcel and Dekker Inc., New York, pp. 309–330.

Copetta, A., Lingua, G., Bardi, L., Masoero, G., Berta, G., Cruz, C., Egsgaard, H., Trujillo. (2007). Influence of arbuscular mycorrhizal fungi on growth and essential oil composition in *Ocimum basilicum* var. Genovese. *Caryologia,* 60: 106–110.

Crews, C.L., Johnson, E.R., Jainer, J.N. (1978). Benefits of growth and development of three woody ornamentals. *Hot.Sci.*13:429–430.

Earanna, N. (2001). VA mycorrhizal association in medicinal plants of Southeastern dry zone of Karnataka and response of *Phyllanthus amarus* and *Withania somnifera* to inoculation with VAM fungi and plant growth promoting rhizomicroorganisms. PhD thesis, Submitted to University of Agricultural Sciences, Bangalore.

Gerdemann, L.M., Nicolson, T. H. (1963). Spore of mycorrhizal Endogone species extracted from soil by wet sieving and decanting. *Trans. Brit. Mycol.Sco.* 46: 235–244.

Gianinazzi, S., Gianinazzi-Pearson, V. and. Trouvelot, V. (1981). Potentialities and procedures for the use of endomycorrhizas with special emphasis on high value crops. In: *Biotechnology of Fungi for Improving Plant Growth* (eds. J.W. Whipps and R.D. Lumsden). Cambridge University Press Cambridge, New York. Pp. 41–54.

Gracy, L., Sailo., Bagyaraj, D.J. (2005). Influence of different AM fungi on growth, nutrition and forskholin content of *Coleus forskohlii*. *Mycological Research.* 109: 795–798.

Gupta, M.L., Janardhanan, K.K (1991). Mycorrhizal association of *Glomus aggregatum* with palmarosa enhances growth and biomass. *Plant and Soil*, 131: 261–264.

Hattingh, M.J., Gray, L.E., Gerdemann, L.W. (1973). Uptake and translocation of 32P labeled phosphate to onion roots by mycorrhizal fungi. *Soil Science*, 116: 383–387.

Hepper, C.M. (1984). Isolation and culture of VA mycorrhizal (VAM) fungi. *In: VA Mycorrhiza* (eds. C.L.I. Powell and D.J. Bagyaraj), CRC Press Inc. Baca Raton, Florida, pp. 95–112.

Jackson, M.L. (1973). Soil Chemical Analysis. Prentice Hall (India) Pvt.Ltd. New Delhi. Pp. 239–241.

Jasper, D.A., Robson, A.D. and Abbott L.K. (1987). The effect of surface mining on the infectivity of vesicular arbuscular mycorrhizal fungi. *Aus. J. Bot.* 35: 641–652.

Jeffries, P. (1987). Use of mycorrhizae in agriculture, *CRC Critical Review of Biotechnology*. 5: 319–357.

Kothari, S.K., Singh Saudan ., Singh, U.B., Kumar Sushil. (1999). Response of bergamot mint (*Mentha citrata*) to vesicular arbuscular mycorrhizal fungi and phosphorus supply. *Journal of Medicinal and Aromatic Plant Sciences*. 21: 990–995.

Krishna, K.R., Bagyaraj, D.J. (1991). Role of vesicular arbuscular mycorrhiza in the uptake of micronutrients by ground nut plants. *Current Research*. 20: 173–175.

Lakshman, H.C. and Patil, G.B. (2004). Effect of AM fungi on *Tinospora cardifolia* Miers. Plants raised from stem cuttings. *J. Microb. World*. 6(1): 27–31.

Lakshman, H.C. and Srinivasulu, Y. (2002). Response of *Guizotia abyssinica* Cass. (Niger) and Linum usitattissimun (Linseed) to VA-mycorrhizal inoculation in an unsterile soil. *Karnatsk. J. Agric. Sci.*15(2) ; 405–407.

Lakshman, H.C. and Geeta. B.P (2004). Influence of AMF (*Glomus fasciculatum*) in relation to nutrient application on the growth of *Acacia auriculifromis* wild. *J.Microb.World*. 6(1), pp.37–40.

Lakshmipathy, R., Gowda Balakrishna., Chandrika, K., Bagyaraj, D.J (2003). Symbiotic response of *Garcinia indica* to VA mycorrhizal inoculation. *Indian Journal of Forestry*. 26: 143-146.

Lambert, D.H., Baker, D.E., Cole, H. (1979). The role of mycorrhizae in the interaction of phosphorus with zinc, copper and other elements. *Soil Science Society American Journal*. 43: 976–980.

Mahadevan, A., Raman, N., and Natarajan, K. (1989). *Mycorrhizae for Green Asia*. University of Madras, Madras. Pp. 351.

Mcgraw, Schenk, N.C. (1981). Effect of two species of vesicular-arbuscular mycorrhizal fungi on the development of Fusarium with tomato. *Mycologia*. 7: 894–897.

Oliveria, V.L.F., De Zambolin, L., Nenes, J.C.L., De Oliveria, V.L. (1995). Growth of eucalyptus seedlings inoculated with mycorrhizal fungi, *Fitopatologia Brasilar*, 20: 164–168.

Pathak, R.K., (2003). Status Report on Genetic Resources of Indian Gooseberry-Aonla (*Emblica officinalis* Gaertn.) in South and Southest Asia.

Phillips, J.M., Hayman, D.S. (1970). Improved procedure for clearing roots and staining parasitic and vesicular Arbuscular mycorrhizal fungi for rapid assessment of infection. *Transaction of British Mycological Society*. 55:158–161.

Piper, C.S. (1950). Soil and Plant analysis interscience Publications, New York. Pp. 91.

Plenchette., Fortin, J.A., furlan, V. (1983). Growth responses of seasonal plant species to mycorrhizae in a soil of moderate p-fertility. I Mycorrhizal dependency under field conditions. *Plant Soil*. 70: 199–209.

Powell, C.L.I. (1976). Development of mycorrhizal infection from *Endogone* spores and infected root segments. *Trans. Br. Mycol. Soc*. 66: 439–445.

Rajan, S.K., Reddy, B.J.D., Bagyaraj, D.J. (2000). Screening of mycorrhizal fungi for their symbiotic efficiency with *Tectona grandis*. *Forest Ecology and Management*. 126: 91–95.

Reeves, F. B., Wagner, D., Moorman, T. and Kiel, J. (1979). The role of endomycorrhizae in revegetation practice in the semi-arid west. I. A comparison of incidence of mycorrhizae in severely disturbed, natural environments. *Amer. J. Bot.*, 66: 6–13.

Sanders, F.E. and Tinker, P.B. (1973). Phosphate flow into mycorrhizal roots. *Pestic.Sci*, 4: 383–395.

Sanders, F.E. and Tinker, P.B. (1971). Mechanism of absorption of phosphate from soil by *Endogone* mycorrhizas. *Nature*, 233: 278–279.

Schenck, N.C. and Perez, Y. (1990). Manual for the identification of mycorrhizal VAM fungi. University of Florida. Gainesville, Florida. Pp. 241.

Snedecor, G.H. and Cochran, W.H (1980). Stastical methods. 7th Ed. Iowa state college Press, Ames, Iowa. USA.

Stathl, P.D., Williams, S.E. and Christenses M. (1988). Efficacy of native vesicular-arbuscular mycorrhizal fungi after severe soil disturbance. *New Phytol.*, 110: 347–354.

Tharun, C., Bagyaraj, D.J., Patil, C.S.P. (2006). Response of *Andrographis paniculata* to different arbuscular mycorrhizal fungi. *Journal of Agricultural Technology*, 2: 221–228.

Ulfath Jaiba, J., Balakrishna, A.N., Bagyaraj, D.J., Arpana, J. (2006). Screening of an efficient AM fungus for inoculating *Piper longum* Linn. *Journal of Soil Biology and Ecology*, 26: 109–116.

Vasanthakrishna, M., Bagyaraj, D.J., Nirmalnath, J.P. (1995). Selection of efficient VA mycorrhizal fungi for *Casuarina equisetifolia* – Second screening. *New Forests*. 9:157–162.

Zak, J.C., Danielson, R.M. and Parkinson D. (1982). Mycorrhizal fungal spore numbers and species occurrence in two amended mine spoils in Alberta. Canada, *Mycologia*. 74: 347–354.

5

Cynobacterial Inoculants for Sustainable Agriculture A Case Study on Three Varieties of Paddy

[1]*M Jayashankara*, [2]*V S Bheemreddy*, [2]*S B Gadi*, [3]*Krishna H Wadder*, [3]*Rekha R Naik* and [3]*H C Lakshman*

INTRODUCTION

In India, for small and marginal farmers, green manuring may be important because of high cost of chemical fertilizers. Moreover, green manuring can also do reclamation of "User lands". In addition to nitrogen, green manures also provide organic matter, N, P, K and minimize the number of pathogenic microorganisms in soil (Dubey, 2004; Gahuikar, 2009).

Role of the cynobacteria in the paddy fields and even water logging condition, they multiply and capable of both

1. Post Graduate Department of Microbiology, Mangalore University, Field March Kariyappa College Campus, Madikeri – 571201. India.
2. Department of Botany, Janata Shikshana Samiti's College, Dharwad – 580004. India.
3. Post Graduate Department of Botany, (Microbiology lab) Karnatak University, Dharwad – 580003, India.

nitrogen fixation and oxygenic photosynthesis. The oxygen sensitivity of nitrogenase requires that these processes be separated in nitrogen fixing cyanobacteria. Heterocystous cyanobacteria achieve this by confining nitrogen fixation to heterocysts while carbon fixation occurs in vegetative cells. While heterocysts organizations such as *Anabaena* are grown on light/dark (L/D) cycles, both CO_2 and nitrogen fixation occur in light (Mackeras, 1990). When the cyanobacterial biomass incorporated in the soil conserves organic matter, nitrogen, phosphorus and moisture and improves the chemical and physical structure of the soil. This characteristic of several strains of cyanobacteria may be very useful in reclaiming saline and alkaline soil. This aspect has been well substantiated and documented under different agroclimatic conditions (Kaushik, 1993; Kaushik and Ummat, 1992; and Bossiba, 1999).

Cyanobacteria can bring about a variety of beneficial effects on the soil. The production of polysaccharides and other gummy substances helps in improving aggregation property of soil by acting as cementing agent. This helps in improved hydraulic conductivity, aeration and water molding capacity of soil. The release of various nitrogen and carbon compounds helps in maintaining or improving soil carbon to nitrogen ration (Kaushik, 1983a and Kaushik, 1983b) leading to improved microbial activity. Cyanobacteria are capable of solubilizing unavailable phosphatic deposits, causes reduction in the soil pH, electrical conductivity, and exchangeable sodium percentage.

Singh (1961) has indicated the possibility of reclamation of salt affected soils with native cyanobacteria. The ultimate success of a bio-ameliorant cyanobacterium in a salt affected soil would be reflected by the successful establishment of the crop with a, moderately high yield (Singh, 1961).

Pot trial experiments were conducted by (Jayaraman and Shanmugasundaram, 1993) to assess the efficacy of *Scytonema lormanii application* as a microbial inoculant in the presence and absence of urea nitrogen. The cyanobacterium was given as a single dose and also in three doses, on the 5th, 35th and 65th day after transplantation of paddy seedlings. Uninoculated, controls with and without the addition of urea were maintained. Cyanobacterial treatments were better compared to control which received neither cyanobacterium nor urea nitrogen. A definite saving of 25% chemical N_2 was found to be possible when cyanobacteria were applied along with chemical N_2 (Jayaraman and Shanmugasundaram, 1993). Application of heterocystous and non-heterocystous cyanobacterial biofertilizers to the rice field clearly showed increase in number of grains and weight of the grains and also enhanced the NPK content of the soil (Joshi, 2007). Immobilized as well as free-living cyanobacterial application was found to be distinctly advantageous over control, as it enhanced significantly the various parameters of growth of rice plants such as shoot and root length, fresh and dry weight of the plants, chlorophyll and protein content over a period of 30 days (Lakshman, 2009). The production of growth promoting substance by cyanobacteria and their direct growth promoting effect on rice crop has been well elucidated by Kannaiyan *et al.* (1997). But, whether these growth promoting substances were hormones, vitamins or amino acids or any other related compounds and their mode of action is not very clear (Sankaram *et al.*,1966).

Fuller and Roger (1952), observed a greater uptake of P by plants from algal materials than from inorganic phosphates when applied in equal amounts. They concluded that: (1) P in algal materials is more available to plants than organic phosphates over longer periods, (2) chemical fixation is not so important a factor with respect to algal material as with respect to inorganic phosphates, and (3) A temporary conversion of available soil or fertilizer P into cell materials by algae may be a desirable process in terms of long-term availability. Algae also have long been known to take up P more than their immediate needs, which may subsequently be released in the form of dissolved organic P (Lean and Nalewajko, 1976).

MASS CULTIVATION OF CYANOBACTERIA

For outdoor mass cultivation of cyanobacterial biofertilizers, the regional specific strains should be used. However, many germplasm collection laboratories have been established by the D.B.T. in different parts of the country for the development of starter inoculum. Mixture of 5 or 6 regional acclimatized strains of cyanobacteria, *e.g.* species of *Anabaena, Aulosira, Cylindrospermum, Gloeotrichia, Nostoc, Plectonema, Tolypothrix* are generally used for starter inoculum. The following four methods are used for mass cultivation : (i) cemented tank method., (ii) shallow metal troughs method, (iii) polythene lined pit method, and (iv) field method. The polythene lined pit method is most suitable for small and marginal farmers to prepared algal biofertilizer. In this method, small pits are prepared in field and lined with thick polythene sheets. Mass cultivation of cyanobacteria is done by using any of the four methods under the following steps:

(i) Prepare the cemented tanks, shallow trays of iron sheets or polythene lined pits in an open area. Width of tanks or pits should not be more than 1.5 m. This will facilitate the proper handling of culture.

(ii) Transfer 2-3 kg soil (collected from open place for 1 m^2 area of the tank) and add 100g of superphosphate. Water the pit to about 10 cm height. Mix lime to adjust the pH 7. Add 2 ml of insecticide *e.g.* malathion to protect the culture from mosquitoes. Mix well and allow settling down soil particles.

(iii) When water becomes clear, sprinkle 100 g of starter inoculum on the surface of water.

(iv) When temperature remains between 35-40° during summer, optimum growth of cyanobacteria is achieved. Always maintain the water level to about 10 cm during this period.

(v) After drying, the algal mat will get separated from the soil and forms flakes. During summer about 1 kg pure algal mat per m^2 area is produced. These are collected, powdered kept in sealed polythene bags and supplied to the farmers.

(vi) The algal flakes can be used as starter inoculum if the same process is repeated. The concept of utilizing BGA as fertilizer for fixing nitrogen in paddy fields was developed by De in India, who recognized that the field fertility was maintained by the proliferation of cyanobacteria in the soil. This principle was later confirmed Arora (1969).

PHYSIOLOGY OF N_2 FIXATION

Cyanobacterial N_2 fixation is carried out by unicellular or filamentous forms with or without heterocysts. The best known examples of nitrogen fixing cyanobacteria are filamentous heterocystous species *(Anabaena, Nostoc)*, which differentiate vegetative cells to heterocysts only when grown in the absence or combined. Heterocyst differentiation is considered as the culmination of nitrogen. Heterocyst differentiation is considered as the culmination of process which begins as a non-developmental response to nitrogen deprivation. The heterocysts, specialized for aerobic fixation of nitrogen are the site of the enzyme complex nitrogenase which catalyzes the conversion of nitrogen to ammonia.

RECENT APPROACH

Kaushik (2002), have reported the solubilization of mussorie rock phosphate by *Calothrix braunii, Tolypothrix ceylonica, Scytonema cincinnatum, Hapalosiphon fontinalis* and *Westiellopsis prolifica*. The cyanobacterial biofertilizers may be applied to the rice fields along with cheaper sources of phosphorus for making phosphorus available to plants, thus allowing the sustainable use of phosphatic fertilizers. Besides the conventional approach of screening natural isolates having mineral phosphate solubilizing phenotype, genetic engineering of non-mineral phosphate solubilizing cyanobacteria to become efficient phosphorus solubilizers is possible. So far, no mineral phosphorus solubilizing genes have been cloned in cyanobacteria.

ADVANTAGES OF CYANOBACTERIAL BIO FERTILIZERS

Biofertilizers are supplement fertilizer for meeting the nutrient needs of crops. They can fix nitrogen under optimum. Thus, a special separation exists between conditions and solubilise, phosphate and potassium from the soil. Also they liberate growth promoting substances and vitamins and help to maintain soil fertility and they suppress the incidence of pathogens and control diseases.

ALGALIZATION

Venkataraman (1975) coined the term '*algalization*' to denote the process of blue green algal culture in field as biofertilizer. He initiated algalization technology in India and demonstrated the way how this technology could be transferred to farmer level that holds small lands (Vekataraman and Shanmugasundaram, 1992).

At present, algalization is being followed in Tamil Nadu and Uttar Pradesh, and tried in Jammu and Kashmir, Andhra Pradesh, Karnataka, Maharastra and Haryana. Algalization is also being practiced in China, Egypt, Philippines and erstwhile U.S.S.R.

In 1990, Department of Biotechnology, Government of India, New Delhi established four major centres in different paddy growing areas of the country for acceleration and extension of works on algal biofertilizers. The programme was launched under Technology Development and Demonstration Projects on Cyanobacterial Biofertilizers in U.P. (Lucknow), Tamil Nadu (Madurai), West Bengal (Calcutta) and New Delhi. The main objectives of the programme were (i) to develop low cost indigenous technology for mass production of cyano bacteria, (ii) to isolate regional specific fast growing and better N_2-fixing strains, (iii) to develop starter inoculum, (iv) to demonstrate the farmers in field, and (v) to study the benefits on both economy and ecology. As a result of these studies cyanobacterial biofertilizer was found very useful, especially for small and marginal farmers of the country with the view point of both economy and ecology (Belwani, 1966).

The starter culture used for multiplication normally is a soil based mixture of *Aulosira, Tylopothrix, Scytonema, Nostoc, Anahaena* and *Plectonema.* The idea of using a mixture of algae is to offset the econologlcal or edaphic dangers to anyone particular strain in a given area. The only shortcoming is that when several forms are used as inocula, the ultimate proportion of individual strains in the soil tends to be unpredictable. Under field conditions this variation is less important, since the establishment of any efficient form will be enough. The standard conditions for propagation of BGA as developed at the Indian Agricultural Research Institute, New Delhi, India are shown. When commercial fertilizer nitrogen is used, the dose can be reduced by one third and supplemented with algae. However, algae can also be used along with high levels of nitrogenous fertilizer.

MATERIALS AND METHODS

In the present investigation, there were three var. of paddy has been selected to understand the effect of cynobacterial and AM fungal noculation of paddy.

Var. Jaya (IET-723)

The scientific name of this variety is IET-723. This paddy variety takes about 130 days to grow it is grown in irrigated areas. It grows upto (82 cm), the grain is long bold and white. It yields 50-60 quintals/hectares. This rice used as food. The rice straw is used as cattle feed and for making strawboards, paper, and mats.

Var. Sonamasuri

Sonamasuri rice is a rice grown largely in the Indian states of Andhra Pradesh and Karnataka. It is considered to be premium quality widely used in making rice porridge during sankranthi or harvest festival in the south. This rice used as food.

Var. Arya

This type rice is common used for cooking. The plant grown within 120 days in irrigated areas. The rice growing seasons depend on climatic conditions and vary in different parts of India. The grain of rice is chiefly used as food. Rice-bran oil for making soaps and cosmetics.

EXPERIMENTAL DESIGN AND TREATMENTS

The starter Nostoc inoculum was collected from Navalur tank, the colour of the pond and identified and puttered. It is situated between 15° 20′ to 15° 20′ North latitude and 75° to 75° 70′ East longitude. The pot experimental was conducted in the net house in the botanical garden, department of Botany, Karnatak University Dharwad, using sterilized sandy, loamy soil to know the effect of cynobacteria on the growth and yield of three rice plants. The experimental soil is sterilized with 5% methyl bromide, and physico chemical characteristics are analyzed as outlined by Jackson (1973). Shown in (Table 5.1). Earthen pots measuring 15 × 15 cm diameter are filled with 3 kg of fine sieved sterilized sandy loamy soil pH 6.4. 10 g of air-dried Nostoc culture were placed 2 cm below the surface of pot mixture before sowing the paddy. The experimental design was a complete randomised block design with 4 replicates of each treatment.

Table 5.1: Showing the List of Nitrogen Fixing Cynobacteria in South Western Part of Karnataka.

Group	Genus	Aerobic species	Anaerobic / microaerobic species
Unicellular forms;	*Gloeocapsa*	1	1
	Gloeothece	4	4
	Synechococcus	0	3
	Xenococcus	0	1
	Pleurocapsa	0	7
Non-heterocystous; filamentous forms	*Oscillatoria*	0	5
	Lyngbya	0	16
	Phormiaium		
	Trichodesmus	1	1
Heterocystous; filamentous forms	*Anabaena*	34	34
	Anabaenopsis	2	2
	Aulosira	1	1
	Calothrix	12	12
	Cylindrospermum	7	7
	Gloeotrichia	4	4
	Nostoc	16	16
	Scytonema	5	5
	Stigonema	1	1
	Tolupothrix	6	6
	Westiella	1	1
	Westiellopsis	1	1

* Mean value of 12 samples.

INOCULANTS USED

There are two treatments given i.e., inoculated with air dried mixed culture of Nostoc and non-algalization (control) to all the three paddy plants. Inoculated consist of the algae

(Nostoc) soil mixture. Each pot received an inoculum of 10 g Nostoc inoculum. The non-mycorrhizal pots received the same quantity of autoclaved soil without inoculum. All the experimental pots are maintained in 5 replicates. Experimental pots are kept free of weeds and irrigated properly. The observations are recorded for the period of 30, 60, 90 days. Plant height, fresh weight of shoot and root were recorded. Shoot and root dry weights were recorded after drying the plants in an oven for 48 hours and then cooling them in a dessicator.

RESULTS AND DISCUSSION

It is clear that important cynobacteria has been recorded in this region of south west part of India (Table 5.1). Soil which was used for the pot experiments had low phosphorus and nitrogen content (Table 5.2). Laboratory experiments were undertaken on three paddy have demonstrated a significant plant height, root length, fresh and dry weight, phosphorus and nitrogen concentration in shoots shown in (Tables 5.3 to 5.5). There was significantly increased plants biomass yield and N, P content in shoots with inoculation nostoc in paddy Jaya Var. compare to non inoculated plants (Table 5.3). In case of paddy Arya Var. root length was increased significantly with increased plant growth in Nostoc inoculated plants (Table 4). However, the inoculation nostoc on paddy var. Sonamasuri have not much influenced growth of plant height, but there was significantly increased concentration phosphorus and nitrogen content in shoot compare to non-inoculated plants (Table 5.4). These data were co-related with each other in all the three paddy varieties with increased plant height (Fig. 5.1) of paddy Var. Jaya, where as biomass production and nutrient uptake in shoots of paddy Var. Arya (Fig. 5.2) and increased plant height and root length in case of paddy Var. Sonamasuri (Fig. 5.3).

Table 5.2: Showing the Physico-Chemical Properties of Soil Used for the Pot Experiments

Sl. No.	Parameters	Results
1.	pH (1:2.5)	8.1
2.	Conductivity (EC) □s / cm	320
3.	Moisture %	4.86
4.	Total organic carbon	1.71
5.	Available Nitrogen %	4.18
6.	Available Phosphorus %	0.52
7.	Available Potassium %	7.94
8.	Magnesium %	0.121
9.	Calcium %	0.472
10.	Copper % (ppm)	0.03
11.	Zinc % (ppm)	3.86
12.	Manganese (ppm)	0.97
13.	Iron (ppm)	8.24

Mean value of 12 samples.

Garden soil three parts mixed with one part of pure sand (2:1).

Effect of growth response on Jaya, Arya and Sonamasuri of three rice plants inoculated with cynobacteria paddy plants had greater growth compared to uninoculated controls. The growth with respect to shoot length, fresh and dry weight of shoot, fresh and dry weight of root and biomass were recorded at one month to over a period of three months, at the end of each month plants are harvested. Root growth of inoculated rice plants increased significantly over the uninoculated control. Results revealed that, there is close correlation in number of rootlets in all experimental plants. Higher number of was recorded in arya followed by jaya.

As it is clear from the table all the plants of Jaya rice, Arya rice and Sonamasuri rice showed, a marked increase of growth in shoot length roots length, fresh and dry weight of root and shoot (Table 5.3 and 5.4) as well as in P content in shoot (Table 5.5 and Fig. 5.2).

Table. 5.3 : Showing the Effect of Nostoc (Cynobacteria) Inoculation with Respect to Shoot Length, Fresh Weight of Shoot, Dry Weight of Shoot, Root Length, Fresh Weight of Root, Dry Weight of Root, P Content in Shoot of Jaya Paddy Plant for 90 Days

Treatment/ duration	shoot length (cm)	Fresh weight of shoot (g/plant)	Dry weight of shoot (g/plant)	Root length (cm)	Fresh weight of shoot (g/plant)	Dry weight of shoot (g/plant)	% P in shoot	% N in shoot
30 days								
N.I.	6.4 a	0.36 c	0.30 c	4.7 b	0.15 b	0.09 c	0.03 c	0.98
I. N.	10.2 a	0.72 d	0.60 b	8.8 bc	0.25 ab	0.19 b	0.08 b	1.92
60 days								
N.I.	8.3 d	0.58 c	0.33 b	6.2 c	0.18 c	0.16 b	0.06 c	1.12
I. N.	20.2 c	0.110 d	0.80 b	10.2 d	0.27 b	0.20 ab	0.10 d	2.06
90 days								
N.I.	16.6 ab	0.70 d	0.40 a	15.2 ab	0.20 cd	0.18 b	0.10 c	1.29
I. N.	24.7 b	0.150 bc	0.99 ab	24.2 d	0.40 a	0.38 a	0.20 d	2.37

N.I. = Non inoculated, I.N. = Inoculated with Nostoc

- Values are mean of replicates
- Mean values followed by same letter within a column do not differ significantly at P = 0.05 by DMRT

Table. 5.4 :Showing the Effect of Nostoc (Cynobacteria) Inoculation with Respect to Shoot Length, Fresh Weight of Shoot, Dry Weight of Shoot, Root Length, Fresh Weight of Root, Dry Weight of Root, P and N Content in Shoot of Arya Paddy Plant for 90 Days.

Treatment/ duration	shoot length (cm)	Fresh weight of shoot (g/plant)	Dry weight of shoot (g/plant)	Root length (cm)	Fresh weight of shoot (g/plant)	Dry weight of shoot (g/plant)	% P in shoot	% N in shoot
30 days								
N.I.	6.2 a	0.28 c	0.26 b	4.1 c	0.01 d	0.08 c	0.04 c	1.05
I. N.	8.2 d	0.50 b	0.46 a	8.2 a	0.19 c	0.16 a	0.10 d	1.89
60 days								
N.I.	8.2 c	0.49 d	0.40 c	6.0 ab	0.13 b	0.10 c	0.06 c	1.14
I. N.	18.5 b	0.80 bc	0.60 d	8.5 b	0.23 ab	0.20 bc	0.12 d	2.42
90 days								
N.I.	13.2 b	0.60 d	0.58 ac	14.1 bc	0.16 dc	0.14 cb	0.08 bc	1.19
I. N.	22.6 ab	0.110 bc	0.105 b	20.2 d	0.36 d	0.30 c	0.21 b	2.26

N.I. = Non inoculated, I.N. = Inoculated with Nostoc

- Values are mean of 5 replicates
- Mean values followed by same letter within a column do not differ significantly at P = 0.05 by DMRT

Table. 5.5 :Showing the Effect of Nostoc (Cynobacteria) Inoculation with Respect to Shoot Length, Fresh Weight of Shoot, Dry Weight of Shoot, Root Length, Fresh Weight of Root, Dry Weight of Root, P and N Content in Shoot of Sonamasuri Paddy Plant for 90 Days

Treatment/ duration	shoot length (cm)	Fresh weight of shoot (g/plant)	Dry weight of shoot (g/plant)	Root length (cm)	Fresh weight of shoot (g/plant)	Dry weight of shoot (g/plant)	% P in shoot	% N in shoot
30 days								
N.I.	3.2 a	0.18 b	0.14 d	3.5 c	0.6 b	0.04 c	0.05 c	0.96
I. N.	6.5 b	0.26 c	0.23 b	6.5 b	0.17 d	0.15 d	0.08 c	1.94
60 days								
N.I.	6.5 bc	0.20 ab	0.16 a	6.1 bc	0.10 ab	0.08 c	0.05 b	1.26
I. N.	8.5 d	0.40 a	0.25 d	9.0 b	0.20 c	0.18 b	0.10 a	2.82
90 days								
N.I.	8.5 a	0.30 b	0.28 ab	12.2 c	0.14 c	0.12 a	0.08 b	1.33
I. N.	16.5 c	0.60 ab	0.49 c	19.5 ab	0.30 a	0.26 d	0.20 b	2.29

N.I. = Non inoculated, I.N. = Inoculated with Nostoc

- Values are mean of 5 replicates
- Mean values followed by same letter within a column do not differ significantly at P = 0.05 by DMRT

The fresh and dry weight of root and shoots of all the three experimental plants had higher in plants over uninoculated control plants.

The maximum root number was recorded in paddy Var. Sonamasuri inoculated with cynobacteria (Nostoc) at 90 days. However, all the cynobacteria inoculated paddy plants showed significantly increased plant growth, P and N contents as compared to the uninoculated (control) plants. The increase of P and N content was very significantly higher in plant jaya followed by closely correlated result was observed in arya and sonamasuri. There is significant increase in shoot phosphorus content in the inoculated plants compared to uninoculated control.

Paddy plants inoculated with Nostoc grow taller than the uninoculated control plants. These results are in agreement with earlier works on cotton, Maize, and soybean grown on unsterilized soil (Cameron and Julion, 1988; Kaushik, 2002). Rice plants inoculated with Nostoc insignificantly increased the thousand-grain weight over uninoculated control. Inoculation of wheat with increased grain yield. Such response are variable and depend upon the initial fertilizer of soil and the/type of crop planted. In unsterilized soil inoculated plants showed higher more rootlets and significant of plant growth compared to uninoculated control plants. This indicates that inoculation with cynobacteria (Nostoc) stimulated paddy growth beyond indigenous. This might be because of low indigenous allowed effectiveness, introduced (Reiwani, 1965; Venkataraman, 1975; Venkataraman and Shanmugasundaram, 1992). The data on growth of shoot, roots dry and fresh weights of Jaya, Arya and Sonamasuri are influenced Nostoc inoculation with 10 g soil and P uptake in shoot of all the experimental plants. These results are in the confirmation with early contributors (Kaushik, 1996; Rajula, 2000). Recently Kannaiyan *et al.*, (1997); Kannaiyan (1999), summarized data from Japan, Thailand, Indonesia and

Philippines and reported that a single rice crop can take from 37–113 kg N/ha from sources other than mineral nitrogen.

CONCLUSION

Cynobacteria is a easily accessable N_2 fixers its role has been well documented in paddy growing fields. AM Fungi aids in P absorption in the rhizosphere of many perennial plants. The inoculation of cynobacteria with AM Fungi brought significant beneficial effect on three var. of paddy. There was a significant increase in shoot growth, root growth, fresh and dry weights of Jaya, Arya and Sonamasuri. This is mainly of inoculation of 10 g of mixed inoculum of AM fungi (*Glomus fasiculatum*) and cynobacteria algalization would greatly play vital role in paddy yield in conventional soils of Indian agriculture.

REFERENCE

Arora, S.K. (1969). The role of algae on the availability of phosphorous in paddy fields.II Riso. 18:135–138.

Bossiba, S. (1999). Cynabacterias for agricultural application. In: International workshop and training course on micro algal biology and biotechnology. Mosonmagyarover, Hungary, June 13–26. Pp. 6

Cameron, H. J. and Julian, G. R. (1988). Utilization of Hydroxyapatite by cyanobactria as their sole source of phosphate and calcium Pl. Soil, 109: 123–124.

Dubey, R. C. (2004). A text book of Biotechnology S. C. Chand and Company Ltd., New Dehli. Pp. 418.

Fuller, W. H., and Roger, R. N. (1952). Utilization of the phosphorus of algal cells as measured by the Neubauer technique. *Soil Sci.* 74: 417–429.

Gahuikar, R.T. (2009). Sustainable agriculture in India Current situation and future needs. International . *J. Agri. Sci.* 5(1):1–7.

Joshi, N.K. (2007). Biotechnology in agriculture. Aavishkar publishers, Distrubuters, Jaipur - India, pp. 454.

Jackson, M.L. (1973). Soil Chemical Analysis, Prentice Hall of India, New Delhi, India, pp. 239.

Jayaraman, S. and Shanmugasundaram, S. (1993). Algal Biofertilizer: Choice species of *Syntonema* Proceedings of the national seminar on Cyanobacterial Research Indian Scene, Jan, 19-21, 1993. Edited by G. Subramanian, Bharathidasan University, Tiruchiraplli, pp. 24–31.

Kannaiyan, S. (1999). Ammonia excretion by the polyurethane foam immobilized N_2 - fixing cyanobacteria for Rice. In: Cyanobacterial and algae metabolism and environmental Biotechnology. Edited by Tasneem Fatma, Narosa Publishing House. New Delhi. pp. 20–27.

Kannaiyan, S., Aruna, S. J., Kumari, Marina Prem S., and Hall, D. O. (1997). Immobilized Cyanobacteria as a biofertilizer for rice crops. *J. Appl. Phycology*. 7:1–8.

Kaushik and Ummat, J. (1992). Reclamation of salt affected soils with blue green algae (cyanobacteria): a technology development. Proc. Natl. Seminar Biofertilizer Technology Transfer. Edited by Gangawne, L. V. Marathwada University, Aurangabad (India). 157.

Kaushik, B. D. (1983a). Effect of native algal flora on nutritional and physico-chemical properties of sodic soil. *J. Biol. Res.* 3: 99.

Kaushik, B. D. (1983b). Amelioration of salt affected so. In : Proceedings of the All India Applied Phycology Congress, Kanpur, India. 60–66.

Kaushik, B. D. (1993). Cyanobacterial research - An IARI Pursuit. Proceedings of the National Seminar On Cyanobacterial Research—Indian Scene, Jan. 19–21, 1993. Edited by G. Subramanian, Bharathidasan University, Tiruchirapalli pp. 32–49.

Kaushik, B. D. (1996). Use of Cyanobacterial biofertilizer in rice cultivation: A technology improvement. In: Cyanobacterial Biotechnology. Proceedings of the International Symposium. September 18–21. Edited by G. Subramanian, B. D. Kaushik, and G. S. Venkataraman Oxford & IBH publishing Co. PVT. Ltd. New Delhi. Pp 211–222.

Lakshman, H.C. (2009). The effect of AM Fungi and Bluegreen algae in improvement of paddy yield. Nat.J.Plant.Sci.113:76–79.

Kaushik, B. D. (2002). Contribution of Cynobacteria to P nutrition in plants. In: Mineral phosphate solubilization. In: proceeding of the 1st National Symposium on phosphate mineral nutrition. Nov. 14 to 15. Eds: A. R. Alagawadi, P. U. Krishna raj and M. S. Kuruvinashetti, U.A.S., Dharwad. Pp. 230.

Lakshman, H. C. (2009). Arbuscular mycorrhizal fungi (AMF) a promising biofertilizer for sustainable plant growth. In: proceedings of (NESCST09-CSIR), April 14–16. NAL Bangalore, pp. 19–22.

Lean, D. R. S. and Nalewajko, C. (1976). Phosphate exchange and organic phosphorus excretion by fresh water algae. *J. Fish Res. Bd. Can.* 33: 1312–1323.

Mackeras, A. H., Youns, B. N., Weir and R.C. Smith, G. D. (1990). Is cyanophycin involved in integration of nitrogen and carbon metabolism in the cyanobacteria *Anabaena cylindrica* and *Gleotheca* grown in light and dark cycles. *J. General Microbiology*. 136: 2049–2050.

Rajula, R. G. and Padmadevi, S. N. (2000). Effect of industrial effluents without and with BGA on the growth and biochemical contents of the seedlings of *Helianthus annuus,* L. Asian Jr. of Microbial. *Biotech. & Env. Sc.* 2: (3–4) 151–154.

Relwani, L. L. (1965). Response of paddy varieties to blue green algae and method of propagation. *Curr. Sci.* 34: 188–189.

Sankaram, A., Mudholkar, M.J. and Sahay, M.N. (1966). Algae in rice production. *Indian Fmg*. 16: 37–38.

Singh, R. N. (1961). Role of blue-greens algae in nitrogen economy of Indian agriculture. Indian Council of Agricultural research, New Delhi, pp. 175.

Venkataraman G. S. and Shanmugasundaram, S. (1992). Algal Biofertilizer Technology for rice. Sankar Printing Press, 271, Goods shed street, Madurai-1.

Venkataraman, G. S. (1975). The role of blue-green algae in tropical rice cultivation. In: Nitrogen fixation by free-living microrganism. Edited by W.D.P. Steward. Cambridge University Press, London. Pp. 207.

6

Arbuscular Mycorrhiza (AMF) Inoculants and Its Role in Sustainable Forestry Production

K Muthuchelian, P Parimala and M Prabhu Inbaraj

INTRODUCTION

The root systems of forest trees and other plants have impressive form, structure, and function. Roots provide plants with the water required for growth, soil minerals essential for metabolism, growth regulators to direct carbon allocation, storage of key food reserves, and physical support. The management and conservation of forest biomes is a recognized priority on a global scale. Integral to the understanding of forest ecology is the rhizosphere and in particular the mycorrhizal symbiotic associations between host plant and fungi. Mycorrhizal relationships are an example of mutual symbioses, involving plants and fungi, whereby both organisms benefit through an exchange of nutrients at the root-soil interface (Smith *et al.*, 1997). Mycorrhizas are an intimate association between plant roots

---

Centre for Biodiversity and Forest Studies School of Energy, Environment and Natural Resources, Madurai Kamaraj University, Madurai - 625 021, India.

and specially adapted mycorrhizal fungi. All trees are completely dependent on mycorrhizal fungi for their mineral nutrition, and in particular phosphorus.

Different types of mycorrhizae occur, distinguished by their morphology and to a certain extent, in their physiology. Two types of mycorrhiza are known today: ecto- and endo-mycorrhizas. Endomycorrhizal plants are typically herbaceous plants, shrubs, many ornamental, fruit and nut trees, vegetables and agronomic crops, and turf grasses. More than 85% of higher plants form endomycorrhizal associations. The ecotomycorrhizal fungi form associations with conifers such as fir and pines, and hardwoods such as birch, beech, eucalyptus, oak, willow, and magnolia. Ectomycorrhiza fungi colonize around 10% of higher plants. Some plants such as eucalyptus will form both endo- and ectomycorrhizal associations.

Generally plants within a plant family tend to form the same kind of mycorrhizas (Table 6.1). Most of the trees that dominate the forests of the Northern Hemisphere including birches (Betulaceae), oaks and beeches (Fagaceae), many softwoods (Pinaceae), European limes (Tiliaceae) and the tropical dipterocarps (Dipterocarpaceae) as well as the Australian eucalypts (Myrtaceae) form ectomycorrhizas. In contrast, the vast majority of cultivated crops, flowering plants, almost all New Zealand natives, ferns, cycads, and some gymnosperms such as North American redwood and cypresses (Taxodiaceae) form arbuscular mycorrhizas. There are a few exceptions such as the Myrtaceae and Rosaceae that contain both ecto and arbuscular mycorrhizal species. A few tree species like the poplars and willows (Salicaceae), some eucalypts and Leptospermum (Myrtaceae) have the best of both worlds as they are able to form both ecto and arbuscular mycorrhizas (Smith and Read, 1997).

Mycorrhizal fungi are a possible keystone guild with regard to nutrient cycling and primary productivity in topical forests. There are almost no studies of the relationship between the diversity of mycorrhizal fungi per se and ecosystem processes, but a number of studies suggest that mycorrhizal fungi may be vulnerable to disturbance, and a reduction in their diversity might reduce the rate of forest recovery. In areas where the upper layer of soil, together with its complement of mycorrhizal fungi, has been eroded away, the capacity of the soil to establish and maintain healthy plant populations can be severely limited (Powell, 1980). In Cameroon, West Africa, complete clearance of forest sites before replanting with native species resulted in a 65% reduction in AM fungal spores; spore populations were less damaged by partial site clearance. The loss of mycorrhizal fungi slows down the reintroduction/ regeneration of trees that are obligately mycotrophic (Janos, 1983). Spread of Arbuscular-mycorrhizal fungi in tropical forests occurs primarily from root-to-root by hyphal growth over distances of millimeters to centimeters because spore production and survival is low (Janos, 1983).

ARBUSCULAR MYCORRHIZAL FUNGI

Arbuscular mycorrhizal fungi (AMF), sometimes referred to as endomycorrhizas, are formed predominantly by the fungal group Glomeromycota (Schüßler *et al.*, 2001). Tropical rainforests are important habitats for arbuscular mycorrhizal fungi (Read, 1994). The tropical rainforests of Xishuangbanna, southwest China are located at the north margin of the tropical zone of Southeast Asia. Since this is a type of transitional vegetation (from tropical to subtropical zone), tropical and subtropical flora coexist; and the habitat contains more species diversity than tropical rainforests of Southeast Asia (Jin and Ou, 1997). It has been found that 434 plant species are condensed in 40000 m^2 (Jin, 1997).

The association is identified by intracellular dichotomously branching haustorial structures called arbuscules

(Figure 6.1), found in the cortical plant root cells and hyphal coils, as well as intercellular hyphal networks and external hyphae that extend into the soil. AM fungi are found on a wide range of host species, predominantly colonizing herbaceous shrubs and tree species (Brundrett *et al.*, 1990). The arbuscular mycorrhizal association was so successful that in the course of evolution, AMF became obligate biotrophs i.e., they can not live without connection to, and carbon supply from, a living host plant. This feature has important implications on the life of AMF in cultivated soils. AMF are found close to plant roots and most of their biomass is in the top 20 cm of the soil (Kabir *et al.*, 1998). AMF appear as networks of fine tubes of a few micrometers in diameter, filled with cytoplasm, and producing spores. These networks are extensive, often with tens of meters per gram of soil (Leake *et al.*, 2004).

AMF is the most abundant kind of mycorrhiza described as 'a universal plant symbioses'. They are found in practically every taxonomic group of plants and the list of species not infected is probably far shorter than the infected ones. Lack of

Fig. 6.1. : Showing an Arbuscular Mycorrhizal Infection Inside a Fine Root (Courtesy of Jim Gerdemann)

Table 6.1: Showing the Occurrence of Arbuscular Mycorrhizal Fungal Species in Some Important Angiospermic Families.

Common name	*Botanical name*	*Plant family*
Akeake	*Dodonaea viscosa*	Sapindaceae
Akiraho	*Olearia paniculata*	Asteraceae
Almond	*Prunus dulcis*	Rosaceae
Apple	*Malus*	Rosaceae
Apricot	*Prunus ameniaca*	Rosaceae
Angelica tree	*Aralia*	Araliaceae
Ash	*Fraxinus*	Oleaceae
Avocado	*Persea americana*	Lauraceae
Bomboo	*Bambusa*	Pooideae
Banana	*Musa*	Musaceae
Barberyy	*Berberis*	Berberidaceae
Bayberry	*Myrica*	Myricaceae
Black locust	*Robinia*	Fabaceae
Blackberry	*Rubus eubatus*	Rosaceae
Box elder	*Acer neugundo*	Aceraceae
Broadleaf	*Griselinia*	Griseliniaceae
Boxwood	*Buxus*	Buxaceae
Buckeye	*Aesculus*	Hippocastaneaceae
Burning bush	*Eunymus*	Celastraceae
Cacoa	*Theobroma cacoa*	Sterculiaceae
Camellia	*Camellia*	Theaceae
Catalpa	*Catalpa*	Bignoniaceae
Cherry	*Prunus avium*	Rosaceae
Chinaberry	*Melia zedarach*	Meliaceae
Coral tree	*Erythrina indica*	Fabaceae
Crabapple	*Malus*	Rosaceae
Cryptomeria	*Cryptomeria japonica*	Taxodiaceae
Cucumber tree	*magnolia acuminata*	Magnoliaceae
Dogwood	*Cornus*	Cornaceae
Fig	*Ficus carica*	Moraceae
Flax, New Zealand	*Phormium tenax*	Agavaceae
Fuschia	*Fuchsia*	Onagraceae
Ginko	*Ginko biloba*	Ginkgoaceae
Gorse	*Ulex europaeus*	Fabaceae
Grapes	*Vitis*	Vitaceae
Hackberry	*Celtis*	Ulmaceae

Contd....

Common name	Botanical name	Plant family
Hilbiscus	*Hilbiscus rosa-sinensis*	Malvaceae
Holly	*Ilex*	Aquifoliaceae
Horse chestnut	*Aesculus*	Hippocastanaceae
Juniper	*Juniperus*	Cupressaceae
Kamahi	*Weinmannia racemosa*	Cunoniaceae
Karamu	*Coprosma robusta*	Rubiaceae
Kauri	*Agathis*	Araucariaceae
Korakia	*Corokia buddleoides*	Cornaceae
Kowhai	*Sophora spp.*	Papilionaceae
Lacebark	*Hoheria populnea*	Malvaceae
Lawson cypress	*Chamaecyparis lawsoniana*	Cupressaceae
Lemonwood	*Pittosporum*	Pittosporaceae
Lyland cypress	*X Cupressocyparis leylandii*	Cupressaceae
Macrocarpa	*Cupressus macrocarpa*	Cupressaceae
Magnolia	*Magnolia*	Magnoliaceae
Mahoe	*Melicytus ramiflorus*	Violaceae
Maples	*Acer*	Aceraceae
Marbleleaf	*Carpodetus serratus*	Carpodetaceae
Mulberry	*Morus*	Moraceae
Olive	*Olea europaea*	Oleaceae
Palms	*Cycad*	Cycadaceae
Papaya	*Carica papaya*	Cariceae
Paulownia	*Paulownia*	Paulowniaceae
Peach	*Prunus persica*	Rosaceae
Pear	*Pyrus communis*	Ropsaceae
Persimmon	*Diospyros*	Ebenaceae
Plum	*Prunus*	Rosaceae
Podocarp	*Podocarpus*	Podocarpaceae
Pohutukawa	*Metrosideros excelsior*	Mytaceae
Privet	*Ligustrum*	Oleaceae
Rain tree	*Koelreuteria elegans*	Sapindaceae
Rata	*Metrosideros*	Myrtaceae
Redwood, coastal	*Sequoia sempervirens*	Taxodiaceae
Redwood, giant	*Sequoiadendron giganteum*	Taxodiaceae
Ribbonwood	*Plaglanthus betulinus*	Malvaceae
Rowan	*Sorbus*	Rosaceae
Sycamore	*Acer*	Aceraceae
Tree-of-heaven	*Alianthus altissima*	Simaroubaceae
Tulip tree	*liriodendron*	Magnoliaceae
Vibumum	*Vibumum*	Caprifoliaceae
Yew	*Taxus spp.*	Taxaceae

host specificity is even more characteristic of this symbiosis than other types known. Studies on AMF fungi conducted during last few decades envisaged their occurrence in a wide variety of hosts, different habitats and variability in their quality and quantity. AMF fungi help in the establishment of the forest trees by increasing the volume of soil exploration and uptake of nutrients particularly phosphorus, increasing disease resistance, improving water absorption, drought tolerance and thus accelerate the ability of the plant species to compete for resources, contributing to efficient recycling of nutrients particularly in acidic soils of tropical areas (Dhar and Mridha, 2003).

AMF FUNGAL DIVERSITY IN INDIA

In India, Bakshi *et al.* (1974) was the first to publish an account of 14 spore types: *Glomus macrocarpum, Glomus macrocarpum* Tul. and Tul. var. *geosporum, G. mosseae, Glomus* sp., *Sclerocystis coremioides* Berk. and Broome, *Sclerocystis* sp., *Gigaspora calospora, Acaulospora* sp., *Endogone gigantean* Nicol. and Gerd. *Endogone microcarpum, Endogone* 1, *Endogone* 2, *Endogone* 3. Gerdemann and Bakshi (1976) reported two new species, viz. *Glomus multicaule* and *Sclerocystis sinuosa.* Bhattacharjee and Mukerji (1982) described the species *Glomus reticulatum* from soils of Bangalore. Bhattacharjee *et al.* (1982) reported the structure and hyperparasitism for *Gigaspora candida* while Bhattacharjee and Mukerji (1982) described the ultrastructure of *Sclerocystis coremioides* sporocarp. Mukerji *et al.* (1983) reported two species of *Glomus,* viz. *Glomus multisubstensum* and *G. delhiense* both from soils of Delhi. Till this date, 102 AMF species have been reported from India. The occurrence of AMF fungi in a natural forest stand was recorded in the Old Delhi Ridge, Saraswati Range of Haryana (Thapar *et al.*, 1996), forest soils and coastal regions of Andhra Pradesh (Manoharachary *et al.*, 1991), Kodayar forest, Tamil Nadu (Ganesan *et al.*, 1991), forest plants of Nilgiris (Raja *et al.*, 1991), coastal tropical forest (Kodikkarai

Reserve Forest) of Tamil Nadu (Raghupathy *et al.*, 1991), Servarayan Hills of Tamil Nadu (Raman *et al.*, 1995), forest soils of Andhra Pradesh (Vijaya *et al.*, 1995) and black pepper grown in the forest soils of Kerala (Lekha *et al.*, 1995). The diversity of AM fungi has also been studied in the coastal regions of Konkan and Servaravan Hills of Tamil Nadu (Gopinathan *et al.*, 1991), Coromandel coast of Tamil Nadu (Raghupathy *et al.*, 1992), coastal sand dunes at Someshwara, Mangalore coast of Karnataka (Kulkarni *et al.*, 1997), coastal sand dunes of the west coast of India (Beena *et al.*, 2000) and western Ghats of Goa (Khade *et al.*, 2003). Sengupta and Chaudhari (1989, 1990) studied the occurrence of AM fungi in *Sueda maritima* (a pioneer mangrove) in terminal, seabound Gangetic delta in West Bengal. Mangrove of Muthupet estuary was surveyed by Selvaraj and Subramanian (1991).

The occurrences of AM fungi in arid and semi-arid regions were studied in Tamil Nadu (Parthipon *et al.*, 1991), deserts (Neeraj *et al.*, 1991), arid zones of Rajasthan (Mohan *et al.*, 1995), and semi-arid grasslands of Maruthamalai hills (off-shoot of the Western Ghats in Peninsular India) (Muthukumar *et al.*, 1995). The diversity of AM fungi in agricultural fields was reported on *Leucaena leucocephala* from Bangalore (Nalini *et al.*, 1987), ornamentals and cultivated plants at Allahabad and adjoining areas (Kehri *et al.*, 1987), crop fields of Konkan and Solapur (Dalal *et al.*, 1995), tea plantation at Nilgiris, Tamil Nadu (Kumaran *et al.*, 1995), pearl millet, maize, wheat, pigeon pea and chick pea in Gwalior (Singh *et al.*, 1996) and different agroclimatic regions of India (Singh *et al.*, 2002). The distribution of AM fungi in stressed ecosystems has also been reported from coal, lignite and calcite mine spoils of India (Mehrotra *et al.*, 1995), Kothagudam coal mine spoil, Andhra Pradesh (Rani *et al.*, 1991), heavy metal polluted soils of Tamil Nadu (Sambadan *et al.*, 1991), petro-effluent irrigated fields, soils polluted with industrial and sewage effluents (Reddy *et al.*, 1995), tannery effluent polluted soils of Tamil Nadu

(Raman *et al.*, 1995) and stressed soils of Bailadila iron ore sites in Madhya Pradesh (Sastry *et al.*, 1999). These studies indicate that the genus *Glomus* is ubiquitous in various ecosystems in India. The distribution of other genera, i.e. *Entrophospora, Gigaspora, Sclerocystis* and *Scutellospora* is limited, indicating greater adaptability of *Glomus* to varied soil conditions

AM FUNGI AND SOIL FERTILITY

AM fungal infection improves plant survival and growth by enhancing the root's ability to absorb moisture, macro-nutrients and micro-nutrients from the soil. Three main components are involved in AMF association: 1) the soil, 2) the fungus and 3) the plant (Figure 6.2). The fungal component involves the fungal structure within the cell of the root and the extraradical mycelium in the soil. The last may be quite extensive under some conditions, but does not form any vegetative structures (Smith and Read, 1997). Its primary function is the absorption of resources from the soil. The *increased* efficiency of mycorrhizal roots versus nonmycorrhizal roots is caused by the active uptake and transport of nutrients by mycorrhizae. Mycorrhizae are described as improving the absortion of several nutrients. AMF have been shown to improve productivity in soils of low fertility (Jeffries, 1987) and are particularly important for increasing the uptake of slowly diffusing ions such as PO_4^{-3} (Jacobsen *et al.*, 1992); immobile nutrients such as P, Zn and Cu (Liu *et al.*, 2002) and other nutrients such as Cadmium (Guo *et al.*, 1996). Joner and Levylal (1997) found that cadmium-tolerant AMF were responsible for uptake, transport and imobilization of cadmium. Copper (Cu) was absorbed and accumulated in the extraradical mycelium of three AMF isolates, as observed in a study with *Glomus spp.*

Indigenous tropical tree species are commonly associated with AM fungi (Moyersoen, 1993). Since it has been hypothesised that P is the limiting element in most tropical rain forests (Vitousek and Sanford, 1986), AMF might be expected to play an important role in P nutrition. However, recent findings show that P does not always limit the growth of tropical rain forest seedlings (Raaimackers and Lambers, 1996). The occurrence of ectomycorrhiza in the tropics on extremely poor soil (Singer and Araujo, 1986) suggests that AMF are particularly beneficial in these habitats.

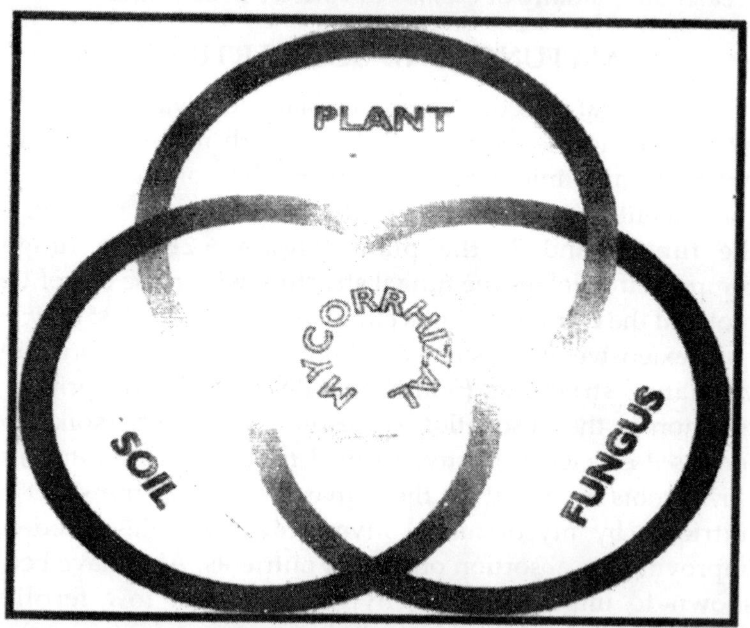

Fig. 6.2 (Brundrett *et al.*, 1996).

PROTECTION AGAINST WATER STRESS

Although most of the work done with AM fungi has concentrated on their effects in plant nutrition, there is an increasing interest also on drought resistance of mycorrhizal plants (Allen *et al.*, 1983). Under drought conditions the uptake of highly mobile nutrients such as NO^{-3} can also be enhanced by mycorrhizal associations (Subramanian and Charest, 1999). Improved P nutrition has been shown to increase in infertile and P fixing soils of the tropics (Dodd, 2000). Mycorrhizal fungi can also improve absorption of N from NH_{4+} -N mineral fertilizers, transporting it to the host plant (Johansen *et al.*, 1993). Its transport and absorption can also increase biomass production in soils with low potassium, Calcium and Magnesium (Liu *et al.*, 2002). The enhanced effect of AM fungi on the uptake of nitrogen and micronutrient uptake may be attributed to two situations. In the first one is mycorrhizal hyphae act as extension to plant root, increasing root surface area and exploring larger soil volume, which will increase the chance of more micronutrient uptake (Osonubi *et al.*, 1991).

PROTECTION AGAINST SALT STRESS

Saline soils occupy over 7% of the earth's land surface and crop production on this area is relatively low. However AM fungi in soil sites affected by salinity may improve salt tolerance and growth (Jain *et al.*, 1989) not only by enhancing the uptake of minerals, particularly P (Graham, 1986), but also by enhancing photosynthetic activity or water use efficiency. Jalaluddin (1993) has observed that seedlings of maize, *Salmalia malabaricum* (*Bombax malabaricum*), *Sorghum* and *Pennisetum* inoculated with *Glomus intraradices* grown in 1% saline soil (16.0 m mho/cm) has increased fresh and dry weights compared with inoculated plants.

Feng-Gu *et al.* (1998) have investigated the influence of VAM fungi on growth and mineral accumulation in *Bromus inermis* at 0, 1, 2, or 3 gm NaCl/Kg soil in a pot experiment. They have concluded that the mechanism of VAM fungi in improving the salt tolerance of *Bromus inermis* is to improve the nutritional condition of the plants by increasing the contents of N and P while decreasing the relative concentration of Na and Cl in the plant. Mandeel and Gul (1999) have made a survey on the occurrence of AM fungi in the rhizosphere soil and root of wild and cultivated plants in coastal habitats in Babrain. Soils were generally highly saline with salt crusts on the top layer, low in total soluble salts, slightly alkaline and poor in organic matter. Only one isolate

G. masseae has been recorded and indicates that salinity may have a detrimental effect upon mycorrhizal establishment and abundance in such habitats.

Feng-Gu *et al.* (1999) have studied the effects of salt stress on AM fungal formation and of inoculation with AM fungi on salt tolerance of plants. Soils with different contents of salt (NaCl) grown with cotton, maize, soybean and melon were inoculated with *Glomes mosseae, G. versforme, Glomus sp.* and *G. caledonium* and it was concluded that the VAM fungi-plant symbiosis might play an important role in the survival of plants grown on saline soil and that inoculation with AM fungi could enhance the production of plants grown on saline soil and reduce the loss of plant yield caused by salt.

PROTECTION AGAINST PATHOGEN

AM fungi have the potential to reduce damage caused by soil-borne pathogenic fungi, nematodes, and bacteria (Borowicz, 2001). In particular, mycorrhizal fungi can provide some protection against root invading primary pathogens (Duchesne *et al.*, 1989). Some examples for woody species include [host/pathogen/ mycorrhizal fungus(i)]: *Castanea/Phytophthora/Laccaria, Hebeloma, Paxillus* (Brazanti *et al.*, 1999); *Picea/ Cylindrocladium/Paxillus* (Morin *et al.*, 1999); *Pinus/Fusarium/ Paxillus* (Farquhar and Peterson, 1991); *Pinus/Phytophthora/ Pisolithus* (Barham *et al.*, 1974); and *Pseudotsuga/Fusarium/Laccaria* (Sylvia and Sinclair, 1983). Further, mycorrhizal plants may be less impacted by pathogenic nematodes than non-mycorrhizal plants (Pinochet *et al.*, 1996), probably because they are better able to maintain a water and nutrient absorbing surface in infected soil.

Carbon distribution in soil may be the major mechanism explaining the relationship between AMF and soil microbial diversity, and the impact of these fungi on the structure of soil microbial communities (Marschner and Baumann, 2003). Soils abundant in AMF are healthier and have been associated with reduced population of soilborne pathogens and disease incidence (St-Arnaud *et al.*, 1995).

ROLE OF MYCORRHIZAE AS A PHYSIOLOGICAL SINK FOR PHOTOSYNTHATES

The AM fungi, the biotrophs, lead a symbiotic life which is characterized by the transfer of nutrients which have been taken up from the soil by the fungus in exchange for carbohydrates provided by the host plant (Fitter, 1998). Mycorrhizae therefore enhance the sink strength of root systems by increasing the consumption of photo assimilates, mainly sucrose, the major transport form of carbohydrates in higher plants.

Sucrose is shown to be cleaved in the host root by a cell wall bound acid invertase. The resulting hexoses are taken up and rapidly metabolized by the fungus (Schaeffer *et al.*, 1997). Schwab and Menge *et al.* (1991) have proposed a carbohydrate based theory pertaining to the control of AM fungal infection. They have proposed that exudation of soluble carbon compounds, assumed to be fungal substrates, into the rhizosphere, is reduced under high P nutrition.

Amijee *et al.*, (1993) have reported that increased P supply from either phosphate addition to soil or AM fungal infection increases the concentration of soluble carbohydrates in leek roots and that the AM fungi behave particularly as a strong physiological sink when there is an excess concentration of sucrose in the host. Plants allocate an estimated 10-20% of net photosynthate to mycorrhizal fungi, although this number can range from 5 to 85% among systems (Allen, 1991). Fitter (1991) has reviewed that as a symbiont, the AM fungi obtain plant carbohydrates for their growth and maintenance. 10% of carbon translocated to the roots pass to the fungal partner.

Thompson *et al.* (1991) have reported that there is a positive correlation between carbohydrate availability in roots and AM fungal formation. The data of Pearson and Schweiger

(1993) suggest that at the onset of sporulation, carbohydrate concentration in the subterranean clover roots is on the decline. Pearson and Jakobsen (1993) have revealed that the concentration of sugar in roots is a balance between supply from the shoot and its consumption. Increase in photosynthesis based on the source sink mechanism should also be taken into account.

Hampp *et al.* (1995) have explained that apoplastic invertase in the mycorrhizae hydrolyses sucrose exuded from host cytoplasm and thus enhances glucose uptake by the fungal tissue. Saito (1995) has also observed that in VAM fungi, hexokinase activity is found to be high in the intraradical hyphae of *Gigaspora margarita,* implying that glucose is used by the fungal hyphae.

Recent NMR study by Shachar-Hill *et al.* (1995) has shown that *Glomus etunicatum* in leek roots is capable of utilizing glucose and synthesized trehalose and glycogen when glucose has been applied exogenously to mycorrhizae. A strong relation between spore numbers and the carbohydrate concentration in roots indicates the direct influence of the host on AM fungi sporulation. Thompson *et al.* (1991) have also revealed that the sensitivity of AM fungi to changes in the carbohydrate concentrations may vary between AM fungal spores and their numbers. Reduced carbohydrate level in roots increases mycorrhizal infection level resulting in an inverse relation (Pearson and Schweiger, 1993).

Johnson *et al.* (1997) have reported that the presence of arbuscules within the roots indicates the nutrient benefit to the host or a nutrient demand of the host, the abundance of hyphae, vesicles or spore numbers indicate a significant carbon sink. Since AM fungi are obligate biotrophs, the host should have a substantial influence on the AM fungi through the regulation of carbon supply.

Harrison (1996) has revealed that glucose is one of the most likely substrates for AM fungi in the symbiotic state and has observed that the expression of a sugar transport gene in the host plant *Medicago truncatula* roots is found to be localized in the areas colonized by *Glomus cersiforme*, implying that carbohydrate supply to the mycobiont is regulated by the host plant.

Salaiman and Saito (1996) have observed that the evolution of $^{14}CO_2$ form onion roots and intraradical hyphae of *Gigaspora marginata* by radiorespirometry after addition of ^{14}C-labelled glucose or sucrose to mycorrhizal roots has been much higher than in the non-mycorrhizal roots but no difference has been found from sucrose. In mycorrhizal roots, the respiration rate from glucose is about twice from sucrose. However there are exceptions, which include studies where no correlation (Amijee *et al.*, 1993) or a negative correlation (Pearson and Schweiger, 1993) has been found between AM fungal colonization levels and root-soluble carbohydrates.

In relation to photosynthate supply to AM fungi, the effect of mycorrhizal colonization on carbohydrate concentration has been well documented, but such results have not shown any consistent tendency. They have not shown if the carbohydrate concentration increases or remains unchanged (Mc Arthur Daj and Knowles, 1993). Schwab *et al.* (1991) have studied extensively the effect of P on root soluble carbohydrates. Other nutrients such as plant nitrogen and potassium have received little attention.

EFFECT OF AM FUNGI ON THE GROWTH OF TREE SEEDLINGS GROWN UNDER NURSERY CONDITION

In reforestation, if the seedlings possess adequate amount of mycorrhizal infection at planting time they can survive and start healthy growth immediately after being transplanted in the field (Janos, 1983). The prevalence of AM fungus suggests that mycorrhizae may be of great importance for plant growth in tropical soils which are relatively poor in mineral nutrients, especially P. The fungal hyphae outside the roots can extract P from longer distance (Mosse and Hayman, 1980).

Mycorrhizal colonization has been observed in *Theobroma cacao* seedlings (Gisela *et al.*, 1990) grown in nursery conditions. AM fungal association forest tree species (Lakshman,1996; Lakshman and Patil, 2006). Puthur *et al.* (1998) have demonstrated that the VAM association (*Glomus fasciculatum* and *Glomus macrocarpum*) help in successful establishment of tissue culture raised plantlets of *Leucaena leucocephala* in the field conditions by alleviating the transplantation stock.

FOREST LAND MANAGEMENT

Studies in a number of ecosystems (Siqueira *et al.*, 1998) show that in climax communities, normally dominated by species heavily colonised by mycorrhizal fungi, disturbance leads to a successional sequence in which re-colonization is initiated by plant species which are non-mycorrhizal or little infected (Read and Birch, 1988). Gap-preferring species thus may have lower rates of mycorrhizal infection than species preferring undisturbed microsites (Onipchenko and Zobel, 2000). However, Brundrett *et al.* (1995) found that for parts of tropical Australia, the activity of mycorrhizal fungi was higher in patches of early-successional vegetation than in undisturbed habitats.

In a study in deciduous tropical forest in Mexico, Allen (1998) concluded that regrowth of vegetation in small gaps was not limited by mycorrhizal fungi, since they were still abundant after tree falls. However, recovery in pastures could be affected by low fungal diversity and dominance of grasses. Jasper *et al.* (1991) found that disturbance of forest and heathland soils decreased colonisation of test plants (clover) compared to disturbance of clover soil. They proposed that a larger number of propagules in the pasture soil may have allowed the pasture soil to maintain infectivity after disturbance. There is a suggestion that ecosystems with a high proportion of grasses and high numbers of AM fungal spores may also be more tolerant of disturbance (Visser *et al.*, 1984).

CONCLUSION

Arbuscular mycorrhiza inoculation has a high potential as a biofertiliser in water stressed environment. The potential benefit of mycorrhizal fungi in rehabilitation of degraded areas by use of agroforestry system is more apparent than ever before. The need to increase food, fibre and fuel wood production to keep pace with the fast growing population in developing countries in the tropics is crucial. The low biomass production of agroforestry tree species in degraded areas can, therefore, be circumvented by the use of mycorrhizal fungi. Use of AM fungi in forestry appears to be more important than in agriculture because in countries like India no large scale provisions exist to irrigate, fertilize and protect the plantation (Chaurasia, 2004).

REFERENCES

Allen, E.B. (1998). Disturbance and seasonal dynamics of mycorrhizae in a tropical deciduous forest in Mexico. *Biotropica* 30: 261–74.

Allen, M.F. (1991). The ecology of mycorrhizae Cambridge University Press, Cambridge, England.

Allen, M.F. and Boosalis, M.G. (1983). Effects of two VA-mycorrhizal fungi on drought tolerance of winter wheat. *New Phytol.* 93: 67–76.

Amijee, F., Stribley, D.P. and Tinker, P.B. (1993). The development of endo-mycorrhizal root and stems. III. Effects of soil phosphorus and fungal colonization on the concentration of soluble carbohydrates in roots. *New Phytol.* 123: 297–306.

Arthur, D. and Knowles, N.R. (1993). Influence of vesicular-arbuscular mycorrhizal fungi on the response of potato to phosphorus deficiency. *Plant Physiology.* 101: 147–60.

Bakshi, B. K. (1974). Mycorrhiza and its role in forestry. P.L. 480 Project Report. Forest Research Institute and Colleges, Dehra Dun. 89.

Barham, R.O., Marx, D.H. and Ruehle, J.L. (1974). Infection of ectomycorrhizal and nonmycorrhizal roots of shortleaf pine by nematodes and *Phytophthora cinnamomi. Phytopathology.* 64: 1260–4.

Beena, K. R., Raviraja, N. S., Arun, A. B. and Sridhar, K. R. (2000). Diversity of arbuscular mycorrhizal fungi on the coastal sand dunes of the west coast of India. *Curr. Sci.* 79: 1459–66.

Bhattacharjee, M. and Mukerji, K. G. (1982). The SEM structure of *Sclerocystis coremiodes. Nova Hedwigia.*36: 101–4.

Bhattacharjee, M., Mukerji, K. G., Tewari, J. P. and Skoropad, M. P. (1982). Structure and hyperparasitism of a new species of *Gigaspora. Trans. Br. Mycol. Soc.* 78: 184–88.

Borowicz, V.A. (2001). Do arbuscular mycorrhizal fungi alter plant-pathogens relations? *Ecology.* 82: 3057–68.

Brazanti, M.B., Rocca, E. and Pisi, E. (1999). Effect of ectomycorrhizal fungi on chestnut ink disease. *Mycorrhiza.* 9: 103–9.

Brundrett, M., Abbott, L.K., Jasper, D.A. and Ashwath, N. (1995). Mycorrhizal associations in disturbed and natural habitats in tropical Australia. p. 34–40. In M. Brundrett, B. Dell, N. Malajczuk and M. Gong (eds). Mycorrhizal Research for Forestry in Asia. ACIAR Proceedings No. 62, Canberra.

Brundrett, M., Beegher, N., Dell, B., Groove, T. and Malajczuk, N. (1996). Working with mycorrhizas in Forestry and Agriculture. ACIAR Monograph. 32.

Brundrett, M., Murase, G. and Kendrick B. (1990). Comparative anatomy of roots and mycorrhizae of common Ontario trees. *Can. J. Bot.* 68: 551–78.

Chaurasia, B. (2004). ENVIS Newsletter: *Himalayan Ecology.* 1.

Dalal, S. and Hippalgaonkar, K.V. (1995). The occurrence of vesicular-arbuscular mycorrhizal fungi in arable soils of Konkan and Solapur. Proceedings of the Third National Conference on Mycorrhizae.In Mycorrhizae: Biofertilizers for the Future (eds Adholeya, A. and Singh, S.). 3–7.

Dhar, P.P. and Mridha, M.A.U. (2003). Status of biodiversity of arbuscular mycorrhizal fungi in different tree species growing in Betagi community forests [J]. *The Chittagong University Journal of Science*, 27: 13–19.

Dodd, J.C. (2000) The role of arbuscular mycorrhizal fungi in agro and natural ecosystems. *Outlook on Agriculture*. 29 (1): 55–62.

Duchesne, L.C., Peterson, R.L. and Ellis, B.E. (1989). The time course of disease suppression and antibiosis by the ectomycorrhizal fungus *Paxillus involutus*. *New Phytol*. 111: 693–8.

Farquhar, M.L. and Peterson, R.L. (1991). Later events in suppression of *Fusarium* root rot of red pine seedlings by the ectomycorrhizal fungus *Paxillus involutus*. *Can. J. Bot*. 69: 1372–3.

Feng, G., Bai, D.S., Yang, M.Q., Li-X.L. and Zhang, F.S. (1999). Effects of salinity on VA mycorrhiza formation and of inoculation with VAM fungi on salt tolerance of plants. *Chinese Journal of Applied Ecology*. 10: 79–82.

Feng, G., Yang, M.Q. and Bai, D.S. (1998). Influence of VAM fungi on mineral elements concentration and composition in *Bromus inermis* under salinity stress. *Acta – Prataculturae – Sinica*. 7: 21–28.

Fitter, A.H. (1988). Water relations of red clover *Trifolium protense* affected by VA mycorrhizal infection and phosphorus supply before and during drought. *J. Exp. Bot*. 39: 595–603.

Fitter, A.H. (1991). Cost and benefits of mycorrhizas – Implications for functioning under natural conditions. *Experiential*. 47: 350–5.

Ganesan, V., Parthipon, B. and Mahadevan, A. (1991). Survey of vesicular-arbuscular mycorrhizae (VAM) in Kodayar Forest, Tamil Nadu, India. Proceedings of the Second Asian Conference on Mycorrhiza (eds. Soerianegara, I. and Supriyanto). 73–5.

Gerdemann, J. W. and Bakshi, B. K. (1976). Endogonaceae of India: two new species. *Trans. Br. Mycol. Soc*. 66: 340–43.

Gisela, U., Rafael, H. and Erasmo, E. (1990). Effects of VA mycorrhiza on the growth of cacao seedlings under nursery conditions in Venezuela. *Plant and soil*. 126: 71–78.

Gopinathan, S., Nagarajan, N. and Raman, N. (1991). Survey of endomycorrhizal spores in the forest of Servarayan Hills of Tamil Nadu, India. In Proceedings of the Second Asian Conference on Mycorrhiza. (eds. Soerianegara, I. and Supriyanto). 274.

Graham, J.H. (1986). Citrus mycorrhizae: Potential benefits and interactions with pathogens. *Hort. Science.* 21: 1302–6.

Guo, Y., George, E. and Marschner, H. (1996). Contribution of an arbuscular mycorrhizal fungus to uptake of Cadmnium and Nickel in bean by maize plants. *Plant Soil.*184: 195–205.

Hampp, R., Schaffer, C., Wallenda, T., Stulten, C., Johann, P. and Einig, W. (1995). Changes in carbon partitioning or allocation due to ectomycorrhiza formation: biochemical evidence. *Can. J. Bot.* 73: 548–56.

Harrison, M.J. (1996). A sugar transporter from *Medicago truncatula*; altered expression pattern in roots during vesicular arbuscular (VA) mycorrhizal associations. *The plant journal.* 9: 491–503.

Jacobsen. I., Abbott. L.K., Robson, A. (1992). External hyphae of vesiculararbuscular mycorrhizal fungi associated with *Trofoluim subterraneum* L. I. Spread of hyphae and phosphorus inflow into roots. *New Phytol.* 120: 371–80.

Jain, R.K., Paliwal, K., Dixon, R.K and Gjerstad, D.H. (1989). Improving productivity of multipurpose trees on substandard soils in India. *J. Forest.* 87: 38–42.

Jalaluddin, M. (1993). Effect of VAM fungus (*Glomus intraradices*) on the growth of sorghum, maize, cotton and *Pennissetum* under salt stress. *Pakistan Journal of Botany.* 25: 2, 215–8.

Janos, D.P. (1983). Tropical mycorrhizas, nutrient cycles and plant growth. In: Sutton SL, Whitmore TC, Chadwick, AC (eds.) Tropical rain forest: ecology and management, Blackwell, oxford. 327–45.

Jasper, D. A., Abbott, L. K. and Robson, D. (1991). The effect of soil disturbance on vesicular-arbuscular mycorrhizal fungi in soils from different vegetation types. *New Phytol.* 118: 471–6.

Jeffries, P. (1987). Use of mycorrhiza in agriculture. *Crit. Rev.Biotechnol.* 5: 319–357.

Jin, Z.Z. (1997). The diversity features of the floristic elements in the tropical rain forest of Xishuangbanna, Yunnan. *Acta Botanica Yunnanica.* 129–148.

Jin, Z.Z. and Ou, X.K. (1997). The diversity features of plant community types in the tropical rain forest vegetation of Xishuangbanna, Yunnan. *Acta Botanica Yunnanica.* 1–30.

Johanssen, A., Jakobsen, I. and Jessen, E.S. (1993). External hyphae of vesicular-arbuscular mycorrhizal fungi associated with Trifolium subterraneum L. 3. Hyphal transport of 32P and 15N. *New Phytol.* 124: 61–8.

Johnson, N.C., Graham, J.H. and Smith, F.A. (1997). Functioning of mycorrhizal associations along the mutualism parasitism continuum. *New phytol.* 135: 575–86.

Joner, E. and Levylal, C. (1997). Uptake of roots by roots and hyphae of Glomusmosseae/Trifolium subterraneum mycorrhiza from soil amenede with high and low concentrations of cadmium. *New Phytol.* 135: 53.

Kabir, Z., O'Halloran, I.P., Widden, P. and Hamel C. (1998). Vertical distribution of arbuscular mycorrhizal fungi under corn (Zea mays L.) in no-till and conventional tillage systems. *Mycorrhiza.* 8: 53–55.

Kehri, H. K., Chandra, S. and Maheshwari, S. (1987). Occurrence and intensity of VAM in weeds, ornamentals and cultivated plants at Allahabad and areas adjoining it. Mycorrhiza Round Table Proceedings of the workshop, JNU, New Delhi. 273–83.

Khade S. W. and Rodrigues B. F. (2003). Occurrence of arbuscular mycorrhizal fungi in tree species from Western Ghats of Goa, India. *J. Trop. Forest Sci.* 15: 320–331.

Kulkarni, S. S., Raviraja, N. S. and Sridhar, K. R. (1997). Arbuscular mycorrhizal fungi of tropical sand dunes of west coast of India. *J.Coastal Res.* 13: 931–6.

Kumaran, K. and Santhanakrishnan, P. (1995). Vesicular-arbuscular mycorrhizal fungi in tea (*Camellia sinensis* (L.) O Kuntz) soil. Proceedings of the Third National Conference on Mycorrhizae. In Mycorrhizae: Biofertilizers for the Future (eds. Adholeya, A. and Singh, S.). 33–7.

Lakshman, H.C. VA mycorrhizal studies in some important timber tree species. Ph.D thesis. Karnatak University, Dharwad India. Pp. 249.

Lakshman, H.C. and Patil. Geeta B. (2006). Screening of AM fungi in sixty forest tree species of Western Ghats of Karnataka Forests. J.Tropical Ecolo.113:79–84.

Leake, J.R., Johnson, D., Donnelly, D.P., Muckle, G.E., Boddy L. and Read, D.J. (2004). Networks of power and influence: the role of mycorrhizal mycelium in controlling plant communities and agroecosystem functioning. *Can. J. Bot.* 82: 1016–45.

Lekha, K. S., Sivaprasad, P., Joseph, P. T. and Vijayan, M. (1995). *Glomus fasciculatum* - a predominant vesicular—arbuscular mycorrhizal fungus associated with black pepper in forest soils of Kerala. Proceedings of the Third National Conference on Mycorrhizae. In *Mycorrhizae: Biofertilizers for the Future* (eds Adholeya, A. and Singh, S.). 81–5.

Liu, A. Hamel, C., Elmi, A., Costa, C., Ma, B. and Smith, D.L. (2002). Concentrations of K, Ca and Mg in maize colonised by arbuscular mycorrhizal fungi under field conditions. *Can. J. Soil Sci.* 82(3): 271–8.

Mandeel, G. and Gul, J. (1999). Occurrence of vesicular-arbuscular mycorrhizae (VAM) in coastal habitats of Bahrain. In Czech. *Mycology*. 52: 1, 69–79.

Manoharachary, C. and Rao, P. R. (1991) Vesicular-arbuscular mycorrhizal fungi and forest trees. In Proceedings of the Second Asian Conference on Mycorrhiza (eds. Soerianegara, I. and Supriyanto), 39.

Marschner, P. and Baumann, K. (2003). Changes in bacterial community structure induced by mycorrhizal colonisation in split-root maize. *Plant and Soil*. 251: 279–289.

Mehrotra, V. S. (1995). Arbuscular mycorrhizal associations in plants colonizing overburdened soil at an opencast coalmine site. Proceedings of the Third National Conference on Mycorrhizae. In Mycorrhizae: Biofertilizers for the Future (eds Adholeya, A. and Singh, S.). 22–9.

Mohan, V. and Verma, N. (1995). Studies on vesicular-arbuscular mycorrhizae association in seedlings of forest tree species in arid zones of Rajasthan. Proceedings of the Third National Conference on Mycorrhizae. In Mycorrhizae: Biofertilizers for the Future (eds. Adholeya, A. and Singh, S.). 52–5.

Morin, C., Samson, J. and Dessureault, M. (1999). Protection of black spruce seedlings against *Cylindrocladium* root rot with ectomycorrhizal fungi. *Can. J. Bot.*77: 169–74.

Mosse, B. and Hayman, D.S. (1980). In: Tropical Mycorrihiza Research. (ed.) Mikola P, Clarendon, Oxford. 213–30.

Moyersoen, B. (1993). Ectomicorrizas y micorrizas vesı´culo-arbusculares en Caatinga Amazo´nica del Sur de Venezuela. *Scientia Guaianae* 3: 82.

Mukerji, K. G., Bhattacharjee, M. and Tewari, J. P. (1983). New species of vesicular–arbuscular mycorrhizal fungi. *Trans. Br. Mycol.Soc.* 81: 641–3.

Muthukumar, T. and Udaiyan, K. (1995). Vesicular-arbuscular mycorrhizae in dicots of a nutrient-deficient semi-arid grassland. Proceedings of the Third National Conference on Mycorrhizae. In Mycorrhizae: Biofertilizers for the Future (eds. Adholeya, A. and Singh, S.). 541– 5.

Nalini, P. A., Byra Reddy, M. S. and Bagyaraj, D. J. (1987). VA mycorrhizal spore types present in the root zone of *Leucaena leucocephala* (LAM) de. Mycorrhiza Round Table Proceedings of workshop, JNU, New Delhi. 129–36.

Neeraj and Verma, A. (1991). Distribution of VAM fungi in the Indian Deserts. In Proceedings of the Second Asian Conference on Mycorrhiza (eds. Soerianegara, I. and Supriyanto). 125.

Onipchenko, V.G. and Zobel, M. (2000). Mycorrhiza, vegetative mobility and responses to disturbance of alpine plants in the Northwestern Caucasus. *Folia Geobotanica and Phytotaxonomica* 35: 1–11.

Osonubi, O., Mulongoy, K., Awotoye, O.O., Atayese, M.O. and Okali, D.U. (1991). Effects of ectomycorrhizal and vesicular-arbuscular mycorrhizal fungi on drought tolerance of four leguminous woody seedlings. *Plant and Soil,* 136: 131–43.

Parthipon, B., Ganesan, V. and Mahadevan, A. (1991). Occurrence of vesicular-arbuscular mycorrhizae (VAM) in semi-arid region of Tamil Nadu, India. In Proceedings of the Second Asian Conference on Mycorrhiza (eds. Soerianegara, I. and Supriyanto). 57–60.

Pearson, J.N. and Jakobsen, I. (1993). Symbiotic exchange of carbon and phosphorus between cucumber and three arbuscular mycorrhizal fungi. *New Phytol.* 124: 481–488.

Pearson, J.N. and Schweiger, P.S. (1993). *Scultellospora calospora* (Nicol and Gerd.) Walker and Sanders associated with subterranean clover: dynamics of colonization, sporulation and soluble carbohydrates. *New Phytol.* 124: 215–219.

Pinochet, J., Calvet, C., Camprubi, A. and Ferandez, C. (1996). Interactions between migratory endoparasitic nematodes and arbuscular mycorrhizal fungi in perennial crop. *Plant and Soil.* 185: 183–190.

Powell, C. (1980). Mycorrhizal infectivity of eroded soils. *Soil Biol Biochem.* 12: 247–50.

Puthur, J.T., Prasad, K.V.S.K., Sharmila, P. and Saradhi, P.P. (1998). Vesicular arbuscular mycorrhizal fungi improves establishment of micropropagated *Leucacna leuocephala* plantlets. *Plant cell Tissue Organ Cult.* 53: 41–47.

Raaimakers, D. and Lambers, H. (1996). Response to phosphorus supply of tropical tree seedlings: a comparison between a pioneer species *Tapirira obtusa* and a climax species *Lecythis corrugata. New Phytol.* 132: 97–102.

Raghupathy, S. and Mahadevan, A. (1991). Vesicular-arbuscular mycorrhizal (VAM) distribution influenced by salinity gradient in a coastal tropical forest. In Proceedings of the Second Asian Conference on Mycorrhiza (eds. Soerianegara, I. and Supriyanto). 91–5.

Raghupathy, S. and Mahadevan, A. (1992). Profile of VA mycorrhizal fungi and nodulation of legumes in the Coromandel coast of Thanjavur District, Tamil Nadu. The International Symposium on Management of Mycorrhizas. In Agriculture, Horticulture and Forestry. 19.

Raja, P., Ravikumar, P. and Mahadevan, A. (1991). Vesicular-arbuscular mycorrhiza (VAM) in the forest plants of Nilgiris, Tamil Nadu, India. Proceedings of the Second Asian Conference on Mycorrhiza (eds. Soerianegara, I. and Supriyanto). 81–9.

Raman, N. and Nagarajan, N. (1995). Incidence of mycorrhizal association in a forest fire site in Servarayans Hills, Tamil Nadu. Proceedings of the Third National Conference on Mycorrhizae. In *Mycorrhizae: Biofertilizers for the Future* (eds. Adholeya, V. and Singh, S.). 100–3.

Raman, N., Sambandan, K., Sahhadevan, C. and Selvaraj, T. (1995). Distribution of vesicular-arbuscular mycorrhizal fungi in tannery effluent polluted soils of Tamil Nadu, India. Proceedings of the Third National Conference on Mycorrhizae. In Mycorrhizae: Biofertilizers for the Future (eds Adholeya, A. and Singh, S.). 168–173.

Rani, D. B. R., Raghupathy, S. and Mahadevan, A. (1991). Incidence of vesicular-arbuscular mycorrhizae (VAM) in coal wastes. Proceedings of the Second Asian Conference on Mycorrhiza (eds. Soerianegara, I. and Supriyanto). 77–80.

Read, D. (1994). Plant-microbe mutualisms and community structure. In: *Biodiversity and Ecosystem Function* (eds. E.D. Schulze and H.A. Mooney). Springer Press, New York: 181–209.

Read, D.J. and Birch, C.P.D. (1988). The effects and implications of disturbance of mycorrhizal mycelial systems. Proceedings of the Royal Society of Edinburgh 94B: 13–24.

Reddy, P. R. P., Reddy, P. J. M., Prakash, P. and Manoharachary, C. (1995). Association of vesicular-arbuscular mycorrhizal fungi in soils polluted with industrial and sewage effluents. Proceedings of the Third National Conference on Mycorrhizae. In Mycorrhizae: Biofertilizers for the Future (eds. Adholeya, A. and Singh, S.). 155–8.

Saito, M. (1995). Enzyme activities of the internal hyphae and germinated spores of an arbuscular mycorrhizal fungus *Gigaspora marginata* Becker and Hall. *New Phytol.* 129: 425–31.

Sambadan, K., Raman, N. and Kannan, K. (1991). Association of VAM fungi with *Casuarina equisetifolia* at different soil types in Tamil Nadu, India. In Proceedings of Second Asian Conference on Mycorrhiza (eds. Soerianegara, I. and Supriyanto). 61–5.

Sastry, M. S. R. and Johri, B. N. (1999). Arbuscular mycorrhizal fungal diversity of stressed soils of Bailadila iron ore sites in Bastar region of Madhya Pradesh. *Curr. Sci.*77: 1095–1100.

Schaeffer, C., Wallenda, T., Hampp, R., Salzer, P. and Hager, A. (1997). Carbon allocation in mycorrhiza. In: Rennenberg, H., Escherich, W., Ziegler, H., eds. Trees-contribution to modern tree physiology. Leiden. The Netherlands. Buckhuya Publishers, 393–407.

Schüßler, A., Schwarzott, D. and Walker, C. (2001). A new fungal phylum, the Glomeromycota: phylogeny and evolution. *Mycol. Res.* 105: 1413–21.

Schwab, S.M., Menge, J.A. and Tinker, P.B. (1991). Regulation of nutrient transfer between host and fungus in vesicular-arbuscular mycorrhizas. *New Phytol.* 117: 387–98.

Schwab, S.M., Menge, J.A. and Tinker, P.B. (1991). Regulation of nutrient transfer between host and fungus in vesicular-arbuscular mycorrhizas. *New Phytol.* 117: 387–98.

Selvaraj, T. and Subramanian, G. (1991). Survey of vesicular-arbuscular mycorrhizae in mangroves of Muthupet Estuary: Ecological implications. In Proceedings of the Second Asian Conference on Mycorrhiza (eds. Soerianegara, I. and Supriyanto). 271.

Sengupta, A. and Chaudhari, S. (1989). Occurrence of vesicular-arbuscular mycorrhiza in *Sueda maritima* (L.) Dumort—A pioneer mangrove of the Chenopodiaceae. *Curr. Sci.* 58: 1372.

Sengupta, A. and Chaudhari, S. (1990). Vesicular arbuscular mycorrhiza (VAM) in pioneer salt marsh plants of the Ganges river delta in West Bengal (India). *Plant and Soil.* 122: 111-3.

Shachar-Hill, Y., Pfeffer, P.E., Douds, D., Osman, S.F., Doner, L.W. and Ratcliffe, R.G. (1995). Partitioning of intermediary carbon metabolism in vesicular - arbuscular mycorrhizal leek. *Physiol. Plant.* 108: 7-15.

Singer, R. and Araujo, I. J. S. (1986). Litter decomposing and ectomycorrhizal Basidiomycetes in an Igapo´ forest. *Plant Systematics and Evolution.* 153: 107-17.

Singh, R. and Adholeya, A. (2002). Biodiversity of arbuscular mycorrhizal fungi (AMF) in different agroclimatic regions of India. IMC 7— Conference on Mycological Advances, 11-17 August, Norway.

Singh, R. and Pandya, R. K. (1995). The occurrence of vesicular-arbuscular mycorrhiza in pearl millet and other hosts. Proceedings of the Third National Conference on Mycorrhizae. In Mycorrhizae: Biofertilizers for the Future (eds. Adholeya, A. and Singh, S.). 56-8.

Siqueira, J.O., Carneiro, M.A.C., Curi, N., Rosado, S.C.S. and Davide. A.C. (1998). Mycorrhizal colonization and mycotrophic growth of native woody species as related to successional groups in southeastern Brazil. *Forest Ecology and Management* 107: 241-52.

Smith, S.E. and Read, D.J. (1997). In *Mycorrhizal Symbiosis,* 2nd edn. Harcourt Brace, London

Solaiman, M.D. and Saito, M. (1996). Use of sugars by intraradical hyphae of arbuscular mycorrhizal fungi revealed by radiorespirometry. *New Phytol.* 136: 533-38.

St-Arnaud, M., Hamel, C., Caron, M. and Fortin, J.A. (1995). Endomycorrhizes VA et sensibilité aux maladies: synthèse de la littérature et mécanismes d'interaction probables. *In* Mycorrhizal Symbiosis Eds. C. Charest and R. Bernier. Orbis, Frelighsburg, Québec, pp. 51-87.

Subramanian, K.S. and Charest, C. (1999). Acquisition of N by external hyphae of an arbuscular mycorrhizal fungus and its impact on physiological responses in maize under drought-stressed and well watered conditions. *Mycorrhiza.* 9: 69-75.

Sylvia, D.M. and Sinclair, W.A. (1983). Phenolic compounds and resistance to fungal pathogens induced in primary roots of Douglas fir seedlings by the ectomycorrhizal fungus *Laccaria laccata. Phytopathology.* 73: 390-7.

Thapar, H. S. and Uniyal, K. (1996). Effect of VAM fungi and Rhizobium on growth of *Acacia nilotica* in sodic and new forest soils. *Indian For.* 122 (11): 1033–9.

Thompson, B.D., Robson, A.D. and Abbott, L.K. (1991). Soil mediated effects of phosphorus supply on the formation of mycorrhizas by *Scutellospora calospora* (Nicol and Gerd.) Walker and Sanders on subtereanean clover. *New Phytol.* 118: 463–9.

Vijaya, T., Kumar, R. V., Reddy, B. V. P., Sastry, P. S. S. and Srivastav, A. K. (1995). Studies on occurrence of endomycorrhiza in some forest soils of Andhra Pradesh. Proceedings of the Third National Conference on Mycorrhizae. In *Mycorrhizae: Biofertilizers for the Future* (eds. Adholeya, A. and Singh, S.). 45–7.

Visser, S., Griffiths, C.L. and Parkinson. D. (1984). Top soil storage effects on primary production and rates of vesicular-arbuscular mycorrhizal development in *Agropyron trachycaulum*. *Plant and Soil.* 82: 51–60.

Vitousek, P. M. and Sanford, R. L. Jr. (1986). Nutrient cycling in moist tropical forest. *Annual Review of Ecology and Systematics.* 17: 137–67.

7

Evaluation of AM Fungal Application on Some Important Hydrocarbon Yielding Plants

R F Inchal and H C Lakshman***

INTRODUCTION

During the last two to three and half decades, increased uses of fertilizer and pesticide have contributed to increase in crop production. But, the unprecedented hike in cost of chemical fertilizers and the present growing concern about the negative effects of various chemicals, which are used in plant production, on environment is promoting the development of sustainable management practices. This situation therefore calls for an intense emphasis on the attempts to control soil and plant associated micro-organisms to lower fertilizer production cost and reduce environmental pollution while assuring high yielding and adaptability of plants to unfavourable conditions. The approach thus would be to inoculate or introduce symbiotic microbes into rhizosphere soil for high crops production.

* Department of Botany S.K. and H.S.K. Science Institute Vidyanagar, Hubli- 580031.

** Post graduate department of Studies in Botany, Microbiology laboratory, Karnatak University Dharwad-580003 India

In a number of developing countries the substitution of imported petroleum derivatives by domestically produced biomass-based engine fuels is currently regarded as a promising strategy for eliminating energy. The main advantage of these petrocrops cultivation is that the latex and resin bearing wild crops grow well with a minimum requirement of water and without much agricultural management. We are so much used to petrol and diesel as transport fuel that we think of converting any renewable source of energy into these fuels, of-course, petroleum fuels are the most convenient for the working of internal combustion engine. Therefore, it would be highly desirable, if we can produce petrol-like hydrocarbons from plants. Considerable work has been done on the cultivation of petroleum plantations (Mukerji *et al.*, 2000).

Many hydrocarbon yielding plants are being considered these days as potential petro-crops yielding biocrude fuel which after various degrees of processing and blending with other fuels can be used for powering internal combustion engines (Smith, 1987). These species can be commercially exploited if the biomass yields are increased by selecting suitable strains and better management of agricultural practices. Association of arbuscular mycorrhizal fungi (AMF) often improves biomass yield (Gerdemann, 1968; Hosamani *et al.*, 2004) by significantly increasing uptake of moisture and other elements. In order to undertake research experiments on productivity of these hydrocarbon-yielding plants and on the effect of AM fungi on its growth and biomass was to be considered. Survey of the literature revealed that there has been no systematic study on the association of AM fungi on hydrocarbon yielding plants.

The intrinsic interest in the study of arbuscular mycorrhizal fungi has been increased since three and half decades due to their beneficial role in the growth of the plants. This fungi is an integral component of majority of the plants in wide ecological range, with mutual dependency between fungus and host for natural function and survival (Neerja and Varma, 1995; Vijaya, 2000). The fungus takes carbohydrates from the host plant and in turn supplies the nutrients. It is established fact that the mycorrhiza in improve plant growth by improved uptake of nutrients especially phosphorus and of the micro-nutrients. The ability of these fungi to produced dramatic response in plant growth is well documented (Chandra, 1992; Lakshman, 2007).

AM fungi are able to bind soil in to semi-stable aggregates, thus improving the structure of the soil. A healthy plant soil system is the backbone of sustainable farming without which sustainable development can never reach its maxima.

MATERIALS AND METHODS

The green house experiments were conducted at the Post Graduate department of Botany, Karnatak University, Dharwad. Geologically it is lying in between 17° 17′ to North latitude 74° 48′ to 76′ East longitude. Using complete randomized design with 4 replications. The experiments were conducted in earthen pots. (25 × 30 cm) filled with sterilized soil to investigate the influence of AMF on five important hydrocarbon yielding plants. 1) *Jatropa curcas* L. 2) *Jatropa gossifolia* L. 3) *Madhuca indica* Gmel (Mahua) 4) *Ricinus communis* L. (Rosa Var.) 5) *Ricinus communis* L. (Mysore local Var.) A brief account of each plants are as follows :

1) *Jatropa curcas* L.

Jatropa curcas is found in a semi wild condition in the vicinity of the villages, scrubby jungles, rock and pillar site etc. It grows rapidly in dry weather conditions. The plant was introduced in India by Portugues as an oil yielding plant. It is one of the promising draught tolerant perennial crop and is adaptable to various kinds of soil environment. The work on improvement of *Jatropa curcas* are very limited. Study of the literature survey revealed that 400 litres of oil could be extracted from the yield per acre.

2) *Jatropa gossifolia* L.

It is a small much branched bushy gregarious shrub naturalized almost throughout India. It is now even being considered a potential substitute for diesel.

3) *Madhuca indica* Gmel (Mahua)

Madhuca indica plant commonly known as Mahua is valued for its flower, fruits and seeds since ancient times. The tree is called 'Madhuca' the spirituous liquor prepared from the flower is Madhu-madhavi. All the parts of mahua tree such as bark, leaf, flower, fruits, seeds and seed oil are used for various purposes. The seed cakes are used as biofertilizers.

4) *Ricinus communis* L. (Castor)

Castor is one of the major oil seed crop of India and infact, India is one the leading producer of Castor crop. The castor export has been fetching 75 crores annually, besides meeting the domestic demand in pharmaceuticals, medicine, industry, soaps, paints, lubricants etc. Inspite of having much potential and scientific technology castor yield/hectare in the country continue to be very low. This crop is grown as Rabi and Karif crops. Two varieties of castor have basically classified in Karnataka i.e. Mysore local and Rosa.

Roots and rhizospheric soil samples were collected from the different localities where these four hydrocarbon yielding plants growing in natural condition. Collected roots were washed thoroughly under tap water and observe directly under stereomicroscope to determine the extrametrical hyphae structure, colour and infected on roots: For quantifying arbuscular mycorrhizal colonization followed the method of Phillips and Hayman (1970). Fine feeder roots (< 2 mm diam) of each collected samples kept in separate polythene bags with proper labeling. Some selected roots were cut into l cm length and stored in alcohol-formaldehyde-acetic acid solution (200ml alcohol + 13ml formaline + 95ml acetic acid). The roots were cleared with 10% KOH, and neutralized with 2% HCl for 3–4 minutes, stained with Tryphan blue mounted in lactophenol. Gently tapped the cover slip to spread root tissue and observed under the compound microscope.

Glomus fasiculatum and *Glomus mosseae* selected for inoculation/purpose. Good quality of these inoculums was multiplied with the germplasm *typhoideum* and *Sorghum bicolor* respectively so as to get sporocarps, spores, freshly colonized root fragments and rhizospheric soils. The seedlings were transplanted to the pots and 50 g of inoculum was provided in the vicinity of root system. Triplicate sets were maintained with suitable controls. Pots were watered on alternate days to maintain sufficient moisture level and plants were fed with Hoagland minus P solution at the rate of 10 ml per pot twice in a month. At the end of each experiment growth parameters such as plant height, shoot, root fresh weight and dry weight, spore number, root-shoot ratio were recorded. Per cent root length colonization was assessed by Gridline intersect method as described by Giovannetti and Mosse (1980). Percentage of P was determined in shoot dry weight. Arbuscular mycorrhizal spores were extracted by wet sieving and decanting technique of Gerdemann and Nicolson (1963). Phosphorus content of the plants were determined calorimetrically by Venadomolybdate-phosphoric yellow colour method as already mentioned earlier. Attempt has been made to extract oil from the seeds of *Jatropa curcas*. The seed oil was extracted by using simple screw press.

The fungal mycelium took bright blue colour stain but the host tissue remains unstained. This gives clear contrast between mycorrhiza and host tissues. The slides are thoroughly scanned for various structural components of Mycorrhiza. (i.e. vesicles, arbuscules, extra metrical chlamydospores etc.)

ASSESSMENT OF COLONIZATION

Assessment of AM colonization of experimental plants were made as described by objective of this study is to determine the percentage, of root colonization as well as intensity of colonization within the root.

The whole root sample is uniformly distributed in a petridish with the help of dissecting needle to eliminate clumping and to enhance light transmission. The number of susceptible root colonization in a sample determines the percentage of AM infection. The roots are cut in to 0.5 cm bits spread on clean glass slides mount it observed under the microscope for the presence or absence of mycelium, vesicles and arbuscules. The percentage of root colonization is calculated after examining number of roots for the presence or absence of colonization. Colonized root samples were grouped as per the classification of (Institute for mycorrhizal research and development USDA fore it service Athens, Georgia (1981). Class 1:0–5 % Class-II 6–26%, Class-III 26–50%, Class IV 51–75%, Class V 75–100%, (Table 7.2) and their components had been examined. Seeds were collected from five different experimental plants from their natural habitats and sown in sterilized soils in a number of pots to obtain healthy seedlings for further experiments. Phosphorus deficient soil samples were used for pot experiments. Soil sieved and mixed with sand at the ratio of 3:1 (3 parts of soil to one part of sand). Physico-chemical characteristic of the soil is given (Table 7.1). Soil was fumigated with Methyl bromide (5%) after six days, these soils were filled to number of earthen pots. Soil physico-chemical characters have been determined following the method of Jackson (1973).

RESULTS AND DISCUSSION

All the experimental plant samples for microscopic observation for AM colonization. The roots were colonized with the arbuscular mycorrhizae. Maceration and anatomical

Table 7.1. Showing the Physico-Chemical Properties of Soil Used for Pot Experiments :

Characteristics	Garden soil + pure sand
Soil moisture (1%)	28.03
pH	6.8
Organic matter (1%)	0.69
E.C. m mohs/cm	0.98
N mg / kg	54.21
P mg / kg	27.16
K mg / kg	72.50
Fe mg / kg	66.14
Zn mg / kg	14.5
Cu mg / kg	11.8
Mg mg / kg	9.3

Each value is the means of 14 samples

studies followed by tryphan blue staining revealed distinctly, stages and different components. Microscopic measurements provided an assessment of the relative abundance of mycelium in roots expressed as the total mycorrhizal root length, the density of hyphae within roots, the length per entry point and the number of entry points, wall thickness, and pattern of outer epidermal cells. Numerous root hairs were seen in the infected roots of *Jatropa curcas* and *Jatropa gossifolia*. Course aseptate hyphal coils were often seen from initial penetration points. AM colonization examined in all the experimental plants, High percent of colonization with profusely branched mycelium were observed in *Jatropa curcas* and *Ricinus cummunis* (64% and 50.7%). 26–50% of mycorrhizal colonization was noticed in *Jatrops gossifolia* and *Madhuca indica*. Colonization was started from the second set of lateral roots from the top; where as young terminal roots had more colonization. But colonization was not been observed in meristematic zone of the roots.

All the five hydrocarbon plants showed the positive response to fungi of *Glomus fasciculatum* and *Glomus mossae*. The overall results have been recorded in 7.4 to 7.8 tables. Soil

Table 7.2. : Showing the Arbuscular Mycorrhizal Fungal Colonization in Five Hydrocarbon Plants

Sl. No.	Plants	No. of Root Samples Examined	Mean Root Colonization Rating	Intensity of Colonization	No. of Localities Pre Seated	No. of Samples Infected	% of Root Segment	% of Root Length Infected Per Sample Mean+SD (Range)
1.	*Jatropa curcas* L.	21	7	5.3±0.00	16	8	62.1±4.2	64.0±0.00
2.	*J. gossifolia* L.	14	4	2.3±1.00	11	6	53.2±2.2 (0–61)	49.7±6.3 (0–72)
3.	*Madhuca indica* Gmel	19	3	1.7±4.2	5	4	44.4±7.1	41.0±3.3
4.	*Ricinus comminis* (Var Mysore local)	31	9	7.5±2.0	14	9	(0–52) 59.5±3.5	(0–59) 67.7±5.6
5.	*Ricinus comminis* (Var. Rosa)	27	8	6.2±2.3	12	7	(0–69) 56.4±4.4 (0–66)	(0–77) 57.2±7.7 (0–63)

Mean colonization rating 1) 0–5%, 2) 6 – 25%, 3) 26 – 50%, 4) 51 – 75%, 5) 76 – 100%
Intensity of colonization 1) small widely scattered, 2) Large colonization, 3) colonization in cortical cells.

used for the green house experiments had low phosphorous content. The mean rating of AM fungal colonization with arbuscules, vesicles and hyphae was assessed. There was increased plant biomass with P content in *Jatropa curcas* (table). There was no much variation in the percent root colonization in *Jatropa gossifolia* (Table 7.5). An increased plant height was recorded in *Madhuca indica* shown in (Table 7.6). Similarly, spore population steadily increased in experimental plants of *Ricinus communis* (variety Rosa). There was a significantly increase in plant height, shoot and root dry weight was observed in *Ricinus communis* (variety Mysore) in mycorrhiza inoculated plant compare to non-inoculated plants. This indicated that the development of arbuscular mycorrhizae was more pronounced in mycorrhizal treated hydrocarbon plants. The percent AM fungal colonization. The experimental plants grown with rhizospheric soil inoculum in glass house condition and roots collected from natural habitat of these plants exhibited the same extent of colonization by AM fungi. The fresh preparation of colonization roots showed the signs of active growth in the cortical region. The plants that showed less percentage of colonization were due to the compactness of tissue system in *Jatropa gossifolia* and *Madhuca indica*. The magnified intersection method gave additional accuracy to measure the fungal components in macerated roots. Magnified intersections were counted in the following categories.

Khan (1979) studied the effect of arbuscular mycorrhizal associations on growth of cereals. Mosse (1977) worked the plant growth responses to AM in *Stylosanthes* spp. Allen *et al.*, (1979) reported the growth response to VAM in non-mycorrhizal *Bouteloua gracilis* in a defined medium. Endomycorrhizae influences the growth of black gram seedlings in flooded soils was studied by Keeley (1980). Growth and yield responses of Cowpea and Maize was recognized by (Gerdemann, 1968; Mosse *et al.*, 1968; Mosse, 1973; Powell and Daniel, 1978, Saif, 1987). Feldman, *et al.*, (1990) demonstrated that every rubber tree has its fungal symbionts under natural conditions. Similarly growth response study conducted on coffee and tea plants by the inoculation of arbuscular mycorrhiza in nursery level (Sieverding and Toro, 1987; McGee *et al.*, 1999). The beneficial effect of inoculation forest tree nurseries with AM fungi to improve plant growth was studied by (Jeffries, 1987; Young 1990; Lakshman, 1996). The effect of plant growth on VA-mycorrhiza was investigated by Powell (1984) reported the influence of VA-mycorrhizae and Zinc on growth of apple seedlings. Smith and Read (1997) studied the plant growth and nutrient uptake in *Azadiracta indica.* The influence of arbuscular mycorrhizae on the growth, and development of hardwood tree species was studied by Kormanik *et al.* (1982). Influence of endomycorrhiza in relation to nutrient application on the growth of *Leucana leucocephala* was reported by Tommer *et al.* (1985). Effects of pre-inoculation with a VA-mycorrhizal fungus on growth of onion transplanted to the field as multi seeded pea modules has been studied by Snellfrove and Stribley (1987).

The maximum shoot weight was recorded in *Jatropa curcas* when the plants were colonized with *Glomus fasciculatum.* This indicates that *G. fasciculatum* have impact on spore germination and root colonization and eventually spore production. The mineral nutrient concentration elevation may be the reason behind over all vigorous growth. The enhanced photosynthetic activity would be due to accumulative effect of higher chloroplast pigmentation and better nutrient uptake. This activity finally resulted in healthier seeds. Similar findings were reported by Sieverding (1991). Interaction studies of both organisms i.e., AM (*Glomus fasciculatum*) and host brought a significant increase in shoot biomass and total P uptake.

The present study brought out to obtain maximum benefit from Arbuscular-mycorrhiza and its compatibility between the fungus host and the both the partners competitiveness was tested. Plant dry weight increased in all the experimental plants over the controls. Similar observation was made by (Tinker, 1975; Santosh, 2004; Devachandra *et al.*, 2008). Root and shoot ratio were gradually reduced with increased days of harvest in all non-mycorrhizal plants over the mycorrhizal

Table 7.3. : Showing the Selected Examples Data Obtained Using the Magnified Intersections Method

Root sub-sample	Microscope	Negative	Number of intersections Arbuscular	Vesicles	Hyphae	Total
1	A	47	7	1	5	60
	B	51	6	0	4	61
	C	44	4	1	4	53
	D	39	5	0	6	50
	Sum	181	22	2	19	224
2	A	23	24	1	4	52
	B	20	15	1	6	42
	C	15	13	1	15	44
	D	21	12	0	3	36
	Sum	79	14	3	28	174

Arbuscular colonization (AC), Vesicular colonization (VC), Hyphal colonization (H C).

Table 7.4. : Showing the Effect of *Glomus fasciculatum* on Growth Response, Biomass Production, Mycorrhizal Colonization, Spore Number and 'P' Uptake in *Jatropha curcas* for 180 Days.

Treatment Duration	Plant Height (cm)	Shoot Fresh wt (g)	Dry wt (g)	Root Fresh wt (g)	Dry wt	Root/shoot ratio (g)	% of colonization	Spore No/ 50 g soil	% of 'P' in shoot
60 days									
NM	12.9±3.2	12.4±3.2	3.7±1.4	0.98±1.3	0.16±0.4	0.16±0.3	0	0	0.05±0.01
M	22.2±4.1	22.2±4.1	10.9±1.4	1.1±4.4	0.86±0.5	0.07±0.4	56.3±4.5	59.4±4.1	0.11±0.2
120 days									
NM	14.5±3.3	14.5±3.3	4.6±1.2	0.98±1.4	0.16±0.1	0.13±0.2	0	0	0.11±0.2
N	37.1±5.5	37.1±5.5	22.3±3.3	1.2±4.5	0.97±0.2	0.04±0.01	61.2±6.8	59.2±6.9	0.14±0.2
180 days									
NM	16.4±3.4	16.4±3.4	5.7±4.2	1.0±1.2	0.65±0.1	0.11±0.2	0	0	0.14±0.2
N	49.2±3.6	49.2±3.6	32.6±3.1	5.5±1.2	0.97±0.3	0.02±0.02	77.4±3.2	63.6±	0.16±0.1

NM = Non mycorrhizal and M = Mycorrhizal values are means of four replicates ± standard error.

Table 7.5 : Showing the Effect of *Glomus fasciculatum* on Growth Response, Biomass Production, Mycorrhizal Colonization, Spore Number and 'P' Uptake in *Jatropha gossifolia* for 180 Days.

Treatment Duration	Plant Height (cm)	Shoot Fresh wt (g)	Dry wt (g)	Root Fresh wt (g)	Dry wt	Root/shoot ratio (g)	% of coloniza-tion	Spore no/ 50 g soil	% of 'P' in shoot
60 days									
NM	7.3±4.4	2.5±2.2	0.96±0.0	0.68±0.0	0.34	0.35			0.05
M	14.6±5.1	6.1±1.0	1.3±1.0	0.92±0.0	0.45	0.34	38.3	51.2	0.10
120 days									
NM	8.2±3.5	14.5±3.3	4.6±1.2	0.98	0.16	0.13			0.11
N	21.1±2.9	37.1±5.5	22.3±3.3	1.2±4.5	0.97	0.04	61.2	59.2	0.14
180 days									
NM	9.1±1.0	16.4±3.4	5.7±4.2	1.0	0.65	0.11			0.14
N	24.2±1.7	7.6±1.3	2.4±5.5	1.3±5.4	0.98	0.40	47.6	61.4	0.15

NM = Non mycorrhizal and M = Mycorrhizal all values are means of four replicates.

Table 7.6. : Showing the Effect of *Glomus fasciculatum* on Growth Response, Biomass Production, Mycorrhizal Colonization, Spore Number and 'P' Uptake in *Madhuca indica* for 180 Days.

Treatment Duration	Plant Height (cm)	Shoot Fresh wt (g)	Dry wt (g)	Root Fresh wt (g)	Dry wt	Root/shoot ratio (g)	% of colonization	Spore No/ 50 g soil	% of 'P' in shoot
60 days									
NM	9.7±5.4	3.8±4.2	1.5±1.1	0.02	0.56	0.37	0	0	0.05
M	16.2±1.1	4.5±2.2	1.9±2.0	1.1±0.0	0.78	0.41	49.3	52.5	0.11
120 days									
NM	10.4±5.2	3.8±4.0	1.6±4.5	0.99	0.56	0.35	0	0	0.07
N	23.3±4.5	4.9±3.2	2.7±2.0	1.5±2.0	0.78	0.28	54.5	52.7	0.12
180 days									
NM	11.6±3.7	4.7±4.3	1.6±5.4	1.00	0.69	0.35	0	0	0.08
N	37.8±4.2	7.8±3.5	3.4±7.2	2.63±3.8	0.81	0.23	54.5	57.1	0.14

NM = Non mycorrhizal and M = Mycorrhizal all values are means of four replicates.

Table 7.7 : Showing the Effect of *Glomus mosseae* on Growth Response, Biomass Production, Mycorrhizal Colonization, Spore Number and 'P' Uptake in *Ricinus communis* (Var. Rosa) for 180 Days.

Treatment Duration	Plant Height (cm)	Shoot Fresh wt (g)	Dry wt (g)	Root Fresh wt (g)	Dry wt	Root/shoot ratio (g)	% of coloniza-tion	Spore No/ 50 g soil	% of 'P' in shoot
60 days									
NM	10.4±1.6	2.6±2.1	1.2±5.2	0.98	0.56	0.46			0.05
M	23.5±3.2	6.2±4.4	2.3±4.2	1.4±0.1	0.99	0.43	42.2	48.5	0.08
120 days									
NM	12.1±5.3	3.1±4.3	1.2±4.4	0.98	0.56	0.46			0.05
N	32.4±7.4	9.4±2.5	5.4±2.3	1.4±1.2	0.47	0.17	47.2	53.3	0.10
180 days									
NM	13.3±3.4	4.1±1.1	2.0±5.1	0.98	0.58	0.58			0.06
N	44.1±5.5	13.2±3.3	5.2±4.3	1.4±1.4	0.97	0.16	51.5	57.5	0.13

NM = Non mycorrhizal and M = Mycorrhizal all values are means of four replicates.

Table 7.8. : Showing the Effect of *Glomus mosseae* on Growth Response, Biomass Production, Mycorrhizal Colonization, Spore Number and 'P' Uptake in *Ricinus communis* (Var. Mysore local) for 180 Days.

Treatment Duration	Plant Height (cm)	Shoot Fresh wt (g)	Dry wt (g)	Root Fresh wt (g)	Dry wt	Root/shoot ratio (g)	% of coloniza-tion	Spore No/ 50 g soil	% of 'P' in shoot
60 days									
NM	11.3±4.1	3.2±1.0	1.1±0.0	1.1±1.0	0.58	0.52			0.04
M	34.2±5.0	8.5±1.4	2.8±1.2	1.4±0.0	0.26	0.34	49.4	54.1	0.07
120 days									
NM	14.1±6.2	13.2±5.5	1.1±2.2	1.0±0.0	0.58	0.52			0.05
N	46.3±7.4	5.5±3.0	1.8±7.1	1.4±1.0	0.97	0.20	51.6	57.3	0.10
180 days									
NM	14.9±1.1	5.5±3.0	1.1±0.0	1.1±0.0	0.59	0.53			0.07
N	59.6±6.6	16.7±5.7	1.9±3.4	1.5±1.1	1.0	0.16	66.2	66.7s	0.14

NM = Non mycorrhizal and M = Mycorrhizal all values are means of four replicates.

plants. This is consistent with the work of (Readhead, 1968; Kormanik *et al.*, 1982 ; Durga *et al.*, 2000). The increased spore population with respect to colonization was not correlated with each other in castor varieties. The observation revealed spore population lowered with higher percent colonization was recorded in *Jatropa gossifolia*, increased dry matter shoots in mycorrhizal plants together with the changes in the distribution of dry matter between root and shoot can be viewed as a feed response to increase nutrient uptake of P in Arbuscular mycorrhiza. The result from this study indicates that potential benefits from VAM in the young seedling production are most encouraging. This technique can be utilized in nursery practices during reforestation programme.

REFERENCES

Aleksandrowicz-Trzcinska, M. (2008). Mycorrhiza of forest seedlings in clear cut regenerations. Sylwan.151 (ll): 3–9

Bryan, C.F. (1963). Endotrophic mycorrhizal influence on yellow poplar seedlings growth. Seedlings growth. Science. 140:1220–1221.

Chandra, S. (1992). VA mycorrhiza Dimensions of its application Indian Phytopath. 45:391–406.

Devachandra, N., Patil, P.B. Swamy, G.S.K. and Durgannavar, M.P. (2008). Screening of different arbuscular mycorrhizal fungi for raising jamun (*Syzygium cumini*) rootstocks. Mycorrhiza News. 20(3): 5–8

Feldmann, F., Jungueria, N.T.V. and Lieberies, R. (1990). Utilization of VA-Mycorrhiza as a factor in integrated plant protection. Agri. Ecosy. And En.zg (1–4) 131–135.

Gerdemann, J.W. (1968). Vesicular arbuscular mycorrhizae and plant growth annual review of Phytopathology. 6:397–418.

Gerdemann, J.W. and Nicolson, T.H. (1963). Spores of mycorrhizal Endogone extracted from soil by wet-sieving and decanting. Trans. Br. Mycol. Soc. 46:235–244.

Giovannetti, M. and Mosse., B. (1980). An evaluation of technique for measuring Vesicular arbuscular mycorrhizal infection in roots. New Phytol, 84: 489-500.

Hosamani, P.A., Lakshman, H. C., Kadam. L.B. and Kolkar, K.P. (2004). Screening of AM fungi and their comparative efficacy on Withania *somnifera* Dunl. Asian. J. Biotech. Environ. Sci.1.3 (2):201–304

Jackson, M. L. (1973). Soil chemical analysis, New Delhi: Prentice Hall, India.

Jefries, P. (1987). Use of mycorrgizae in Agriculture. Critical Review. Biotechnology. 5; 319–357.

Keeley, J.E. (1980): Endomycorrhizae influence growth of blackgram in flooded soils. Amer. J. Bot. 67(1).

Khan, A.G. (1979). The effect of Vesicular arbuscular mycorrhizal association on growth of cereals. I. Effects as Maize growth. New Phytol. 71. 71:613.

Kormanik, P. P., Schultz, R. C. and Bryan, W.C. (1982). The influence of Vesicular arbuscular mycorrhizae on the growth and development of eight hard wood tree species forest. Forest Sci. 28(3): 531–539.

Lakshman, H. C. and Bheemreddy, V.S., (2007). Microbial inoculants for *Capsicum annum* J. Micro World. 1.8 (2): 184–186.

Lakshman, H.C and Patil. Geeta. B. (2006). Efficacy of tree species to AM fungi inoculation at nursery stage. Microbial World. 13: 72–76.

Lakshman, H.C. (1992). Development and response of *Terminalia bellerica* to inoculation Vesicular arbuscular mycorrhizae. Tropical Forestry. 79:109–114.

Lakshman, H.C. (1996). Vesicular arbuscular mycorrhizae studies in some economically important tree species. Ph.D. thesis Karnatak University Dharwad, India.

Lakshman, H.C. (1994). Arbuscular mycorrhiza in hydrophytic and xerophytic plants. Nat. Con. 73:114–118.

Lambert, D. H., Baker, D.F. and Cloe, H. (1979). The role of mycorr-hizae in the interaction of phosporus with zinc, copper and other elements. Soil. Science. Soc of America. J. 43: 976–980.

McGee, P.A., Torrisi, V. and Pattinison, G.S. (1999). The relation between density of Glomus mosseae propagules and the initiation and spread of arbuscular mycorrhizal in cotton roots. Mycorrhiza. 9(4): 221–225.

Mikola, P. (1980) (Ed). Tropical Mycorrhizal research. Oxford press. Oxford. 270 pages.

Mohan, V. Singh, Y.P. (1997). Endomycorrhizal association of *Acacia* in nurseries and plantation of Indian arid zones. Indian Forester. 1223: 323–330.

Mohan, V., Veerma N. Singh, Y.P. (1995). Distribution of VAM fungi in nurseries and plantations of Neem (*Azardicta indica*) in arid zone of Rajasthan. India forester. 121 :1069–1076.

Mosse, B. (1973). Plant growth responses to Vesicular arbuscular mycorrhiza IV. In soil given additional phosphate. New Phytol. 42:127–136.

Mosse, B. (1977). The role of mycorrhizae in legume nutrition as marginal soil in exploiting the Legume-Rhizobium symbiosis. Tropical Agriculture (Ed. By J.M. Vinecnt, A.S. Whiteny and J. Bose), pp. 192–275. College of Tropical Agriculture. University of Hawai, U.S.s Miscellaneous.

Mosse, B. (1981). Vesicular arbuscular mycorrhiza research for Trophical Agriculture Hawaii Institute of Tropical Agriculture and Human Resources, Research Bulletin, 194–196.

Mukerji, K. G., Chamola, B.P. and Singh, J. (2000): Mycorrhizal Biology. Kluwer Academic Publishers New York, London, pp. 338.

Nirmala, C.B., Rajalakshmi G. and Chandrakarthick.(2009). Plant biotechnology. MJP publishers. Chennai, pp. 382.

Parvathi, K., Venkateshwarlu, K. and Rao, A. S. (1984). Occurence of VA-mycorrhizas on different legumes in laterite soil. Curr. Sci. 53: 1254–1255.

Powell, C.V. (1984). Field inoculation with VAM fungi in : VA Mycorrhiza, Powell C.L. and Bagyaraj, J.D. (EDS.) 1984. CRC press, Florida, USA. 205–222.

Powell, C.Vm and Daniel. (1978). Mycorrhizal fungi stimulate uptake of soluble and insoluble phosphates fertilizer from a phosphate deficient soil.

Read, D. 1. (1991). Mycorrhizas in ecosystems. Experientia. 47: 376–391.

Red Head, J.F. (1968). Mycorrhizal association in some Nitrogen forest trees. Trans. Brt. Mycol. Soc. 15: 377–387.

Reena, J. and Bagyaraj, D.J. (1990). Growth stimulation of *Tamarindus indica* by a VA-Mycorrhizal fungi. World Journal of Microbiology and Biotechnology. 6 (1): 59–63.

Saif, R.S. (1987). Growth responses of tropical forage plant species to Vesicular arbuscular mycorrhizal. I. growth and mineral uptake and mycorrhizal dependency. Plant and soil. 97: 25–35.

Santosh (2004). Enhancement germination, growth of Mango root stocks using bioformulations. U. A. Science journal 12:7–9

Sieverding, E. (1991). Arbuscular mycorrhizae management in tropical agro systems. Zusammenarbeit. 11: 75–81.

Sieverding, E., and Toro, T.S. (1987) (a). Growth of coffee and Tea plants in nurseries inoculated with different VAM fungal species. In Sylvia, D M., Hung, L. L., and Grahm, J.H. (eds) Mycorrhiza in the next decade, practical applications and research priorities, 58. Proceedings of the Seventh Nacon-IFAS, University of Florida. Gainesville.

Smith, K. R. (1987). Biofuels and health a global review, New York Plenum Press, pp. 452.

Smith, S. and Read, D. J. (1997). Mycorrhiral Symbiosis London:s Academic Press pp. 584.

Stribley, D.P. (1987). Mineral nutrition PP.60–70. *In:* Ecophysiology of VA-Mycorrhizal plants *.Ed.,* R.Safir.C.R.C press, Boca Roton, Florida.

Thapar, H.S. and Khan, S.N. (1973). Studies on Endomycorrhizal in some forest species Indian Nat. Sci.Acad. 39: B (6): 687–694.

Thapar, H.S., Vijayan, A.K. and Unyal, K. (1993). Vesicular arbuscular mycorrhizal associations and root colonization in some important tree species. Indian Forester 18 (3): 207–212.

Tinker, P. B. (1975) (b). Effect Vesicular arbuscular mycorrhizal as higher plants. In: Endomycorrhizas (Eds. E. Sanders, B. Mosses and P.B.Tinker). pp 353–371. Academic press, London and New York.

Tinker, P.B. (1975). Effect of Arbuscular mycorrhiza on higher plants. Sym. Society for Exp. Bio. 29:325–349.

Vijaya, J., Srivasuki, K. (1996). Performance of VAM fungi in composition to Superphoshate application in *Tamarindua indica*

Yasid, M.S., Lee, S.S., Lapeyreee,E. (1996): Mycorrhizal inoculation of *Hopea odorata* (Dipterocarpaceae) in nursery. Journal of Tropical Forest Science. 9 (2): 276–278.

Young chiu-chung (1990). Effect of phosphate solubilizing bacteria and vesicular arbuscular mycorrhizal fungi on the growth of tree species in subtropical-tropical soils. Soil Science plant nutrition 36 (2): 225–235.

8

Arbuscular-Mycorrhizal Technology for Cassava (*Manihot esculenta* Crantz.)L.

Deiveekasundaram, M and P Muralikannan*

INTRODUCTION

Arbuscular mycorrhizal (AM) fungi are ubiquitous soil microorganisms. These fungi have great potential to enhance plant growth and soil aggregation. (Fitter, 1989; Read, 1991). The benefits of mycorrhizal fungi in experimental condition with individual fungi are generally related to the rate and extend of Mycorrhiza formation (Abbott and Robson, 1991 and Graham *et al.*, 1991). In the field, the timing of mycorrhiza formation is also likely to be important for benefits to plant growth (Sanders and Fitter, 1992). Although there are more than 150 species of arbuscular mycorrhizal fungi (Morton, 1988), relatively little is known about the way in which they coexist in soil community to form mycorrhizal root system in field soil.

Department of Microbiology, Faculty of Agriculture, Annamalai University, Annamalai nagar - 608 002, Tamilnadu
*Email:-deiveekasundaram@yahoo.co.in

AM fungi is formed by non-septate phycomycetous fungi belonging to the genera *Glomus, Gigaspora, Acaulospora* and *Sclerocyctis* in the family Endogonaceae of the order Mucorales. These fungi are non-specialized in host range, yet are apparently obligate symbionts (Gerdemann, 1968). They depend on their host plants for deriving energy source viz., carbohydrate. They can be recognized by the irregular coenocytic hyphae, which ramify within the cells of the root cortex and form vesicles (Storage pouches) and arbuscules (absorptive structure).

Mycorrhizal infection enhances plant growth by increasing nutrient uptake via increase in the absorbing surface or by mobilizing sparingly available nutrient sources or by excretion of chelating compounds or coenzymes (Marschner and Dell, 1994). Competition between the fungus and the roots for photosynthates is the main factor responsible for the large shoot: root dry weight ratio in favour of the shoots in mycorrhizal plants (Berta *et al.*, 1990).

A close correlation exists between root colonization and active hyphae within the roots (or) those attached to the root surface (Abbott and Robson, 1985). The growth enhancement effects of root infection with mycorrhizal fungi are caused by increase in P absorption particularly from sparingly soluble P sources (Bolan *et al.*, 1987). When root exploration is restricted, up to 80 per cent of plant phosphorus can be delivered by the external hyphae to the host plants over a distance of more than 10 cm from the root surface (Li *et al.*, 1991a). The P acquisition by the external hyphae is related to a. formation of polyphosphate in the hyphae and thus maintaining low internal P concentration and b. increase in acid phosphatase activity of mycorrhizal roots (Dodd *et al.*, 1987). It is well documented that the increase in plant growth resulting from AMF symbiosis is usually associated with increased nutrient uptake by the hyphae from the soil (Rhodes and Gerdemann, 1980). The increased ability of plants to absorb phosphates

(Mosse, 1973) and possibly other elements such as nitrogen, potassium, calcium, zinc, iron and manganese (Gray and Gerdemann, 1973) are the instances to cite. About 53 per cent of the total P uptake, 24 per cent of N uptake and 10 per cent of total K uptake in mycorrhizal plant is attributed to the uptake and delivery by the external hyphae (George *et al.*, 1992). The uptake of 25 per cent of Zn and 62 per cent of Cu is attributed to the external hyphae of mycorrhizal roots (Lambert and Weidensanl, 1991). The increased nutrient uptake by mycorrhizal plants is attributed to either one or more of the following interactions between the Vesicular-arbuscular mycorrhizal fungi and host plant :

1. Hyphal transport of the nutrients
2. Modifying the transpiration rate, and
3. Altering the rhizosphere microflora

Cassava (*Manihot esculenta* Crantz.), one of the major tuber crops is one among the group of plants inhabited by AM fungi. It is grown in more than 80 countries of the humid tropics of the world and forms the most subsistence crops for farmers. Cassava possesses a high potential for yielding large amount of food per unit area and also it is an efficient producer of calorie (135 calories/100 g fresh tuber) compared with other cereal crops. In India, cassava ranks first in area (2.35 lakh ha) and production (5.4 m tonnes) of fresh tuber. Cassava is highly mycotrophic and there exists variability in harbouring AM fungi in its root system. (Sundaram and Muralikannan, 1997)

In recent years, it is recognized as a subsidiary food-cum-industrial crop. Cassava is used in starch and starch based industries, manufacturing sago, liquid glucose, dextrin, vitamin C, gums, high fructose syrup and potable alcohol. Cassava tubers are a good source of carbohydrate (32.4g/100g of fresh tuber), minerals and vitamins. Cassava leaves are rich in protein (20.6–36.40 per cent by dry weight), calcium and vitamin C.

Cassava demands high amount of nutrients especially phosphorus for higher productivity. With the hike in fertilizer price due to fertilizer decontrol, the total area of cultivation of cassava in India has declined. It is essential to focus on the alternate means of improving productivity and reducing the fertilizer input cost of cassava crop. The AMF spore population and root colonization by AMF fungi were found to be poor in cassava soils of coastal Tamilnadu, although cassava was a highly mycorrhizal dependent crop (Sundaram and Muralikannan 1997).

MYCOTROPHY OF CASSAVA

The chlamydospore population of different AM fungi in the soils of coastal Tamilnadu ranged from 90–170 per 100 kg of soil and percent root colonization ranged from 38.85 to 56.50. Cassava soils of Bahur, Pinnalur, Puduchattram, Vadalur, Kurinchipadi, Varakalpattu, Vallampadugai, Sivapuri, Sirkali, Vaitheeswaran koil and Mayiladuthurai recorded spores of 90 to 110.110 and that of Poondiyakuppam, Panruti, Kullanchavadi, Kollidam and Tindivanam recorded 110 to 120 spores, while Pondicherry, Cuddalore, Poombuhar and Villupuram soils recorded 121 to170 spores. The cassava soils of these different locations were belonged to any one of the textural group's viz., loam, sandy loam, loamy clay, and clay loam and clay soil. The nutrient status is either low or medium in cassava soils of Tamilnadu, it is all the more important to manage mycorrhizal status of cassava (Sundaram and Sunder, 1992).

Negative correlation existed between mycotrophy and soil N, P, K contents. The increased levels of N, P and K reduced the dependency of plant root to mycorrhizal fungi. The mycorrhizal dependency of root system is more pronounced under 'P' limited condition. Hence the variation in the levels of 'P' between samples formed the determining factor for mycorrhizal condition of cassava in different locations.

Arbuscular mycorrhizal (AM) fungi occur in almost all soils (Mosse *et al.*, 1981). Soil chemical and physical properties change with the depth and are known to affect distribution of AM fungi (Bethlenfalvay, 1984). The AM spore population in cassava soils of coastal Tamilnadu decreased with increase in depth up to 30 cm(Sundaram and Muralikannan, 1997). The surface layer (10–15 cm) of different soil types viz., loam, clay, sandy loam, loamy clay and sand respectively recorded 110, 101, 90, 88 and 84 spores per 100g of soil. The bottom most layer (75–90 cm) recorded the least AM spore number of 4, 3, 1, 3 and 1 per 100 g of soil with loam, clay, sandy loam, loamy clay and sand.

Organic matter content of tropical soil decreased with increase in depth. The root proliferation is generally limited to top soil where the nutrient concentration is more due to mineralization. The limited availability of nutrients and organic matter contents are the reasons for poor mycorrhizal status at deeper layers.

Growth response to mycorrhizal symbiosis depends on three major components viz., the plants, the mycorrhizal fungi and the soil environment. A positive response of plant to mycorrhizal colonization of root during early growth stages of the host is highly essential (Tholkappian and Sundaram, 1987). In cassava, the host response to AMF varied with different soil types. The percentage root infection in different soil types increased with age of the cassava plant. The dependency of cassava to mycorrhizal condition positively correlated with the age of the crop, as the tuber formation and development proceeds and the plant require more phosphorus for the accumulation of starch.

BIODIVERSITY OF ARBUSCULAR MYCORRHIZAL FUNGI IN CASSAVA

The vesicular-arbuscular mycorrhizal fungi usually more numerous and diverse in cultivated soil than natural vegetation soil (Mosse and Bown, 1968). The predominant AM fungus associated with cassava, sweet potato and *Coleus* were found to be members of the genus, *Glomus* and *Gigaspora* (Potty, 1982). *Glomus mossea, G. manihotis, G. fasciculatum, Acaulospora* sp. were isolated from cassava fields (Sieverding, 1982; Kennedy and Sundaram, 2001). The information on the distribution and frequency of occurrence of AM fungi in different soil types of coastal Tamilnadu revealed that no generalization could be made on the distribution of AMF species in relation to soil characteristics except that the members of this genus *Glomus* found to be more predominant in all the soil types, which accounted for 45 to 60 per cent of total AM fungi. Cassava grown in all soil types recorded higher mycorrhizal colonization with maximum length of root (Potty, 1980 & 1982).

FACTORS INFLUENCING MYCORRHIZAL DEPENDENCY IN CASSAVA

Plants having the thick root system were considered to be more dependent on mycorrhizal association. Symbiotic status is very important for plant mineral nutrition and plant health, but the degree of mycorrhizal dependence differ with plant species (Hayman, 1986). Root characteristics are known to influence the inherent capacity of plant to absorb soil P and thus influence the dependency of plant species on AM fungi (Manjunath and Habte, 1991, no specificity existed between varieties and AM fungal species tested. *G. fasciculatum* was found to be the most effective AM fungi for all the cassava varieties, followed by *Glomus mossae, Acaulospora laevis* and *Gigaspora margarita* based on the different parameters studied. Among the different varieties viz., the varieties Co-1, M-4, MVT-1 and H-165. The variety H-165 was the most preferred partner to different Vesicular-arbuscular mycorrhizal species. No specificity of preference could be observed between cassava varieties and AM fungal species in the coastal soils of Tamilnadu. However, Sieverding (1991) reported the crop specificity to AM fungal species in cassava soils. The level of AM fungi association may be related to the plant inherent factor such as root morphology, physiology, metabolism and plant growth rate. Plant growth response results from a greater ability of mycorrhizal roots to acquire more phosphorus from the soil (Santhanakrishnan, 1986 and Kodie, 1991).

Inoculation of *G. fasciculatum, G. mossae* and *A. laevis* as pure culture was found to be better than mixed inoculum culture of these different AM fungi. The root infection rating was the highest when the inoculum contained *G. fasciculatum* as pure culture. It is also of interest to observe that wherever *G. fasciculatum* was one of the components in the mixed culture inoculum, the effectiveness was better than those that do not have *G. fasciculatum* component. These findings indicate that G. *fasciculatum* is able to compete effectively in mixed culture inoculum to increase the colonization and growth of cassava in coastal soils.

SCREENING VA MYCORRHIZAL FUNGI FOR EFFICIENCY

Glomus fasciculatum, G. mossae, Acaulospora laevis and *Gigaspora margarita* when screened for its efficiency on cassava by determining the root infection per cent and AMF spore population in soils under sterile and non-sterile conditions, AMF infection ratings were more in non-sterile than in sterile soil. The response of cassava in terms of root colonization by AM fungi was highest with *G. fasciculatum,* closely followed by *G. mossae, Acaulospora laevis* and *Gigaspora margarita.* The more pronounced colonization and AM spore population in non-sterile soil may be to the additional benefit obtained due to the synergism between AM and mycorrhization helper bacteria (Brown and Carr, 1984). The spread of native inoculum with the plant system was comparatively slower than that of plants infected with introduced ones (Puppl and Riess, 1987).

EFFECTIVENESS OF INTRODUCED VERSUS INDIGENOUS VAMYCORRHIZAL FUNGI

The inoculation with introduced species of *G. epigacum* and *G. mossae* stimulated the plant growth of cassava, increased the dry matter production, plant phosphorus, plant K content and root infection percentage better than that of plants infected with native AM fungi (Garcia and Almeida, 1987). In the cassava soils of coastal Tamilnadu, the introduced cultures viz., *Glomus fasciculatum, G.mossae, Acaulospora laevis* and *Gigaspora margarita* were found to be more effective in improving the growth parameters viz., root length, number of roots, per cent root colonization, spore population and phosphatase activity in the rhizosphere of cassava than the indigenous isolates. However, among the introduced cultures of AM fungi *G. fasciculatum* was found to be more effective than other AM fungi tested. Both indigenous and introduced cultures of AM fungi considerably increased the chlamydospore production. The positive response of cassava plants to both indigenous and the introduced AM fungi are evinced from augmented root length, shoot length and higher root infection percentage by AM fungi. The soil application of both indigenous and introduced AM fungal inoculum increased the phosphatase activity in the rhizosphere of cassava. However, the phosphatase activity was higher with the introduced AM fungi than that observed with indigenous fungi. The increased enzyme activity observed in the mycorrhizal cassava may be due to root exudates pattern mediated stimulation of microorganism and microbiological activity. The soil enzymes activity reflects as an index for the total microbial activity of a given soil.

FUNCTIONS OF ARBUSCULAR MYCORRHIZAL FUNGI AS INFLUENCED BY AGRONOMIC PRACTICES

The principal function of AM fungi is to enhance the nutrient uptake. This phenomenon would have increased the plant dry weight (Geetha, 2001).

Levels of Phosphorous

The 'P' uptake response of cassava to AMF inoculation was maximum when the available 'P' was medium in the soil. The uptake of K, Ca and Mg of cassava was more with Vesicular-arbuscular mycorrhiza fungal roots (Sieverding and Toro, 1988 and Clark, 1997). In the coastal soils of Tamil Nadu, cassava inoculated with *G. fasciculatum*, recorded a nitrogen uptake ranging from 623.26 to 1121.38, phosphorus uptake ranging from 76.46 to 123.238 and Potassium uptake ranging from 957.21 to 1357.15 mg plant^{-1} (Sundaram and Muralikannan, 1997). The higher uptake of N, P and K by *G. fasciculatum* in inoculated roots may be due to its infective capacity and the proliferation of external mycelium, which considerably increased soil volume explored for nutrient uptake. The major transfer of phosphorus from the fungus to the plant, occurs in those root cells which contain arbuscules however root internal hyphae can release phosphorus to the host takes place in a process of interchange with carbon compounds from the host to the fungus.

Micronutrients

Arbuscular mycorrhizal fungi infection in the plants increased the uptake of micronutrient such as Zinc (Zn), Copper (Cu), Sulphur (S), Boron (B) and Molybdenum (Mo) and proved to be transported to the host plant (Swaminathan and Verma, 1975).

Soil fumigation

The four soil Sterilants viz., formaldehyde, dazomat, metam sodium and methyl bromide decreased the cassava root colonization by arbuscular mycorrhizal fungi and their spore population in the rhizosphere soil. The degree of inhibition varied with the types of soil fumigants used. Formaldehyde treated soil exerted the least influence on the root Colonization by AM fungi and their spore production. The inhibitory effect of the soil fumigants on AMF colonization was in the order of methyl bromide > metam sodium > dazomat > formaldehyde. The plants growing in fumigated soil were exposed to biotic and abiotic factors, which differ from those of untreated soil with AM fungi.

Higher 'P' level decrease the cell membrane permeability for carbohydrate thus may suffer through inhibited photosynthate support from plants. The mechanism of P limiting root infection is more likely to be associated with the permeability of the root cell membranes rather than with any changes in the root carbohydrate content. Under low P nutrition, increased root cell membrane permeability leads to increased loss of metabolites, sufficient to sustain the growth of the mycorrhizal fungus during pre and post infection. The per cent root colonization by AM fungi increased with the level of P application up to 75 kg P_2O_5 thereafter, reduced proportionately to the level of P application. The AM fungal infection rating also depended on the soil texture, soil P level and soil conditions. It is recognized that it is not soil P that regulates mycorrhizal infection, but the amount of P available to the plant (Menge *et al.*, 1978 and Sanders, 1975). There is scope for saving of P fertilizers to cassava with AM fungal inoculation technology.

BENEFICIAL EFFECT OF AM FUNGI UNDER STRESS

The soil quality and related problem could affect the efficiency of AM fungi. Further, the plants associated with AM fungi could tolerate low pH to sustain the plant growth. Tolerance and ability of plant to grow in different pH may be associated with the level of AM fungi colonization with roots and adoptability of mycorrhizal roots to different pH conditions (Koslowsky and Boerner, 1989). *Glomus mossae* and *Gigaspora margarita* were found in the soil pH>5.5 (Sieverding, 1991) showing appreciable mycorrhizal colonization.

Drought tolerance is complex of many morphological, physiological and biochemical characteristics. AMF association greatly improves water use efficiency of mycorrhizal plant under drought condition (Bethlenfalvay, 1992). Rapid accumulation of water is a metabolic consequence of water deficiency AM inoculations contribute towards the improvement in water uptake in cassava plants under moisture stress condition. AM hyphae ramifying into the soil are likely to increase the absorbing area for water uptake and able to by-pass the dry zone (Ney and Tinker, 1977). Arbuscular mycorrhizal infection showed to cause a reduction in water potential and stomatal resistance to water stress condition (Saif *et al.*, 1981). Drought resistance is complex of many morphological, physiological and biochemical characteristics. Certain chemical constituents are considered for indicator for drought resistance. Reduction of proline content in mycorrhizal cassava plants that may implies a lower physiological water stress in mycorrhizal plant. Higher amount of relative leaf water content in mycorrhizal plants is due to altered hormonal balance in leaf that is high gibberellins activity in mycorrhizal plant leaf (Allen *et al.*, 1980). The mycorrhizal cassava plants are not affected by created moisture stress because AM fungi hyphae may function as physical bridge and that they maintain the contact between root and the soil water. This function helps in the preservation of water and the transpiration may be maintained. The drought resistance mechanism in mycorrhizal plant improved. AMF inoculation technology in the drought prone areas of coastal Tamilnadu helped to overcome moisture stress condition and to sustaining the yield of the rainfed cassava varieties.

Organic amendments

The effectiveness of *G. fasciculatum* on cassava variety increased with the application of different organic amendments viz., vermicompost, FYM and pressmud. The vermicompost supported the highest root infection percent of 88.60 followed by 80.60 per cent with FYM and 78.60 per cent with pressmud. The effect of organic manure (or) organic amendments on VAM activity may be due to the indirect effect of increased biological activity by providing the carbon sources and there by increasing the microbes associative with AM activity. These indirect effects can be attributed to improved soil conditions such as water holding capacity, aggregate formation and nutrient composition in mycorrhizal roots. The organic amendments increased the activity of soil bacteria. The root associated bacteria excrete polysaccharides around the root system which play a vital role in the interactions and inter relationships between microorganism between AM fungi and rhizosphere bacteria. AM fungi were reported to be vehicle for soil rhizobacteria (Bianciotto *et al.*, 1996). This phenomenon helps in increasing the root colonization in organic matter amended soils. Higher root colonization percent lead to augmentation in the nutrient uptake, growth and yield of cassava in the coastal soils of Tamil Nadu.

Pesticide chemicals

The chemicals that affect the mycorrhizal fungi and mycorrhizal formation are of serious concern in crop productivity. Population dynamics of AM fungi in fields are inversely correlated to the quantity of applied fungicides. The fungicide such as captafol, chloroneb, ethazole, triadimenol, metalaxyl and thiram, increased AM root infection. It is more probable that these fungicides may change the root exudation pattern of the host plant, which probably explains its positive effect on AM formation (Schwab *et al.*, 1982). The fungicide viz. captan, mancozeb and copper oxy chloride significantly decreased the mycorrhizal roots in cassava at higher doses of application. Application of the fungicides mancozeb and captan @ 50 mg kg^{-1} of soil affected the root colonization by *G. fasciculatum,* the fungicides mancozeb, captan and copper oxychloride increased the total root length, infected root and spore production by AM fungi. The constituents observed in mycorrhizal plants were considered as factor for inducing resistance of mycorrhizal cassava plants towards the root pathogens.

INTERACTIONS BETWEEN AM FUNGI AND SOIL BACTERIA

Mycorrhizal fungi improved plant growth and enhanced microbial activity in the rhizosphere of mycorrhizal plants (Bagyaraj and Menge, 1978). Much more improved growth in plants observed, when inoculum contains AM endophytes along with *Azotobacter chroococcum*. The additional benefit obtained may be due to the synergistic effect of the AM and mycorrhization helper bacteria (MHB) (Brown and Carr, 1984). The presence of *Azospirillum brasilense* with the AM fungus, *G. fasciculatum* improved the root colonization and yield of tuber crop (Kandasamy *et al.*, 1988). The percent root colonization and AM spore population increased with, *P. fluorescens, A. brasilense* and *Azotobacter chroococcum*. The diazotrophic bacterium, *A. brasilense* with AM showed the greatest benefit to cassava by improving the plant shoot dry weight and P uptake followed by associations with *P. fluorescens* and *A. chroococcum*. The combined inoculation of *Azospirillum brasilense* with *G. fasciculatum* exerted the maximum beneficial effect on root colonization (88.00) and AM spore production (309.60/100 g) in cassava. Inoculation of diazotrophic bacteria with AM fungal increased mycorrhizal colonization by 40% and sporulation up to 168% compared to the AMF alone inoculated cassava plant (Balota *et al.*, 1995). Inoculation of AM with PGPR enhanced the spore germination and root infection by Mycorrhiza due to the production of enzymes/hormones like substances which helped to increase the cell wall plasticity and in turn fungal infection (Requena, 1998). AM interaction with *Azospirillum* showed positive effects on plant production, probably via hormonal effects. However, the beneficial effect of AM fungi on diazotothrophs was quite possibly an indirect one. The increased photosynthetic rate resulted in the increased growth and altered the root exudates pattern, which stimulated the bacteria that are associative with AM fungi.

AM FUNGAL EFFECT ON GROWTH AND YIELD

The beneficial effect of AM to plant growth is highly pronounced in marginal soil. The low mobility of P in soils leads to a P depletion zone in the rhizosphere of crop plants (Schwab *et al.*, 1982). Mycorrhizal root enhanced the uptake of P by plant roots by providing a large absorbing surface and also due to the external hyphae of AM fungi mediated translocation to the host plant root. However, host plants must provide carbohydrate to AM fungi for development and growth.

The different growth and yield parameters recorded by AM inoculated cassava at 50 kg P_2O_5 level were on par with the values recorded at 100 kg P_2O_5 level indicating that a saving of 50 per cent of P by adopting AM inoculation technology. The P content of AM inoculated cassava at 50 kg P_2O_5 with SSP was on par with 100 kg P_2O_5 with AM inoculation.

CONCLUSION

Mycorrhizal status of cassava soils of coastal Tamil Nadu was found to be poor and needs AMF inoculation for effective mycorrhizal cassava symbiosis. *G. fasciculatum* was found to be the most effective AM fungi for the cassava variety tested viz., CO-1, M-4, MVT-1 and H-165. The cassava variety H-165 was found to be the most responsive for AM fungal inoculation. For better management of AM fungi in cassava soils, the variety H-165 and the AM fungus *G. fasciculatum* was identified as suitable for coastal soils of Tamilnadu.

AM fungal inoculation effect on cassava as pure culture was better than mixed cultures. Soil fumigation reduced the mycotrophy of cassava, since the mycorrhization bacteria considerably reduced. Introduced cultures of AM fungi were found to be more effective than indigenous cultures to cassava varieties in coastal soils of Tamil Nadu.

The effectiveness of AM fungi to cassava was highly significant in sandy loam soil. *G. fasciculatum* inoculation can help in increasing the drought tolerance of cassava and can with stand up to 20-30 per cent moisture holding capacity (MHC).

Vermicompost was the most ideal organic amendment to increase the mycotrophy of cassava and the spores of AM fungi in cassava soils. The fungicides, viz, mancozeb, captan and copper oxychloride at 50 mg kg^{-1} of soil affected the mycorrhizal colonization, fungal population while bacterial and actinomycete population remains unaffected or slightly stimulated at lower concentrations.

The biochemical defense mechanism of cassava towards soil born pathogens can be accelerated by AMF inoculation. Co-inoculation effect of AM fungi with diazotrophs on cassava was found to be synergistic and increased the mycotrophy, growth and yield of cassava.

The level of saving of P fertilizer to cassava was 50 per cent with SSP and 25 per cent with RP. Application of 50 kg of P_2O_5 ha^{-1} as SSP (or) 75 kg P ha^{-1} as RP has been worked out to be as optimum for effective cassava-mycorrhizal symbiosis. Mycorrhizal inoculation effect (MIE) was the maximum at 75 kg P_2O_5 level. The mycorrhizal colonization significantly reduced at 100 kg P_2O_5 ha^{-1}. The P response ratio was highest at 50 kg P_2O_5 level and decreased at 75 & 100 kg P_2O_5 level to cassava.

REFERENCES

Abbott, L.K. and Robson, A.D. (1985). Formation of external hypae in soil by four species of vesicular mycorrhizal fungi. *New Phytol.* 99:245-255.

Abort, L.K. and Robson, A.D. (1991). Factors influencing the occurrence of VAM mycorrhizal fungi. *Agri. Ecos. Environ.* 35: 121–150.

Allen, M.F., Moore T.S. and Christensen M. (1980). Phytohormone changes in *Bouteloua gracilis* infected by vesicular-arbuscular mycorrhizae. I. Cytokinin increases in the host plant. *Can. J. Bot.* 58: 371–374.

Bagyaraj, D.J. and Menge J.A. (1978). Interaction between a VAM and Azotobacter and their effects on rhizosphere microflora and plant growth. *New Phytol.* 80 : 567–573.

Balota, E.L., Lopes, S.L., Hungria, M. and Dobereiner, J. (1995). Interaction and physiological effect of diazotrophic bacteria and arbuscular mycorrhizas on cassava. *Pesquisa Agropecuaria Brasileir.* 30(11) : 1335–1345.

Berta, G., Fusconi, A., Trotta, A., and Scannerini, S. (1990). Morphogenetic modifications induced by the mycorrhizal fungus *Glomus* strain E3 in the root system of *Allium porrum* L. *New Phytol.* 114: 207–215.

Bethlenfalvay, G.J., Dakessian, S. and Pacovsky, R.S. (1984). Mycorrhizae in a Southern California desert. Ecological implications. *Can J. Bot.*, 62: 519–524.

Bianciotto, V., D., Minerdi, S., Perotto. and Bonfante, P. (1996). Cellular interaction between arbuscular mycorrhizal fungi and rhizobacteria. *Proto plasma.* 193(1/4): 123–131.

Bolan, N.S., Robson, A.D. and Barrow, N.J. (1987). Effects of Vesicular Arbuscular mycorrhizae on the availability of iron phosphates to plant. *Plant and Soil.* 22, 401–410.

Brown, M.E. and Carr, G.R. (1984). Interaction between *Azotobacter chroococcum* and VAM and their effect on plant growth. *J. Appl. Bacteriol.*, 56: 429–437.

Clark, R.B (1997). Arbuscular mycorrhizal adaptation spore germination, root colonization and host plant growth and mineral acquisition at low pH. *Plant and Soil.* 192(1): 15–22.

Dodd, J.C., Burton, C.C., Burns, R.G. and Jeffries, P. (1987). Phosphotase activity associated with the roots and the rhizosphere of plants infected with vesicular arbuscular mycorrhizal fungi. *New phytol*, 21, 163–171.

Fitter, A.H. (1989). The role and ecological significance of vesicular arbuscular mycorrhizal in temperate ecosystems. *Agric. Ecosys. Environ.*, 29: 137–151.

Geetha, S. (2003). Studies on the mycotrophy of certain medinal herbs and its antimicrobial property. M.sc(Ag) Thesis, Annamalai Univ., Annamalainagar.

George, E., Haussler, K.V., Vetterlein, D., Gorgus, E. and Marschner, H. (1992). Water and nutrient translocation by hypae of *Glomus mossae*. *Can. J. Bot.*, 70, 2130–2137.

Gerdemann, J.W. (1968). Vesicular-arbuscular mycorrhiza and plant growth. *Annu. Rev. Phytopathol.*, 6: 397–418.

Garcia, L.F. and De Almeida, R.T. (1987). Behaviour of cassava cv. Olha Verde regarding inoculation with vesicular arbuscular mycorrhizal fungi. *Ciencia Agronomica*, 18(1): 1–8.

Graham, J.H., Eissenstat, D.M. and Drovillard, D.L. (1991). Relationship between a plant mycorrhizal dependency and rate of vesicular arbuscular mycorrhizal colonization. *Functional Ecol.*, 56: 773–779.

Gray, L.E. and Gerdemann, J.W. (1973). Uptake of Sulphur-35 by vesicular arbuscular mycorrhizal. *Plant and Soil*, 39: 687–689.

Hayman, D.S. (1986). VA mycorrhizae in field crop system. In: Safir, G.R. (ed.) Ecophysiology of VA mycorrhizal plants. CRC Press, Boca Raton, Fla., pp. 171–192.

Kandasamy, D., Palanisamy, D., and Oblisamy, G. (1988). Screening of germplam of sweet potato for VAM fungal occurrence and response of the crop to the inoculation of VAM fungi and *Azospirillum*. *J. Root Crops*, 14(1): 37–42.

Koide, R.T. (1991). Nutrient supply, nutrient demand and plant response to mycorrhizal infection. *New Phytol.*, 117: 365–386.

Koslowsky, S.D. and Boerner, R.E. (1989). Interactive effects of aluminium phosphorus and mycorrhizae on growth and nutrient uptake of *Panicum virgatum* L. (Poaceae), Environ. Pollution, 61: 107–125.

Lambert, D.H. and Weidensanl, T.C. (1991). Element uptake by mycorrhizal soybean from sewage sludge-treated soil. Soil Sci. *Am.J.*, 55: 393–398.

Li, X.L., E. George, H. Marschner, Zhang and Jun-Ling. (1997). Phosphorus acquisition from compact soil by hyphae of a mycorrhizal fungi associated with red clover (*Trifolium pratense*). Can. J. Bot., 75(5): 723–729.

Li, X.L., George, E. and Marschner, H. (1991). Phosphorous depletition zone in VA mycorrhizal white clover in a calcareous soil. Plant and Soil, 136, 41–48.

Marschner, H. and Dell, B. (1994). Nutrient uptake in mycorrhizal sybiosis. In: Management of mycorrhizal in Agriculture, Horticulture and Forestry (Eds. Robson, A.D., Abbott, L.K. and Malajezuk, N. Kluwer academic publishers, p. 238.

Manjunath, A. and Habte, M. (1991). Root morphological characteristics of host species having distinct mycorrhizal dependency. *Can. J. Bot.*, 69: 671–676.

Menge, J.A., Sterile, D., Bagyaraj, D.J. Johnson, E.L.V. and Leonard, R.T. (1978). Phosphorus concentration in plants responsible for inhibition of mycorrhizal infection. *New Phytol.*, 80: 575–578.

Mosse, B. (1973). Plant growth responses to vesicular arbuscular mycorrhizal. IV. In soil given additional phosphate. *New Phytol.*, 72: 127–136.

Morton, J.B. (1988). Taxonomy of VA-mycorrhizal fungi calsssifictaion, nomenclature and identification. Mycotaxon., 32: 267–324.

Ney, P.H. and Tinker, S. (1977). Solute movement in the soil root system. University of California Press, U.S.A.

Potty, V.P. (1980). Plant microbe interrelationship in tuber crops. Ann. Report, CTCRI, Trivandrum, India, pp. 149–151.

Potty, V.P. (1982). Annual Report, Central Tuber Crops Research Institute, Trivandrum, Kerala, India, pp. 146–148.

Puppl, G. and Riess, S. (1987). Role and ecology of VAM mycorrhizae in sand dunes. Angew. Bot., 61: 115–125.

Read, D.J. (1991). Mycorrhizal om eepsustes, Experientia, 474, 376–390.

Requena, N. (1998). Mycorrhizal interaction in the rhizosphere. In : Microbes for Health, Wealth and Sustainable Environment, Varma, A. (Ed.), Malhotra Publishing House, New Delhi, 725–738.

Rhodes, L.H. and Gerdemann, J.W. (1980). Nutrient translocation in vesicular arbuscular mycorrhiza. In: Cellular Interactions in symbiosis and parasitism. (Eds) C.B. Cook, P.W. Pappas and E.D. Rudolph. The Ohio state University Press, Columbus.

Saif, G.R., Boyer, J.S. and Gerdeman, J.W. (1981). Mycorrhizal enhancement of water transport in soybean. Science, 172: 581–585.

Sanders I.R. and Fitter, A.H. (1992). The ecology and functioning of vesicular and arbuscular mycorrhizal in co-existing grass land species I-seasonal patterns of mycorrhizal occurance and morphology. *New phytol.*, 1204, 517–524.

Santhanakrishnan, P. (1986). Studies on the dual inoculation of VA - mycorrhizal and Rhizobium on ground nut (Arachis hypogaea L.). Ph.D. Thesis, Tamil Nadu Agrl. Univ. Coimbatore, p. 260.

Schwab, S.M., Johnson, E.L.V. and Menge, J.A. (1982). Influence of Simazine in formation of vesicular arbuscular mycorrhizal in *chenopodium quinona* wild. Plant and Soil, 64: 283.

Sieverding, E. (1991). Vesicular-arbuscular mycorrhiza management in tropical agrosystems. Deutsche. Gesellschaft Technisuhu Zusammonarbeit (GTZ) GMBH, Eschborn, p. 371.

Sieverding, E. and Toro, T.S. (1986). Evaluation cauntitatia uqualitative de hongos formadores de micorriza vescul arbuscular en la region de mondomo. Sue los Eavatoriales, 16: 122–124.

Sundaram M.D. and Sundar, S.K. (1992). Response of groundnut (Arachis hypogea L.) VRI 1 to inoculation with VA-mycorrhizal fungi as fertilizers. Proc International Symposium on Management of Mycorrhizas in Agriculture, Horticulture and Forestry, University of Western Australia from 28th Sep. to 2nd Oct. 1992.

Sundaram, M.D. and Muralikannan, P. (1997). Technology development and use of VAM fungi for Horicultural crop-Cassava. ICAR, ADHOC Scheme Final Report.

Swaminathan, K. and Verma, B.C. (1975). Symbiotic effect of VA-mycorrhizal fungi on the phosphate nutritional of potatoes. Proc. Indian Acad. Sci., 85(B): 310–318.

9

Positive Response of *Syzygium cumini* Lam. to Different Bioinoculation in the Nursery

[1]*R H Ratageri*

INTRODUCTION

The beneficial effect of AM fungal association on growth of plants is well known and there are many reports on effect of growth of nursery plants. There are reports proving that AM fungi not only improved the plants growth but also improved the water relation of citrus plant (Allen, 1982). AM are known to increased nutrients and water absorption through improved absorption area (Alkaraki, 1998). It also increased host resistance to natural stress such as drought, cold, salinity etc. (Ahhott and Rabson, 1983). Thus AM fungal inoculations have tremendous potential because of their ability to improve the efficiency and sustainability of crop yield especially in low 'P' soils. Hence, it is very much important to select proper AM fungi and manage their potential for good and low cost crop productivity technology.

Through AM fungi are not host specific but some studies

1. P. G. Department of Studies in Botany (Microbiology Lab) Govern-ment Science College, Chitradurga -577 501 India.

have indicated the host preferences of native AM isolates of mycorrhizal fungi (Bagyaraj *et al.*, 1988) thus suggesting the need for selecting efficient. A mycorrhizal fungi for specific host (Anilkumar *et al.*, 2007).

Syzygium cumini Lam is a large evergreen tree with much densely branched and appears like crown. The tree is very useful as its leaves are lopped for fodder. The fruits are edible and very nutritious. The wood is moderately hard and is fairly durable used for building purposes and preparation of agricultural implements (Balakrishnan, 1996). It is one of the promising drought tolerant perennial adapted to varied soils. The main advantage of cultivation of this plant grows well with a minimum requirement of water and without much agricultural management (Lakshman, 1996). Therefore, it would be highly desirable to grow these plants, when the biomass yield is increased by inoculating suitable biofertilizers containing live cells of phosphorus absorbing (AMF), P solublizers (Phosphobacteria) and nitrogen fixing micro-organisms can aids in minimizing our dependence on inorganic fertilizers, which are very wide application and also be assure a pollution free environment (Rodriguez *et al.*, 2003).

The role of biofertilizers has already documented on different annual crops, but the information in pertaining to its applications on perennial is very scanty. However, in forest nursery, few research reports are available in recent days, to demonstrate that biofertilizers stimulate the growth of many tree species (Huang *et al.*, 1885; Dela Cruz *et al.*, 1988; Lakshman, 1992 & 1996; Singh, 2001) thereby the survival rate of planted seedlings are increased. Hence, this study was undertaken to assess the effect of biofertilizers on *Syzygium cumini* seedling at nursery level.

MATERIALS AND METHODS

A total of 10 × 25 seeds were collected from 10-year-old plant *S. cumini.* Seeds were surface sterilized in 2% sodium hyprochlorite, later on washed in water and sown in earthen pots contained 3 kg of sterilized sand loam soil, in (1:1) ratio (loam soil and pure sand). The experiments were set up in a completely randomized design and replicated thrice. The earthen pots were inoculated with biofertilizers, viz., (i) AMF (arbuscular mycorrhizal fungi), (ii) Phosphobacteria, (iii) *Azotobacter*, (iv)AMF+ phosphobacteria, (v) AMF+*Azotobacter* (vi) AMF+ phosphobacteria + *Azotobacter* uninoculated seedlings were maintained as control.

Biofertilizer inoculation was prepared with a base of peat soil 200 g of *Azotobacter* and phosphobacteria were weighed and mixed with 3 kg of well-decomposed and powdered FYM separately. *Azotobacter* and phosphobacteria were inoculated to seeds with peat based culture (Sattar and Gaur, 1989). 25 g mixed inoculum AMF *Glomus fasciculatum* were applied to each pots at 5 cm depth near the root zone. The AMF inoculum consists of 238 chlamydospores, hyphae from the pots culture of Sudan grass (*Sorghum bicolour* L. Moench. Var Sudanse), which was infected with *G. fasciculatum*.

Six seedlings were selected in each treatment at random and observed initially 60 and 120 days after inoculation. The effect of biofertilizers on root length, shoot length, stem diameter, leaf length was measured. The per cent AMF colonization calculated by following the by procedure (Phillips and Hayman, 1970) chemical analysis was determined following the method (Jackson, 1973). The results were subjects to analysis of variance and tested for significant differences ($p<0.5$) after Panse and Sukhatme (1967).

RESULTS AND DISCUSSION

A significant increases in number of leaves 48.3 cm (80.9%) leaf length 11.8 cms (68.8%) for 120 days (Table 9.1), and shoot length 27.4 cm (69.0%), root length 29.1 cm (45.2%) (Table. 9.2). Shoot dry weight 31.3 g (75%) in 120 days (Table 9.3) and N.P.K. mg content of shoots was achieved after dual inoculating the seedlings with arbuscular mycorrhizal fungi and *Azotobacter*. These results supporting the early workers contribution (Shiva *et al.,* 2002) in *Azadirachta indica* seedlings in nursery stage. Many beneficial microorganisms that are used as biofertilizers colonize in the rhizosphere and stimulate the plants growth by providing required nutrients. They are important contributors to organic matter turnover and nutrient cycling in agro-forestry ecosystem (Watrud and Seidler, 1998). The association may also regulates the physiological process in fixation substances, increased availability of mineral nutrients and protection of plants from pathogens besides increasing the availability of nutrient elements at the root zone (Suresh and Baygyaraj, 2002; Sannikova, 2003).

An increased 31.9 g dry weight over control was evident in seedlings inoculated with combined Azotobactor, phosphobacteria and arbuscular mycorrhizal fungi (Table 9.3). Similar increase in biomass production was recorded due to arbuscular-mycorrhiza fungal inoculation on *Terminalia spp.* (Lakshman, 1992) and on *pongamia pinnata* (Venkatesh *et al.,* 1998). In the present study individuals inoculation of AMF also increased biomass relatively higher over the controls though not to the tune of combined microbial inoculation. The increase in the seedling biomass production may be strongly correlated with increased accumulation of N due to *Azotobacter* (Siddiqui *et al.,* 1993) and P due to AMF and Phosphobacteria (Venkatesh *et al.,* 1998). Tripartite inoculation of *Azotobaceter* phosphobacteria and AMF registered significantly increased shoot and root length (Table 9.2). And Nitrogen, Phosphorus, Potassium and Magnesium content in those plants received all the bifertilizers (Table 9.4). The relative enhanced growth of shoots might be attributed to the role of phosphobacteria in enhancing the availability of P in the soil (Clarkson, 1985) and making it available for metabolic activity in the seedlings. The increased root system in the present investigations is in agreement with the results of

Table 9.1. : Effect of Biofertilizers on Number of Leaves and Leaf Length of *Syzygium cumini* in the Nursery at 60 and 120 Days Seedlings

Biofertilizers	Numbers of leaves Months after inoculation				Leaf length (cm)			
	60 days		120 days					
		Mean		Mean	60 days	Mean	120 days	Mean
Control	18.3	17.5	34.1	32.1	5.8	5.6	8.3	7.2
Phosphobacteria (P)	40.1	38.8	41.2	39.8	6.9	6.6	9.3	8.5
AMF	41.2	39.1	38.4	37.2	7.7	7.5	9.7	8.8
Azotobacter(A)	43.5	41.2	48.7	43.5	7.4	7.1	10.2	9.3
AMF+P	45.9	42.6	54.4	51.2	7.8	7.5	11.3	10.4
AMF+A	46.1	44.7	56.5	54.3	8.2	8.0	10.4	9.6
AMF+A+P	48.5	46.3	61.2	58.1	9.4	9.2	12.3	11.8
Mean	40.5	38.4	42.0	45.1	7.6	7.4	10.2	9.4

	CD(P=0.05)	CD (P= 0.05)
Biofertilizers(B)	2.16	0.27
Growth stages(GS)	1.40	0.53
B × GS	3.72	0.71

Table 9.2. : Effect of Biofertilizers on Shoot Length, Root Length and Per Cent of AMF of Colonization of *Syzygium cumini* in the Nursery at 60 and 120 Days Seedlings

Biofertilizers	Shoot length (cm)				Root length (cm)				% of AMF colonization	
	60 days		120 days		60 days		120 days			
		Mean		Mean		Mean		Mean	60 days	Mean
Control	7.4	7.1	9.8	8.8	19.3	18.6	22.0	21.2	-	-
Phosphobacteri (P)	9.2	9.0	14.3	13.9	23.5	22.3	25.1	24.0	41.7	39.6
AMF	8.5	8.4	10.7	9.5	20.4	19.7	24.3	23.0	59.6	57.4
Azotobacter(A)	7.9	8.1	9.6	9.2	22.3	21.8	25.4	24.2	38.4	37.6
AMF+P	9.6	9.3	13.5	12.9	26.1	25.0	29.6	28.1	67.1	63.5
AMF+A	10.5	9.7	20.6	19.8	26.7	25.2	28.2	27.6	62.2	55.1
AMF+P+A	12.3	12.0	31.0	29.2	27.8	27.0	31.1	31.7	71.3	68.4
Mean	9.3	9.0	15.6	14.7	23.7	22.6	26.7	25.5	56.5	53.6

	CD(P=0.05)	CD (P= 0.05)	CD (P= 0.05)
Biotertilizers(B)	0.51	NS	0.45
Growth stages(GS)	0.29	0.07	0.61
B X GS	0.93	0.09	0.13

Table 9.3. Effect of Biofertilizers on Shoot Dry Weight of *Syzygium cumini* in the Nursery at 60 and 120 Days Seedlings

Biofertilizers	Shoot dry weights g^{-1} days after inoculation			
	60 days	Mean	120 days	Mean
Control	5.80	3.90	10.4	8.62
Phosphobacteria (P)	8.60	6.30	13.26	12.81
AMF	11.20	9.14	18.75	16.47
Azotobacter(A)	10.40	9.60	16.53	15.28
AMF+P	21.60	19.41	24.35	23.62
AMF+A	22.81	20.92	26.14	24.83
AMF+P+A	23.54	22.23	33.30	31.54
Mean	19.71	18.8	24.3	22.9

	CD(P=0.05)
Biofertilizers (B)	1.463
Growth stages (GS)	1.029
B × GS	2.876

Table 9.4. : Effect of Biofertilizers on N, P, K and Mg Content in Shoots of *Syzygium cumini* in the Nursery at 60 and 120 Days seedlings

Biofertilizers	Nutrients N.P.K. Mg(%) content in shoots							
	N%		P%		K%		Mg%	
		Mean		Mean		Mean		Mean
Control	1.9	1.7	0.11	0.10	1.06	1.02	0.11	0.11
Phosphobacteria (P)	2.12	2.10	0.22	0.21	1.11	1.10	0.12	1.12
AMF	2.11	2.11	0.25	0.23	1.21	1.19	1.12	1.12
Azotobacter(A)	1.13	2.11	0.20	0.18	0.8	0.7	0.12	0.12
AMF+P	2.12	2.10	0.21	0.20	1.10	1.9	0.12	0.12
AMF+A	2.17	2.14	0.21	0.20	1.21	1.19	0.13	0.12
AMF+P+A	2.19	2.15	0.28	0.24	1.19	1.15	0.15	0.14
Mean	2.11	1.81	0.21	0.19	0.99	0.96	0.12	0.12

	CD(P=0.05)	CD (P= 0.05)	CD(P=0.05)	CD(P=0.0)
Biofertilizers (B)	0.25	0.025	0.09	0.032
Growth stages (GS)	0.26	0.026	0.09	0.05
B × GS	0.62	0.051	0.21	0.07

(Marchner and Dell, 1994) in tea seedlings. The possible reason for increasing the root lengths in the combined inoculation is that a synergistic interaction between these beneficial micro-organisms might have resulted in enhanced growth and nutrient uptake.

In the present study, N.P.K. concentrations were higher due to phosphobacteria+*Azotobacter*+AMF inoculation. Several scientists have reported the positives and significant enhancement of biofertilizers on the uptake of nitrogen. Similar findings are reported by early workers. (Reena and Bagyraj, 1990; Rahnel *et al.*, 1993).

The results of the present study are in strong support of combined inoculation of the *Syzygium cumini* seedlings with AMF, *Azotobacter* and phosphobacteria. Therefore, this kind of work may be recommended that seedlings produced for large-scale afforestation should be subjected to such an inoculations treatment. This could ensure better plant growth and survival to maintain plants sustainability in new area.

CONCLUSION

Experiments were conducted to understand the response of *Syzygium cumini* biofertilizer inoculation viz., VMF phosphobacteria and Azobacter individually and conjointly under nursery conditions. The non-inoculated seedling become control. The results indicated on enhancement in number of leaves, leaf length, total dry matter of shoot, increased shoot length, root length and N, P, K, contents in shoots has been recorded relative significantly higher over the control.

REFERENCES

Abbott, L. K., Robson, A.D. and Hall, I. R. (1983). Introduction of Vesicular arbuscular mycorrhizal fungi into agricultural soils, Aust. J. Agric. Res. Vol. 34. Pp. 741-749.

Al-Karaki, G. N. (1998). Benefit, cost and water use efficiency of arbuscular mycorrhizal durum wheat grown under drought stress, mycorrhiza, V018, pp. 41–45.

Allen, M.F. (1982). Influence of vesicular-arbuscular mycorrhizae on water movement through Boutelous gracilis (H.B.K) Lag. ex steud. New phytologist Vol. 91, pp.191–196.

Anil Kumar. A., Shukla, S., Hashmi, and Tewari, R. K. (2007). Effect of trees on colonization of intercrops by vesicular-arbuscular mycorrhiza in agroforestry systerms. Indian Journal of Agricultural Sciences. Vo. 77(5), pp. 291–298.

Bagyaraj, D.J., Manjunath, A. and Govinda Rao., Y. S. (1998). Mycorrhizal inoculation effect on different crops. Journal of Soil Biology and Ecology. Vol. 8, pp. 98–103.

Balkrishnan, L. (1996). The roles of renewable in meeting domestic energy needs. Natural Energy Review, 1:14–25.

Clarkson, D.T. (1985). Factors affecting mineral nutrient acquisition with VAM by plants Ann. Rev. plant physiology 36:77–115.

Jackson, J.L. (1973). Soil and plant chemical analysis. Prentice Hall of India (Pvt.) Ltd. New Delhi.

Lakshman, H. C. (1992). Development and response of VA-mycorrhizal fungi in *Terminalia bellerica* Roxb. J. Tropical Forests, 8(11):179–182.

Lakshman, H. C. (1996). VA-mycorrhizal studies in some important timber tree species Ph.D. thesis Karnatak University, Dharwad, India.

Marchner, H. and Dell, B. (1994). Nutrient uptake in mycorrhizal symbiosis pl. soil 159:89–102.

Nogueira, M. A. and Cardosa, E. J. B. N. (2002). Microbial interactions as manganese availability and uptake by soy bean Pes Agros Brasiliera, 37(11): 1605–1612.

Phillips, J.M. and Hayman D.S. (1970). Improved procedure for clearing and staining parasitic and vesicular arbuscular mycorrhizal fungi for rapid assessment of infections. Trans Brit. Mycol Soc. 55:158–161

Rachel, E. K., Reddy, S.R. and Reddy, S. M. (1993). Effect of different mycorrhizal treatments on the growth and yield of sunflower, India J. Microbial Ecol. 3:99-103.

Reena, J. and Bagyaraj, D. T. (1990). Response of *Acacia nilotica* and *Calhandra calathyrus* to different VAM. Arid soils Research and Rehabilitation, 4(4): 261-268.

Rodriques, L.A., Martins, M.A. and Salomao, H.S.H.B. (2003). Use of mycorrhizas and *Rhizobium* in intercropping system of Eucalyptus and Sestania-II Phosphorus uptake and efficiency of use and phosphate fractions R. B. de Ciencia Do Solo, 27(4):593-599.

Sattar, M.A. and Gaur, A. C. (1989). Effect of vesicular arbuscular mycorrhiza & phosphate dissolving micro-organisms, on the yield and phosphate uptake of Lentil Lens esculentamoench. Thai. J. Agri. Sci. 22:129-136.

Siddiqui, S. A., Anamika, N. and Shukla, N.P. (1993). The rate of *Azotobacter* inoculation in mulberry cultivations. Indian For. 119(7): 557-563

Venkatesh, Mallika, V.K., Vanangamudi, K.T., Parthiban and Ravichandran (1998). Effect of Biofertilizers on seedling growths, Physiology nodule production and VAM colonization in pangamiapinnata (L) Pierreo In: Bio fertilizers and Biopesticides PP1-7 Ed. A. M. Deshmukh Technoscience Publications Jaipur, India.

10

Effect of Arbuscular Mycorrhizae in Improvement of *Solanum indicum* Linn. A Vulnerable Medicinal Plant

*Hosamani P A, **Lakshman H C

INTRODUCTION

"Eat leeks in March and wild garlic in May, and all the year after the physicians may play," -Traditional Welsh rhyme (Tayler, 1987). "An apple a day keeps the doctors away," -Traditional American rhyme. Finding healing powers in plants is an ancient idea. People of all continents have long applied poultices and imbibed infusions of hundreds, if not thousands, of indigenous plants, dating back to prehistory. India is a veritable emporium of medicinal plants. It is bestowed with rich natural resources, diverse ecological conditions and a long practice of traditional system with their ethnic diversity and ancient civilization. The remarkable fact

* Department of Botany, Bangurnagar Degree College Dandeli-581 325, India

** P.G.Department of Botany, Karnataka University Dharwad -580 003 India Email: pahosamani@rediffmail.com

is that in traditional system of Indian medicine Ayurveda, Siddha, Unani and Tibetian, there are hundreds of texts on pharmacy that specifically deal with plants and plant products. Most of the plants are found in the tropics growing in wild condition and mainly used by herbal doctors and ayurvedic vaidyas, who refer to them as "jadibutis". The history of use of plants in medicine probably dates back to the origin of human race itself and the medicinal values of plants is recognized by every civilization on this earth. One of the oldest treaties of Indian medicine is Charaka Samhita (1000 BC) in which 340 medicinal drugs of plant origin are described. Plant products hold high promises as the therapeutic agents. Totally 13,000 plant species are known to have been used as drugs throughout the world (Tayler, 1994). Nearly 25–30% of the drugs currently used are of plant origin. The World Health Organization (WHO) estimate that up to 80% of the world population is still rely on plants for their primary health care. Plants have contributed more than 7000 different compounds used for heart drugs, laxatives, anti-cancer agents, hormones, contraceptives, diuretics, antibiotics, decongestants, analgesics, anesthetics, ulcer treatments and anti parasitic compounds till today. Extensive screening of plants is being carried out to identify chemical compounds that may provide new treatment for human diseases.

Use of medicinal plants is advantageous over synthetic drugs as latter has low cost and less side effect and little risk of disorders. Presences of active compounds like alkaloids, glucosides, volatile compounds etc., justify the use in treatment of many diseases. It has been observed that some natural products suffer from certain disadvantages like, they are available in minute quantities in source material, poor solubility, stability, absorption and failure to reach the demand. Presently the need to study medicinal plants in details from various points of view is generally well recognized.

Importance of AM association

Frank (1885) coined the term "Mycorrhiza" which literally means "fungus root". There are two main groups of mycorrhizae namely Ectomycorrhizae and Endomycorrhizae. Endomycorrhizae is again subdivided into three types (i) Vesicular arbuscular mycorrhizae (VAM) (ii) Ericaceous mycorrhizae and (iii) Orchidaceous mycorrhizae, but most important from agricultural point of view is VAM, which is an obligatory parasite, belonging to the order Glomales of Zygomycotina (Morton and Benny, 1990). VAM fungi are distinguished by producing branched, aseptate, fine hyphal structures called as "arbuscules" and globular structures called as "vesicles" in the region of the root cortex of host plant. These fungi are now recognized as the most widespread in occurrence in various plants and under different agro-climatic conditions covering a broad ecological range (Mosse, 1973 and Allen, 1991). VAM fungi also moderates the adverse effect of higher root temperature, improves plant water relation, soil structure, have greater tolerance to heavy metal effect, salinity and lower incidence of soil borne plant diseases, in addition they also improve the N_2 fixing capacity. Because of their widespread occurrence in nature and their benefits, researchers in several disciplines are interested in studying mycorrhizae.

The arbuscules are the sites of exchange of nutrients between the fungus and host cell, vesicles are apical or intercalary swellings of hyphae, which contain lipids and are reserve organs of the fungus. However, in some cases vesicles were not formed and they are known as "Arbuscular Mycorrhiza" (AM) coined by Allen (1991). Plants absorb phosphorus from soil leading to depletion of phosphorus in the zone of about 1-2 mm around the roots (Lewis and Kirk, 1967; Bhat and Nye, 1974). But mycorrhiza has the capacity to extend the hyphae far beyond the depletion zone and helps in absorption and translocation of phosphorus, in return the fungus obtains nutrients from the host plant. Apart from phosphorus it also helps in better utilization of other microelements such as Zn, Cu, Mn, Fe etc.

In recent days there is an increasing demand for medicinal plants and their products. Very little experiments have been conducted to understand the effect of VAM fungi, which can improve the growth and quality of medicinal plants. Therefore, the present investigation was undertaken to evaluate the efficacy of four different indigenous VAM fungi (*G. fasciculatum, G. mosseae G. macrocarpum* and *G. aggregatum*) on plant growth, biomass production, per cent colonization, spore number and "P" uptake in *Solanum indicum* Linn. medicinal plant.

The aim of mycorrhizal research is to utilize AM fungi to accelerate the growth and productivity of plants and to reduce the need of fertilizers and water. The growing necessity to understand the diversity of life contained within a gram of soil as well as intricacies of lining these living entities to terrestrial life in general, provided, the impetus to work on AM association in medicinal plants. The literature survey clearly indicates that, there is much to be digged out from the soil about AM, their association and their role with reference to the productivity of medicinal plants. However, the studies are very meager on medicinal plants using AM inoculation to understand growth response, biomass production and "P" uptake and changes in the chemical constituents of AM inoculated medicinal plants. Therefore, there is need of research in improving the production of native medicinal plant drugs in relatively shorter period and lower expense by utilizing a potential biofertilizer like AM fungi, and the present investigation was undertaken to evaluate the efficacy of four different indigenous AM fungi (*G. fasciculatum, G. mosseae G. macrocarpum* and *G. aggregatum*) on plant growth, biomass production, per cent colonization, spore number and "P" uptake in *Solanum indicum* medicinal plant.

MATERIALS AND METHODS

The experiments were conducted in pot under glass house using sterile red sandy, loamy soil to understand the effect of AM fungi on the growth and yield of *Solanum indicum* Linn. medicinal plant. Seeds were collected from nurseries. The experimental soil was sterilized with 5% methyl bromide and its physico-chemical characteristics were analyzed as outlined by Jackson (1973). Before sowing the seeds were surface sterilized with 2% sodium hypochlorite and washed in distilled water for 2–3 times.

Mycorrhizal Inoculum production: Pot cultures of the VAM fungi were multiplied in sterilized sand: soil (1:1), according to (Sreenivasa, 1992). Different strains of VAM such as *Glomus fasciculatum, G. mosseae G. macrocarpum* and *G. aggregatum* were selected. The VAM fungi-*Glomus fasciculatum* was multiplied by using onion *(Allium cepa)* as host plant, *G. mosseae* was multiplied on Jawar (*Andropogan sorghum*), *G. macrocarpum on Pennisetum typhoidium,* and *G. aggregatum on* Foxtail millet *(Setaria italica*). 15 gram of air dried VAM inoculum of these cultures was given to the soil as a thin layer 2 cm below the soil surface in each pot. The inoculum consists of 3 g root bits plus 12 g rhizospheric soil of host plant with hyphae and sporocarps (114 Chalmydospore/50g soil approximately). Ruakura plant nutrient solution (Harley and Smith1983) without phosphorus was given to the seedlings at the interval of 15 days. There were two treatments i.e., inoculated and non-inoculated (control) and were maintained in triplicates. Experimental pots were kept free of weeds, insects, and pests and irrigated properly. The plants were harvested intervals of 30, 60 and 90 days (days after sowing) and the parameters like plant height, stem diameter, leaf area, dry weight of shoot and "P" uptake were recorded. The spore count was carried out by wet sieving and decanting method (Gerdemann and Nicolson, 1963). The percent of colonization was determined according to Phillips and Hayman (1970). The percentage of colonization was calculated by the following formula.

$$\text{Per cent of root colonization} = \frac{\text{No. of root bits shows colonization}}{\text{Total number of root bits observed}} \times 100$$

The plant height was measured from ground level to tip of the plant and expressed in centimeters. Leaf area was recorded by graph method and stem diameter was measured one cm above the soil surface using vernier callipers. The uptake of 'P' in shoots was determined according to (Jackson, 1973). For each harvest experimental plant shoot and root was oven dried at 70°C until a constant weight was obtained to determine the dry weight and was expressed in terms of grams.

RESULTS AND OBSERVATIONS

In general, mycorrhizal inoculation resulted in a significant increase in plant height, stem diameter, dry weight, percent colonization, spore number and leaf area in all the medicinal plant seedlings over uninoculated (control) seedlings (Table 10.1). The present study with an objective of screening for an efficient AM fungi for *Solanum indicum* Linn. medicinal plant has resulted in varied plant growth responses to different AM fungi. The response of the medicinal plants to inoculation with different AM fungi varied. AM colonization in inoculated experimental plants was low in early days of inoculation i.e. 30 days, but steadily increased after 60 days. However it was observed that after 90 days of inoculation there was increased in colonization and spore number in all the AM inoculated *Solanum indicum* Linn. medicinal plants. The growth rate of inoculated plants increased after 60 days than of uninoculated control plants. The hyphae, arbuscles and vesicles were predominant sign of infection in all medicinal plants. *Solanum indicum* significant increase in plant growth i.e. height, stem diameter, dry weight, leaf area, "P" content and mycorrhizal parameters like percent colonization and spore number was observed in the plants inoculated with *Glomus macrocarpum* followed by *Glomus mossae* among the inoculated treatments. Inoculation of *Glomus aggregatum* significantly increased the percent colonization and spore count as compared to *Glomus fasciculatum*. All the mycorrhizal plants showed higher 'P' content in the shoots over the uninoculated control in all stages of harvest. The highest 'P' uptake was seen in the *Solanum indicum* plants inoculated with *Glomus macrocarpum*, followed by *Glomus mossae, Glomus aggregatum* and *Glomus fasciculatum*.

Table 10.1. Efficiency of Different VAM Inocula on Plant Height, Stem Diameter, Leaf Area, Dry Weight of Shoot and Root, Per Cent Colonization, Spore Number and 'P' Uptake of *Solanum indicum*

VAM inocula	Plant height (cm)	Stem diameter (cm)	Leaf area (cm^2)	Dry weight of shoot (g)	Dry weight of root (g)	Percent coloni-zation	Spore No./ 50 g of soil	Percent of 'P' uptake in shoot
30 Days NM	16.56 e	2.10 d	60.33 e	6.85 e	0.44 e	00.00 e	00.00 d	00.04 c
G.f	22.33 d	3.30 c	71.66d	8.23 d	1.25 d	61.20 d	115.66 c	00.14 b
G.m	24.36 b	3.53 b	95.66 b	10.82 b	1.55 b	70.56 b	120.33 c	00.16 b
G.ma	29.50 a	4.63 a	112.66 a	15.22 a	2.54 a	76.30 a	150.00 a	00.18 a
G.ag	23.33 c	3.26 c	81.33 c	9.026 c	1.38 c	64.90 c	118.66 c	00.15 b
60 Days NM	32.10 e	3.30 d	64.00 e	7.48 e	0.87 e	00.00 e	00.00 e	00.08 e
G.f	41.13 d	3.70 c	81.33 d	8.26 d	2.18 d	63.10 d	185.33 d	00.17 c
G.m	48.56 b	4.33 b	99.33 b	11.43 b	2.54 b	72.36 b	237.66 b	0.19 b
G.ma	52.20 a	5.36 a	121.00 a	16.72 a	3.52 a	77.36 a	271.66 a	0.22 a
G.ag	44.16 c	3.76 c	89.00 c	9.23 c	2.38 c	65.20 c	191.33 c	0.15 d
90 Days NM	50.86 e	3.40 d	72.66 e	8.14 d	1.44 e	00.00 e	00.00 e	00.12 e
G.f	65.46 d	3.76 c	88.00 d	9.28 d	3.18 d	64.30 d	276.66 d	00.18 d
G.m	71.70 b	4.23 b	105.00 b	12.40b	3.54 b	74.20 b	306.00 b	00. 22 b
G.ma	81.53 a	6.06 a	132.33 a	17.63 a	4.52 a	78.26 a	326.00 a	00.25 a
G.ag	68.53 c	3.93 c	100.33 c	10.11 c	3.39 c	70.90 c	290.00 c	00.20 c

NM- Non mycorrhizae ; *G.f—Glomus fasciculatum* ; *G . m—Glomus macrocarpum*; *G.ag—Glomus aggregatum*; *G.m-Glomus mossae*. Mean values followed by the same letter within a column do not differ significantly at $P = 0.05$ by DMRT.

In the present study AM status was considerably higher in all inoculated treatments compared to the control plants. Overall results in the present study *Glomus macrocarpum* caused higher plant height, stem diameter, dry weight, leaf area, "P" content and mycorrhizal parameters like percent colonization and spore number in the plants compared to other AM fungi and it can be considered an efficient AM fungus for *Solanum indicum.*

DISCUSSION

Arbuscular mycorrhiza (AM) association with plants will increase uptake of mineral nutrients, especially phosphorus in conditions of low 'P' availability (Bagyaraj, 1989; Manjunath *et al.*, 1998), reduce susceptibility of plants to certain pathogens (Giovannetti, 1990), alter the water relations and photosynthetic activity of the plant (Brown and Bethlenfalvay, 1988), increase seed production (Koide *et al.*, 2000) and improve health and vigor of the seedlings (Reena and Bagyaraj, 1990).

The medicinal plant *Solanum indicum* grown in the presence of AM showed a general increase in plant height, stem diameter, leaf area, dry weight, "P" content and mycorrhizal parameters like per cent colonization and spore number as those grown in soils without AM fungus (Table 10.1). The enhanced plant growth in *Solanum indicum* was obtained inoculated with *Glomus macrocarpum.* Several workers have observed the effects of AM fungi on plant growth (Abbott *et al.*, 1989; Michelsen and Rosendahl, 1990; Aggangan *et al.* and Jaize-Vega and Azcon 1995). Dhilon (1992) suggested that species might result in selection procedures, which favor certain specific fungus host species combination. Plant biomass is an important parameter, which directly reflects the efficiency of a particular fungus. In the present study the enhanced plant biomass in *Solanum indicum* was obtained inoculated with *Glomus macrocarpum.* The increase in shoot dry weight was up to four times the dry weight of uninoculated plants. Further Pope *et al.* (1983) in their experiment observed a 200 percent increase in total dry weight of *Platanus occidentalis* inoculated with *Glomus fasciculatum.* Several workers have observed the effects of VAM fungi on plant biomass (Nemec, 1970; Aggangan *et al.*, 1990; Sivaprasad *et al.*, 1990; Mandal and Kaushik 1994; Jaize-Vega and Azcon 1995; Mulla, 2002; Hemalatha and Selvaraj, 2003 and Yudhvir, 2004).

The total photosynthretic area expressed as the leaf area was significantly enhanced leaf area in *Solanum indicum* was obtained inoculated with *Glomus macrocarpum.* Enhanced leaf area because of mycorrhizal inoculation has been observed in *Trifoliate orange* and *Troyer citrange* (Vinayak and Bagyaraj, 1990). Murugan and Selvaraj (2003), observed enhanced leaf area in *Cichorium intybus* L. medicinal plant inoculating different native arbuscular mycorrhizal fungi. Also enhanced leaf area because of mycorrhizal inoculation has been observed in *Asparagus racemosus, Ocimum basilicum, Andrographis paniculata,* and *Becopa monnieri* (Yudhvir, 2004).

The "P" content was more pronounced in plants inoculated with *Glomus macrocarpum* in *Solanum indicum* and AM fungi are known to improve plant growth mainly through increased uptake of P and other nutrients (Jeffries, 1987). Abbott and Robson (1982) defined efficiency as the ability of the fungus to increase plant growth in a phosphate deficient soil. This depends on the hyphae in soil and extensive root colonization to absorb P from soil. Mallesha, Padmavathy and Bagyaraj (1994) reported that species and strains of AM fungi differed to the extent by which they increased nutrient uptake and plant growth. Hence some researchers suggested the need for selecting efficient VAM fungi that can be used for inoculating different plants (Bagyaraj, Byra Reddy and Nalini 1989; Mallesh and Bagyaraj, 1994). Hemalatha *et al.* (2003) and Murugan *et al.* (2003) also increased "P" content in medicinal plants.

In present study mycorrhizal parameters like percent root colonization and extra matricular spore count were higher in plants inoculated with AM. The influence of *Glomus macrocarpum* was highest on percent colonization and spore number in *Solanum indicum.* Similar observations were also made by Reena and Bagyaraj (1990) in their studies with *Calliandra calothrysus* inoculated with 13 different VA mycorrhizal fungi. So naturally the fungus having higher root colonization will be better adapted and absorb more nutrients and thus better growth. Rani (2001), Mulla (2002), Mulani (2002) and Lakshmipathy *et al.* (2003) observed higher colonization and up take of more nutrients in medicinal plants.

The results from this study indicated that, potential benefits could be obtained from the AM fungi in production of medicinal plants for their better use in future. It is concluded from the present experiments that the inoculation of medicinal plant *Solanum indicum* inoculated with *Glomus macrocarpum* seems to be a great boon to the botanist to minimize manure dosage in improving the growth, biomass production and the chemical constituents in medicinal plants by inoculating biofertilizers like AM fungi.

REFERENCES

Abbott, L.K. and Robson, A.D. (1982). Infectivity of vesicular arbuscular mycorrhizal fungi in agricultural soils. Aust.J.Agric.Res.33: 1049–1059.

Aggangan, N.C., Lorilla, E.B. and Cruz, R.E., de La. (1990). Screening for effectiveness of four VA mycorrhizal fungi on Acacia auriculiformis in fertile soil. Sylvatrop 1987 Publication 12(3&4): 133–140.

Allen, M.F. (1991). The ecology of mycorrhizae, Cambridge University, New York, p. 184 .

Bagyaraj, D.J., (1989). Mycorrhizas. In, Trophical rain forest ecosystems(Eds.)Leith,H. and Werger,M.J.A. Elsevier science publishers, B.V. Amsterdam. Pp 537–546.

Bhat, K.S. and Nye, P.H. (1974). Diffusion of phosphate to plant roots in soil. II. Uptake along the roots at different times and effect of different levels of phosphorus. Plant Soil, **41:** 365–382.

Dhillon, S.S. (1992). Host endophyte specificity of vesicular arbuscular mycorrhizal colonization of *Oryza sativa* L. at the preplant stage in low or high phosphrous soil. Soil Biol. Biochem., **24:**405–411.

Gerdemann, J.W. and Nicolson, T.H. (1963). Spores of mycorrhizal endogone species extracted from the soil by wet sieving and decanting. Transaction of British Mycological Society, 46: 235–244.

Giovannetti, M. (1990). Mycorrhizas as agents of biological control of agricultural plant pathogens. Informatore, Fitopathologica,40:17–20.

Hemalatha,Mand Selvaraj,T. (2003). Association of AM fungi with Indian borage (*Plectranthus amboinicus*(Lour.) Spenge) and its influence on growth and biomass production. Mycorrhiza News 15(1):18–21.

Jackson, M.L. (1973). Soil chemical Analysis. Prantice Hall of India Pvt. Ltd, New Delhi.

Jaizme-Vega,M.C. and Azcon, R. (1995). Responses of some tropical and subtropical cultures to endomycorrhizal fungi. Mycorrhiza 5(3):213–217.

Jeffries, P. (1987). Use of mycorrhiza in agriculture. Critical review in biotechnology. **5:**319–357.

Koide, R.T., Goft, M.D. and Dickie, I.A. (2000). Component growth efficiencies of mycorrhizal and non-mycorrhizal plants. New Phytologist. **148(1):** 163–168.

Lakshmipathy, Balakrishnagouda and Bagyraj, (2003). VA Mycorrhizal colonization patteren in RET medicinal plants in different parts of Karnataka. Asian Jr.of Microbiol.Biotech.Env.Sc. Vol.**5**,No(**4**):505–508.

Lewis, D. G. and Kirk, J.P. (1967). Phosphate diffusion in soil and uptake by plants II. ^{32}P movement and uptake by plants as indicated by 32p autoradiography. Plant and Soil. **26:** 445-453.

Mallesha, B.C.,Padmavathy, R.T. and Bagyaraj, D.J. (1994). Response to tomato to different VA mycorrhizal fungi. Journal of Soil Biology and Ecology **14:**36–40.

Mandal, B.S. and Kaushik, J.C. (1994). Effect of VAM inoculation of growth and development of *Acacia nilotica* (L.) Wild. ex. del. Annals of Biology (Ludhiana) **10:**264–267.

Manjunath, G.T.S. (1998). Incidence of vesicular arbuscular mycorrhizae in medicinal plants of MPC area, Sandur. M.Phil Dissertation, Gulbarga.Uni. P.15–23.

Manjunath, G.T.S. (1998). Incidence of vesicular arbuscular mycorrhizae in medicinal plants of MPC area, Sandur. M.Phil Dissertation, Gulbarga.Uni. P.15–23.

Michelson, A. and Rosendahl, S. (1990). The effect of VA mycorrhizal fungi, phosphorous and drought stress on the growth of Acacia nilotica and Leucaena leucocephala seedlings. Plant Soil, **124:**7–13.

Morton, J.B. and Benny, G.L. (1990). Revised classification of arbuscular mycorrhizal fungi (Zygomycetes): a new order Glomales, two suborders, Glomineae and Gigasporineae, and two new families, Acaulosporaceae and Gigasporaceae, with an amendation of Glomaceae. Mycotaxon. 37: 471–491.

Mosse, B. (1973). Plant growth responses to vesicular arbuscular mycorrhizae. IV. In soil given additional phosphate. New Phytologist. 72: 127–136.

Mulani, R.M. R. Prabhu R. and Dinkaran, M. (2002). Occurance of VAM in the roots of Phyllanthus fraternus Webster. Mycorrhiza News **14(2):**11–14.

Mulla, F.I. (2003). VA-mycorrhizal studies in some important medicinal plants. Ph.D. thesis Karnatak University Dharwad, India.

Murugan, R. and Selvaraj. T (2003). Reaction of Kashini (Cichorium intybus L.) to different native arbuscular mycorrhizal fungi. Mycorrhiza News **15(1):**10–13.

Phillips, J.M. and Hayman, D.S. (1970). Improved procedures for clearing roots and staining parasite and vesicular arbuscular mycorrhizal fungi for rapid assessment of infection. Transaction of British Mycological Society. **55**:158–160.

Pope, P.E., Chaney, W.R., Rhoides and Woodhead, S.H. (1983). The mycorrhizal dependency of four hardwood tree species, Can.J. Bot., **61:412**–417.

Rani,V. and Bhaduria, S. (2001), Vesicular arbuscular mycorrhizal association with some medicinal plants growing on alkaline soils of Mainpuri district, Uttar Pradesh. Mycorrhiza News **13(2)**:12–13.

Reena, J. and Bagyaraj, D.J. (1990). Response of *Acacia nilotica* and *Culliandra calothyrus* to different VA mycorrhizal fungi. Arid Soil Research and Rehabilitation.**4**:261–268.

Sreenivasa, M.N. (1992). Selection of efficient vesicular arbuscular mycorrhizal fungus (*Capsicum annum* Linn). Scientia Horticulture. **50:** 53–58.

Sreenivasa, M.N. and Bagyaraj, D.J. (1988). Effect of culture, age and pruning on mass production of the VA mycorrhizal fungus *Glomus fasciculatum*. Pertanika 11(1): 143–145.

Tayler, V.E. (1987). The new honest herbal. Geroge F. Stickley Co., Philadelphia,Pa.

Vinayak, K. and Bagyaraj, D.J. (1990a). Selection of efficient VA mycorrhizal fungi for *Trifoliant orange*. Biological Agriculture and Horticulture, **6**:305–311.

Vinayak, K. and Bagyaraj, D.J. (1990b). Vesicular arbuscular mycorrhizae screened for *Troyer citrange*. Biological and Fertility of soils, **6**: 305–311.

K.Bhoon, Y. (2004). Mycorrhizal inoculation of some medicinal plants field trials. Agrobios Newsletter, Vol.3 No.4:15–16.

11

Symbiotic Response of *Pisum sativum* L. to Different Arbuscular Mycorrhizal Fungi

[1]*V S Bheemreddy*

INTRODUCTION

Indian soil, in particular, is mostly deficient in phosphorus (P). Arbuscular mycoprrhizal (AM) fungi are ubiquitous and form obligate symbiotic association with most families of plants. VA-mycorrhizal fungi have been shown to be beneficial to plants in better utilization of soil phosphorus (Mosse, 1981). Inoculation of soil with AM fungi are known to increase the uptake of P, Cu and Zn, there by increase the dry matter yield of several legumes, cereals and horticultural crops (Krishna and Bagyaraj, 1985; Lakshman, 1992). Although plant mycorrhizal association is considered to be the current field of interest in bio-fertilizer technology, very little or no information is available on AM fungal association with *Pisum sativum* and its impact on plant growth and nutrient uptake.

Pisum sativum is an important legume. It grows well in waste lands and leached soils with low fertility. It is under

1. Department of Botany, JSS College, Vidyagiri, Dharwad – 580 004, India.

such situation the mycorrhizal dependency of the host plant become more evident and pronounced (Chung *et al.*, 1989). In the present work, experiments were undertaken on the effect of different mycorrhizal ioslates for *Pisum sativum*.

MATERIALS AND METHODS

The AM fungi used in this study were maintained in a glass house as pot cultures using sterilized pure sand: garden soil mix (1:1 v/v) as the substrate and Gninea grass (*Panicum maximum* Jacq) as the host. The hyphae, spores present in host soil and root segments present in the substrate served as the inoculum (487 chlomydosspores/50 g host soil). It was estimated using the most probable number technique (MN) as outlined by Porter (1979) [4]. Healthy seeds of *Pisum sativum* were sown for earthen pots measuring 25 × 25 cm holding 4 kg of sterilized sand : Soil : Composite in the ratio of 1:1:0.25. Twenty five gms mixed inoculum was placed 2 cm below the surface of each pot one seed per pot was sown and plants were maintained for 12 days in a poly house.

The plant height and stem girth was recorded at 40, 60, 80 and 100 days after sowing. Plants were harvested after 100 days of sowing. However, data obtained after 100 days of sowing only is presented in this paper. Stem girth was measured by vernier callipers. The harvested plant samples were oven dried constantly at 120°C for 12 hrs and total dry weight was determined. Per cent mycorrhizal root colonization was determined following the gridline intersect method (Giovannetti and Mosse, 1980), after clearing 10% kg/h and staining the root with acid fuchsion. The AM fungal spore number per 50g/soil was determined by wet-sieving and decanting technique (Gerdemann and Nicolson, 1963). Phosphorus content of shoots were estimated by Vanado molybdate phosphoric acid yellow colour method (Jackson, 1973).

RESULTS AND DISCUSSION

Pisum sativum plants responded quite well to inoculation with vesicular arbuscular mycorrhizal fungi. Plants height, stem girth and total shoot biomass was more in AM fungi inoculated plants compared to uninoculated plants (Table 11.1). Stem dry weight was significantly more in plants inoculated with *Gigaspora decipiens, Glomus fasciculatum* and *Glomus mosseae* compared to uninoculated plants. Such increase in plant growth and Biomass of *Pisum sativum* because of AM fungi inoculation, similar results were recorded by (Lakshmipathy *et al.*, 1999) on cashew nut root stock. Role of VAM fungi in improving the stem girth and plant biomass is well documented in tropical plantation crops like citrus and mango and an forest tree species (Vinayak and Bagyaraj, 1990; Lakshman, 1996).

Table 11.1: Effect of Mixed Inoculum with Different VAM Fungi on Content Height, Stem Girth and Total Biomass of *Pisum sativum* (for 100 Days).

Different VAM fungi treatment	Plant height (cm)	Stem girth (cm)	Stem biomass g/plant	Total shoot weight g/plant
Uninoculated (control)	24.6	0.59	1.63	3.86
Acaulospora undulata	43.5	0.84	4.50	9.38
Gigaspora decipiens	50.2	1.00	5.72	14.26
Glomus botryoides	43.8	0.78	4.63	9.47
Glomus fasciculatum	52.1	1.02	6.34	14.80
Glomus intraradices	44.7	0.85	4.53	9.55
Glmous leptonica	45.8	0.86	4.74	9.35
Glomus mosseae	51.3	1.05	5.68	14.35
Scutellospora calospora	44.1	0.87	4.58	9.16
Scutellospora persica	43.5	0.85	4.76	9.71
LSD at 5%	2.05	0.03	0.11	1.04

Inoculated plants generally had higher percent of P (Table 11.2) and it was significantly more plants inoculated with *Gigaspora decipiens, Glomus fasciculatum* and *Glomus mosseae* compared to uninoculated plants. These mycorrhizal (AMF) root colonization and spore number in root zone soil were significantly more in plants inoculated with *Glomus fasciculatum* and *Glomus mosseae* (Table 11.2). It is well established fact that AMF fungi help in greater uptake of P by plants. Especially *Glomus fasciculatum* and *Glomus mosseae* resulted high phosphorus content (Fig. 11.1). An extensive and well distributed hyphae proliferation in the soil together with the root system facilitate absorption of phosphorus scavenging of larger volume of soil thus aiding better plant growth (Hayman, 1983; Asghari *et al.*, 2005). Vigarous plant growth and better stem girth of *Pisum sativum* plants due to vesicular arbuscular mycorrhization helps in early establishment seedlings and promote growth and phosphorus with biomass production (Fig. 11.1). In order to promote and accelerate succession in restoration processes and to develop sustainable revegetation practices. Similar results of increased root colonization and enhanced P uptake inresponse to vesicular arbuscular mycorrhizal inoculation in some tropical plantation crops and tree species has been reported by several

Table 11. 2: Effect of Mixed Inoculum with Different VAM Fungi on Mycorrhizal Root Colonization, Spore Number in Root Zone Soil and P Content of Shoot of *Pisum sativum* (for 100 Days).

Different VAM fungi treatment	VAM root colonization	VAM spores/ 50 g soil	P content in shoot (%)
Uninoculated (control)	-	-	0.10
Gigaspora decipiens	62.52	198	0.51
Glomus botryoides	45.06	141	0.39
Glomus fasciculatum	57.41	173	0.48
Glomus intraradices	44.08	143	0.41
Glomus leptonica	49.54	139	0.37
Glomus mosseae	61.70	168	0.50
Scutellospora calospora	35.42	147	0.34
Scutellospora persica	38.53	152	0.42
LSD at 5%	8.51	50.24	0.08

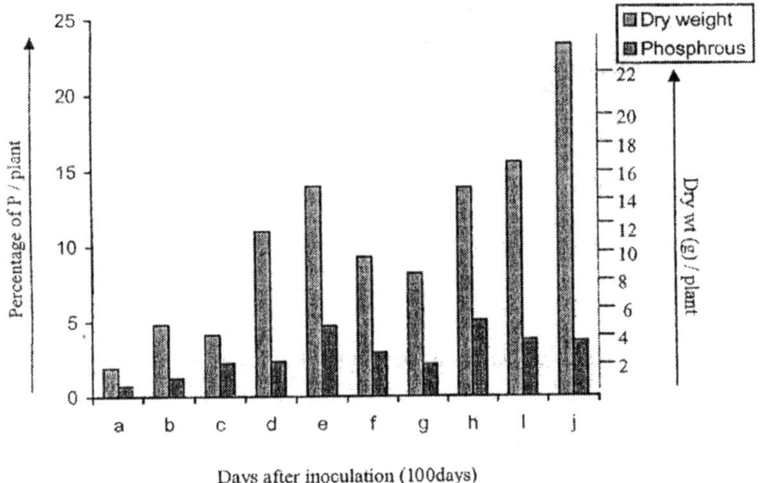

Fig. 11.1. Figure Showing the Effect of Mixed Inoculum with Different VAM Fungi on Plant Dry Weight, with Phosphorous Content of *Pisum sativum*

a= Control	b= *Acaulospora undulata*
c= *Gigaspora decipiens*	d= *Glomus botryoides*
e= *Glomus fasciculatum*	f= *Glomus intraradices*
g= *Glomus leptonica*	h= *Glomus mosseae*
i= *Scutellospora calospora*	j= *Scutellospora persica*

workers (Cardoso et al., 1986; Suman and Bagyaraj 1996; Lakshman, 2001). From the present study, it can be concluded that the best AMF fungus for inoculating *Pisum sativum* nursery seedling is *Glomus fasciculatum*.

REFERENCES

Cardoso, E. J. B.N., Hutunes, V., Da Selveria, A. P. D. and De dliverira, M.H.A., Revista Braseleira de Eienei do solo. (1986). **10**, 25–30.

Chung, R.S.Su., Yu-Chiou and Chang, Tse-Chon. (1989). Effect of VAMF on the growth and uptake of phosphorous and zinc of vorn plant under different phosphorous sources. J.chin.agri>chem.Soc. 27:469–504.

Gerdemann, J. W. and Nicolson, T. H. (1963). Spore of mycorrhizal endogone species extracted from soil by wet sieving and decanting. Trans. Brit Mycol. Soc., 1963, **46,** 235–244.

Giovannetti, M. and Mosse, B. (1980) An evaluation of techniques for measuring vesicular arbuscular mycorrhizal infection in roots. New Phytol New Phytol. 84: 489–500.

Hayman, D. S., Can J. Bot., 1983, **61,** 944–963.

Jackson, M.L. (1973). Soil Chemical Analysis, Prentice Hall of India, New Delhi, India, pp. 239.

Krishna, K. R. and Bagyaraj, D. J. (1985). Phosphatases in the rhizosphere soil of mycorrhizal and non-mycorrhizal goundnut. J. Soil. Biol. and Biochem. 52: 81–85.

Lakshman, H. C. (1992). Response and development of Terminallia bellerica Roxb. Infected with VA-mycorrhiza Tropical forest, 109:81–86.

Lakshman, H. C., (1996), VA-mycorrhizal studies in some important timber tree species Ph.D. thesis, Karnatak University, Dharwad, India.

Lakshman, H. C., J. Trophical Ecol., 2001, **42(2)**, 283–286.

Lakshmipathy, R., Balakrishna, A. N. and Bagyaraj, D. J., (1999), In: proc. Frontiers in fungal biotechnology and plant pathogen relations. (Ed). C. Manoharachary, Jan. 16–18 p. 90–93.

Mosse, B.(1981). Vesicular-arbuscular mycorrhizal research for tropical agriculture. Institute and Human Resource, Research Bulletin, **194,** 1–82.

Verma, Neeraj. Kaur, S and Verma, A.K. (1990). Effect of soil inoculation with endomycorrhiza fungi on growth of *Cyamopsis tetragonaloba*. In: Mycorrhiza symbiosis and plant growth . Proc. 2nd NaH.On Mycorrhiza, Bangalore (Eds. D.J. Bagyaraj and A. Manjunath) pp. 82–83.

Porter, W. M. (1979). 'The most probable number' method for enumerating infective propagules of vesicular arbuscular mycorrhizal fungi in soil. Aust.J.Soil. Res., 17:515–519.

Suman, D. A. and Bagyaraj, D. J., In : Proc. HWFRO Symp. On Impact of diseases and insect pests in tropical forests (Eds.) K.S.S. Nair, J.K. Sharma and R. V. Varma Pub. By KFSI Peechi, Kerala, p. 246–251.

Vinayak, K. and Bagyaraj, D. J. (1990). Selection of efficient VA mycorrhizal fungi trifoliate orange. Biol. Agri. Hort., 6:303–311.

12

Selection of Suitable AM Fungus to *Artocarpus heterophyllus* Lam. A Fruit/Timber for an Ecofriendly Nursery

[1]*H C Lakshman*

INTRODUCTION

The accentuating demand for food, fodder, fiber and fuel has resulted in the emergence of the fungal symbionts, the mycorrhizae. Thereafter, scientists of agriculture, horticulture and forestry started exploiting mycorrhizae to augment the productivity of crop, forestry and ecosystem (Harley, 1989; Lakshman *et al.*, 2007).

AMF (arbuscular mycorrhizal fungi) are highly evolved, mutualistic associations with plant roots. It is estimated that 90% of plant species characteristically form mycorrhizae. AMF have shown to improve plant productivity in soils of low fertility, increase plant vigour, and reduce transplant injury. There has been a shift in the horticultural scenario, which necessitates raising the rootstocks organically (from the

1. Post Graduate Studies in Botany, (Microbiology Lab), Karnatak University, Dharwad-580 003, India.

nursery itself) to ensure higher grafting, take, better growth, and in particular more stem girth. The present study was carried out to select AMF most suitable for *Artocarpus heterophyllus*.

Arbuscular mycorrhiza fungal association plays an important role in P cycling and uptake of phosphate by plants. Mycorrhizal plants are therefore adapted to cope with nutrient deficient situation or prevent pathogenesis by other organisms. AM fungi will not prove effective with all plants with then plants have little or no mycorrhizal dependency (Wood, 1992; Lakshman, 1996). Variation in mycorrhizal dependency limits the potentiality of the fungus in improving the plant growth performance.

These arbuscular mycorrhizal fungi are not host specific although evidence is growth that certain AM fungi may form preferential association with certain host plants. Several investigations indicated that even in unsterile soils plants do respond to inoculation with efficient strains (Manjunath *et al.*, 2001; Gracy *et al.*, 2005). Numerous experiments have been conducted and confirming that growth of mycorrhizal plants is favoured in soil with low to moderate fertility (Mukerji and Kapoor 1987 ; Lakshman, 2007). The jackfruit (*Artocarpus heterophyllus*) is a species of tree of the mulberry family (Moraceae) native to parts of South and Southest Asia. It is well suited to tropical lowlands. The jackfruit has played a significant role in the Indian agriculture (and culture) from time immemorial. Archeological findings of India have revealed that jackfruit was cultivated in India 3000 to 6000 years ago. The main products from this tree are fruits, timber, fodder, latex. Many parts of the plant including the bark, roots, leaves, and fruit are attributed with medicinal properties. The tree can provide many environmental services. It is highly wind tolerant and therefore makes a good component in a windbreak. The objective of this study is to select suitable AM fungus on most promising plant *Artocarpus heterophyllus* at nursery, to improve its growth, biomass production and nutrients uptake.

MATERIALS AND METHODS

This investigation was carried out under nursery conditions in a green house. The soil used in this study was collected from an uncultivated field from a depth of 0–15 cm and has been classified as fine kaolinitic isohyperthermic, kanhaplustalfs. The soil pH was 6.8 (1:10 soil to water extract ratio), and it contains 2.4 ppm available phosphorus (extractable NH_4F + Conc. HCl) and an indigenous AM mycorrhizal population of 247 spores/50 ml of soil. Seedlings raised in the nursery, and before transplanting in to earthen pots measuring (20–25 cm) in diameter containing 4 kg sterilized mixture (soil : sand : FYM) in the ratio of 2 : 1 : 2 (v/v/v basis).

The AM fungal species used in this study were isolated, identified and pure cultures were maintained in glass house. These fungi were multiplied using sterlized sand soil mix (1:1 v/v) the substrate and *Sorghum vulgaris* Pers (Johnson grass) as the host, After 30 days, the growth, roots of *Sorghum vulgaris* Pers (Johnson grass) were severed and the substrate containing hyphae, spores and root bits was air dried and used as the inoculum. The inoculum potential (IP) of each culture was estimated adopting the most probable number (MPN) method as outlined by Porter (1979).

Each earthen pot containing the potting mixture, with and without AM inoculum, as the treatment may be and were planted with *Artocarpus heterophyllus* seedlings. Each treatment had 4 replications and was maintained in green house and watered regularly so as to maintain the field capacity of the soil. Plant nutrient solution without phosphate was added to the earthen pots at the rate of 50 ml per earthen pot once in 15 days (Harley and Smith, 1983).

Plant height was measured from soil surface to the growing tip of the plant. Stem girth was measured 1 cm above the soil surface using vernier callipers on 120 and 210 DAP. However, only observations recorded on day 210 are presented in this paper. The plants were harvested 210 DAP. At harvest, leaf area was recorded using a leaf area meter. Dry biomass was determined after drying the plant sample at 60°C to a constant weight in a hot air oven. Soil samples (50 ml) were collected from each earthen pot and subjected to wet sieving and decantation method as outlined by Gerdemann and Nicolson (1963) to estimate the population of spores. Fine terminal feeder roots were stained using 0.05% trypan blue as described by Philips and Hayman (1970) and the per cent root colonization was estimated by adopting the gridline intersect method (Giovannetti and Mosse, 1980). Estimation of soil aggregates (<50 μm size), which indirectly denotes the extent of external hyphae is soil, was done as described by Van Bavel (1980).

Phosphorus content in the plant tissue was determined by employing the Vanado molybdate phosphoric acid yellow colour method (Jackson, 1973). Atomic absorption spectrophotometer was employed to estimate zinc and copper contents of the plant samples, using respective hallow cathode lamps. Bio-volume index, a measure of the total volume of a seedling, and quality index, which reflects the quality of a seedling, were determined using the formula given by Hatchell (1985). Data thus generated were subjected to statistical analysis of completely randomized block design and the means were separated by Duncan's Multiple Range Test (DMRT) (Duncan, 1955).

RESULTS AND DISCUSSION

In general, mycorrhizal inoculation resulted a significant increase in plant height, stem girth, plant biomass and plant phosphorus content of seedlings of *Artocarpus heterophyllus*.

Host preference among AM fungi has been reported by earlier workers (Pankow *et al.*, Vasanthakrishna *et al.*, 1995). Hence the need for selecting efficient AM fungi that can be used for inoculating different mycotrophic plants has been stressed (Jeffries, 1987). The present study with an objective of screening for an efficient AM fungus for seedlings of *Artocarpus heterophyllus* has also resulted is varied plant growth responses with stem girth (Fig. 12.1) to different AM fungi.

Nutritional status of seedlings of *Artocarpus heterophyllus*, considering the phosphorus, zinc and copper contents, was significantly high in those raised in soil inoculated with arbuscular mycorrhizal fungi (Table 12.1). Seedlings raised in the presence of *Glomus macrocarpum* showed an increase of 124%, 110% and 152% in tissue phosphorus, zinc and copper contents, respectively, as compared to seedlings raised in the absence of inoculation. The extent of increase in the plant phosphorus, zinc and copper contents varied among the fungi studied with seedlings grown in the presence of *Glomus*

Table 12.1. Effect of Soil Inoculation with Arbuscular Mycorrhizal Fungi on Phosphorus, Zinc and Copper Contents of *Artocarpus heterophyllus* at 210 Days.

AM fungi	P content (mg/plant)	Zn content (mg/plant)	Cu content (mg/plant)
Acaulospora mellea	33.0 f	293.9 f	140.9 f
Gigaspora albida	38.5 c	417.1 c	191.3 c
Glomus fasciculatum	28.3 g	269.6 g	127.8 h
Glomus mosseae	37.8 d	379.2 d	136.1 g
Glomus microcarpum	30.9 f	331.5 e	159.7 e
Glomus macrocarpum	43.4 a	479.3 a	233.0 a
Glomus tenue	44.1 b	446.7 b	206.3 b
Scutellospora aurigloba	35.9 d	421.4 c	179.6 d

Means followed by the same letter with a column do not differ significantly at P=0.05 by DMRT

macrocarpum containing a significantly higher content of these nutrients, followed by those grown in the presence of *G. tenue*. Such a variation in the plant nutrient status in relation to the fungal species and other tree species is well documented (Saif, 1987; Bagyaraj *et al.*, 1989; Reena and Bagyaraj, 1990).

In all the growth phases, uninoculated control registered least growth of rootstocks (Silva and Sigueira 1991; Bassangowda 2005), citrus (Vinayak and Bagyaraj 1990; Rilling and Mummey, 2006) and sapota (Sreeramulu, Gowda and Bagyaraj, 1998). This increased growth of vegetative parameters is the consequence of increased root proliferation via increased nutrients and water uptake due to mycorrhizal inoculation.

It is well known that the enhanced nutritional status of a plant manifests in its improved growth (Jeffries, 1987). Plants grown in the presence of arbuscular mycorrhizal showed a general increased in the plant growth parameters like plant height, stem girth, leaf area and total dry weight as against those grown in soils uninoculated with AM fungus. *G. macrocarpum* significantly enhanced the plant height as compared to all other treatments except for *Acauolspora meallea*. However, seedlings raised in the presence of *Gigaspora albida* had a significantly higher stem girth compared to all other treatments excepting *G. macrocarpum*. The total photosynthetic area expressed as the leaf area was significantly more in plants grown in the presence of *Gigaspora albida*. This increased leaf area and enhanced nutrient content in seedlings colonized by *Gigaspora albida* have probably results in significantly higher biomass observed compared to other treatments. The enhancement in growth and nutritional status is also related to the percent root colonization apart from several soil and environmental factors.

Considering the percent root colonization observed in this study, *Glomus tenue* inhabited a significantly higher percentage of roots compared to other AM fungi (Table 12.3).

Table 12.2Effect of Soil Inoculation with Arbuscular Mycorrhizal Fungi on Plant Height, Stem Girth, Leaf Area and Plant Dry Weight of Artocarpus heterophyllus at 210 Days.

AM fungi	Plant height (cm)	stem girth (mm)	Leaf girth (cm/plant)	Dry weight (g/plant)
Acaulospora mellea	46.1 cde	6.7 d	1222.9 e	16.5 e
Gigaspora albida	52.0 ab	8.9 b	1721.4 b	19.2 b
Glomus fasciculatum	48.1 ef	7.9 b	1156.3 f	17.1 f
Glomus mosseae	48.0 ef	8.8 ab	1379.5 d	18.4 cd
Glomus microcarpum	45.2 cde	8.5 b	1260.0 e	16.6 e
Glomus macrocarpum	52.7 cde	8.2 a	1995.6 a	21.7 a
Glomus tenue	54.3 a	7.1 c	1589.1 b	23.2 a
Scutellospora aurigloba	53.9 bc	7.2 c	1428.8 c	22.8 bc

Means followed by the same letter with a column do not differ significantly at P=0.05 by DMRT

Table 12.3Effect of Soil Inoculation with Arbuscular Mycorrhizal Fungi on Percent Root Colonization, Spore Number in Root Zone Soil and Percent Aggregation of Rhizosphere Soil of *Artocarpus heterophyllus* at 210 Days.

AM fungi	Root colonization (%)	Spore number/ 50 ml soil	Percent aggregation of soil
Acaulospora mellea	46.9 de	195 c	49 f
Gigaspora albida	72.0 a	170 e	51 c
Glomus fasciculatum	53.9 ef	179 d	51 b
Glomus mosseae	55.5 d	207 c	56 c
Glomus microcarpum	48.7 f	181 d	37 g
Glomus macrocarpum	69.3 c	291 a	52 ab
Glomus tenue	77.2 b	256 b	52 d
Scutellospora aurigloba	73.2 b	163 e	59 a

Notations as in Table 1.

Means followed by the same letter with a column do not differ significantly at P=0.05 by DMRT

However, spore numbers were highest in soil samples inoculated with *Glomus fascilatum*, indicating the better proliferating ability of this fungus with *Artocarpus heterophyllus* as the host. This capacity of *Glomus fasciculatum* to sporulate and hence multiply is of great significance as it will not only increase the colonization of the roots further but also improve the mycorrhizal potential of the soil to which it would be transplanted. The ability of this species to improve considerably the growth and nutrient content of *Artocarpus heterophyllus* plants and to sporulate in higher numbers despite a significantly lower colonization level compared to *Gigaspora albida, Scutellospora aurigloba* and *Aculospora mellea* it is further emphasizes that *G. macrocarpum* utilizes the carbon sources of *Artocarpus heterophyllus* seedlings efficiently, an observation which supports the work of Abbott and Robson (1985). Mycorrhizal fungi are also implicated in improving the soil structure by increasing the soil aggregation by their hyphae (Miller and Jastrow, 1992). Soil aggregation is a measure of the amount of extramatrical hyphae, which is in turn related to the efficiency of the fungus (Reena and Bagyaraj, 1990). This observation was further strengthened by the present study as mycorrhizal fungi used in this study significantly improved the aggregation of soil compared to the uninoculated treatment (Table 12.3). Soil aggregation was highest in soil inoculated with *Glomus mosseae* but it did not significantly differ from the soils incoulcated with *Glomus microcarpum*.

Further, seedlings raised in the presence of *Glomus macrocarpum* showed a greater biovolume index and quality index compared to all other treatment (Table 12.4) and this increase was to an extent of 68% and 66.7%, respectively, over those seedlings raised in soil uninoculated with AM fungus. Such high values of biovolume index and quality index indicate a sturdier stem and a proportionate top dry weight compated to seedling dry weight, qualities which are desirable among nursery seedlings (Pankow *et al.*, 1991).

Table 12.4.: Effect of Soil Inoculation with Arbuscular Mycorrhizal Fungi on Bivolume Index and Quality Index of *Artocarpus heterophyllus* at 210 Days.

AM fungi	Biovolume index	Quality index
Acaulospora mellea	2473.5	2.1
Gigaspora albida	3579.1	3.0
Glomus fasciculatum	3094.0	2.4
Glomus mosseae	3411.5	2.8
Glomus microcarpum	3344.0	2.6
Glomus macrocarpum	3789.4	3.3
Glomus tenue	3073.6	2.8
Scutellospora aurigloba	2887.2	2.1

Notations as in Table 12.1. Means followed by the same letter with a column do not differ significantly at P=0.05 by DMRT

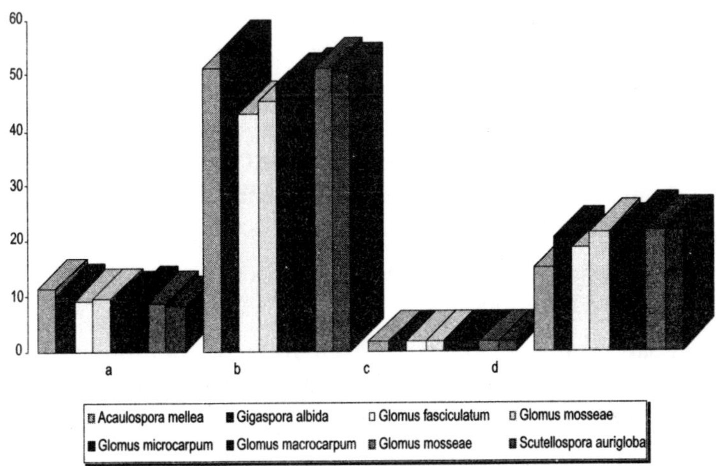

a = non-inoculated plant height; b = inoculated plant height; c= non-inoculated tuber girth; d= inoculated tuber girth

Fig. 12.1.: Showing the Efficacy Different AM Fungi on *Artocarpus heterophyllus* with Regard to Plant Height and Dry Weight for 210 days.

Hence it can be concluded that *Artocarpus heterophyllus* seedlings show varied responses to different AM fungi *Glomus macrocarpum* confers maximum growth benefits compared to all other fungi used in this study. Further, considering the ability to proliferate in greater numbers, utilize plant carbon sources efficiently, enhance nutrient status and growth of *Artocarpus heterophyllus* seedlings and also improve the soil structure, *Artocarpus heterophyllus* seedlings raised in the presence of *Glomus macrocarpum* can possibly established and perform better when planted in degraded and impoverished lands.

REFERENCES

Abbott, L. K., Robson, A.D. (1985). Formation of external hyphae in soil by four species of vesicular-arbuscular mycorrhizal fungi. New Phytol. 99, 245–255.

Bagyaraj, D. J., Byra Reddy, M.S., Nalini, P.A. (1989). Selection of an efficient inoculant VA mycorrhizal fungus for *Leucanena* for Ecol. Manage. 27, 791–801.

Gerdemann, J.W., Nicolson, T.H. (1963). Spores of mycorrhizal *Endogone* species extracted from soil by wet-sieving and decanting. Trans. Br. Mycol. Soc. 46, 235–244.

Giovannetti, M. and Mosse, B., (1980). An evaluation of techniques to measure vesciular-arbuscular infection in roots. New Phytol. 84, 489–500.

Gracy, L. Sailo and Bagyaraj, D. J. (2005). Influence of different AM fungi on growth, nutrition and forskolin content of *Coleus forskohlii* Mycol. Res. 109: 795–798.

Harley, J. L. (1989). The significance of mycorrhizae mycological Res. 92: 129–139.

Harley, J.L. and Smith S.E. (1983). Mycorrhizal Symbiosis, Academic Press, London, UK, pp. 245.

Jackson, M.L. (1973). Soil Chemical Analysis, Prentice Hall of India, New Delhi, India, pp. 239.

Jeffries, P. (1987). Use of mycorrhizae in agriculture. Crit. Rev. Biotechnol. 5, 319–357.

Lakshman, H. C. (1996). VA mycorrhizal studies on some important timber tree species Ph.D. thesis Karnatak Universiy, Dharwad (India). Pp. 255.

Lakshman, H. C. (2009). Arbuscular mycorrhizal fungi (AMF) a promising biofertilizer for sustainable plant growth. In: proceedings of (NESCST09-CSIR), April 14–16. NAL Bangalore, pp. 19–22.

Manjunath, V.G., Patil, C. P. Swamy, G.S.K and Patil, P. B. (2001) . Effect of different VAM fungi an growth parameters of papaya cv. sunset solovar J. Maharashtra Agri. Uni. 26(3): 269–271.

Mukerji, K. G. Chamola, B.P. and Singh, Jagjit (2000). Mycorrhizal biology. Kluwer academic, NewYork, London and Moscow. Pp. 340.

Pankow, W., Boller, T. and Wiemken, A. (1991). Structure function and ecology of mycorrhizal symbiosis expermentia. 47: 391–394.

Philips, J.H., Hayman, D.S., (1970). Improved procedures for clearing roots and staining parasitic and vesicular-arbuscular mycorrhizal fungi for rapid assessment of infection. Trans. Br. Mycol. Soc. 55, 158–161.

Porter, W.M. (1979). The 'Most probable number' method for enumerating propagules of VAM fungi in soil. Aust. J. Soil Res. 17, 515–519.

Reena, J. and Bagyaraj, D.J., (1990). Response of *Acacia nilotica* and *Calliandra calothrysus* to different VA mycorrhizal fungi, Arid Soil Res. Rehabil. 4, 261–268

Rilling, M.C. and Mummey. D.L. (2006). Mycorrhizas and soil structure, New Phytologist 171 (1): 41–53.

Saif, S. R. (1987). Growth responses of tropical forage plant species to vesicular-arbuscular mycorrhiza I. Growth, mineral uptake and mycorrhizal dependency. Plant and Soil. 97, 25–35.

Sreeramulu, K. R., Gowda, V. N. and Bagyaraja D. J. (1998). Influences of VA mycorrhiza on establishment and growth of Khirni (*Manilkara hexantra* (Roxb.) Dub) seedlings. Current Research, University of Agricultural Sciences, Bangalore. 27(3): 50-52.

Vasanthakrishna, M., Bagyaraj, D. J. and Nirmalnath, J.P. (1995). Selection of efficient VA mycorrhizal fungi for *Casuarina equisetifolia* - second screening. New. For. 9, 157-162.

Vinayak, K., Bagyaraj, D.J. (1990). Vesicular arbuscular mycorrhizae screened for Troyer citrange. Biological and Agriculture Horticulture, 6:303-311.

Wood, T. (1992). Mycorrhizal fungi: Challenges for commercialization. In Handbook of applied mycology. Eds. D.K. Arora, R. P. Elander and K. G. Mukerji Marcel Dekker Inc. New York. Pp. 823-847.

INTERACTION BETWEEN AM FUNGI AND OTHER BENEFICIAL MICRO-ORGANISMS

13

Synergistic Effect Between Arbuscular Mycorrhizal Fungi and *Rhizobium* in Yield and Growth Response of Soybean

Shweta J Sabannavar[1], *H C Lakshman*[1] *and Roopa K Jambotkar*[2]

INTRODUCTION

Pulses and legumes are next to cereals in importance as source of human food. Being rich in protein, they are considered an essential ingredient of a predominantly cereal-based diet. In India, pulses form a very important part of dietary in all the states. Pulses and legumes, which contain more protein than any other vegetable product, are nearer to animal flesh in food value. Besides protein, the content of which varies generally from 20 to 30% carbohydrates and fats are also present in them. Being leguminous crops possessing root nodules, they fix atmospheric nitrogen and thus add up

1. Post Graduate Department and Research in Botany (Microbiology Laboratory), Karnatak University, Dharwad -580 003, India (Email id: shwetsab@gmail.com)

2. J.S.S's Shri Manjunatheshwar P.U.Science College, Vidyagiri, Dharwad-580005

to 30 kg of N per hectare to the soil. Pulses and legumes also contain more calcium and phosphorus than cereals and are a good source of vitamin B.

The area under pulse crops in India is about 17% of the total cultivated area in the country. The pulses and legumes, which belong to the sub-family Papilionaceae, of the great family Leguminosae, have been cultivated and used for centuries all over the world. Not only the seeds, but all other parts of the plant are rich in protein and, therefore legumes are very valuable field and forage crops. They are of short duration and can be easily grown. Further, they fit in very well in crop rotation and crop mixtures. On account of their ability to fix atmospheric nitrogen, they are regarded as restorer of soil fertility. Besides, they serve as excellent fertilizer when ploughed under the soil.

Successful establishment of Nitrogen Fixing Trees (NFTs) in plantation programmes is possible through the production of high quality tree seedlings at nursery level. *Rhizobium* inoculation of NFTs at seedling stage helps in production of healthy stock in the nursery capable of growing successfully when transplanted in field (Ferrera-Cerrato and Villerias, 1985; Rensberg and Strijdom, 1985). The advantage of NFTs includes their ability to fix atmospheric nitrogen, and their potential to transfer some of the fixed nitrogen to succeeding and associated crops. Therefore the contribution of biological nitrogen fixation to the improvement of soil fertility and the influence of NFTs to soil nitrogen management should be given attention (Danso *et al.*, 1992). *Samanea saman* (Jacq.) Merr. is a fast growing avenue tree growing extensively as a shade tree along the road side, large parks and also used in various afforestation programmes (Islam, 1990). Being a leguminous tree, it enhances soil fertility through its association with effective rhizobia and proficient litter fall. The capacity of *Samanea saman* to form nitrogen fixing symbiosis with rhizobial bacteria has been recognized for many years but little information is available on the diversity of rhizobial strains that associate with this species. This necessitates the screening of efficient *Rhizobium* isolate, which would be helpful in selecting the strains to be used in nurseries for production of healthy seedling growth and production plantations.

Phosphorus plays a key role in the balance nutrition of plants. The native soil phosphorus is mostly unavailable because of its low solubility. It's deficiency in soil is met by the addition of phosphatic fertilizers but the efficiency of utilization of the added phosphorus by plants is also very low. According to Mosse (1973), 75% of the phosphorus applied to the crops is not utilized by them. The role of AM fungi in stimulating plant growth, enhanced nutrients and water uptake on waste land is widely recognized (Hiremath *et al.*, 2007). Mycorrhizal inoculation helps in mobilization of unutilized phosphorus and *Rhizobium* helps in fixation of nitrogen resulting in a more efficient acquisition of these nutrients by plants.

While the exploitation of legume, *Rhizobium* symbiosis can provide an economic source of available nitrogen (Ahmad, Saxena *et al.*, 2002) the use of arbuscular mycorrhizal (AM) fungi can improve phosphorus uptake and ultimately plant growth and yield (Barea, and Azcon-aguilar, 1983; Bass 1990; Rodrigues *et al.*, 2003). Phosphorus deficiency is an important limiting factor in nitrogen fixation and legume production (Jacobson, 1985). Legume inoculated with both rhizobia and arbuscular mycorrhizal fungi benefit from 'P' uptake and generally have greater N_2 fixation, nodule mass, N and P content, and yield, than legumes inoculated only by *Rhizobium* (Thiagarajan *et al.*, 1991).

In recent years, quite a good number of studies have been conducted, with interaction between arbuscular mycorrhizal fungi and *Rhizobium* on legumes (Bethlenfalvay, Milford, Brown and Stafford 1987; Lakshman, 1999) and showed that the growth and yield of nodulating soybean increased after inoculation with *Glomus mosseae* in sterilized soil. Inoculation on different crop plants with arbuscular mycorrhizal fungi and *Rhizobium* was found to have synergistic beneficial effect on the nodulation, nitrogen fixation and the plant growth (Mosse, 1977 b; Baas, 1990; Hazarika, Dubey and Phookam 2000; Sampath Kumar and Ganesh Kumar, 2003). The genetic elements for nodulation and autoregulation systems in legumes may be involved in the accommodation of root associated microbes other than rhizobia (Seishi *et al.*, 2008). The present study was undertaken with the objective to assess the response of Soybean (*Glycine max* (L.) Merr.) variety DSb-1 to dual incoluation with arbuscular mycorrhiza and *Rhizobium* and the effect of each organism in the earthen pots using sterilized soils.

MATERIALS AND METHODS

The physico-chemical characteristics of the soil used for pot experiments were tested in soil testing laboratories at Jalavahini Management Private Limited, Dharwad district, India. The results are as follows: Soil was sandy loam with pH 8.1, Conductivity (EC) 320uS/cm, Moisture 4.86%, Total Organic Carbon 1.71%, Available Nitrogen 0.08%, Available Phosphorus 4.52%, Available Potassium 7.94%, Magnesium 0.121%, Calcium 0.472%, Copper 0.03ppm, Zinc 3.86 ppm, Manganese 0.97 ppm, and Iron 8.24 ppm. The soil was steam sterilized for 1hr on 2 consecutive days. The seeds of Soybean (*Glycine max* (L.) Merr.) variety DSb-1 was collected from Soybean research station, University of Agricultural Sciences, Dharwad, India. Seeds were germinated in experimental pots. After germination, uniform seedlings were made one per pot. There were four inoculation treatments:

(1) Uninoculated control

(2) Inoculated with *Rhizobium phaseoli*

(3) Inoculated with the arbuscular mycorrhizal fungus *Glomus fasciculatum* (Thaxt) and Gerd and Trappe.

(4) Inoculated with *Rhizobium* and *Glomus fasciculatum.*

The *Rhizobium* inoculation was done by treating Soybean variety DSb-1 seeds with a peat based culture 10^{-8} cells ml^{-1} before sowing. The *Glomus fasciculatum* was maintained for 12 weeks in greenhouse conditions using Jowar (*Sorghum vulgare* L.) as host for mass multiplication in 30 cm diameter pots containing sterilized sand-soil mix (1 : 1 v/v) was used as inoculum. A thin layer of inoculum (*Glomus fasciculatum*) was placed 2 cm below the soil surface consisted 10 g roots and soil contained 100–150 chlamydospores was inoculated into the earthen pots. A control without bio-inoculant was maintained. There were three replications for each treatment and plants were watered every alternate day. At 40 and 80 days of sowing, growth of the plant height, number and dry weight of nodules, root and shoots was measured. Phosphorus content in the shoot was determined colourimetrically by the Vanadomolybdate yellow colour method of Jackson (1973). Total nitrogen content was determined by the microkjeldhal method (Bremner, 1960). Percentage of mycorrhizal infection in the roots was determined by the root slide technique (Nicolson, 1960) after clearing the roots with 10% KOH and stained with 0.05% trypan blue (Philips and Hayman, 1970).

$$\text{Root colonization (\%)} = \frac{\text{No. of colonized segments}}{\text{Total No. of segments examined}} \times 100$$

The number of *Glomus fasciculatum* spores in the rhizospheric soil was determined by the wet sieving and decanting technique (Gerdemann and Nicolson, 1963) and grain yield were also recorded at the time of harvest.

RESULTS AND DISCUSSION

The data on number of root, dry weight and nitrogen content in nodules of the Soybean variety DSb-1 roots as influenced by *Rhizobium* and mycorrhizal inoculation is given in Table 13.1. The effect of *Glomus fasciculatum* and *Rhizobium phaseoli* on Soybean variety DSb-1 at 40 days is shown in (Figure 13.1). There was no nodulation observed in uninoculated control plants. Similarly, the plants inoculated with the *Glomus fasciculatum* and *Rhizobium* nodulated well. The number, size, dry weight and nitrogen content of nodules in plants inoculated with mycorrhiza (*Glomus fasciculatum*) and *Rhizobium* were significantly greater than those of the plants inoculated with only *Rhizobium*. The number of root nodules favourably increased at 80 days than 40 days after inoculation of *Rhizobium* or *Rhizobium* + *Glomus fasciculatum*. Plants inoculated with *Glomus fasciculatum* showed significantly greater mycorrhizal spores in the rhizospheric

Table 13.1. : Dry Weight and Nitrogen Content of the Soybean Variety DSb-1 root Nodules as Influenced by Inoculation with *Glomus fasciculatum* and *Rhizobium phaseoli*

Treatment	Number of nodules per plants		Dry wt. of Nodule per plant (gm)		Nodule nitrogen per plant (%)	
	40*	80*	40*	80*	40*	80*
Control	0.66± 0.33c	1.00± 0.00c	0.01± 0.00c	0.08± 0.03b	0.00± 0.00b	0.11± 0.00c
Glomus fasciculatum	1.33± 0.66c	2.00± 0.57c	0.20± 0.05c	0.23± 0.08b	0.03± 0.01a	0.05± 0.00b
Rhizobium phaseoli	5.00± 0.57b	11.33± 1.76b	1.27± 0.00b	3.79± 0.12a	0.02± 0.00a	0.07± 0.12b
Glomus fasciculatum+ Rhizobium phaseoli	12.00± 0.57a	17.00± 0.57a	1.85± 0.29a	5.0± 0.77a	0.03± 0.00a	0.12± 0.12a

40* = 40 days after sowing; 80* = 80 days after sowing

Means followed by the same letter are not significantly different at the 0.05 level of confidence. Values represent the mean ± SD.

Fig. 13.1. : Showing Effect of *Glomus fasciculatum* and *Rhizobium phaseoli* on Soybean Variety DSb-1 at 40 days. 1- Control, 2- *Glomus fasciculatum*, 3- *Rhizobium phaseoli*, 4- *Glomus fasciculatum* and *Rhizobium phaseoli*.

soils compared to the uninoculated control plants and plants which received only *Rhizobium* (Table 13.2). The increased spore number were more striking, when the plants were 80 days old than 40 days. The percentage of mycorrhizal infection was higher in the plants which were inoculated with *Glomus fasciculatum and Rhizobium.*

The height of the plant, dry weight of shoots above ground, the grain yield and grain weight are presented in Table 13.3. The differences in the dry weight of the plant shoots were not significant in 40 days old plants. The dual inoculation of *Glomus fasciculatum* and *Rhizobium phaseoli* produced significantly greater plant biomass production compared to uninoculated control produced and plants inoculated with either *Glomus fasciculatum* or *Rhizobium phaseoli*. The grain yield of Soybean variety DSb-1 treated with *Rhizobium* + *Glomus* was significantly greater than that of uninoculated control, and plants inoculated with *Rhizobium* and *Glomus* alone. The nitrogen and phosphorus content of

Table 13.2. : AM Fungal Spores in Soil and Percent of Mycorrhizal Infection of Soybean Variety DSb-1 as Influenced by Inoculation with *Glomus fasciculatum* and *Rhizobium phaseoli*

Treatment	Spore count / 50 gm Soil		Percent of AMF infection (%)	
	40*	80*	40*	80*
Control	0.00± 0.00c	0.00± 0.00c	0.00± 0.00c	0.00± 0.00c
Glomus fasciculatum	57± 0.57a	90.66± 2.90a	46± 0.57b	62.0± 0.57b
Rhizobium phaseoli	43± 1.52b	57.0± 0.57b	0.00± 0.00c	0.00± 0.00c
Glomus fasciculatum+ Rhizobium phaseoli	42± 0.57b	92.0± 0.57a	51.6± 1.45a	67.0± 0.57a

40* = 40 days after sowing; 80* = 80 days after sowing

Means followed by the same letter are not significantly different at the 0.05 level of confidence. Values represent the mean ± SD.

Table 13.3. : Dry Weight of Shoot of the Soybean Variety DSb-1 as Influenced by Inoculation with *Glomus fasciculatum* and *Rhizobium phaseoli*

Treatment	Height of plant (cm)		Dry weight of shoot per plant (gm)		Grain yield per pot (gm)	
	40*	80*	40*	80*	40*	80*
Control	9.00± 0.57c	21.0± 0.57c	1.03± 0.00d	1.82± 0.01d	1.22± 0.00c	1.99± 0.00c
Glomus fasciculatum	11.0± 0.57b	25.0± 1.15b	2.11± 0.00b	3.14± 0.01b	1.82± 0.00b	2.02± 0.00c
Rhizobium phaseoli	9.00± 0.57c	20.0± 0.57c	1.22± 0.00c	2.32± 0.05c	1.81± 0.01b	2.12± 0.00b
Glomus fasciculatum +Rhizobium phaseoli	14.00± 0.57a	30.33± 0.88a	2.41± 0.00a	3.86± 0.02a	2.00± 0.08a	3.07± 0.03a

40* = 40 days after sowing; 80* = 80 days after sowing

Means followed by the same letter are not significantly different at the 0.05 level of confidence. Values represent the mean ± SD.

the above ground shoots at two different intervals are presented in (Table 13.4 and Figure 13.2). There were not

Table 13.4. : Nitrogen and Phosphorus Content of the Shoot of Soybean Variety DSb-1 as Influenced by Inoculation with *Glomus fasciculatum* and *Rhizobium phaseoli*

Treatment	Total nitrogen content per shoot (%)		Total phosphorus content per shoot (%)	
	40*	80*	40*	80*
Control	0.00± 0.00b	0.03± 0.00b	0.02± 0.00b	0.03± 0.01 c
Glomus fasciculatum	0.02± 0.00b	0.03± 0.01b	0.04± 0.01ab	0.13± 0.00b
Rhizobium phaseoli	0.02± 0.01b	0.03± 0.00b	0.04± 0.00a	0.14± 0.01b
Glomus fasciculatum + *Rhizobium phaseoli*	0.07± 0.00a	0.13± 0.00a	0.06± 0.02ab	0.23± 0.01a

40* = 40 days after sowing; 80* = 80 days after sowing

Means followed by the same letter are not significantly different at the 0.05 level of confidence. Values represent the mean ± SD.

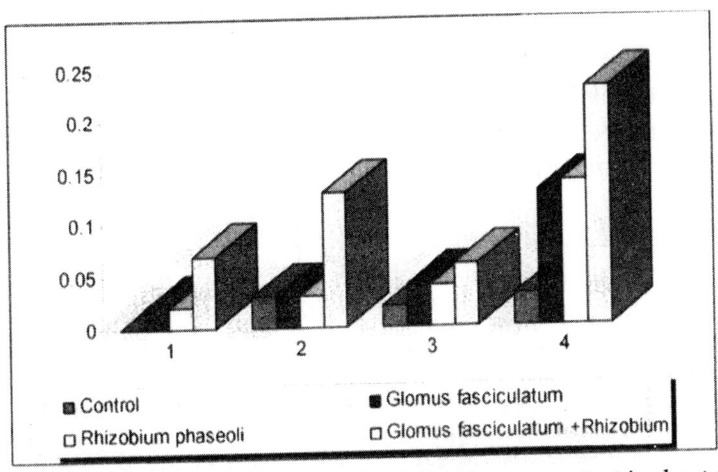

1-Nitrogen content in shoot (at 40 days), 2- Nitrogen content in shoot (at 80 days),

3- Phosphorus content in shoot (at 40 days), 4- Phosphorus content in shoot (at 80 days).

Fig. 13.2. : Showing total Nitrogen and Phosphorus Content in Shoot of Soybean Variety DSb-1 by Inoculation with *Glomus fasciculatum* and *Rhizobium phaseoli*.

much significant differences in 40 days old plants. But in 80 days old plants the nitrogen and phosphorus content of the plants that received *Glomus* and *Rhizobium*, was significantly higher compared to the uninoculated control and plants inoculated with *Glomus* or *Rhizobium* alone. Data were subjected to analysis of variance and mean values were compared using the Duncan's multiple range test (at the 0.05 level), as recommended by Snedecor and Cochran (1980).

Improvement by inoculation with vesicular arbuscular mycorrhiza to the host plant and its growth yield under green house conditions and in soils with low phosphorus has been helpful to obtain better results even by earlier workers (Mosse 1977a; Smith *et. al.*, 1979). Similar results are obtained in the present work, where the sterilized soil used for pot experiments had low phosphorus content. But Manjunath *et al.* (1984) and Thiagarajan (1991) have worked on soybean and cowpea by using unsterilized soil and they concluded that the increased biomass production, nodulation, N and P uptake in mycorrhiza with *Rhizobium* inoculated plants was due to the presence of indigenous arbuscular mycorrhizal fungi. The present investigation supports the work of (Ross and Harper, 1970; Azcon and Rubio, 1990) those two major elements, nitrogen and phosphorus which can be supplied to the host plant by means of symbiotic association. There is also growing evidence that certain endophytes form preferential association with certain host plants (Mosse, 1977b). However, the present investigations do not agree with their findings. Arbuscular mycorrhizal fungi is non-specific to host plants. Therefore inoculation of the introduced endophyte is more efficient to boost in improving the plant growth than the indigenous fungi, *Glomus fasciculatum* (Lakshman, 1999). In present study the inoculation of *Glomus fasciculatum* to Soybean variety DSb-1 which resulted in the increased plant height, dry weight, total phosphorus content of shoot, nitrogen content in nodules and the grain yield compared to the uninoculated control plants.

The better response of Soybean variety DSb-1 to *Rhizobium phaseoli* inoculation could be attributed to the presence of Soybean variety DSb-1 specific rhizobial strain inoculation, in the sterilized soil pot experiments. The better plant growth in 40 days was due to *Rhizobium* inoculation alone when compared to uninoculated control plants and was not statistically significant. While the grain yield at the time of 80 days harvest was significantly greater, suggesting that the crop derived more benefit from fixed atmospheric nitrogen in later stages of plant growth. Smith and Daft (1977) and Tilak (1993) pointed out that nodule activity of soybean reached the maximum during pod filling and did not decline until pod filling was complete. Similar process may be expected in Soybean variety DSb-1 plants of the present work. This investigation brings out very clearly that arbuscular mycorrhizal fungi (*Glomus fasciculatum*) can greatly increase nodulation and nitrogen fixation on Soybean variety DSb-1 plants grown in pots inoculated with *Rhizobium phaseoli*. Similar results were obtained in Beans using fumigated soil conducted in pot experiments, with double inoculation of endomycorrhiza+*Rhizobium* (Lakshman, 2001; Sampathkumar and Ganeshkumar, 2003) with a single inoculation of *Rhizobium* alone. While the principal effect of mycorrhiza on nodulation is undoubtedly phosphate mediated (Asini *et al.*, 1980), there may be other secondary effects, possibly of hormonal nature (Mosse, 1977b; Peterson and Farquhar, 1994; Andrade, Azcon and Bethlenfalvay, 1995).

CONCLUSION

There is a synergistic or additive effect in introduced arbuscular mycorrhizal fungi and inoculation of *Rhizobium*, resulted in greater yield and biomass production. This type of investigation could be suggested to grow nodulated plants in poor agricultural soils to increase soil fertility and effective strains of rhizobia for treating seeds. Therefore, the present study suggests that an efficiently introduced arbuscular mycorrhizal fungus with *Rhizobium* inoculation can contribute to the efficiency of such system, especially in phosphorus deficient soils.

ACKNOWLEDGEMENT

Authors are indebted to U.G.C, New Delhi for awarding UGC research fellowship in Science and financial support.

REFERENCES

Ahmad, M. H., Eagleshman, A. R. J., Hassouna. (1981). Examining the pot trial for inoculant use with cowpeas in West African soils. *Tropical soils*. 58: 325–335.

Andrade, G., Azcon, R. Bethlenfalvay, G.J. (1995). A Rhizobacterium modifies plant and soil responses to the mycorrhizal fungus *Glomus mosseae*. *Applied Soil Ecology*. 2(3): 195–202.

Asini, Gianinazzi-Pearson, V.Gianinazzi, S. (1980). Influence of increasing soil phosphorus levels an interaction between vesicular-arbuscular mycorrhizas and Rhizobium in soybean. *Canadian Journal of Botany*. 58: 2200–2206.

Azcon, R. and Rubio, R. (1990). Interaction between different VA-mycorrhizal fungi *Rhizobium* strains on growth and nutrition of *Medicago sativa*. *Agriculture Ecology and Environment*. 29:5–9.

Baas, R. (1990). Effect of *Glomus fasciculatum* and isolated rhizosphere micro-organisms on growth and phosphorus uptake of *Plantago major* sp. *Pleiosperma*. *Plant soil*. 124:187–193.

Barea, J. M. and Azcon-aguilar, C. (1983). Mycorrhizas and their significance in nodulating nitrogen fixing plants. *Advances in Agronomy*. 36: 1–54

Bethlenfalvay, G. J., Milford, S., Brown, K. L. M. and Stafford, A. E. (1987). *Glycine Glomus - Rhizobium* symbiosis. *Plant Physiology*. 85: 115–119.

Bremner, J. M. (1960). Determination of nitrogen in soil by Kjeldahl method. *Journal of Agricultural Science*. 55: 11.

Danso, S.K.A.; Bowen, G.D. and Sanginga, N. (1992). Biological nitrogen fixation in trees in agro-ecosystems. *Plant and Soil,* 147: 177–196.

Ferrera-Cerrato, R. and Villerias, S.J. (1985). The VA endomycorrhiza and its effect on the development of three arborus legumes In: Proc. Of the sixth North American Conference on Mycorrhizae. Bend, Oregon, USA. pp. 328.

Gerdemann, J. N. and Nicolson, T. H. (1963). Spores of mycorrhizal Endogone species extracted from soil by wet sieving and decanting. *Transaction of British mycological Society*. 46: 235–244

Hazarika, D. K., Dubey L.N. and Phookam, A. K. (2000). Effects of vesicular mycorrhizal fungi and *Rhizobium* on growth and yield of green gram *(Vigna radiata* L.). *Journal of Mycology and pathology*. 30(3): 424–426.

Hiremath, S.G. and Lakshman, H.C. (2007). Survey of arbuscular mycorrhizal fungi in plants growing in coastal beaches and their effect on *Solanum nigram* L. *Nature Environment and pollution Technology*. 6: 81–84.

Islam, M. (1990). The Flora of Majuli. Published by Bishen Singh and Mahendra Pal Singh, 23 A, Connaught Place, Dehra Dun. pp.122.

Jackson, M.L. (1973). Soil Chemical analysis. TS Delhi Prentice-Hall of India Pvt. Ltd. pp. 498.

Jacobson, I. (1985). The role of Phosphorus in nitrogen fixation by Young Pea Plants (*Pisum sativum). Physiology Plantaram.* 64:190–196.

Lakshman, H. C. (1999). Dual inoculation of Vesicular and *Rhizobium* is beneficial to *Pterocarpus marsupium* Roxb. *Journal of Environment and Conservation*. 5(2): 133–136.

Lakshman, H.C. (2001). Effect of introduced *Glomus fasciculatum* on field bean (*Dolchos Lab Lab,* L.). *Indian Journal of Environment and Ecoplanning*. 113(41): 11–17.

Manjunath, A., Bagyaraj, D.J. and Gopalgowda, H. H. (1984). Dual inoculation of VAM and *Rhizobia* is beneficial to Leucaena. *Plant and Soil*. 78: 445–449.

Mosse, B. (1973). Advances in the study of vesicular arbuscular mycorrhizal. *AM. Rev. Phytophtho.* 11:171–176.

Mosse, B. (1977b). The role of mycorrhiza in legume nutrition on marginal soils. In; *exploiting the legume- Rhizobium symbiosis in Tropical Agriculture* (ed. by J.M Vincent, A.S. Whitney and J. Bose), pp. 275–292. College of Tropical Agriculture, University of Hawaii, U.S.A., Miscellaneous Publications No. 145.

Nicolson, T. H. (1960). Mycorrhiza in Gramineae. II. Development in different habitats particularly sand dunes. *Trans. Brit. Mycol. Soc.* 43: 132.

Peterson, R. C. and Farquhar, M. C. (1994). Mycorrhiza integrated development between roots and fungi. *Mycologia.* 86:311–326

Phillips, J. M. and Hayman, D. S. (1970). Improved procedures for clearing roots and staining parasitic and vesicular arbuscular mycorrhizal fungi for rapid assessment of infection. *Transaction of British mycological Society.* 53:158–161.

Rensberg Jansen-van, H. and Strijdom, B.W. (1985). Effectiveness of *Rhizobium* strains used as inoculants after their introduction in soil. *Applied Environment Microbiology,* 49: 127–131.

Rodrigues, L. A., Martin, M. A. and Salomao, M. S. (2003). Use of mycorrhizas and *Rhizobium* in intercropping system of *Eucalyptus* and *Sesbania* II. Phosphorus uptake and efficiency of use and phosphate - fractions. *Revista Brasileira de Eiencia Do Solo.* 27(4): 593–599.

Ross, P. and Harper, J. A. (1970). Effect of Endogone mycorrhiza on soybean yields. *Phytopathology.* 60: 1552–1556.

Sampathkumar, G. and Ganeshkumar, A. (2003). Effect of AM fungi and Rhizobium on growth and nutrition of *Vigna mungo* L. and *Vigna ungniculata* L. *Mycorrhiza news.* 14(4): 15–18.

Saxena, A. K., Shende, R. and Tilak, K. V. B. R. (2002). Interaction of arbuscular mycorrhiza with nitrogen fixing bacteria. In: Rhizosphere and soils. Eds. A. K. Sharma and B. N. Johri Oxford IBH publishing Co. (Pvt.) Ltd. New Delhi. pp. 311.

Seishi Ikeda, Lynn Esther E. Rallos, Takashi Okubo, Shima Eda, Shoko luaba, Hisayuki Mitsui, and Kiwamu Minamisawa. (2008). Microbial community analysis of field grown soybeans with different nodulation phenotypes. *Applied & Environmental Microbiology*. 74: 5704–5709.

Smith, E., Nicholas, D. J. D. and Smith, F. A. (1979). Effect of early mycorrhizal infection on nodulation and nitrogen fixation in *Trifolium subterraneum* L. *Aus. J. of Plant Physiol.* 65: 305–315.

Smith, S. E. and Daft, M. J. (1977). Interaction between growth phosphate content and nitrogen fixation in mycorrhizal and non-mycorrhizal *Medicago sativa. Aust. J. of Plant Physiol.* 4: 403–413.

Snedecor, G.M. and Cochran, W.G. (1980). Statistical Methods, 7th edn. Iowa State College Press, Ames, IA.

Thiagarajan, T. R., Ames, R. N. and Ahmad, M. H. (1991). Response of cowpea *(Vigna ungniculata)* to inoculation with co-selected vesicular-arbuscular mycorrhizal fungi and *Rhizobium* strains in field trails. *Canadian Journal of microbiology*. 38:573–576.

Tilak, K.V.B.R. (1993). Associative effects of vesicular-arbuscular mycorrhiza with nitrogen fixers. *Proceedings of the Indian National Science Academy*. 59:325–332.

14

Importance of AM Fungi and *Rhizobium* Inoculants on Some Important Leguminous Tree Seedlings

Romana M Mirdhe and H C Lakshman

INTRODUCTION

The depletion of forest cover leading to scarcity of wood, fodder and timber for human beings and animals, coupled with increasing wasteland formation, is a major concern of developing countries. Leguminous trees and shrubs have been suggested as a renewable source of fuel and wood products and are known to have symbiotic association with *Rhizobium* and vesicular arbuscular mycorrhizal (AM) fungi (National Academy of Sciences, 1977, 1979).

Woody legumes are useful for revegetation of water-deficient ecosystems that have low availability of Nitrogen, Phosphorous and the other nutrients. The scarcity of available phosphorous and the imbalance of trace elements in desertified ecosystems actually limit legume establishment

Post Graduate Department of Studies in Botany, Microbiology Laboratory
Karnatak University, Dharwad-580 003, INDIA.
Email id: romeebot@gmail.com

and nitrogen fixation. But, it was found that AM fungal association increase the establishment of legume. Moreover woody legumes exhibit a considerable degree of dependence on mycorrhiza to thrive in stress situations. Arbuscular mycorrhiza is by far more widespread in nature and the most common natural association markers with the nodulated nitrogen fixing legumes. After forming symbiotic association with legume roots, AM fungi develops an extra radical mycelium that links the root and the soil environment and helps the plants to use soil nutrients more efficiently, particularly those that diffuse slowly towards the root surface, such as phosphate and trace elements. In addition, the symbiosis enhances the ability of plant of become established and cope with stress situations (nutrient deficiency, soil disturbance) which are typical in desertified situations.

Most of the land plants require Arbuscular mycorrhizal fungi (AMF) to achieve maximum growth in nutrient deficient soils by an increased uptake of soil phosphorus (P) (Barea and Azcon-Aguilar, 1983). These fungi may have no effect on plant growth in the presence of applied phosphorus in phosphorus deficient soils or when phosphorus supply is not limiting growth of non mycorrhizal plants (Aaron and Ahmed, 1987). Legumes need phosphorus for optimum growth nodule formation and nitrogen (N) fixation (Mosse, 1977). Crush (1974) reported increased nodulation growth in *Centrsema pubescens* and *Stylosantnes guyanensis*. Benificial responses to AM fungi inoculation in sterilized in natural soils have also been obtained on French bean (Daft and El-Giahmi, 1976), Peanut (Krishna and Bagyaraj 1984).

Although man and higher plants live immersed in atmosphere which has 78% nitrogen, they ironically are unable to use that element in its gaseous form. Instead they must rely on other sources to provide them with fixed nitrogen in either in oxidized or reduced form which is suitable for further metabolism. From an agronomic point of view the most important of these organisms are bacteria of the genus *Rhizobium* which live symbiotically within root nodules of leguminous plants (Phillips *et al.*, 1978). Leguminous trees and shrubs are currently being suggested as nodulation and the resources of fuel and wood (Hogberg and Kvarnstrom, 1982; Salo, 1985; Lakshman, 1996).

Albizzia lebbeck (L) Benth is native to India, Burma and it is used to cure many ailments, *Dalbergia sisso* is usually found in arid and semi arid regions are renowned for their fast growth rate even on marginal lands and are popular lands for use in land rehabilitation. *Pongamia pinnata* (L)., is native to humid and sub-tropical environment. Juices from plant are antiseptic. *Tamarindus indica* L. leaves are used for curing malarial fever. It is used in ayurvedic medicine.

Literature survey revealed that very limited studies have been carried between *Rhizobium,* and AM Fungi on tree species, but little information is available as to its symbiotic relationship. Therefore, this paper reports the growth response of four leguminous tree seedlings *Albizzia lebbeck* (L) Benth., *Dalbergia sisso* Roxb., *Pongamia pinnata*(L)., *Tamarindus indica* L., to AM fungal inoculation with *Rhizobium* in pot experiments in unsterilized soil.

Maximum growth of legumes can be achieved by inoculation with compatible and effective strains. The beneficial effect of inoculation with *Rhizobia* and AMF on plant growth in tree species is meager. Hogberg and Kvarnstrom (1982) reported fixation of 110 kg N.ha^{-1} year equivalent by an effective strain of *Rhizobium* with *Dalbergia sisso.* Enhanced growth owing to *rhizobium* and mycorrhizal association has been reported by Roskoski *et al.* (1987) in *Dalbergia sisso* seedlings. The objective of the present study was to evaluate the ability of different isolates of *Rhizobium* to nodulate *Albizzia lebbeck* (L) Benth., *Dalbergia sisso* Roxb., *Pongamia pinnata* (L). and *Tamarindus indica* L .

Ferrere–Cerrato and Villerias (1985) reported that nitrogen content increased higher in plants inoculated with *Glomus* species than in those inoculated only with *rhizobium*. The largest nitrogen content corresponded to the treatment with *Glomus* and *Rhizobium*. Selviraj (1998) reported that in *G. fasciculatum* and *rhizobium* inoculated *Prosopis juliflora* treated with tannery effluent showed a higher dry weight, increased rate of photosynthesis, higher protein content, increased sugar, lipid and amino acid levels, increased activity of enzyme catalase, peroxidase, phosphatase, nitrate reductase, increased level of growth hormones, higher N, P, K and Ca, Fe, Co, Mo levels in root and shoot, than their single inoculations. Also translocation of heavy metals such as Cd, Cr and Zn are highly restricted in extra radical hyphae of AM fungi with dual inoculated treatments, when compared with the single inoculations. Thus the combined symbiotic associated of AM fungi with *rhizobium* are more to the plants when compare with their single symbiotic association.

MATERIALS AND METHODS

Experiments were conducted in earthen pots with unsterile soil, maintained at Department of Botany, Karnatak University, Dharwad. The used sandy loam soil with pH 6.9, Phosphorus deficient (2.6 ppm) available P extracted with NH_4 and HCL) and practically devoid of *rhizobium* nodulation. *Dalbergia sisso* Roxb had an indigenous AMF population 239/50g soil. Earthen pots measuring (25 × 30 cm) in diameter. All the four leguminous tree seeds were surface sterilized with 2% pot hypochlorite and germinated in pots containing sterile sand. One week seedlings were transplanted to already arrange experimental pots. Unsterile soil infested without any inoculation. The following treatments were adapted.

1. Unsterile soil without any inoculation.
2. Unsterile soil infested with *Glomus fasciculatum*

3. Unsterile soil infested with *Rhizobium species*
4. Unsterile soil infested *Glomus fasciculatum* and *Rhizobium*

The *Rhizobium* inoculation was done by treating the seeds with a peat based culture before sowing. Mycorrhizal inoculation was done by placing the seeds over a thin layer of mycorrhizal inoculam (10 g mixed inoculum) at the time of sowing. The mycorrhizal inoculation consists of roots and soil from a pot of culture of *Sorghum bicolor*. Moench. var sudnese (Sudan grass) which was infected with *Glomus fasciculatum* and grown for four months. The inoculum contained hyphae, vesicles, and chlymadospores of *Glomus fasciculatum*. There were four replicates for each treatment. At interval of 180 days after sowing, observation on plant weight, total dry matter, percent root colonization, spore, number was recorded per 50 g soil, phosphorus content of shoot was determined calorimetrically by the Vanadomolybolate phosphoric yellow colour method outlined by Jackson (1973). Total nitrogen was determined by the microkjeldhal method (Bremmer, 1960). Percentage of mycorrhizal root colonization of was determined by the root slide technique (Nicolson, 1960), after clearing the roots with 10% KOH and stained with 0.05% trypan blue in lacto phenol (Phillips and Hayman, 1970). The number of *Glomus fasciculatum* spores in the mycorrhizosphere soil was determined by the wet sieving and decanting technique (Gerdemann and Nicolson, 1963).

RESULTS AND DISCUSSION

The presence of both mycorrhizal fungi and N_2 fixing bacteria stimulated the greatest above ground mean growth however; this growth was not significantly greater than *Glomus*-inoculated plants.

The below ground portion of the plants responded differently to the various treatments tested. The control plants had limited root development, compared with the other treatments (Table 14.1). In general, plant height was proportional to the number of leaves. Dual inoculation with both symbionts resulted in significantly P *Rhizobium* but response was not significantly different from plants infected with *Glomus*. Separate inoculations with *Rhizobium* and *Glomus* produced similar responses in plant height and both treatments produced taller than control.

Inoculation with *Glomus, Rhizobium,* or a combination of *Glomus* and, Rhizobium produced significantly (P = 0.05) more leaves than unioculated control plants. Dual inoculation with *Glomus* and, *Rhizobium* produced more leaves than the control and *Rhizobium* treatments, but there were no differences between plants inoculated with *Glomus* when compared with plants inoculated just with *Rhizobium*. This result is supported with other above ground measurements and confirms similar findings (Bagyaraj *et al.*, 1979). This may be explained by the increased level of plant available N supplied by the symbiotic bacteria present or the formation of nodules. The increased level of available N may have caused a proliferation of roots as a result of N concentrating in the root cells and hastening cell division and elongation. Total root production was significantly (P = 0.05) lesser when *Glomus* was present, as compared with *Rhizobium*. Perhaps this response is a result of mycorrhizal fungi increasing the surface area and thus the efficiency of roots to absorb nutrients, therefore reducing the amount of root material required the plant. There is much evidence in the literature that the presence of mycorrhizae decreases the root and shoot ratio by increasing above ground production and possibly by reducing the heed for below ground production. Daft and El-Giahm (1975), Crush (1974), reported reduced root: shoot ratios when *Glomus* was present, as compared with Rhizobium-inoculated plants. The root : shoot ratios calculated in the present study indicate that ratios decrease with the presence of mycorrhizal fungi.

The presence of *Glomus* or *Rhizobium* has been shown to have a positive influence on the P and N content of infected plants: Mosse 1977 and Bagyaraj *et al.*, 1979). Present results in this study of *Dalbergia sisso* Roxb., show that *Rhizobium* significantly increased N content and *Glomus* significantly increased P content. This response may be a result of very low plant-available P levels in the soil. Phosphorus has been reported by Jacobson (1985) to stimulate root development to a greater extent than shoot. The magnitude of root production by plants inoculated with *Rhizobium*. These plants produced significantly more root mass than all other treatments. Tables (14.1–14.4) show the effect of *Glomus* and *Rhizobium* on experimental plants *Albizzia lebbeck, Dalbergia sisso Pongamia pinnata, Tamarindus indica.* There is an increase in plant height, Dry weight of shoot, dry weight of root, Number of nodules, Nitrogen content, percent colonization, phosphorous content, spore number with dual inoculation of *Glomus* + *Rhizobium,* Fig. (A-D) shows the macerated root sections of the experimental plants, Fig. E(1–3) shows controlled and inoculated plants of *Pongamia pinnata,* Fig. F(1–2) shows the nodulation in controlled and inoculated plant of *Pongamia pinnata,* Fig. G (1–2) shows controlled and mycorrhiza inoculated plant of *Tamarindus indica* and Fig. H (1–3) shows controlled, mycorrhiza and mycorrhiza + *Rhizobium* inoculation in *Tamarindus indica.*

Nitrogen levels in shoot materials were significantly (P = .05) higher in *Rhizobium* inoculated plants when compared with plants that with non infected (control) or infected with *Glomus.* Presence of both symbionts increased the N level in the aboveground plant material only slightly over the single inoculation with *Rhizobium*. The response of inoculated plants in accumulating P in shoot material is also recorded. *Rhizobium*-inoculated plants represent, respectively, 21 and 22% increase in P content in above ground foliage. Plants infected with *Glomus* had higher levels of P in than *Rhizobium* inoculated plants. Dual inoculation resulted in the highest

Table 14.1. : Showing Effect of AMF Inoculation and *Rhizobium* with Regard to Plant Growth, Dry Weight of Shoot, Dry Weight of Root, Number of Nodules, % of AMF Colonization and P Uptake in Shoots of *Albizzia lebbeck* at 180 Days

Treatment	PH (cm)	DWS (gm)	DWR (gm)	No. Nod	SN	PC	P %	N %
Control	14.7	3.4	2.2	10.1	30.5	55.3	0.02	26.42
AMF	34.8	7.8	12.5	15.4	45.41	71.2	0.08	45.41
Rhizobium	31.7	6.5	9.8	13.6	43.6	73.6	0.06	42.1
Rhizobium +AMF	48.5	9.3	14.6	17.9	49.4	74.5	0.14	49.4
CD 0.05 %	15.11	6.21	4.12	11.01	5.11	0.43	3.34	45.0

PH = plant height ; DWS= Dry weight of shoot ; DWR= Dry weight of root; No. Nod= Number of nodules; SN= spore number; PC= percent colonization

P = phosphorous content ; N= Nitrogen content

Table 14.2. : Showing Effect of AMF Inoculation and *Rhizobium* with Regard to Plant Growth, Dry Weight of Shoot, Dry Weight of Root, Number of Nodules, % of AMF Colonization and P Uptake in Shoots of *Dalbergia sisso* at 180 Days

Treatment	PH (cm)	DWS (gm)	DWR (gm)	No. Nod	SN	PC	P %	N %
Control	11.6	2.9	1.2	12.1	40.2	65.0	0.09	21.46
AMF	35.1	3.8	1.9	16.5	48.41	68.2	0.12	38.49
Rhizobium	32.9	4.5	1.7	15.6	45.9	75.8	0.11	45.7
Rhizobium +AMF	48.5	5.4	2.6	18.0	52.0	79.5	0.19	48.1
CD 0.05 %	11.11	6.29	5.11	17.01	8.11	68.9	0.56	36.8

PH = plant height; DWS= Dry weight of shoot; DWR= Dry weight of root; No. Nod= Number of nodules; SN= spore number; PC= percent colonization

P = phosphorous content; N= Nitrogen content

Table 14.3. : Showing Effect of AMF Inoculation and *Rhizobium* with Regard to Plant Growth, Dry Weight of Shoot, Dry Weight of Root, Number of Nodules, % of AMF Colonization and P uptake in Shoots of *Pongamia pinnata* at 180 Days

Treatment	PH (cm)	DWS (gm)	DWR (gm)	No. Nod %	SN %	PC	P	N
Control	14.6	4.3	2.2	3.5	11.2	33.9	0.09	0.05
AMF	25.1	6.7	2.6	4.3	15.4	45.2	0.15	0.10
Rhizobium	22.2	6.4	2.4	4.1	14.9	47.1	0.17	0.15
Rhizobium +AMF	27.7	7.7	2.9	4.8	17.7	48.0	0.19	0.18
CD 0.05%	9.11	6.28	4.12	11.01	7.11	0.43	0.67	0.78

PH = plant height; DWS= Dry weight of shoot; DWR= Dry weight of root;
No. Nod= Number of nodules; SN= spore number; PC= percent colonization
P = phosphorous content; N= Nitrogen content

Table 14.4. : Showing Effect of AMF Inoculation and *Rhizobium* with Regard to Plant Growth, Dry Weight of Shoot, Dry Weight of Root, Number of Nodules, % of AMF Colonization and P Uptake in Shoots of *Tamarindus indica* at 180 Days

Treatment	PH (cm)	DWS (gm)	DWR (gm)	No. Nod	SN	PC	P %	N %
Control	15.7	3.9	2.4	2.8	15.8	48.1	0.4	0.21
AMF	24.7	4.7	3.6	3.5	17.6	54.8	0.9	0.30
Rhizobium	23.1	4.2	3.4	3.2	16.4	51.1	0.6	0.27
Rhizobium +AMF	29.4	5.1	4.2	3.8	19.2	57.6	0.16	0.32
CD 0.05%	11.11	5.29	5.18	17.01	7.11	0.43	0.24	0.45

PH = plant height; DWS= Dry weight of shoot; DWR= Dry weight of root;
No. Nod= Number of nodules; SN= spore number; PC= percent colonization
P = phosphorous content; N= Nitrogen content

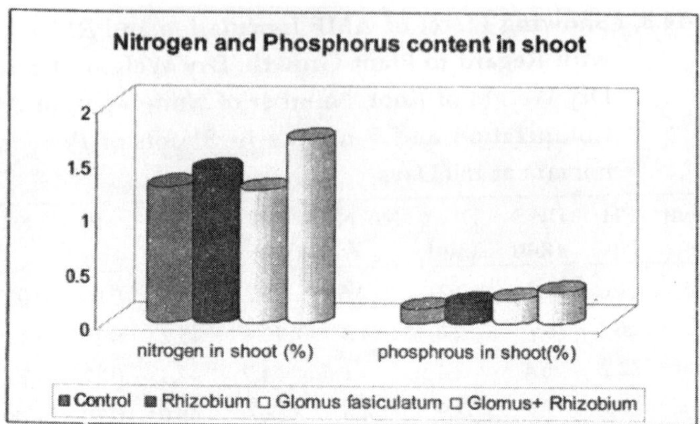

Fig 14.1.:Showing the Effect of AM Fungus and *Rhizobium* Interaction with Regard to Nitrogen and Phosphorous Content in Shoot of *Dalbergia sisso*

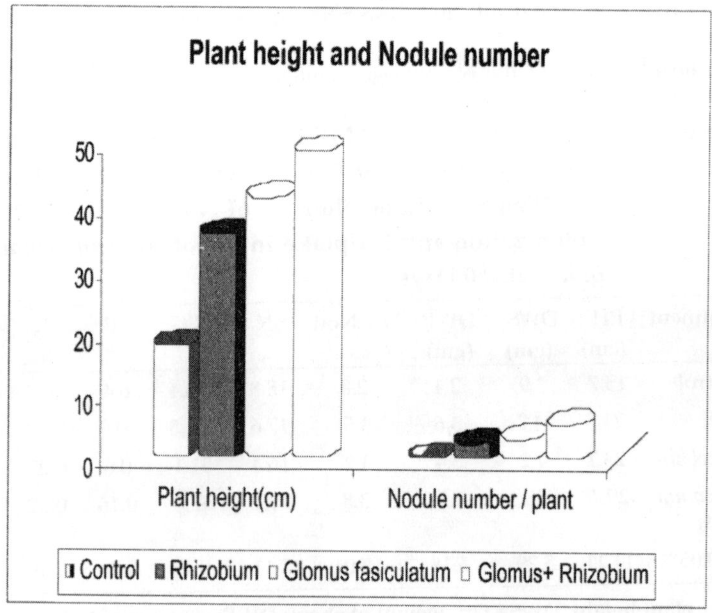

Fig 14.2.:Showing the Effect of AM Fungus and *Rhizobium* Interaction with Regard to Plant Height and Nodule Number of *Dalbergia sisso*

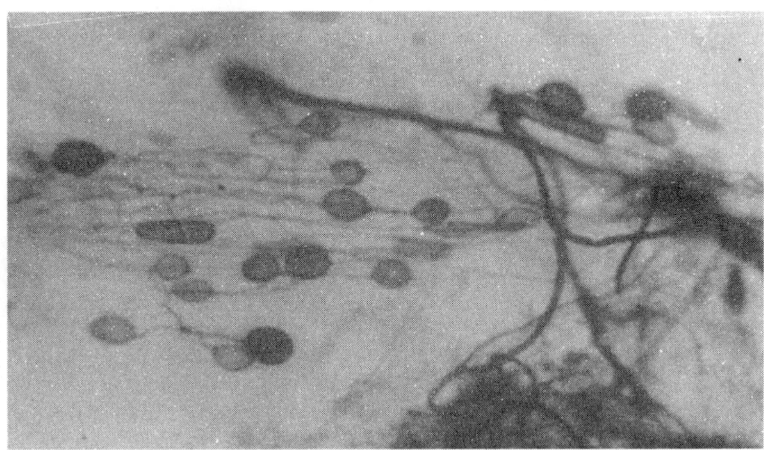

Fig. A. :Macerated Root Section of *Albizzia lebbeck* Showing Group of Vesicles, Hyphae and Arbuscules after AMF Inoculation (Not in Scale)

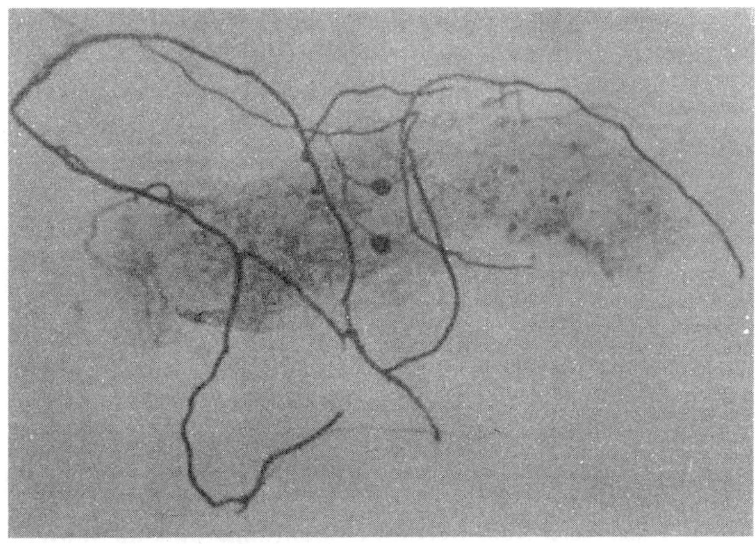

Fig. B. : Macerated Root Section of *Albizzia lebbeck* Showing the Hyphae is Separated from the Root (not in Scale).

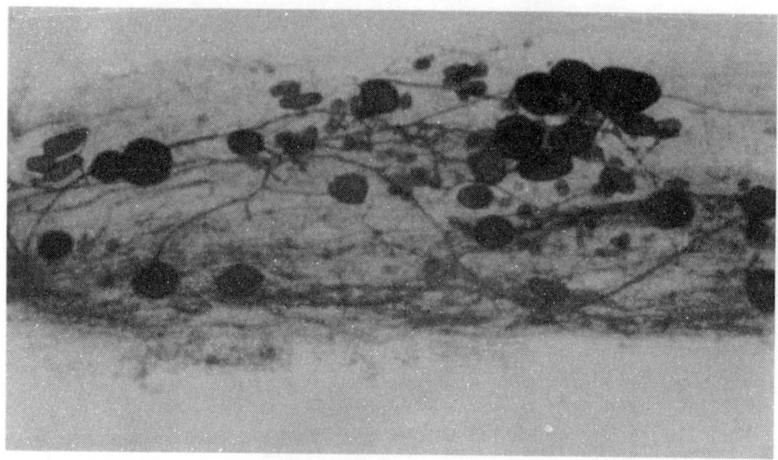

Fig. C.: Macerated Root Section of *Dalbergia sisso* Showing Group of Vesicles, Hyphae and Arbuscules after AMF Inoculation (Not in Scale)

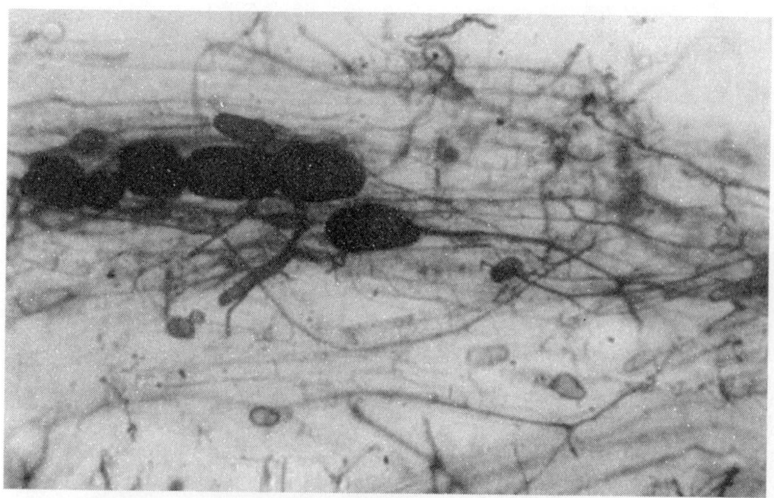

Fig D.: Macerated Root Section of *Pongamia pinnata* Showing Group of Vesicles, Hyphae and Arbuscules after AMF Inoculation (Not in Scale)

Fig E. : 1. *Pongamia pinnata* Plant Inoculated with AMF +*Rhizobium*

2. *Pongamia pinnata* Plant Inoculated with Only AMF

3. *Pongamia pinnata* Plant Uninoculated (Control)

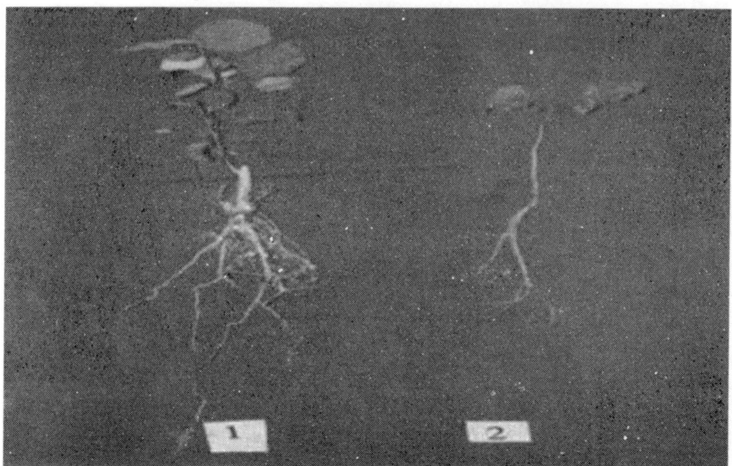

Fig F. : 1. *Pongamia pinnata* Showing Root Nodules in Root System after Inoculation with AMF + *Rhizobium*

2. *Pongamia pinnata* Showing Root Without Root Nodules in Control

Fig. G. 1.: Non-inoculated Plants of Tamarindus indica (Control)
2. Mycorrhiza Inoculated of Tamarindus indica

Fig. H. : 1. Non-inoculated Plants of *Tamarindus indica* (control)
2. Mycorrhiza Inoculated of *Tamarindus indica*
3. Inoculated with Mycorrhiza+ *Rhizobium* (*Tamarindus indica*)

P level. In unsterilized soils, the positive growth responses could in part be due to the low inoculum potential or effectiveness of the introduced mycorrhizal fungi (Lakshman and Ratageri, 2005). Dual inoculation of VA mycorrhizae and *Rhizobium* beneficial to *Pithecollobium dulce*, Roxb., has also shown that *Glomus fasciculatum* was very efficient strain at stimulating uptake, plant growth and nodulation of several legumes in many unsterile, tropical acid soils. While the principal effect of mycorrhiza may have other secondary effects, possibly of hormonal nature (Mosse, 1977b). In conclusion the present study brings out that an effective AM fungus *Glomus fasciculatum* with *Rhizobium* inoculation could contribute to the efficiency of such a system, especially in phosphorous deficient soils, even though native endophytes may be present.

CONCLUSION

The dual inoculation of AM fungi and *Rhizobium* play important role in the production of rootlets in the all four legume trees *Albizzia lebbeck, Dalbergia sisso, Pongamia pinnata* and *Tamarindus indica*. The results clearly demonstrated that *Glomus fasciculatum* inoculation significantly improved in phosphorous uptake in all four leguminous tree species as the *Rhizobium* influenced in the uptake of nitrogen. However, the inoculation of *Glomus fasciculatum* with *Rhizobium* brought out optimum 22% increase in phosphorous in above ground foliage. Therefore, it may be concluded, in unsterilized soil, the positive growth response could result due to low inoculum potential. Thus, the introduced efficient strain of AM fungi with *Rhizobium* could bring promising nodulation in unsterile tropic soils of leguminous trees.

REFERENCES

Aaron, S., and Ahmed, M. H. (1987). Growth and Survival of Cowpea rhizobia in bauxite silt loam and sand clayloam soils. *FEMS microbial. Ecol.*45: 77–83.

Bagyaraj, D.J., Manunath ,A. and Patil, R.B. (1979),. Trans. Brit. Mycol. soc. 72: 164-165.

Barea, J.M., and Azcon., Aguilar, C. (1983). Mycorrhiza and their significance in nodulation nitrogen plants. *Adv. Agron*. 36: 1–54.

Bremmer, J.M. (1960). Determinations of nitrogen soil by Kjeldahl method. *Journal of Agricultural science* 55, 11.

Crush, J R. (1974). Plant growth responses to AM fungi .VII. Growth and nodulation of herbage legumes. *New Phytol*. 73: 743–749.

Daft, M.J. and EL-Ggrahmi, A A. (1976). Studies on nodulation and Mycorrhizal peanuts. *Ann. Appl . Biol*. 83: 273–276.

Ferrera-Cerrato, R. and Villerias, S.J. (1985). The VA endomycorrhiza and its effect of development of three arborous legumes, in:Proceeedings of the sixth North American Conference on Mycorrhizae held at Bend, Oregon, USA, pp. 328.

Gerdemann, J.W. (1973). VA mycorrhizae. In development and functions of roots Ed by J.B.Torry and D.T. Clarkson. pp. 575–591. Academic Press, London.

Hogberg, P. and Kvarnstrom, M., (1982). Nitrogen fixation by the woody Legume *Leucaena leucocephala* in Tanzania. *Plant soil*. 66:21–28.

Jacobson, I. (1985). The role of phosphorous in nitrogen fixation by young pea plants (*Pisium sativum*). *Physol Plant*. 64:190–196.

Jackson, N.E., Franklin, R.E. and Miller, R.H. (1972). Effects of vesicular arbuscular mycorrhizae on growth and phosphorous content of three Agronomic crops. Proceeding of the soil science, Society of America, 36: 64–73.

Krishna, D.R. and Bagyaraj. D.J. (1984). Growth and nutrient uptake of peanut inoculated with the mycorrhizal fungus *Glomus fasciculatum* compared with non-inoculated ones. *Plant and soil*.77: 405–408.

Lakshman, H.C. (1996). VA-mycorrhizal studies in some important timber tree species. Ph.D. thesis, Karnataka University, Dharwad, India. pp. 255.

Lakshman, H.C. and Ratageri. (2005). Dual inoculation of VA mycorrhizae and *Rhizobium* beneficial to *Pithecollobium dulce* Roxb. *J. Microbial world*. 7(2):163–168.

Mosse, B. (1977). The role of mycorrhiza in legume nutrition on marginal soils. In: Exploiting the legume–*Rhizobium* symbiosis in tropical Agriculture. Eds. H. J. Vincent and A S Whitney and J. Bose. College of Tropical of Agriculture university of Hawaii, U.S.A. Miscellaneous Publications No.145. pp 275–292

Nicolson, T. H. (1974). Vesicular-Arbuscular mycorrhiza—a universal plant symbiosis. Sci.Prog. Oxford 55, 561–581.

Phillips, D.A., Rains, D.W., Valentine, R.C. and Huffaker, R.C. (1978). Enhanced biological nitrogen fixation. *Cereal foods world.* 23(1):26–26.

Salo, L. (1985).Competition between strains of *Rhizobia* for nodulation of *Leucaena leucocephala* Masterthesis. Uni-Arizona Tucson. AZ. pp. 109.

Selviraj, T. (1998). Studies on mycorrhizal and *rhizobial* symbiosis on tolerance of tannery effluent treated *Prosopis julifera,* Ph.D thesis, University of Madras, Chennai, India,. pp. 208.

15

The Effect of Indigenous Introduced AM Fungi and its Interaction Between *Rhizobium* for Improving of *Cajanus cajan* (L.) Millsp. (Pigeon Pea)

[1]*H C Lakshman*

INTRODUCTION

The success of green revolution depends upon the availability of fertilizers, high yielding variety of seeds, improved agronomic practices and timely availability of water. Bioinoculants are the preparations containing cells of micro-organisms which may be nitrogen fixers, phosphate solubilizers, sulphur oxidisers or organic matter decomposers. In short, they are called bio-inoculants which on supply to plants improve their growth and yield. Biofertilizers are of environment friendly, non-bulky and low cost agricultural inputs. Micro-organisms can be used as biofertilizers to stimulate plant growth by providing necessary nutrients as a result of their colonization at the rhizosphere (Rilling and Mummey, 2006). The high protein content is related with the

1. Post Graduate Deptt. Studies in Botany, (Microbiology Lab), Karnatak Univ., Dharwad-580 003, India. Email id: dr.hclakshman@gmail.com

presence of root nodules of legume containing nitrogen fixing bacteria.

In many tropical soils scarcity of available phosphate is an important factor which affects both plant growth and nodulation. It is well known in practice that an adequate phosphate supply is necessary for satisfactory nodulation and symbiotic nitrogen fixation (Lakshman, 2009). Nodules often contain 2-3 times more phosphorus per unit dry matter. The rate of fixation per g of nodule can be greatly increased by phosphorus.

Cajanus cajan (L.) Millsp. (Pigeon pea) is one of the important crops of India. Since, it is generally grown as mixed crop. It is grown in almost all states of the country. Seeds of this Pigeon is good source of human food. The green leaves and plant tops may be used as a animal fodder. The husks of pods and seeds are regarded as the valuable cattle feed. The dry plants left after threshing are used for making baskets. Research on this plant is very meager. Therefore, the investigation has been undertaken by inoculating AMF with Rhizobium for its yield improvement.

MATERIALS AND METHODS

Field survey work was carried in different locations of field bean grown centers of Dharwad (Agriculture University of Sciences). The geographical location of studied area situated in between 15°20 to 15°28 North latitude and 75° to 75°70 East longitude. Minimum eight rhizospheric soil samples and twenty-five roots were collected from each individual plants growing in the fields. A summary of the physico-chemical analytical details of the field soil consist of pH: 6.8, organic matter: 1.72(%), total Nitrogen: 810 (ppm). Total phosphorus: 491(ppm) and total potassium: 563 (ppm). Randomly selected root samples were cut into 1cm segments and cleared with 10% KOH and strained with 0.05% tryphan blue in lactophenol, following the procedure of Phillips and Hayman (1970). The percentage of root colonization was determined according to Giovanetti and Mosse (1980). The number of AMF spores per 50 g. Soil was calculated by adapting the procedure of wet-sieving and decanting technique [Gerdemann and Nicolson (1963)]. Identification of AM fungal species was done following the keys suggested by (Morton 1988; Schenck and Perez, 1990).

The soil used in pot experiments was phosphorus deficient 2.4 ppm available extracted with NH_4F+HCL sandy loam, pH 6.8. Two days old seedlings were transplanted in pots containing 5 Kg soil sterilized in 5% methyl bromide. Four triplicate pots were maintained in glass house at 25–27 temp., watered alternatively. *Rhizobium* inoculation was done by treating the field bean seeds with a peat based culture before sowing. Mycorrhizal inoculum (introduced/ indigenous) was applied to the planting hole in the corresponding pot and it consists of spores, hyphae and infected fragments thoroughly homogenized and divided into similar aliquots. The inocula containing mycorrhizal spores of *Glomus mosseae* and *Glomus fasciculatum* approximately 148 AM spores/ 50 g soil were taken. There were five inoculation treatments : 1) Non-mycorrhizal (control), 2) Inoculated with *G. mosseae* introduced AM Fungus 3) Inoculated with *G. fasciculatum* indigenous AM Fungus 4) Inoculated with *G.mosseae* + *rhizobium* 5) Inoculated with (*G. fasciculatum* + *rhizobium*). At two intervals, i.e., 45 and 60 days after sowing observation on the number and dry weight of nodules and dry weight of shoots were taken. Bean yield and 100-seeds weight were recorded at the time of harvest. Phosphorus content of the shoot was determined colorimetrically by the vanadomolybdate / phosphoric-yellow colour method cut lined by Jackson (1973). Total nitrogen determinations were made by the microkjeldahl method (Bermner, 1960). Percentage of mycorrhizal colonization of roots, clearing of roots and the number of *G. fasciculatum* spores in the surrounding the roots were determined according to the procedures as already mentioned earlier.

RESULTS AND DISCUSSION

Different arbuscular-mycorrhizal fungal spores were isolated and identified, in different field bean grown sites. The number of nodules per each plant was recorded. The results showed that the alkaline soil pH 7.6 don't influence much in (the production of nodules, though there was many mycorrhizal spores associated in their rhizoshperic soils (Table 15.1). Fourteen AMF spores were isolated and identified. The most dominated genera and species were listed against each site. Important VAM-spores were i.e. *Glomus mosseae, Gigaspora margrita, Acaulospora aurigloba, scutellospora geospora, Glomus velum, Glomus monosporum, Glomus constrieturn, Glomus aggregatum, Glomus macrocarpum, Gigaspora globosa, Glomus citricolum, Glomus fasciculatum, Gigaspora margarita, and Sclerocystis clavispora.* The percentage of mycorrhizal colonization was varied in different sites. And thus the intensity of mycorrhizal colonization in the field (sites) was not much as that observed under glass house. This may be due to the pattern of root growth in the field, which might have easily outgrown with different mycorrhizal fungi. Plants inoculated with *Glomus fasciculatum* plus *Rhizobium phaseoli* showed a significantly greater plant height (six and half fold). Per cent mycorrhizal colonization and spore number was higher compared to the non-mycorrhizal plants, and the plants received only introduced *G. mosseae* plus *Rhizobium phaseoli* (Table 15.2). These increased plant height were more striking when the plants were 60 days old than 45 days. The dry weight of shoots, bean yield and seed weight was also greater in plants inoculated with indigenous *G. fasciculatum* plus *Rhizobium phaseoli* (Table 15.3). Number of nodules per plant, nodule dry weight and nitrogen content in the nodules (Table 15.4). The differences in dry weight of the shoots were not significant in 45 days old plants. Dual inoculation with indigenous *G. fasciculatum* and *Rhizobium* produced very significantly increased biomass production compared to non-mycorrhizal plants and plants inoculated with introduced *G.*

mosseae, indigenous *G. fasciculatum* alone. The 100 bean weight of field bean seeds, was significantly greater in plants inoculated with *Rhizobium* plus *G. fasciculatum*. Nitrogen and phosphorus content of the shoot at two different intervals are presented in (Table 15.5). No significant differences could be seen in 45 days old-plants. In 60 days old plants the nitrogen and phosphorus content of the plant which received indigenous *G. fasciculatum* plus *Rhizobium phaseoli* were significantly higher compared to these plants, inoculated with *G. mosseae* plus *Rhizobium phaseoli* than non-mycorrihzal (control) plants and the plants which received only introduced *G. mosseae* or *G. fasciculatum* plus *Rhizobium phaseoli*.

Improvement by inoculation with Arbuscular mycorrhiza in phosphate deficient soil under glass house experiments as Legumes obtained by earlier workers (Crush, 1974; Redente and Reeves, 1981; Cruz *et al.*, 1988; Banwarilal *et al.*, 1990; Singh, 2000). The present studied results suggest that soil conditions that favour the development of *Rhizobium* also increase the AMF colonization and spore population. This may reflect a correlation between Arbuscular mycorrhizae and *Rhizobium*. The intensity of mycorrhizal colonization was varied from 25 to 100% shown (Table 15.1) compared to the glass house experiments. This may be due to the pattern of root growth in the field sites, which might have easily out-grown associated with different endophytes. Barea, Escadero and Azcon Gdeaguilar (1980) worked on *Medicago sativa* and they concluded, the indigenous and introduced endophytes co-operate together to assist the growth of legumes. This is probably due to experiments which might not be conducted in sterilized soil, so the native endophytes in unsterilized soil could be deduced from calculation of growth. However, the experimental data of the present study suggests that indigenous mycorrhiza *(G. fasciculatum)* influence much favourable mycorrhizal colonization, number of spores and nodulation. But, the best response was obtained with combined or dual inoculation of indigenous *G. fasciculatum* +

Table 15.1. : Showing AM Fungi Associated with Field Bean Grown in the Fields of University of Agricultural Sciences, Dharwad

Sites	pH	Nodulation Number/plant	Category of colonization range/ Recorded VAM fungi / 25 g. soil
L1	6.8	09.3	*[a] *G. mosseae, Gigaspora margrita, Acaulospora aurigloba, Scutellospora geospora*
L2	6.9	11.4	*[b] *G. Glomus mosseae, Glomus velum, Glomsus monosporum*
L3	6.4	13.1	*[a] *Glomus constrietum, Glomus aggregatum, Scutellospora geospora*
L4	7.3	0.74	*[a] *Glomus mosseae, Glomus aggregatum, Scutellospora geospora*
L5	7.2	09.2	*[c] *G. mosseae, Glomus margarita, Aculospora denticulata*
L6	7.4	05.4	*[b] *Glomus macrocarpum, Gigaspora globosa, Glomus citricolum*
L7	6.6	12.5	*[a] *Scutellospora reticulata, Acaulospora constritum, G. mossea, G. velum*
L8	6.8	10.3	*[b] *Gigaspora reticulatum, Glomus microcarpum, Gigaspora reticulatium*
L9	6.9	08.4	*[a] *Glomus mosseae, Glomus fasciculatum*
L10	7.3	06.1	*[a] *Glomus fasciculatum, Gigaspora margrita, Sclerocystic clavispora.*
L11	7.1	06.5	*[a] *Glomus fasciculatum, Glomus aggregatum, Acaulospora aurigloba*
L12	6.7	12.4	*[a] *Glomus mosseae, Sclerocystis clavispora, Glomus aggrigatum.*
C.D. at 5% 5.1 6.53			

*Colonization range : a: 0%–25%, b:25%–50%, c:50%–75%, d: 75%–100%

Table 15.2. : Showing Plant Height, Shoot Dry Weight, Per Cent VAM Colonization and Spore Number as Influenced by Introduced *G. mosseae* Indigenous *Glomus fasciculatum* and *Rhizobium Phaseoli* on Pigeon Pea for 45 to 60 days

Treatment	Plant height (cm)		% VAM colonization		VAM spores / 50g. soil	
	45 days	60 days	45 days	60 days	45 days	60 days
NM	7.9 ± 1.0	10.2 ± 1.1				
G. mosseae	13.7 ± 0.0	18.4 ± 0.1	44.2 ± 4.3	45.2 ± 5.0	64.3 ± 1.0	69.0 ± 2.4
G. fasciculatum	24.1 ± 1.0	26.2 ± 2.0	58.3 ± 2.1	61.5 ± 2.0	67.2 ± 2.2	71.3 ±1.0
G.m.+Rhizobium	27.2 ± 5.0	38.6 ± 3.2	59.2 ± 5.1	61.7 ± 5.2	68.5 ± 3.1	72.3 ± 4.2
G.f.+Rhizobium	42.3 ± 1.2	64.1 ± 3.0	72.6 ± 2.3	76.1 ± 3.2	101 ± 4.1	105 ± 2.0
C. D. at 5%	7.3	10.2	14.1	14.7	21.2	24.0

Table 15.3. : Showing Shoot Dry Weight, Bean Yield and 100 Seed Weight as Influenced by Inoculation with *G. mosseae* Indigenous *Glomus fasciculatum* and *Rhizobium phaseoli* on Pigeon Pea for 45 Days.

Treatment	Shoot Dry weight plant(g)		Bean Yield / Plant (g)		100 seeds weight (dry)	
	45 days	60 days	45 days	60 days	45 days	60 days
NM	9.8 ± 0.0	1.98 ± 1.0	5.2 ± 3.0	5.91 ± 4.0	1.3 ± 0.0	2.2 ± 1.0
G. mosseae	2.1 ± 1.0	3.3 ± 1.1	23.4 ± 1.0	42.6 ± 4.3	8.5 ± 3.2	14.2 ± 2.4
G. fasciculatum	3.0 ± 0.0	5.8 ± 2.0	32.1 ± 2.2	58.2 ± 2.1	11.5 ± 7.0	21.2 ± 5.2
G.m.+Rhizobium	7.3 ± 1.2	8.3 ± 5.2	73.5 ± 3.0	76.2 ± 4.1	28.6 ± 5.5	29.0 ± 1.0
G.f.+Rhizobium	10.0 ± 0.0	17.6 ± 2.4	126.2 ± 3.3	132.1 ± 5.2	38.1 ± 3.2	44.5 ± 5.1
C. D. at 5%	4.3	5.3	11.3	13.1	5.5	14.2

Table 15.4. : Number, Dry Weight and Nitrogen Content Root Nodule as Influenced by Inoculation with Introduced *G. mosseae* Indigenous *Glomus fasciculatum* and *Rhizobium phaseoli* on Pigeon Pea for 45 to 60 days.

Treatment	Shoot Dry weight plant(g)		Bean Yield / Plant (g)		100 seeds weight (dry)	
	45 days	60 days	45 days	60 days	45 days	60 days
NM						
G. mosseae		3.1 ± 0.0		0.92 ± 0.0		2.12
G. fasciculatum	1.3 ± 0.0	4.2 ± 3.0	0.76 ± 0.0	1.6 ± 2.2	1.81	3.24
G.m.+Rhizobium	2.1 ± 0.0	9.7 ± 5.0	0.87 ± 0.0	3.9 ± 3.0	3.6	5.71
G.f.+Rhizobium	6.3 ± 1.2	17.5 ± 4.1	3.1 ± 2.2	5.8 ± 1.4	8.69	12.14
C. D. at 5%	3.0	5.2	0.12	0.30	0.06	0.06

Table 15.5. : Showing Nitrogen and Phosphorus Content Shoot as Influenced by Inoculation with Introduced *G. mosseae* indigenous *Glomus fasciculatum* and *Rhizobium phaseoli* on Pigeon Pea for 45 to 60 days

Treatment	Shoot Dry weight plant(g)		Bean Yield / Plant (g)	
	45** days	60* days	45** days	60* days
NM	7.30	10.64	2.16	9.33
G. mosseae	14.31	17.21	8.25	13.72
G. fasciculatum	21.51	31.10	11.89	11.84
G.m.+Rhizobium	41.42	59.72	15.30	17.33
G.f.+Rhizobium	52.20	151.32	26.37	41.64
C. D. at 5%	0.00	0.12	0.07	0.07

**: Not significant, G.m. = *Glomus mosseae,* G.f. = *Glomus fasciculatum*

* : Significant at $p = 0.05$

Rhizobium phaseoli than introduced *G. mosseae* + *Rhizobium phaseoli,* as increased nodule number, dry weight of shoots, increased bean yield and significant increase on total uptake of P and N by the host plants (Table 15.3 to 15.5). Such a significant increase was compared with non-mycorrhizal plants and the plants inoculated with *G. mosseae* or *G. fasciculatum* alone. These results are consistent with other workers (Rao and Tilak, 1986; Shivaprasad and Rai, 1991; Lakshman, 1999). Table 15.4 showed, the gradual increased mycorrhizal colonization in those plants inoculated with *G. mosseae* (introduced) and *G. fasciculatum* (indigenous) irrespective of *Rhizobium* inoculation. It is clear that Modulation by *Rhizobium* species depends on the adequate mycorrhization. It is obvious that the introduction of efficient strain (indigenous mycorrhiza) that may co-operate with introduced Arbuscular mycorrhizal fungi, might lead to the improvement of soils after several harvests. Manjunath *et al.* (1984) have reported a closely similar growth response in *Leucaena leucocephala* when native mycorrhizal fungi were inoculated with *Rhizobium phaseoli* on mugbean plants.

CONCLUSION

Hence, it can be concluded that field bean inoculation with arbuscular mycorrhizal fungi in phosphorus deficient soils was most successful, since it not only improved plant growth and nutrition, but also enhanced the activity of *Rhizobium* applied as inoculant. It is also observed that arbuscular mycorrhizal fungi not penetrate the nodule tissue directly and therefore influence of the rhizobial activity through altered root or rhizosphere environment. Hayman (1986) has emphasized that their studies dealing with mycorrhizas of nitrogen fixing legumes should be taken into account the total dynamics of the system, so that practical applications can be reached.

REFERENCES

Azcon, G., de Aguilar, C., Azcon, R. and Barea, J.M. (1979). Endomycorrhizal fungi and Rhizobium as biological fertilizers for *Medicago sativa* in normal cultivation. *Nature.* 279 : 325–327.

Banwarilal, Adholeya, Alok Kapoor Anita and Khanna, S. (1990). Does dual inoculation really, help increase of biomass of tree Legumes. In: mycorrhizal symbiosis and plant growth (eds.) D.J. Bagyaraj and A. Manjunath proceedings of the National Conference on mycorrhiza. Feb.14–16. Pp 97–98.

Barea, J.M. and Azcon-Aguilar, C. (1985). Mycorrhizas and their significance in nodulating nitrogen-fixing plants. Adv. Agron. 36:1–54.

Bremner, J.M. (1960). Determination of nitrogen in soil by Kjeldahl method. *Journal of Agricultural Science.*

Crush, J.R. (1974). Plant response to vesicular-arbuscular mycorrhiza VII. Growth and nodulation of some herbaceous Legumes. *New phytol.* 73: 743–749.

Cruz, R.E., Delimanalo, M.Q., Aggangan, N.S. and Tambalo, J.D. (1988). Growth of three legume trees inoculated with VA-mycorrhizal fungi and *Rhizobium. Plant and soil.* 108: 111–115.

Gerdemann, J.M. and Nicolson, T.H. (1963). Spores of mycorrhizal endone species extracted from soil by wet sieving and decanting. *Trans. Brit. Mycol. Soc.* 46: 235–244.

Giovanetti, M. and Mosse, B. (1980). An estimation of techniques for measuring vesicular arbuscular mycorrhizal infection in roots. *New phytol.* 64:489- 500.

Hayman, D.S. (1986). Mycorrhizae of nitrogen fixing legumes. *MIRCEN. J.* 2: 121–145.

Jackson, M.L. (1973). Soil chemical analysis Prentice Hall of India, New Delhi.

Lakshman, H.C. (1999). Dual inoculation with Rhizobium and G. mosseae to enhance biomass production and nitrogen fixation in *pterocarpus marsupium.* Nat. Con. 97(1): 105–111.

Manjunath, A., Bagyaraj, D. J. and Gopala Gowda, H. S. (1984). Dual inoculation with VA mycorrhiza and Rhizobium is beneficial to *Leucaena. Plant soil.* 56: 131–135.

Morton, J.B. (1988). Taxonomy of VA-mycorrhizal fungi, classification, nomenclature and identification. *Mycotaxon*. 32: 267–324.

Phillips, J.M. and Hayman, D.S. (1970). Improved procedure for clearing roots and staining plastic and vescular-arbuscular mycorrhizal fungi for rapid assessment for infection. *Trans. Brit. Mycol. Soc.* 55:158–161.

Rao and Tilak (1986). Journal of Microbıology 29: 649-652.

Redente, E.F. and Reeves, F.B. (1981). Interaction between vesicular arbuscular mycorrhiza and Rhizobium and their effects on sweet gum tree growth. *Soil. Sci.* 132(6): 410–415.

Rilling, M.C and Mummey, D.L. (2006). Mycorrhizas and soil structure. New Phytol.171(1): 41–43.

Schenck, N.C. and Perez, Y. (1990). Manual for the identification of VA-mycorrhizal fungi. Synergistic publications. Gainesville. Pp. 287.

Singh, S. (2000). Interaction between VA-mycorrhiza and beneficial micro organisms. *Mycorrhiza news*. 12(3): 1–5.

Sivaprasad and Rai, P.V. (1991). Synergistic association between *Glomus fasiculatum* and Rhizobium species and its effect on pigeonpea *cajananus cajana*. *J. Agri. Sci.* 61(2): 97–101.

16

Interaction Between Arbuscular Mycorrhizal Fungi and *Rhizobium* and their Effects on *Acacia mangium* Willd

M Jayashankara and H C Lakshman***

INTRODUCTION

Acacia mangium L. is an important plant growing upto 1-2 m of an height, distributed in tropical and subtrophical countries. Medicinally roots are used in skin diseases, leaf and fruits are anthelimintic. Seeds contain sennosides, N-memorphidine, anthraquinone, glucoside and anthroquinones. Therefore seeds used in diabetic treatment.

While the exploitation of legume—*Rhizobium* symbiosis can provide an economic source of available nitrogen (Ahmad, Eagleshman and Hassling, 1981; Saxena *et al.*, 2002) the use of arbuscular mycorrhizal (AM) fungi can improve phosphorus uptake and ultimately plant growth and yield

* Department of Microbiology, Mangalore University, P. G. Centre, F. M. Kariyappa College, Madekeri -571201, India

** Post Graduate Department and Research in Botany (Microbiology Lab.) Karnatak University Dharwad -580 003, India

(Barea, and Azcon-aguilar, 1983; Bass 1990; Rodrigues *et al.*, 2003). Phosphorus deficiency is an important limiting factor in nitrogen fixation and legume production (Jacobson, 1985). Legume inoculated with both rhizobia and arbuscular mycorrhizal fungi benefit from 'P' uptake and generally have greater N_2 fixation, nodule mass, N and P content, and yield, than legumes inoculated only by *Rhizobium* (Thiagarajan *et al.*, 1991).

In recent years, quite a good number of studies have been conducted, with interaction between arbuscular mycorrhizal fungi and *Rhizobium* on legumes (Bethlenfalvay, Milford, Brown and Stafford 1987; Lakshman, 1999) and showed that the growth and yield of nodulating soyabean increased after inoculation with *Glomus mossae* in sterilized soil. Inoculation on different crop plants with arbuscular mycorrhizal fungi and *Rhizobium* was found to have synergistic beneficial effect on the nodulation, nitrogen fixation and the plant growth (Mosse, 1977b; Hayman 1986; Baas, 1990; Hazarika, Dubey and Phookam 2000; Sampathkumar and Ganeshkumar, 2003). The present study was undertaken with the objective to assess the response of *Acacia mangium* to dual inoculation with arbuscular mycorrhiza and *Rhizobium* and the effect of each organisms in the earthen pots using sterilized soils.

MATERIALS AND METHODS

Seeds of *Acacia mangium* were surface sterilized in 1% sodium hypochlorite and germinated in sterile sand. Two weeks old seedlings were selected for uniformity and transplanted singly into 15 × 15 cm pots containing 2 kg sandy loam soil and pure sand in the ratio of initial pH of 6.7, organic matter content of 1.4% and an available P content of 2.6 ppm extracted with NH_4 and HCL and air dried, pulverized, passed through a 4 mm sieve and sterilized with 2.5% methyl bromide. There were four inoculation treatments: a) Uninoculated control b) Inoculated with *Rhizobium phaseoli*

c) Inoculated with the arbuscular mycorrhizal fungus *Glomus fasciculatum* (Thaxt) and Gerd and Trappe d) Inoculated with *Rhizobium* and *G. fasciculatum.*

Rhizobium inoculation was done by treating *Acacia mangium* seeds with a peat based culture 10^{-8} cells ml^{-1} before sowing. Mycorrhizal inoculation was done by placing the seeds over a thin layer of mycorrhizal inoculum mycorrhizal consisted of chopped root bits and the soil from a pot culture of Sudan grass, *Sorghum bicolor* (L.) Moench var. Sudanense which was infected by *Glomus fasciculatum* and grown for 4 months at the time of sowing. The inoculum contained hyphae, vesicles, 142 chlamydospores per 50 g soil and arbuscules of *G. fasciculatum.* There were three replications for each treatment and plants were watered every alternate day. At two intervals i.e., after 40 and 80 days of sowing, growth of the plants by height, the number and dry weight of nodules, root and shoots was measured. Phosphorus content of the shoot was determined calorimetrically by the vanadomolybdate/ phosphoric-yellow colour method of Jackson (1973). Total nitrogen content was determined by the microkjeldahl method (Bremner, 1960). Percentage mycorrhizal infection of the roots was determined by the root slide technique (Nicolson, 1960) after clearing the roots with 10% KOH and stained with 0.05% trypan blue (Philips and Hayman, 1970). Number of *G. fasciculatum* spores in the soil surrounding the roots were determined by the wet sieving and decanting technique (Gerdemann and Nicolson, 1963). The grain yield and 100 grain weight were also recorded at the time of harvest.

RESULTS AND DISCUSSION

Data on root nodule number, dry weight and nitrogen content of the *Acacia mangium* roots as influenced by *Rhizobium* and mycorrhizal inoculation is given in Table 16.1. There was no nodulation observed in uninoculated control plants. Similarly, in those plants inoculated with the *G. fasciculatum* plants treated with *Rhizobium* nodulated well. The number, size, dry weight and nitrogen content of nodules in plants inoculated with mycorrhiza *(G. fascipulatum)* and *Rhizobium* were significantly greater than those of the plants inoculated with only *Rhizobium*. The number of root nodules favourably increased at 80 days than 40 days after inoculation of Rhizobium or *Rhizobium* + *Glomus fasciculatum* (Fig. 16.1). Plants inoculated with *Glomus fasciculatum* showed significantly greater mycorrhizal spores in the rhizosphere soils compared to the uninoculated control plants and plants which received only *Rhizobium* (Table 16.2). These increased spore number were more striking, when the plants were 80 days old than 40 days. The percentage of mycorrhizal infection is higher in the plants which were inoculated with *Glomus fasciculatum*.

The height of the plant, dry weight of shoots above ground, the seeds yield and 100-seeds weight are presented in Table 16.3. The differences in the dry weight of the plant shoots were not significant in 40 days old plants. The dual

Table 16.1. Dry Weight and Nitrogen content of the *Acacia mangium* Willd. Root Nodules as Influenced by inoculation with *Glomus fasciculatum* and *Rhizobium phaseoli*

Treatment	Number of nodules per plants		Dry wt. of Nodule per plant (g^{-1})		Nodule nitrogen per plant (mg^{-1})	
	40*	80*	40*	80*	40*	80*
Control	0d	0d	0d	0d	0d	0d
Glomus fasciculatum	3c	6.00c	0.06c	0.10c	1.22c	1.51c
Rhizobium phaseoli	7b	14.33b	0.13b	0.17b	2.07b	2.15b
Glomus+Rhizobium	18b	21.00b	0.24b	0.26b	3.51b	5.61c

40* = 40 days after sowing; 80* = 80 days after sowing

Significant at p=0.05 values not followed by identical letters in each vertical column are significantly different.

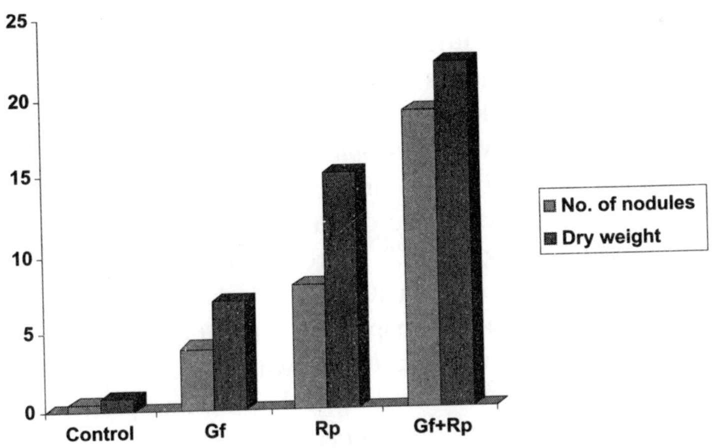

Fig 16.1. : No. of Nodules and Dry Weight as Infl uenced by Inoculation with *Glomus fasciculatum* and *Rhizobium phaseoli* in *Acacia mangium* Willd. for 80 Days

Table 16.2. AM Fungal Spores in Soil and Percent of Mycorrhizal Infection of *Acacia mangium* Willd. Root Nodules as Influenced by Inoculation with *Glomus fasciculatum* and *Rhizobium phaseoli*

Treatment	Spore count / 50 g. soil		Percent of AMF infection	
	40*	80*	40*	80*
Control	0d	0d	0d	0d
Glomus fasciculatum	55.00c	74.00b	70.33b	81.00b
Rhizobium phaseoli	72.33b	62.33c	50.00c	53.00c
Glomus+Rhizobium	82.00a	92.00a	75.00a	87.66a

40* = 40 days after sowing; 80* = 80 days after sowing

Significant at p=0.05 values not followed by identical letters in each vertical column are significantly different.

inoculation of *Glomus fasciculatum* and *Rhizobium phaseoli* produced significantly greater plant biomass production compared to uninoculated control produced and plants inoculated with either *Glomus fasciculatum* or *Rhizobium phaseoli*. The seed yield of *Acacia mangium* treated with

Rhizobium alone or *Rhizobium* + *Glomus* was significantly greater than that of uninoculated control plants and plants inoculated with *Glomus* alone (Table 16.3). The 100 grain weight of *Acacia mangium* seeds was significantly greater in plants inoculated with *Rhizobium* with or without *Glomus fasciculatum*. The nitrogen and phosphorus content of the above ground shoots at two different intervals are presented in Table 16.4 and figure 16.2, there was not much significant differences could be seen in 40 days old plants. In 80 days old plants the nitrogen and phosphorus content of the plants that received *Glomus* plus *Rhizobium*, was significantly higher compared to the uninoculated control plants and plants which received only *Glomus* or *Rhizobium*.

Improvement by inoculation with arbuscular mycorrhiza to the host plant and its growth yield under green house conditions and in soils low in phosphorus has been helpful to obtain better results even by earlier workers (Mosse, 1977a; Smith *et al.*, 1979). Similar results are obtained in the present

Table 16.3. : Dry Weight of Shoot of the *Acacia mangium* Willd. as Influenced by Inoculation with *Glomus fasciculatum* and *Rhizobium phaseoli*

Treatment	Height of plant (cm)		Dry weight of shoot per plant (g^{-1})		Grain yield per pot (g^{-1})	50 Grain wt (g^{-1})
	40*	80*	40*	80*	80*	80*
Control	9.74d	13.74d	0.55d	1.95d	1.46d	1.64d
Glomus fasciculatum	21.30c	32.25c	1.35c	2.76c	41.22c	14.82c
Rhizobium phaseoli	25.47b	44.20b	1.54b	2.95b	77.94b	17.54b
Glomus+Rhizobium	38.31a	52.81a	1.63a	4.75a	95.77a	19.37a

40* = 40 days after sowing; 80* = 80 days after sowing

Significant at p=0.05 values not followed by identical letters in each vertical column are significantly different.

Percentage infection values did not tend themselves to statistical analysis.

Table 4. Nitrogen and Phosphorus Content of the Shoot of *Acacia mangium* Willd. as Influenced by Inoculation with *Glomus fasciculatum* and *Rhizobium phaseoli*

Treatment	Total nitrogen content per shoot (mg)		Total phosphorus content per shoot (mg)	
	40*	80*	40*	80*
Control	2.05d	2.18d	1.05d	1.11d
Glomus fasciculatum	38.12c	74.71b	4.56c	14.37a
Rhizobium phaseoli	42.03b	96.11a	4.73b	11.27b
Glomus+Rhizobium	47.15a	57.11c	5.02a	10.21c

40* = 40 days after sowing; 80* = 80 days after sowing

Significant at p=0.05 values not followed by identical letters in each vertical column are significantly different.

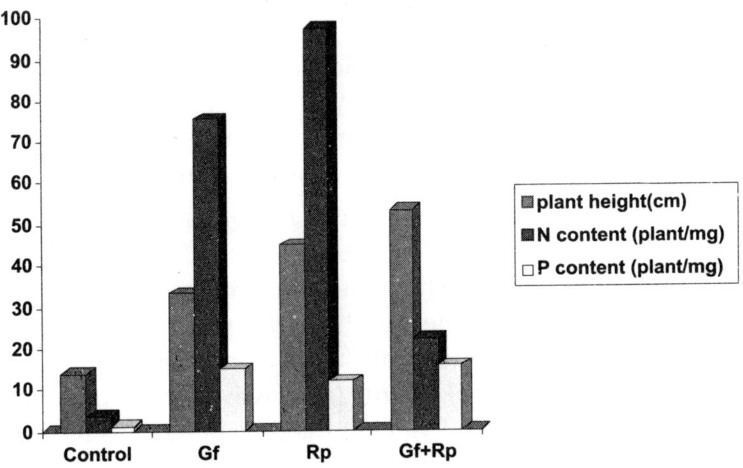

Fig 16.2. Plant Height, Total Nitrogen Content and Total Phosphorus content Per Shoot as Influenced by Inoculation with *Glomus fasciculatum* and *Rhizobium phaseoli* in *Acacia mangium* Willd. for 80 Days

work, where the sterilized soil used for pot experiments had low phosphorus content. But Manjunath *et al.* (1984) and Thiagarajan (1991) have worked on soyabean and cowpea by using unsterilized soil and they concluded that the increased biomass production, nodulation, N and P uptake in mycorrhiza with *Rhizobium* inoculated plants was due to the presence of indigenous arbuscular mycorrhizal fungi. The present investigation supports the work of (Ross and Harper, 1970; Azcon and Rubio, 1990) those two major elements, nitrogen and phosphorus, can be supplied to the host plant by means of symbiotic association. There is also a growing evidence that certain endophytes form preferential association with certain host plants (Mosse, 1977b). However, the present investigations do not agree with their findings. An Arbuscular mycorrhizal fungus is non-specific to host plants. Therefore inoculation of the introduced endophyte is more efficient to boost in improving the plant growth than the indigenous fungi, *G. fasciculatum* (Lakshman, 1999). In present study the inoculation of *G. fasciculatum* to *Acacia mangium* with resulted in the increased plant height, dry weight, total phosphorus content of shoot, nitrogen content in nodules and the grain yield compared to the uninoculated control plants.

The better response of *Acacia mangium* to *Rhizobium phaseoli* inoculation could be attributed to the presence of *Acacia mangium* specific rhizobial strain inoculation, in the sterilized soil pot experiments. The better plant growth in 40 days was due to rhizobium inoculation alone when compared to uninoculated control plants and was not statistically significant. While the seed yield at the time of 80 days harvest was significantly greater, suggesting that the crop derived more benefit from fixed atmospheric nitrogen in later stages of plant growth. Smith and Daft (1977) and Tilak (1993) pointed out that nodule activity of *Acacia mangium* reached the maximum during pod filling and did not decline until pod filling was complete (Fig. 1 and 2). Similar process may be expected in *Acacia mangium* plants of the present work. This investigation brings out very clearly that arbuscular mycorrhizal fungi *(G. fasciculatum)* can greatly increase nodulation and nitrogen fixation on *Acacia mangium* plants grown in pots inoculated with *Rhizobium phaseoli.* Similar results were obtained beans in pot experiments, which were conducted in fumigated soil, with double inoculation of endomycorrhizal plus *Rhizobium* (Lakshman, 2001; Sampathkumar and Ganeshkumar, 2003) with a single inoculation of *Rhizobium* alone. While the principal effect of mycorrhiza on nodulation is undoubtedly phosphate mediated (Asini *et al.,* 1980), there may be other secondary effects, possibly of hormonal nature (Mosse, 1977b; Peterson and Farquhar, 1994; Andrade, Azcon and Bethlenfalvay, 1995), and early nodullins by *Rhizobium* (Brewin, 2004).

It is a common practice to grow nodulated plants on poor agricultural soils to increase soil fertility and effective strains of *Rhizobia* are often used for treating seeds. Therefore, the present study suggests that an efficiently introduced arbuscular mycorrhizal fungus can contribute to the efficiency of such system, especially in phosphorus deficient soils, even though indigenous (native) endophytes may be present.

REFERENCES

Ahmad, M. H., Eagleshman, A. R. J. and Hassouna. (1981a). Examining the pot trial for inoculant use with cowpeas in west African soils. *Tropical soils.* 58: 325–335.

Andrade, G., Azcon, R. and Bethlenfalvay, G. J. (1995). A Rhizobacterium modifies plant and soil responses to the mycorrhizal fungus *Glomus mossae. Applied Soil Ecology.* 2(3): 195–202.

Asini, Gianinazzi-Pearson, V. and Gianinazzi, S. (1980). Influence of increasing soil phosphorus levels an interaction between vesicular-arbuscular mycorrhizas and Rhizobium in soyabean. *Canadian Journal of Botany. 58:* 2200–2206.

Azcon, R. and Rubio, R. (1990). Interaction between different VA-mycorrhizal fungi *Rhizobium* strains on growth and nutrition of *Medicago sativa. Agriculture Ecology and Environment.* 29:5–9

Baas, R. (1990). Effect of *Glomus fasciculate* and isolated rhizosphere micro-organisms on growth and phosphorus uptake of *Plantago major* ssp. *pleiosperma. Plant soil.* 124:187–193.

Barea, J. M. and Azcon-aguilar, C. (1983). Mycorrhizas and their significance in nodulating nitrogen fixing plants. *Advances in Agronomy.* 36: 1–54

Bethlenfalvay, G. J., Milford, S., Brown, K. L. M. and Stafford, A. E. (1987). *Glycine Glomus - Rhizobium* symbiosis. *Plant Physiology.* 85: 115–119.

Bremner, J. M. (1960). Determination of nitrogen in soil by Kjeldahl method. *Journal of Agricultural Science.* 55: 11.

Gerdemann, J. N. and Nicolson, T. H. (1963). Spores of mycorrhizal Endogone species extracted from soil by wet sieving and decanting. *Transacation of British mycological Society.* 46: 235–244

Hayman, D. S. (1986). Mycorrhiza of nitrogen fixing legumes. *MIKCEN.l)l:* 319–357.

Hazarika, D. K., Dubey L.N. and Phookam, A. K. (2000). Effects of vesicular mycorrhizal fungi and *Rhizobium* on growth and yield of green gram *(Vigna radiata* L.). *Journal of Mycology and pathology* 30(3): 424–426.

Jackson, M. L. (1973). Soil Chemical analysis. Prentice Hall of India Pvt. Ltd. pp. 498.

Jacobson, I. (1985). The role of Phosphorus in nitrogen fixation by Young Pea Plants (*Pisum sativum*). *Physiolog run Planta.* 64:190–196.

Lakshman, H. C. (1999). Dual inoculation of Vesicular and *Rhizobium* is beneficial to *Pterocarpus marsupium* Roxb. Journal of Environment and Conservation. 5(2): 133–136.

Lakshman, H. C. (2001). Effect of introduced *G. fasciculatum* on field bean (*Dolchos Lab Lab,* L.) *Indian Journal of Environment and Ecoplanning.* 113(41): 11–17.

Manjunath, A., Bagyaraj, D.J. and Gopalgowda, H. H. (1984). Dual inoculation of VAM and *Rhizobia* is beneficial to *leucaena. Plant and Soil.* 78: 445–449.

Mosse, B. (1977b). The role of mycorrhiza in legume nutrition on marginal soils. In; *Exploiting the legume- Rhizobium symbiosis in Tropical Agriculture* (ed. by J.M. Vincent, A.S. Whitney and J. Bose), pp. 275–292. College of Tropical Agriculture, University of Hawaii, U.S.A., Miscellaneous Publications No 145.

Myers, N. (1988). A wealth of wild species store house of Human Welfare, USA West view press.

Nicolson, T. H. (1960). Mycorrhiza in Gramineae. II. Development in different habitats particularly sand dunes. *Trans. Brit. Mycol. Soc.* 43 : 132.

Peterson, R. C. and Farquhar, M. C. (1994). Mycorrhiza integrated development between roots and fungi. *Mycologia.* 86:311–326

Phillips, J. M. and Hayman, D. S. (1970). Improved procedures for clearing roots and staining parasitic and vesicular arbuscular mycorrhizal fungi for rapid assessment of infection. *Transacation of British mycological Society* 53: 158–161.

Rodrigues, L. A., Martin, M. A. and Salomao, M. S. (2003). Use of mycorrhizas and *Rhizobium* in intercropping system of *Eucalyptus* and *Sesbania* II. Phosphorus uptake and efficiency of use and phosphate-fractions. *Revista Brasileira de Eiencia Do Solo* 27(4): 593–599.

Ross, P. and Harper, J. A. (1970). Effect of Endogone mycorrhiza on soyabean yields. *Phytopathology.* 60: 1552–1556.

Sampathkumar, G. and Ganeshkumar, A. (2003). Effect of AM fungi and Rhizobium on growth and nutrition of *Vigna mungo* L. and *Vigna ungniculata* L. *Mycorrhiza news.* 14(4): 15–18.

Saxena, A. K., Shende, R. and Tilak, K. V. B. R. (2002). Interaction of arbuscular mycorrhiza with nitrogen fixing bacteria. In : Rhizosphere and soils. Eds. A. K. Sharma and B. N. Johri Oxford IBH publishing Co. (Pvt.) Ltd. New Delhi. pp. 311.

Smith, E., Nicholas, D. J. D. and Smith, F. A. (1979). Effect of early mycorrhizal infection on nodulation and nitrogen fixation in *Trifolium subterraneum* L. *Aus. J. of Plant Physiol.* 65: 305–315.

Smith, S. E. and Daft, M. J. (1977). Interaction between growth phosphate content and nitrogen fixation in mycorrhizal and non-mycorrhizal *Medicago sativa. Aust. J. of Plant Physiol.* 4: 403–413.

Thiagarajan, T. R., Ames, R. N. and Ahmad, M. H. (1991). Response of cowpea *(Vigna ungniculata)* to inoculation with co-selected vesicular-arbuscular mycorrhizal fungi and *Rhizobium* strains in field trails. *Canadian Journal of Microbiology.* 38:573–576.

Tilak, K.V. B. R. (1993). Associative effects of vesicular-arbuscular mycorrhiza with nitrogen fixers. *Proceedings of the Indian National Science Academy.* 59:325–332.

Young, C. C., Juang, T.C. and Chao, C. C. (1998). Effects of *Rhizobium* and Vesicular-arbuscular mycorrhiza inoculation, symbiotic nitrogen fixation and soyabean yields in subtropical-tropical fields. *Biology and fertility Soil.* 6: 165–196.

17

Beneficial Effect of Solubilizing Bacteria and Arbuscular Mycorrhizal Fungi on Two Fiber Yielding Plants

Sandeepkumar K and H C Lakshman

INTRODUCTION

Soil fungi and bacteria participate in the process of agricultural importance which includes nitrogen fixation, mineralization, pesticide decomposition, disease resistance, production of growth promoters; antibiotic production, soil system processes, soil formation and other biological processes similarly the phosphate solubalizing bacteria have a capacity to solubilize inorganic/mineral phosphate. Tricalcium phosphate solubilization was demonstrated by soil bacteria in liquid and solid medium for the first time in 1903. Since then many bacteria have been added to the list of P-solubilizers. The important factors influencing P-solubilization under *in vivo* conditions include type of phosphate, soil conditions, environment, plant genotypes and the type of PSB, and it release most of the organic acids during

Post Graduate Department of studies in Botany (Microbiology laboratory)
Karnatak University Dharwad-580003. Email: sandeep.vam@gmail.com

growth plants has been reported (Draff *et al.*, 1974). The PSB are involved not only in P-solubilization but also in release of phytohormones, nitrogen fixation, mineralization and mobilization of other nutrients.

Arbuscular mycorrhiza is the most dominant fungal association in agricultural and horticultural cropping systems. These fungi represent an important component by their ubiquitous in the soil microbial biomass and their direct involvement in essential processes at the plant-soil interface (Bagyaraj, 1984). They belong to the order Glomales and class Zygomycetes (Morton and Benny, 1990). They produce two distinct structures known as vesicles and arbuscular, occurring intracellular that interface with the host cytoplasm and increase the surface area that aids exchange of nutrients between the host and the fungal partner (Lakshman, 1992).

The benefits of AM fungi in agricultural ecosystems are now widely known. Increase in P uptake and growth response due to AM fungi are well documented (Mosse *et al.*, 1981; Hayman, 1983). The increased growth of plants inoculated with AM fungi is not only attributed to improved phosphate uptake but also better availability of other diffusion-limited nutrients like Zn, Cd, Fe and Cu (Krishna and Bagyaraj, 1991).

Studies using ^{32}p labeled tricalcium phosphate and super phosphate by Raj *et al.* (1981) have clearly recorded that AM fungi do not solubilize unavailable form of P but still enhance the P uptake. This was attributed to absorption of P from a larger surface area. Tinker (1982) found better P uptake in mycorrhizal plants when soil was amended with rock phosphate observed that higher number of PSB survived for longer period of time in the rhizosphere of mycorrhizal roots, thus further confirming the observations of Barea *et al.* (1975). They also reported that a Pseudomonas sp. and an *Agrobacterium* sp. having the capacity of phosphate solubilization inoculated on to seeds or seedlings maintained a high population of these bacteria in the rhizospheres of mycorrhizal than non-mycorrhizal roots of lavender and maize. Further, they also found that inoculation of lavender plants with arbuscular mycorrhiza and phosphate solubalizing bacteria recorded increased plant dry matter and P uptake in soils amended with rock phosphate. The increase in PSB could be due to better uptake of nutrients by mycorrhizal plants and better root exudation. They later found that the inoculated PSB also produced growth hormones, which probably increased plant growth. Gopalakrishna (1980) found a synergistic interaction between *Glomus fasciculatum* and the phosphate solubalizing *Aspergillus niger* and *Penicillium funiculosum*, with each organism stimulating the growth of the other, with consequent improvement in growth and P, K, Zn and Mn nutrition of finger millet.

Recent research on plant nutrition through PSB and AM fungi has that these organisms play an important role in uptake nutrient from the marginal soils. Research in the last three and half decades has established that dual inoculation of phosphate solubalizing bacteria and AM fungi inoculation stimulates plant growth. Phosphate solubalizing bacteria soubilize insoluble P and help plants to absorb and translocate more soluble phosphate (Azcon-Angliar *et al.,* 1986). However no research work was directed on *Corchorus capsularis* L. and *Hibiscus cannabinus* L. two important fiber yielding plants hence, the present study was under taken to understand the interaction between phosphate solubilizing bacteria and AM fungi on the growth and P uptake of in green house.

MATERIALS AND METHODS

Corchorus capsularis L. (Jute). Annual plant growing to 3–5 m at fast rate the plant prefers light sandy and medium loamy soils. A fiber obtained from the stem, it is the main source of jute plant. Leaves raw cooked young leaves are added to salads whilst older leaves are cooked as a pot herb tea. Dried leaves are used as appetizer, carminative, demulcent, laxative; stimulant and stomachi infusion is used in treatment of dysentery. A decoction of the roots and unripe fruits is used in the treatment of dysentery. *Hibiscus cannabinus* L. (Pundi). The plant prefers light sandy, medium loamy moist soil. Fiber stands, which are 1.5–3 metres long, are used for making rope, cordage, canvas, sacking, carpet backing, nets, table cloths etc. Young leaves cooked used as a potherb or added to soups, seed are roasted or ground into a flour and made a kind of cake. Roots it is edible but very fibrously oil is obtained from seeds. The juice of the flowers mixed sugar and black paper is used in the treatment of biliousness with acidity the seeds are aphrodisiae promote increase in weight. In ayurvedic medicine, they use in the treatment of dysentery and bilious, blood and throat disorder.

The Green house experiment were conducted in the Post Graduate Department of Botany, Karnatak University Dharwad. *Bacillus polymyxa*, a phosphate solubilizing bacterium, was procured from the Department of Agril. Microbiology U.A.S Dharwad. The AM fungus *Glonus fasciculatum* was multiplied with Sudan grass (*Sorghum valgare* var. sudaneese., as a host plant. Soil based AM fungal inoculation was established and maintained in pot culture at poly house department of Botany.

Two fiber yielding plants seedlings were raised in earthen pots containing 6 kg of sterilized garden soil. The soil separately used in the experiment was a sandy loam a pH of 6.7, EC: 0.13 mmhas/cm, organic carbon: 0.38%, available N: 199 kg/ha, available K:204 kg/ha and available P:4.6 kg/ha. One week old seedlings of *Corchorus capsularis* L. and *Hibiscus cannabinus* L. were inoculated with PSB and AMF, around the root system. PSB (0.5 mg) and mycorrhizal colonized chopped root bits (5 g) and 5 g soil of the host plants which consisted of spores (approximately 150/250 g soil) and external hyphae was used as inoculation singly or in combination. The treatments were as :

1. Un-inoculated (control)
2. *Bacillus polymyxa* alone
3. *Glomus fasciculatum* alone
4. *Bacillus polymyaxa* + *Glomus fasciculatum*

The following observation was recorded on 60 days old seedlings growing in green house. Plant growth parameters such as Plant height, dry weight of shoot and P uptake in shoots was determined. The per cent of AM colonization of *Corchorus capsularis* L. and *Hibiscus cannabinus* L. of roots were estimated according to Phillips and Hayman (1970). The extra metrical chalamydospores were isolated by adopting wet sievings an decanting technique outlined by Gerdemann and Nicolson (1963). Phosphorus in shoots of both the plants was estimated according method outlined by Jackson (1973).

RESULT AND DISCUSSION

The data in treatment of *Corchorus capsularis* L and *Hibiscus cannabinus* L. Plant height, dry matter and P uptake recorded at 60 days of plant growth in both sterilized and unsterilized soils are presented in Table 17.1 and 17.2. *Corchorus capsularis* L plant biomass, presented results of *Corchorus capsularis* L. plant growth, dry matter, of shoot and single single inoculation either AM fungi or PSB did not contribute significantly in enhancing dry matter of shoot or root. However at 60 days harvest, enclosed dry weight of *Corchorus capsularis* L. may be due to increased solubilization P from PSB the activated process of subsequently increased uptake of P by the plant shown in plate-I. Data in Tables 3 and 4 indicate a significant reduction at 60 days in *Hibiscus cannabinus* L. Even though AM fungus inoculation effectively enhanced P and N in shoots of *Hibiscus cannabinus* L. was found to be

Table 17.1. : Effect of PSB and AMF Inoculation on Growth, Dry Matter, and P Uptake of 60 Days Old *Corchorus capsularis* L. in Sterilized Garden Soil

Treatments	Plant height (cm)	Shoot dry weight (g/pl)	% Root colonization	Spore count /50g soil	Root Dry weight (g/plant)	P uptake in shoot (ppm)
Un-inoculated (Control)	10.2	4.1	-	-	0.911	41
PSB	27.2	9.8	25.5	34.2	2.144	99
AMF	27.8	9.9	45.4	58.3	7.188	96
PSB+AMF	44.2	15.0	47.8	60.0	9.234	107
C.D. at (0.05%)	2.21	0.31	0.6	0.01	0.11	0.07

Table 17.2. : Effect of PSB and AMF Inoculation on growth, Dry Matter, and P Uptake of 60 days Old *Corchorus capsularis* L. in Unsterilized Garden Soil

Treatments	Plant height (cm)	Shoot dry weight	% Root colonization (g/pl)	Spore count /50g soil	Root Dry weight (g/plant)	P uptake in shoot (ppm)
Un-inoculated (Control)	7.8	2.7	-	-	0.211	30
PSB	21.5	8.5	21.5	29.0	0.341	62
AMF	24.2	8.2	38.6	41.1	0.356	54
PSB+AMF	41.3	12.6	43.6	52.0	0.375	99
C.D. at (0.05%)	2.11	0.15	0.4	0.01	0.10	0.05

Table 17.3. : Effect of PSB and AMF Inoculation on Growth, Dry Matter, and P Uptake of 60 Days old *Hibiscus cannabinus* L. in Sterilized Garden Soil

Treatments	Plant height (cm)	Shoot dry weight	% Root colonization (g/pl)	Spore count /50g soil	Root Dry weight (g/plant)	P uptake in shoot (ppm)
Un-inoculated (Control)	12.3	5.1	-	-	0.991	45
PSB	29.1	10.4	27.2	36.2	2.544	102
AMF	29.8	10.7	50.3	60.5	7.122	100
PSB+AMF	46.6	17.2	48.6	62.4	9.242	121
C.D. at (0.05%)	2.81	0.45	0.7	0.01	0.12	0.08

Table 17.4. : Effect of PSB and AMF Inoculation on Growth, Dry Matter, and P Uptake of 60 days Old *Hibiscus cannabinus* L. in Unsterilized Garden Soil

Treatments	Plant height (CM)	Shoot dry weight (g/pl)	% Root colonization	Spore count /50g soil	Root Dry weight (g/plant)	P uptake in shoot (ppm)
Un-inoculated (Control)	8.1	2.9	-	-	0.235	35
PSB	23.2	8.8	22.1	30.2	0.358	65
AMF	26.3	8.9	40.0	43.2	0.408	58
PSB+AMF	42.3	13.5	44.1	53.5	0.452	100
C.D. at (0.05%)	2.45	0.23	0.5	0.01	0.11	0.04

Fig. 1

quite higher at 60 days harvest (Shown in Fig. 17.1) there by fungus to grow and produce sufficient amount of phosphate dissolving organic acids which in turn solubilizing P fro PSB (Goenadi *et al.*, 2000). Jones Nirmalnath and Sreenivasa (1993) reported higher plant P concentration in sunflower, when *Glomus fasciculatum and P. striata* was inoculated Prathibha *et al.* (1994) reported beneficial effect of tripartite symbiosis between *Azosprillium sp.* and *Pseudomonas striata* or *Glomus fasciculation* in terms of higher counts of rhizosphere microflora and growth of cotton. Chandrasheker *et al.* (1995a and b) reported AM fungi mediated P effect on growth and yield of sun flower under field conditions. Similar bebefits have been derived in tree species from synergistic interactions between P-solubilizers and AM fungi (Lakshman, 1996; Gurumurthy and Sreenivasa, 2000; Lakshman *et al.*, 2004).

AM colonization, spore population was greatly improved by simultaneous inoculation of PSB and AMF. This kind of variation has been observed in both inoculated and non-inoculated plants is mainly due to microbial interaction (Linderman, 1988; Kadam and Lakshman, 2004).

The bacterium *Bacillus polymyxa* in combination with *Glomus fasciculatum* gave a significant increase in plant height, per cent AM colonization, spore population and P uptake in sterilized than unsterilized soil. The response was higher due to mixed inoculation than single inoculation indicating the synergistic effect of two organisms. The results indicated that dual inoculation could be used for better growth and biomass production of plants enhancement of nutrient uptake (Bagyaraj,1984). Similar interaction between AM fungi and certain rhizospheric microorganisms results of earlier workers (Barea *et al.*, 1975: Sattar and Gaur, 1989) on lavender, maize and lentils.

REFERENCES

Azcon-Anglir, Gianinazzi-Pearson, V., Fardean,J.C. and Gianinazzi, S. (1986). Effect of vesicular arbuscular mycorrhizal fungi and phosphate solubilizing bacteria on nutrition of soybean in a neutral calarious soil amended with $^{32}P_{45}$ Ca-tricalcium phosphate. *Pl. Soil,* 93: 3–15.

Bagyaraj, D.J. (1984). Biological interaction with VAM fungi. In: VA mycorrhiza (Eds.) Powell, C.L. and Bagyaraj, D.J. CRC Press, Boca Raton, Florida, pp.131-153.

Barea, J.M., Azcon, R. and Hayman, D.S. (1975). Possible synergistic interaction between endogone and phosphate soluibilizing bacteria in low phoaphate soil. In: *Endogone and Phosphate* (Eds. F.E. Sanders, B. Mosse and P.B. Tonker) Academic Press, London, New York, pp. 409–418.

Daft, M.J. and Nicolson, T.H. (1990). The Arbuscular mycorrhizae in plants colonizing coal wastes in scotland. New Phytol. 73:1129–1138.

Chandrashakhar, C.P., Patil, V.C. and Sreenivasa M.N. (1995a). VA mycorrhiza mediated P effect on growth and yield of sunflower at different P levels. *Pl. Soil,* 176: 325–328.

Gerdemann, J.W. and Nicolson, T.H. (1963). Spores of mycorrhizal endogone species extracted from soil by wet sieving and decanting. *Trans. Brit. Mycol. Soc.,* 46: 235–244.

Goenadi, D.H., Siswanto, and Sugiarto, Y. (2000). Bioactivation of poorly soluble phoshate rocks with a phosphorus-solubilizing fungus. *Soil Sci. Soc.* Am. *J.,* 64:927–937.

Gopalakrishna, N.N. (1980). Interaction between VA mycorrhiza and phosphate solubilsing fungi and their effect on rhizosphere microflora and growth of fingher millet. M.Sc (Agri) Thesis, University of Agriculture Science, Banglore, India.

Gurumurthy, S.B., and Sreenivasa, M.N. (2000). Effect of VAM fungus and P-solubilizers on Shisham. *Myforest.,* 35: 73–80.

Hayman, D.S. (1983). The physiology of vesicular arbuscular endomycorrhizal symbiosis. *Can. J . Bot.,* 61: 944–963.

Jackson, M.L. (1973). *Soil chemical analysis.* Prentice Hall of India (P) Ltd. New Delhi. Pp. 498.

Jones Nirmalnath, P. and Sreenivasa, M.N. (1993). Response of sunflower to the inoculation of VA mycorrhiza and or phosphate solubilizing bacteria in black clayey soil. *J. Oil Seeds. Res.,* 10: 86–92.

Krishna, K.R. and Bagyaraj, (1991). Role of arbuscular mycorrhiza in the uptake of micronutrients by groundnut plants. Curr. Res., 20:173–175.

Lakshman, H.C., Srinivasalu, Y., Patil, G.B. and Hosamani, P.A. (2004) synergistic effect between PSB and VAM on growth and uptake of "P" of Clitoria ternate L. in proceedings of mineral phosphate solublization . No.14–16 UAS, Dharwad pp159–166.

Lakshman, H. C. (1996). Possible way nutrient distribution by weeds colonized with VA-mycorrhiza world weeds, 11: 72–74.

Lakshman, H.C. (1992). Response and development of Terminallia bellerica Roxb. Infected with VA-mycorrhiza. Tropical forest, 109: 81–86.

Laxman B.kadm and Lakshman (2004). Survey of AMF in some important crop plants and the effect of introduced *Glomus fasciculatum* on Fooxtail millet. Nature Env. And Pollution Teconology, Vol.3(3): pp. 377–381.

Morton, J.B. and Benny, G.L. (1990). Revised classification of arbuscular mycorrhizal fungi (Zygomcetes): A new order Glomales, two new families, *Acaulosporaceae* and *Gigasporaceae* with an emendation of *Glomaceae. Mycotaxon.,* 37:471–491.

Phillips, J.M. and Hayman, D.S. (1970). Improved procedure of clearing and staining of vesicular arbuscular mycorrhizal fungi for rapid assessment of infection. *Trans. Brit. Mycol. Soc.,* 55: 158–161.

Prathibha, C.K., Alagawadi, A.R. and Sreenivasa, M.N. (1994). Rhizosphere microflora and growth of cotton as influenced by combined inoculation of *Azospirillum and Pseudomonas striata or Glomus fasciculatum, J. Soil. Biol. Ecol.,* 14: 11–16.

Raj, J., Bagyaraj, D.J. and Manjunath, A. (1981). Influence of soil inoculation with vesicular arbuscular mycorrhiza and phosphate dissolving bacterium on plant growth and $_{32}$P-uptake. Soil boil. Biochem., 13: 105.

Sayyar, M.A. and Gaur, A.C. (1989). Effect of vesicular arbuscular mycorrhiza and phosphate dissolving microorganisms on the yield and phosphate uptake of *Lentil, Lens esculenta* Moench. Thai. J. Agri. Sci., 22: 129–136.

Tinker, P.B. (1982). Mycorrhizas: The present position. In: Whither soil Research Trans. 12th International Congress Soil Science, New Delhi, pp. 150.

18

Intereactions Between Azotobacter, Pseudomonas and Arbuscular Mycorrhizal Fungi on Two Varieties of Finger Millets

Geeta B Patil and H C Lakshman

INTRODUCTION

The most important components of the P cycle are mineralization, solubilization and immobilization. The processes of organic P mineralization and mineral P solubilization in soils are of great economic importance for P nutrition of crop plants. P solubilizing microorganisms play a major role is making the nutrient available to plants, specially in alluvial soils.

Several, heterotrophics and chemoautotrophic bacteria have the capacity to solubilize insoluble mineral phosphates. Many genera of bacteria such as *Achromobacter, Agrobacterium, Bacillus, Pseudomonas, Serratia* and several others have been reported to solubilize varying quantities of phosphorus depending on the efficiency of the strains. The most efficient

---

P. G. Department of Botany (Microbiology Lab) Karnatak University, Dharwad -580 003 India.

and dominant solubilizers belong to genera Bacillus and pseudomonas.

The rhizosphere is a heterogeneous, continuous and natural habitat in which different types of interactions occur between soil microbes and plants. The beneficial plant-microbe interactions in the rhizosphere are the primary determinants of plant health and soil fertility (Jeffries *et al.*, 2003). Concerted efforts are being made worldwide to develop nutrient use—efficient crop cultivars responsive to biofertilizers to increase crop yield and also to maintain good soil health (Manske *et al.*, 1995). Van der Heijden *et al.* (1998) indicated that plants might select the arbuscular mycorrhizal fungus (AMF) and Vandenkoornhuyse *et al.* (2002) demonstrated that distinct AMF communities are associated with different plant hosts. Synergistic positive interactions have been reported between AMF and plant growth-promoting bacteria (PGPB) such as nitrogen fixers, fluorescent pseudomonads and sporulating bacilli (Hameeda *et al.*, 2007). Plant growth can be stimulated in response to root colonization by bacteria referred to as plant growth-promoting rhizobacteria (Kloepper *et al.*, 1980; Ryu *et al.*, 2003, 2004a,b).

Millet is one of the most important oilseed plants. Millet oil has an excellent nutritional, medicinal, cosmetic and cooking quality for which it is known as 'the queen of oils'. These potent antioxidants are also called 'the seeds of immortality'. With growing health consciousness, the international demand and export of Millet is continuously increasing (Duhoon *et al.*, 2004). Agri-export is an important source of foreign exchange in India. The modernization of agriculture has resulted in the extensive use of chemical fertilizers, indiscriminate use of pesticides, and disruption and degradation of agro-ecosystems. Excessive use of agro-chemicals increases pollution, decreases soil fertility and leads to nutrient imbalance. Integrated use of organic and inorganic fertilizers in a balanced proportion for sustainable production of Millet was emphasized by USDA (1980), Pannase *et al.*, (1995), Tiwari *et al.* (1995), Hegde (1998) and Deshmukh *et al.*, (2002a). Therefore, studies are undertaken to optimize the yield through biofertilizers. The aim of this study was to determine the role played by bacteria associated with AMF in the interaction of AMF with its plant hosts. Plant species play an important role in regulation of the composition and diversity of AM fungal and bacterial communities due to different spectra of their root exudates (Grayston *et al.*, 1998; Eom *et al.*, 2000, Marschner *et al.*, 2001, Johnson *et al.*, 2003). On the other hand, both AMF and bacteria play an important role in the development of the plant community (Eom *et al.*, 2000).

MATERIALS AND METHODS

Soil and plant material

The soil physico-chemical characteristics used for pot experiments were estimated as per Jackson (1973). Per cent of organic matter was determined according to Piper (1950). Electric conductivity is measured using Bridge meter and pH by 1 : 1 (w/v) soil to water ratio. The experimental soil was sandy loam with pH 7, organic matter 0.84 %, nitrogen 1.41 mg kg^{-1}, potassium 2.41 mg kg^{-1}, phosphorus 0.18 mg kg^{-1}, zinc 2.02 mg kg^{-1}, copper 1.04 mg kg^{-1}, magnesium 1.42 mg kg^{-1}, E.C. 10.17 m.mho cm^{-1}. The soil was steam sterilized for 1 h on two consecutive days. Two varieties of Finger millet namely DS 1 and E 8 seeds were collected from the seed production unit, University of Agricultural Sciences, Dharwad, India. Seeds were germinated in small plastic cups containing sterilized soil. After germination, uniform seedlings were transplanted (one per pot).

AM fungi and bacterial isolates

The two AM fungal spores *Glomus fasciculatum* (Thax. sensu Gerd.) Gerd. and Trappe and *Acaulospora laevis* Gerd. &

Trappe were maintained in a poly house using Jowar (*Sorghum vulgare* L.) as host for mass multiplication in 30 cm diameter pots containing sterilized sand-soil mix (1:1 v/v) and was used as inoculum. The bio-fertilizers like *Azotobacter chroococcum* (Ac) and *Pseudomonas fluorescens* (Pf) were collected from the Microbiology Laboratory, UAS, Dharwad, India.

Pot experimental set-up

Pots (15 cm diameter) amended with air-dried sterilized soil-sand mix (3:1 v/v) ratio was used for the experiments. Before transplanting DS 1 and E 8 varieties of Millet seedlings, a thin layer of inoculum—*G. fasciculatum* and *A. laevis* respectively—was placed 2 cm below the soil surface. Ten grams of the inoculum consisted of roots and soil contained 100–150 chlamydospores as single inoculation. In dual inoculation, 3 ml of culture suspension containing 10^{10} cells ml^{-1} of Ac and Pf was inoculated on the inoculum as AMF + Ac and AMF + Pf and in triple inoculation AMF + Ac + Pf was inoculated. A control without bio-inoculants was maintained. A summary of treatments is given in Tables 1 and 2 of both the varieties. After a week, old healthy germinated seedlings of both the varieties were transplanted as a single plant per pot. The experiment was completely randomized with three replications per treatment of the two varieties. The pots were exposed to sunlight and received 10 ml of Hoagland nutrient solution minus phosphate, once in 15 days. To maintain moisture pots were watered every alternate day.

The plants were uprooted periodically after 30, 60 and 90 days. The fresh and dry weight of the shoot and root was recorded. The shoot length, root length and the number of capsules were also recorded. The AM fungal spores were counted after removing them from the soil by wet sieving and decanting (Gerdemann and Nicolson, 1963). The percentage of mycorrhizal infection of the roots was determined (Phillips and Hayman, 1970). Phosphorus content in the shoots was determined in terms of percentage in the plant sample (Jackson, 1973).

Statistical analysis

Data were subjected to analysis of variance and mean values were compared using the Duncan's multiple range test (at the 0.05 level), as recommended by Snedecor and Cochran (1980).

RESULTS AND DISCUSSION

Response to AMF, Ac and Pf on Finger millet var. DS 1 and E 8 had beneficial effect on growth. In the DS 1 variety, single inoculation of AMF (G. *fasciculatum)* was not much effective than double inoculation of AMF (G. *fasciculatum)* with Ac or Pf, whereas the triple inoculation of G. *fasciculatum,* Ac and Pf was very much effective than the other treatments. Certain bacterial treatments significantly modified the extent of mycorrhizal colonization. Thus the percentage of mycorrhizal colonization by G. *fasciculatum* was increased when co-inoculated with Ac and Pf.

With regard to shoot development, AMF (G. *fasciculatum)* alone and dual inoculation of G. *fasciculatum* and Ac or Pf were most effective inoculation treatments (Table 18.1 and Fig. 18.1). The combination of G. *fasciculatum,* Ac and Pf produced the greatest shoot growth, phosphorus content, number of capsules than the other treatments. Root and shoot weight were especially enhanced by inoculation with G. *fasciculatum,* Ac and Pf. The highest value of root length was observed when G. *fasciculatum* was co-inoculated with the nitrogen-fixing bacteria Ac and Pf. Thus their latter treatments showed the highest shoot values (Fig. 18.1). DS 1 variety plants inoculated with G. *fasciculatum* alone or in combination with Ac had greater shoot length and root length than their respective non-mycorrhizal controls, but the greatest increase was in plants inoculated with G. *fasciculatum* + Ac + Pf.

Table 18.1. : Shoot Length (SL), Fresh Weight of Shoot (FWS), Dry Weight of Shoot (DWS), Root Length (RL), Fresh Weight of Root (FWR), Dry Weight of Root (DWR), Number of Flowers (NF), Number of Capsules (NC), Per Cent Colonization (PC), Spore Number (SN) and Phosphorus Uptake (P Uptake) in B5J Variety as Influenced by Different Treatments

	SL (cm)	FWS (g)	DWS (g)	RL (cm)	FWR (g)	DWR (g)	NF	NC	PC (%)	SN	P uptake (%)
30 days											
CN	29.00 ± 0.57 e	2.33 ± 0.00 c	0.26 ± 0.00 e	13.00 ± 1.00 c	0.21 ± 0.00 e	0.05 ± 0.00 e	1.00 ± 0.00 c	0.00 ± 0.00 d	0.00 ± 0.00 d	0.00 ± 0.00 c	0.00 ± 0.00 d
Gf	41.00 ± 0.57 d	3.20 ± 0.00 d	0.72 ± 0.00 c	15.00 ± 1.15 b	0.42 ± 0.00 d	0.06 ± 0.00 d	2.66 ± 0.66 bc	1.33 ± 0.33 c	46.00 ± 1.52 c	64.66 ± 8.11 a	0.10 ± 0.00 c
Gf + Ac	47.66 ± 0.88 b	4.40 ± 0.04 c	0.54 ± 0.00 d	17.00 ± 0.57 a	0.64 ± 0.00 c	0.12 ± 0.00 c	6.66 ± 0.88 a	3.00 ± 0.57 b	56.33 ± 0.88 b	62.00 ± 1.15 b	0.13 ± 0.00 b
Gf + Pf	45.00 ± 0.57 c	5.10 ± 0.00 b	0.86 ± 0.00 b	16.00 ± 0.57 b	0.66 ± 0.00 b	0.11 ± 0.00 c	3.66 ± 0.88 b	1.66 ± 0.33 c	52.66 ± 2.18 b	62.33 ± 1.85 b	0.11 ± 0.00 c
Gf + Ac + Pf	53.00 ± 0.57 a	8.65 ± 0.04 a	1.52 ± 0.00 a	19.00 ± 0.00 a	1.04 ± 0.00 a	0.24 ± 0.00 a	7.00 ± 0.57 a	4.66 ± 0.44 a	87.33 ± 0.33 a	75.66 ± 0.66 a	0.20 ± 0.00 a
60 days											
CN	41.00 ± 0.57 e	3.68 ± 0.00 e	0.76 ± 0.00 e	15.66 ± 0.33 b	0.51 ± 0.00 e	0.08 ± 0.00 e	3.00 ± 0.57 d	0.33 ± 0.33 d	0.00 ± 0.00 d	0.00 ± 0.00 c	0.00 ± 0.00 d
Gf	52.00 ± 0.57 c	4.14 ±0.03 d	0.97 ± 0.03 d	18.00 ± 0.57 b	0.98 ± 0.00 d	0.32 ± 0.00 d	6.66 ± 0.66 c	2.66 ± 0.33 c	51.00 ± 0.57 c	75.00 ± 2.88 a	0.11 ± 0.00 c
Gf + Ac	55.00 ± 0.57 b	8.52 ± 0.00 b	1.66 ± 0.00 b	23.66 ± 0.88 a	1.32 ± 0.00 b	0.63 ± 0.00 b	8.66 ± 0.33 ab	4.66 ± 0.33 b	63.00 ± 1.15 b	76.33 ± 1.45 a	0.15 ± 0.00 b
Gf + Pf	48.00 ± 0.57 b	5.53 ± 0.00 c	1.21 ± 0.00 c	21.66 ± 1.20 a	1.28 ± 0.00 c	0.41 ± 0.00 c	8.33 ± 0.33 b	2.66 ± 0.33 c	62.00 ± 1.15 b	66.66 ± 1.45 b	0.14 ± 0.00 b
Gf + Ac + Pf	64.66 ± 0.88 a	10.61 ± 0.00 a	2.64 ± 0.00 a	23.66 ± 0.33 a	1.53 ± 0.00 a	0.83 ± 0.00 a	10.00 ± 0.00 a	6.00 ± 0.57 a	91.33 ± 1.33 a	77.00 ± 0.57 a	0.22 ± 0.00 a
90 days											
CN	53.00 ± 0.57 e	4.16 ± 0.03 e	1.07 ± 0.00 e	19.66 ± 0.66 d	0.93 ± 0.00 e	0.41 ± 0.00 e	4.66 ± 0.33 d	6.00 ± 0.57 d	0.00 ± 0.00 d	0.00 ± 0.00 d	0.11 ± 0.00 d
Gf	61.00 ± 0.57 d	5.10 ± 0.00 d	1.31 ± 0.01 d	22.00 ± 1.00 cd	1.28 ± 0.00 d	0.52 ± 0.00 d	8.00 ± 0.57 c	8.33 ± 0.33 c	68.66 ± 0.33 c	81.33 ± 1.20 a	0.13 ± 0.00 c
Gf + Ac	73.00 ± 0.57 b	9.34 ± 0.00 b	1.73 ± 0.00 b	26.00 ± 1.15 b	1.83 ± 0.00 a	0.78 ± 0.00 b	10.66 ± 0.66 b	12.00 ± 0.57 b	77.66 ± 0.88 b	77.66 ± 0.88 b	0.16 ± 0.00 b
Gf + Pf	67.00 ± 1.45 c	6.38 ± 0.00 c	1.42 ± 0.00 c	24.66 ± 0.66 bc	1.35 ± 0.00 c	0.64 ± 0.00 c	14.33 ± 0.88 a	11.00 ± 0.57 b	73.66 ± 1.12 e	68.33 ± 1.20 c	0.15 ± 0.00 b
Gf + Ac + Pf	80.66 ± 0.66 a	13.62 ± 0.00 a	2.92 ± 0.00 a	34.00 ± 1.15 a	1.73 ± 0.00 b	1.24 ± 0.00 a	16.00 ± 0.57 a	18.33 ± 0.33 a	90.66 ± 5.33 e	83.00 ± 1.52 a	0.35 ± 0.00 a

CN, control; Gf, *Glomus fasciculatum*; Ac, *Azotobacter chroococcum*; Pf, *Pseudomonas fluorescens*; Values represent mean ± SD. Mean values followed by the same letter are not significantly different at the 0.05 level of confidence.

Table 18.1. : Shoot length (SL), fresh weight of shoot (FWS), dry weight of shoot (DWS), root length (RL), fresh weight of root (FWR), dry weight of root (DWR), number of flowers (NF), number of capsules (NC), per cent colonization (PC), spore number (SN) and phosphorus uptake (P uptake) in B5J variety as influenced by different treatments

	SL (cm)	FWS (g)	DWS (g)	RL (cm)	FWR (g)	DWR (g)	NF	NC	PC (%)	SN	P uptake (%)
30 days											
CN	26.66 ± 0.88 c	2.67 ± 0.00 c	0.33 ± 0.00 e	14.00 ± 1.15 c	0.20 ± 0.00 e	0.04 ± 0.00 e	3.66 ± 0.88 b	0.00 ± 0.00 c	0.00 ± 0.00 e	0.00 ± 0.00 c	0.05 ± 0.00 c
Al	36.66 ± 0.83 b	3.63 ± 0.00 d	0.44 ± 0.00 d	15.33 ± 0.88 bc	0.54 ± 0.00 c	0.10 ± 0.00 b	4.33 ± 0.88 ab	0.33 ± 0.33 c	55.00 ± 0.57 d	52.66 ± 2.66 b	0.12 ± 0.00 b
Al + Ac	43.66 ± 0.88 a	5.76 ± 0.00 b	0.78 ± 0.00 b	17.33 ± 0.66 ab	0.62 ± 0.00 b	0.10 ± 0.00 b	5.33 ± 0.88 ab	1.66 ± 0.33 b	65.00 ± 0.57 b	64.00 ± 3.21 a	0.12 ± 0.00 b
Al + Pf	36.33 ± 1.45 b	3.83 ± 0.00 c	0.54 ± 0.00 c	14.33 ± 1.20 bc	0.49 ± 0.00 d	0.09 ± 0.00 c	5.00 ± 1.52 ab	0.33 ± 0.33 c	58.66 ± 0.88 c	56.66 ± 1.76 b	0.12 ± 0.00 b
Al + Ac + Pf	46.66 ± 1.45 a	8.72 ± 0.00 a	1.12 ± 0.00 a	20.33 ± 0.88 a	1.24 ± 0.00 a	0.22 ± 0.00 a	7.33 ± 0.33 a	2.66 ± 0.33 a	76.00 ± 0.57 a	66.00 ± 1.15 a	0.19 ± 0.00 a
60 days											
CN	47.00 ± 0.57 d	5.53 ± 0.00 e	1.21 ± 0.00 e	18.33 ± 0.33 e	0.78 ± 0.00 e	0.19 ± 0.00 e	3.66 ± 0.88 d	2.66 ± 0.33 c	0.00 ± 0.00 e	0.00 ± 0.00 c	0.07 ± 0.00 e
Al	51.00 ± 0.57 c	5.58 ±0.00 d	1.73 ± 0.00 d	19.66 ± 0.33 d	0.98 ± 0.00 d	0.24 ± 0.00 d	5.66 ± 1.20 d	2.00 ± 0.57 c	63.00 ± 1.52 d	73.33 ± 3.17 a	0.13 ± 0.00 d
Al + Ac	58.66 ± 0.33 b	9.60 ± 0.00 b	2.05 ± 0.00 b	25.00 ± 0.57 b	1.70 ± 0.00 b	0.38 ± 0.00 c	12.66 ± 0.33 c	6.66 ± 0.33 ab	76.00 ± 1.15 b	56.66 ± 1.45 b	0.17 ± 0.00 b
Al + Pf	51.00 ± 0.57 c	8.53 ± 0.00 c	1.83 ± 0.00 c	21.00 ± 0.00 c	1.33 ± 0.00 c	0.56 ± 0.00 b	15.33 ± 0.33 b	5.66 ± 0.33 b	71.66 ± 0.88 c	60.00 ± 4.35 b	0.15 ± 0.00 c
Al + Ac + Pf	68.33 ± 0.33 a	11.61 ± 0.00 a	3.23 ± 0.00 a	1.83 ± 0.00 a	1.83 ± 0.00 a	1.06 ± 0.00 a	18.33 ± 0.33 a	7.66 ± 0.33 a	80.33 ± 0.88 a	62.33 ± 1.76 b	0.22 ± 0.00 a
90 days											
CN	53.00 ± 0.57 d	5.63 ± 0.00 e	1.73 ± 0.00 e	22.00 ± 0.57 d	1.09 ± 0.00 e	0.26 ± 0.00 e	6.00 ± 0.57 e	4.66 ± 0.33 d	0.00 ± 0.00 d	0.00 ± 0.00 d	0.09 ± 0.00 e
Al	60.00 ± 0.00 c	8.71 ± 0.00 d	1.96 ± 0.00 d	23.66 ± 0.88 cd	1.16 ± 0.00 d	0.41 ± 0.00 d	9.66 ± 0.33 d	5.66 ± 0.33 cd	72.33 ± 1.45 c	66.66 ± 1.45 b	0.15 ± 0.00 d
Al + Ac	65.00 ± 0.57 b	10.34 ± 0.00 b	3.23 ± 0.00 b	28.00 ± 0.57 b	1.80 ± 0.00 b	0.63 ± 0.00 b	15.00 ± 0.57 c	8.33 ± 0.33 b	81.66 ± 0.88 b	62.00 ± 1.52 c	0.21 ± 0.00 b
Al + Pf	61.00 ± 0.57 c	9.30 ± 0.00 c	2.21 ± 0.00 c	24.66 ± 0.33 c	1.40 ± 0.00 c	0.59 ± 0.00 c	16.66 ± 0.33 b	6.00 ± 0.57 c	76.66 ± 2.84 c	61.33 ± 0.88 c	0.18 ± 0.00 c
Al + Ac + Pf	82.00 ± 0.57 a	13.80 ± 0.00 a	4.12 ± 0.00 a	41.00 ± 0.57 a	3.32 ± 0.00 a	1.23 ± 0.00 a	20.33 ± 0.33 a	11.33 ± 0.33 a	91.00 ± 0.57 a	72.66 ± 1.85 a	0.24 ± 0.00 a

CN, control; Gf, *Acaulospora laevis*; Ac, *Azotobacter chroococcum*; Pf, *Pseudomonas fluorescens*; Values represent mean ± SD. Mean values followed by the same letter are not significantly different at the 0.05 level of confidence.

The phosphorus content and the number of capsules were more efficient in AMF (G. *fasckulatum),* Ac and Pf than single and double inoculation. A comparison between the dual inoculations revealed that G. *fasckulatum* and Ac was more efficient in all the growth parameters than G. *fasckulatum* and Pf.

In the E 8 variety, inoculation with *A. laevis* + Ac + Pf resulted in increased shoot and root development, the fresh and the dry weight of the shoot and root, the number of capsules and P uptake. As regards the percentage of colonization, the infection percentage of the mycorrhiza was enhanced when inoculated with Ac and Pf compared with the single inoculation of A. *laevis* and dual inoculation of *A. laevis* and Ac or Pf (Table 18.2, Fig. 18.2). In dual inoculation of A. *laevis* and Ac, the response was greater in all the growth parameters than the combined inoculation of *A. laevis* and Pf. The response within the varieties was almost similar but only with the dry weight of the root it differed—DS 1 showed greater response in root dry weight than the E 8 variety.

Millet production is scanty despite its importance as an oilseed crop in the past decades. Plant root systems have complex and functional characteristics (Rengel, 1997). Efficient water and mineral uptake is closely related to roots and their growth (Kapulnik *et al.*, 1985). The stimulation of root biomass and total root length for plants grown in single, dual and triple inoculations was observed in our studies, may be due to the result of the cumulative effects of favourable plant-AMF-microbe interactions.

It is well known that the magnitude of plant response to any microbial inoculation is greatly affected by the P and N content of the soil (Paula *et al.,* 1992). In particular, P deficiency has been described as a main factor in restricting not only plant development but other biological processes such as biological nitrogen fixation, owing to the high requirement of P (as ATP) for the nitrogen fixation process

(Singleton *et al.*, 1985; Israel, 1987). The analysis of the phosphorus is present in lesser available forms, and this could be a factor limiting survival of bacteria. In this context, the role of the AM fungi as phosphorus suppliers to the plant appears to be of great relevance. Thus in both varieties DS 1 and E 8, inoculation with AM fungi, Ac and Pf produced the highest effect on either plant growth or nutrient uptake, together with a noticeable increase in mycorrhizal root colonization. Bacteria may affect the root cell walls, and thereby increase the susceptibility of the plant tissue to fungal penetration. *Azospirillum brasiliense* produces pectolytic enzymes *in vitro* which soften the root cell walls in the soil (Umali-Garcia *et al.*, 1980). This mechanism was suggested by Mosse (1962) for increased colonization of roots of several plants by Endogone species in the presence of *Pseudomonas* species. It was also suggested by Meyer and Linderman (1986) to explain the enhanced colonization of clover roots by an unidentified AM fungus in the presence of *Pseudomonas putida*. Azcon (1987) reported that the growth of emerging mycelium from *Glomus mosseae* spores was enhanced in the presence of PGPB. The nodulating bacteria, e.g. *Frankia, Rhizobium* and *Bradyrhizobium*, generally form synergistic interactions with the AM fungi. It is believed that AM symbiosis reduces phosphate stress for the plant, which is beneficial for the N_2-fixing nitrogenous system of the bacteria, resulting in enhanced fixation and improved N status of the plant. Despite frequent reports of mutualistic influences, antagonistic relationships between the micro-symbionts have also been found (Bethlenfalvay *et al.*, 1985).

The beneficial effect of *Azotobacter* and AMF on plant growth might be through nitrogen fixation or through the production of growth-promoting substances. Brown (1974 and 1975) observed no nitrogenous activity in the rhizosphere of young *Paspalum notatum* plants inoculated with *Azotobacter paspali* and pointed out that the bacterium frequently does not fix the improved plant growth by the production of growth-regulating substances.

Singh (1992) observed that inoculation of nitrogen-fixing bacteria (Ac) and phosphorus solubilizing bacteria (*Pseudomonas striata, Bacillus polymyxa*) enhanced sporulation, root volume and percentage VAM root colonization of *Pennisetum padicillatum* in the presence of *Glomus macrocarpum.* Synergistic interaction effects of PSB, *Enterobacter agglomerans* and AM fungus *Glomus etunicatum* increased growth of tomato plant as reported by Kim *et al.,* 1998. This is clear from our results where AMF in both varieties appeared to be more effective in plant growth and the number of capsules when inoculated with Ac and Pf and even promoted increased mycorrhizal colonization. Both AMF and bacteria play an important role in the development of the plant community (Eom *et al.,* 2000). The enhanced plant growth may result directly from microbial production of plant growth regulators, including indole-3-acetic acid (IAA) (Lambrecht *et al.,* 2000, O'Donnell *et al.,* 2003, Bano and Musarrat, 2004). The IAA produced by the microbes colonizing the seed or root surfaces is proposed to act in conjunction with endogenous IAA *in planta* to stimulate cell proliferation and elongation and also enhance the host's uptake of minerals and nutrients from the soil (Patten and Glick 2002), Suzuki *et al.,* (2003). *Azotobacter* excretes phytohormones, which improve growth of plant roots, and AMF may solubilize phosphorus from surrounding areas and make it available to roots (Behl *et al.,* 2003). Some AMF spore-associated bacteria (AMB) such as *Bacillus pabuili* have the ability to enhance AMF root colonization (Xavier and Germida 2003) and can also improve plant growth (Artursson *et al.,* 2006). Gurumurthy and Sreenivasa (1996) have found higher population of free living nitrogen fixers and P solubilizers in the rhizosphere of chilli inoculated with *G. macrocarpum* compared with control. Thus there is a synergistic or additive beneficial effect on Millet. The results confirmed the feasibility of substituting chemicals in the form of fertilizers and pesticides by organic/natural resources. Besides overcoming the major problem of pesticide residue, the Millet produced with bio/natural inputs will get a premium in the international market. This in turn will reflect higher prices in the domestic market and ultimately higher profit to the growers. The Millet seed so produced will be a perfect health food with no adverse effects on the human body.

REFERENCES

Artursson, V., Finlay, R. D. and Jansson. J. K. (2006). Interactions between arbuscular mycorrhizal fungi and bacteria and their potential for stimulating plant growth. Environ. Microbiol. 8, 1–10.

Azcon, R. (1987). Germination and hyphal growth of *Glomus mosseae in vitro:* effects of rhizosphere bacteria and cell-free culture media. Soil Biol. Biochem. 19, 417–419.

Behl, R. K., Sharma. H., Kumar. V., and Narula. N., (2003). Interactions amongst mycorrhiza, *Azotobacter chroococcwn* and root characteristics of wheat varieties. J. Agron. Crop Sci. 189, 151–155.

Bethlenfalvay, G. J., Brown, M. S. and Stafford. A. E. (1985). *Glycine-Glomus-Rhizobium* symbiosis. 11. Antagonistic effects between mycorrhizal colonization and nodulation. Plant Physiol. 79, 1054–1058.

Brown, M. E. (1974). Seed and root bacterization. Annu. Rev. Phytopathol. 12, 181.

Brown, M. E. (1975). Rhizosphere microorganisms - opportunists, bandits or benefactors. In: N. Walker, ed. Soil Microbiology, pp. 21–38. Halsted Press, New York.

Deshmukh, M. R., Jain, H. C., Duhoon, S. S. and Goswami. U. (2002a). Integrated nutrient management in Millet for Kymore plateau zone of M.P. J. Oilseeds Res. 19, 73–75.

Eom, A. H., Hartnett, D. C. and Wilson. G. W. T. (2000). Host plant species effects on arbuscular mycorrhizal fungal communities in tallgrass prairie. Oecologia 122, 435–444.

Gerdemann, J. W. and Nicolson. T. H. (1963). Spores of mycorrhizal endogene species extracted from the soil by wet sieving and decanting. Trans. Br. Mycol. Soc. 46, 235–244.

Giller, K. E. and Cadisch, G. (1995). Future benefits from biological nitrogen fixation: an ecological approach to agriculture. Plant Soil 174, 255–277.

Grayston, S. J., Wang S., Campbell. C. D. and Edwards, A. C. (1998). Selective influence of plant species on microbial diversity in the rhizosphere. Soil Biol. Biochem. 30, 369–378.

Gurumurthy, S. B. and Sreenivasa, M. N. (1996). Response of chilli to different inoculum levels of *Glomus macrocarpum* in two soil types of Karnataka. Karnataka J. Agric. Sci. 9, 154–159.

Hameeda, B., Srijana. M., Rupela. O. P. and Reddy, G. (2007). Effect of bacteria isclated from composts and macrofauna on sorghum growth and mycorrhizal colonization. World J. Microbiol. Biotechnol. 23, 883–887.

Hegde, D. M. (1998). Integrated nutrient management for production sustainability of oilseeds. A review. J. Oilseeds Res. 15, 1–17.

Jackson, M. L. (1973): Soil Chemical Analysis. Prentice Hall, New Delhi.

Jeffries, P., Gianinazzi, S. Perotto, S. Turnau, K. and Barea, J.M. (2003). The contribution of arbuscular mycorrhizal fungi in sustainable maintenance of plant health and soil fertility. Biol. Fertil. Soils 37, 1–16.

Johnson, D., Booth. R. E., Whiteley. A. S., Bailey. M. J., Read. D. J., Grime. J. P. and Leake. J. R (2003). Plant community composition affects the biomass, activity and diversity of microorganisms in limestone grassland soil. Eur. J. Soil Biol. 54, 671–677.

K|epper, J. W., Leong. J., Teintze, M. and Schroth. M. N. (1980). Enhancing plant growth by siderophores produced by plant growth-promoting rhizobacteria. Nature 286, 885–886.

Kapulnik, Y., Okon, Y. and Henis, Y. (1985). Changes in root morphology of wheat caused by Azospirillum inoculation. Can. J. Microbiol. 31, 881–887.

Kim, K. Y., Jordan, D., and. Dowald. G. A. (1998). Effect of phosphate solubilizing bacteria and VA-mycorrhizae on tomato growth and soil microbial activity. Biol. Fertil. Soils 26, 79–89.

Lambrecht, M., Okon. Vande Broek Y, and Vanderley-den. J. (2000). Indole-3-acetic acid: a reciprocal signaling molecule in bacteria plant interactions. Trends Microbiol. 8, 298–300.

Manske, G. G. B., Luttger, A. B., Behl., R. K. and Vlek, P. L. G. (1995). Nutrient efficiency based on VA mycorrhizae (AMF) and total root length of wheat cultivars grown in India. Angew. Bot. 69, 108–110.

Marschner, P., Yang, C.H., Lieberei. R. and Crowley. D. E. (2001). Soil and plant specific effects on bacterial community composition in the rhizosphere. Soil Biol. Biochem. 33, 1437–1445.

Meyer, J. R. and Linderman. R. G. (1986). Response of subterranean clover to dual inoculation with vesicular-arbuscular mycorrhizal fungi and a plant growth-promoting bacterium, *Pseudomonas putida*. Soil Biol. Biochem. 18, 185–190.

Mosse, B. (1962). The establishment of vesicular-arbuscular mycorrhizae under aseptic conditions. J. Gen. Microbiol. 27, 509–520.

Pannase, S. K., Thakur. N. S. and Sawarkar. S. D., (1995). Sustainable yield of rainy Millet oilseed in Satpura plateau zone of M.P. Indian J. Agron. 40, 323–324.

Patten, C. L., and Glick, B. R. (2002). Role of *Pseudomonas putida* indole acetic acid in development of the host plant root system. Appl. Environ. Microbiol. 68, 3795–3801.

Paula, M. A., Urquiaga, S., Siqueira. J. O. and Dobereiner, J. (1992). Synergistic effects of vesicular-arbuscular mycorrhizal fungi and diazotrophic bacteria on nutrition and growth of sweet potato *(Ipomoea batatas)*. Biol. Fertil. Soils 14, 61–66.

Phillips, J. M. and Hayman. D. S. (1970). Improved procedure for clearing roots and vesicular arbuscular mycorrhizal fungi for rapid assessment of infection. Trans. Br. Mycol. Soc. 55, 158–161.

Rengel, Z. (1997). Root exudation and microflora populations in rhizosphere of crop genotypes differing in tolerance to micronutrient deficiency. Plant Soil 196, 255–260.

Singh, C. S. (1992). Mass inoculum production of vescicular arbuscular (VA) mycorrhizae: II. Impact of nitrogen fixing and P-solubilizing bacterial inoculation on VA mycorrhizae. Zentralbl. Mikrobiol. 147, 503–508.

Singleton, P. W., Abdel-Magid, H. M. and Tavares. J. W. (1985). Effects of phosphorus on the effectiveness of strains of *Rhizobium japonkum*. Soil Sci. Soc. Am. J. 49, 61–63.

Snedecor, G. M. and Cochran. W. G. (1980). Statistical Methods, 7th edn. Iowa State College Press, Ames, IA.

Suzuki, S., He, Y. and Oyaizu, H. (2003). Indole-3-acetic acid production in *Pseudomonas fluorescens* HP72 and its association with suppression of creeping bentgrass brown patch. Curr. Microbiol. 47, 138–143.

Tiwari, K. P., Namdeo. K. N., Tomar, R. K. S. and Raghu, J. S. (1995). Effect of macro and micro nutrients in combination with organic manure in the production of Millet. Indian J. Agron. 40, 134–136.

Umali-Garcia, M., Hubbell, D. H., Gaskins, M. H. and Dazzo, F. B. (1980). Association of Azospirillum with grass roots. Appl. Environ. Microbiol. 39, 219–226.

Xavier, L. J. C. and Germida. J. J. (2003). Bacteria associated with *Glomus clarum* spores influence mycorrhizal activity. Soil Biol. Biochem. 35, 471–478.

19

Combined Inoculation of Arbuscular Mycorrhizal Fungi and Azotobacter Beneficial to *Glycine max* Merr. (Soyabean)

[1]*A K Venu*

INTRODUCTION

Soil provides the matrix for the biological processes involved in nutrient cycling. Among the biological processes involved in the rhizoplane, the unique role of symbiotic bacteria and the AM fungi which ensure fixation and mobilization, and availability to nitrogen and phosphorous to plants have been well recognized (Marchner, 1995). It is well-established fact that the AM always prefer certain host exhibiting maximum symbiotic response and increased the growth and yield of crop mainly through improved uptake of nutrients (Allen, 1991). A few scientists have observed wide variation among and within different species on AM fungi in their ability to promote plant growth (Read, 1996). This led to the concept of host preference by AM fungi (Mosse, 1973). Hence it is always better to select an efficient AM fungus for a

1. Department of Microbiology Mangalore University, P. G. Centre, F. M. Kariyappa College, Madekeri, India

particular host-soil-climate combination to harness maximum benefits. This study was aimed to find the response of *Glycine max* to inoculation *of Glomus macrocarpum, G. fasciculatum* and *Azotobacter* in unsterile soil either singly or in combinations.

MATERIALS AND METHODS

Present investigation carried out in sterilized soil of polyhouse during 2006 to 2007 using aromatic plant seedling Soyabean (*Glycine max* Merr.). The soil used for experiment contained Organic carbon 0.86%, pH 6.8, available N 0.74%, available P_2O_5 2.8 kg/ha, available 91 kg/ha, Electrical conductivity (EC) was 0.17 ohm^{-1} and native AM spore population averaging 87 spores/50 g soil. Seedlings, used in the experiment were grown on sterile soil and 30 seedlings were transplanted in earthen pots measuring 25 x 15 cm diameter. Soil-root-cultures of *Glomus macrocarpum* and *G. fasciculatum* were cultivated on maize.

Roots using mixture of soil : sand. : FYM (1:1:1). The cultures containing clamydospores (96–112 spore/50 g soil) and root segments of maize colonized by particular AM fungus were used as mycorrhizal inoculum. Application of AM inoculum was 129 g soil/ plant when single species was used and 5 g soil/plant when two species used in combination in soil with seedling roots. Azotobacter as per treatment was 10g/plants. The experiment was arranged in a completely randomized design with five replications. The five treatments included and uninoculated control. *G. mosseae, G. fasciculatum* and *Azotobacter.* Observations were recorded, 30 and 60 days after planting. Observation such as chlorophyll content, plant height, number of leaves, size of stem, and fresh weight of bulb were recorded. The fresh weight of plant and leaves was recorded immediately after harvest. The dry weight was determined after drying the plant at 80 C for 48 hrs. The root samples of each treatment were collected, processed and stained in 0.05% cotton blue in lactophenol (Phillips and Hayman, 1970). Per cent root colonization was calculated using the method of Giovanetti and Mosse (1980). The nitrogen and phosphorus from shoot and bulb of onion were determined following the method of Jackson (1973). Estimation of chlorophyll was carried out using the method of Arnon (1949).

RESULTS AND DISCUSSION

There was more than 6 folds increase in total chlorophyll content 30 days after inoculation of plants with *G. macrocarpum*, *G. fasciculatum* and *Azotobacter chrococcum* compared to uninoculated plants of Soyabean.

In inoculated plants with *G. fasciculatum* or with *G. macrocarpum* and *G. fasciculatum* also total chlorophyll content increased. However, 60 days after inoculation the total chlorophyll content dropped gradually (Table 19.1) while control plants showed a continued small increase. Increase in chlorophyll content in AM fungi inoculated plants were reported by Devi and Reddy (2004) in groundnut and Shivaputra *et al.* (2004) in papaya. Plants treated with *G. fasciculatum* and Azotobacter in combination recorded rapid increase in dry weight from 30 to 60 days after inoculation. A similar increase in dry weight was observed when inoculated with *G. macrocarpum* and *G. fasciculatum* or *G. fasciculatum* or *Azotobacter* (Table 19.1).

Significant increases were recorded in growth parameters number of leaves, plant dry weight, stem diameter, fresh and dry weight of leaves (Table 19.2). The inoculation of *G. macrocarpum*, *G. fasciculatum* and *Azotobacter* increased plant height 55.35 cm after 30 days to 74.25 cm after 60 days Similarly in the same treatment, maximum number of functional leaves increased from 6 and 12 and leaf area from 51.20 to 90.64 cm^2. Similar findings were recorded by (Krishna

Table 19.1. Effect of AM Fungi and *Azotobacter* Inoculation Showing Chlorophyll Content and Fresh and Dry Weight of Shoots *Glycine max* Merr

AM	Chl a	Chl b	Total chl	Chl a	Chl b	Total chl	Shoot fresh wt (g)		Shoot dry wt (g)	
	30 DAP			60 DAP			30 DAP		60 DAP	
Control	0.170	0.126	0.113	0.213	0.088	0.311	3.11	11.13	0.74	1.32
Gm	0.236	0.129	0.381	0.260	0.128	0.380	9.69	24.17	0.76	1.32
Gf	0.317	0.281	0.587	0.243	0.098	0.342	13.74	32.68	1.23	3.52
Gm+Gf	0.269	0.19	0.439	0.252	0.147	0.386	8.01	41.12	0.92	4.63
Gm+Gf+B.P	0.371	0.414	0.782	0.244	0.173	0.423	14.25	71.03	1.24	7.24
CD(*P*=0.05)	0.07	0.12	0.23	0.012	0.033	0.046	3.02	27.62	0.25	2.72

Gb - *Glomus macrocarpum*, Gf = *Glomus fasciulatum*, B.P=*Azotobactor chrococcum*

Table 19.2. : Effect of AM Fungi and *Azotobacter* Inoculation on Leaves, Plant, VAM Colonization N and P Status *Glycine max* Merr

AM	Leaf (no)		Leaf area		Plant height (cm)		VAM colonization (%)		N (%) in shoot/plant		P (%)in shoot/plant	
	(40 d)	(80 d)	(40 d)	(80 d)	(40 d)	(80 d)	(40 d)	(80 d)	(40 d)	(80 d)	(40 d)	(80 d)
Control	3	6	30.4	35.7	49.11	46.21	-	-	2.05	3.11	0.06	0.08
Gm	5	10	28.01	51.20	44.52	59.30	49.10	73.10	3.42	3.72	0.08	0.11
Gf	5	10	34.8	77.00	51.4	71.14	61.0	62.50	3.55	3.80	0.12	0.12
Gm+Gf	7	10	30.15	84.32	46.32	74.10	61.0	78.4	3.68	4.62	0.13	0.15
Gm+Gf+B.P	7	12	35.12	91.64	55.35	74.25	67.3	84.22	3.71	3.86	0.13	0.18
CD (*P*=0.05)	1	12	3.18	22.15	5.05	13.10	31.06	41.03	0.71	0.82	0.02	0.02

Gb - *Glomus macrocarpum*, Gf = *Glomus fasciulatum*, B.P=*Azotobactor chrococcum*

and Bagyaraj, 1982) in Lady's finger and (Indi *et al.*, 1989) in brinjal recorded significant increase in height number of leaves and overall growth of plants due to inoculation with AM fungi. It is evident from (Table 19.2) root colonization 30 days after inoculation was higher in combined inoculation with *G. macrocarpum, G. fasciculatum* and *Azotobacter*. After 60 days max. root colonization was found this treatment. Clamydospore count in rhizosphere soil 60 days after inoculation was also the highest on inoculation of onion plants with *G. macrocarpum, G. fasciculatum* and *Azotobacter* (187 spore/50 g soil). *G. fasciculatum* and *G. macrocarpum* also showed high levels (147– 161 spores/50 g oil). Inoculation with G. *mosseae* or *G. fasciculatum* alone was also effective (152 spore/50 g soil). This indicated that AM fungi used in inoculations were efficient in colonization as reported earlier in *Ablemoschus esculentus* (Krishna and Bagyaraj, 1982). In the present investigation, AM species were proved to be the best for the growth of *Glycine max* when used in combination. All AM inoculated plants showed significantly higher uptake of N and P over control (Table 19.2). Plants inoculated with G. *macrocarpum, G. fasciculatum* and *Azotobacter* recorded significantly after 30 days the max. 0.13 P% uptake (3.71% N in plant). *G. macrocarpum,* and *G. fasciculatum* and *G. fasciculatum, G. fasciculatum* and *G. macrocarpum* also showed high uptake of N and P. After 60 days of inoculation, the maximum N uptake was with *G. macrocarpum, G. fasciculatum* and *Azotobacter* was 3.86% N in plant. Plants inoculated with *G. macrocarpum, G. fasciculatum* and *Azotobacter* in combination showed a max. P uptake (0.18%). Inoculation with *G. macrocarpum* and *G. fasciculatum, G. fasciculatum* or *G . macrocarpum* also led to increased P uptake.

The increased uptake of N and P could be attributed to the increased root absorption area induced by AM through an efficient symbiosis with the host and by assimilation and translocation of N (Yao and Li, 1999). It is well known fact that P uptake by plant is improved by AM fungal association (Shivputra *et al.*, 2004). Differential ability of inoculated AM fungi stimulating P uptake was also reported in different crops viz. on Lady's finger (Krishna and Bagyaraj 1982) and and Garlic (Wani and Konde 1998). There were significant differences in stem diam, fresh and dry weight of leaves following inoculations with AM fungi (Table 3). Inoculations with any AM in general increased bulb weight and size and the best treatment was inoculation made with *G. macrocarpum, G. fasciculatum* and *Azotobacter* in combination. Similar studies have been conducted earlier (Lakshman, 2000). Ramananda and Sreenivasa (2000), reported that the fresh and dry weight of *Glycine max* L. inoculated with different AM fungi increased significantly. An increase in stem diam is attributed to the increased dry matter accumulation. In conclusion, the results confirm that appropriate strains of AMF and *Azotobacter* inoculum would help in increased biomass production and nutrient uptake in *Glycine max* plants.

Table 19.3. : Effect of AM fungi inoculation on leaf weight and stem size of *Glycine max* Merr.

Treatment	Leaf fresh wt (g)		Leaf dry wt (g)		Stem size (cm)	
	(30 d)	(60 d)	(30 d)	(60 d)	(30 d)	(60 d)
Control	1.0	2.9	0.189	0.431	0.6	1.13
Gm	1.3	4.10	0.381	0.473	0.8	1.64
Gf	2.7	6.62	0.302	0.442	0.8	1.72
Gm+Gf	2.1	7.72	0.576	0.761	0.9	1.78
Gm+Gf+B.P	2.6	8.44	0.605	0.891	0.9	1.84
CD at 0.05%	0.88	2.45	0.158	0.210	0.14	0.23

Gb - *Glomus macrocarpum*, Gf = *Glomus fasciulatum*, B.P=*Azotobactor chrococcum*

REFERENCES

Allen, F. (1991). Ecophysiology of mycorrhiza, Oxford Press, London, pp. 249.

Arnon, D. I. (1949). Copper enzymes in isolated chloroplast. Polyphenol oxidase in *Beta vulgaris*. *Plant Physiol*. 24:1–15.

Devi, M. C. and Reddy, M. N. (2004). Effect of arbuscular mycorrhizal fungi and Rhizobium association on chlorophyll content of groundnut (*Arachis hypogaea* L.), *Mycorrhiza News*. 16(1): 15–17.

Giovanetti, M. and Mosse, B. (1980). An evaluation of technique for measuring vesicular-arbuscular mycorrhizal infection in the plants. *New Phytol* 95:69–82.

Hattingh, M. S., Gray, L. E. and Gerdemann, J. W. (1973). Uptake and translocation of 32p. labelled phosphate to onion roots by endomycorrhizal fungi. *Soil Sci* 16:383–387.

Indi, D. V., Konde, B. K., Wani, P. V. and Kale, P. N. (1989). Response of brinjal genotypes to mycorrhizal inoculation. In: *Biofertilizer transfer Technology*- (Ed. Gangwane L.V.). Assoc. Publ. Co. New Delhi. Pp. 223–229.

Jackson, M. L. (1973). *Soil Chemical Analysis*. Prentice Hall of India Pvt., New Delhi, pp.370–387.

Krishna, K. R. and Bagyaraj, D. J. (1982). Effect of vesicular arbuscular mycorrhize and soluble phosphate on *Ablemoschus esculenlus* (L.) Moeach. *Plant Soil* 64(2); 209–213.

Lakshman, H. C. (2000). Dual inoculation of *Rhizobium* and AM fungi improve seedling of *Pterocarpus marsupium*, *J. Env. Ecol. Cons.* 171:205–210.

Marchner, H. (1995). Mineral nutrition of higher plants. Academic press, London, pp. 198.

Mosse, B. (1973). Advances in the study of vesicular arbuscular mycorrhiza. *Ann. Rev. Phytopathol* 11:171–196.

Panse, V. G. and Sukhatme, P. V. (1957). *Statistical Methods for Agricultural workers*. ICAR Publ. New Delhi. pp. 253–256.

Phillips, J. M. and Hayman, D. S. (1970). Improved procedures for clearing roots and staining parasitic vesicular arbuscular mycorrhizal fungi for rapid assessment of infection. *Trans. Brt. Mycol. Soc.* 55:158–161.

Ramananda, T.V. and Sreenivasa, M. N. (2000). Influence of different AM fungi on growth of onion. *J. Maharashtra. Agric. Univ.* 25(1):93–94.

Read, J. W. (1996). Mycorrhizal function. Academic press. *New York*. pp. 91.

Sandersen, P., Raja, N.U., Gunasekaran, P. and Lakshamanan, M. (1988). Studies on Nitriate reduction by VAM fungal spores. *Curr Sci*, 57(2):84–85.

Shivaputra, S. S., Patil, C. P., Swamy, G. S. K. and Patil, P. B. (2004). Cumulative effect of VAM fungi and vermicompost on nitrogen, phosphorus, potassium and chlorophyll content of papaya leaf. *Mycorrhiza News*. 16(2): 15–16.

Wani, P. V. and Konde, B. K. (1998). Response of garlic to inoculation with vesicular arbuscular mycorrhizae. *J. Maharashtra. Agric. Univ.* 23(I):35–38.

Yao, Q. and Li, D. (1999). Influence of VA mycorrhiza on juice components and arid colour of Citrus fruit. *J. fruit. Sc.* China. 16: 38–42.

20

Effect of AM Fungi and PGPR to Improve Growth and Yield of *Solanum Tuberosum* L. (Potato)

[1]*S B Gadi*

INTRODUCTION

AMF (arbuscular mycorrhizal fungi) is widely distributed in agro-ecosystems (Smith and Read, 1997), forming symbiotic associations with the roots of plants. They play an important role in plant mineral nutrition and plant health (Barea, 2000; Azcon *et al.,* 2001). These fungi have a wide range of application in sustainable low input agricultural systems. The use of AMF may contribute to reducing chemical fertilizer inputs and sustaining plant productivity in agriculture (McGonigle, 1988).

The contribution of PGPR (plant growth promoting rhizobacteria) in phytostimulation, phytoremediation, and biofertilization is well documented (Goldstein 1986; Barea, 2000). *Azotobacter is* regarded as a broad-spectrum inoculant as it could be used for inoculating wide variety of crops such as wheat, rice, sorghum, barley, potato, sugarbeet, cotton, maize, etc. (Dhillion *et al.,* 1980; Rai and Gaur, 1982).

---

1. Department of Botany, JSS College, Vidyagiri, Dharwad –580 004 India

The present study, therefore, was conducted under net house conditions using unsterilized soils, to test the effectiveness of the introduced AMF and plant growth promoting rhizobacteria in the presence of their indigenous counterparts on *Solanum tuberosum* grown at different soil nitrogen sources.

MATERIALS AND METHODS

Soil

Test soil was collected from rice fields located near Karwara South Canara district of Karnataka. It contained 0.037% phosphorus (total), 0.109% nitrogen (total), and 1.3 ppm potassium with pH 5.70, and moisture content of 38%. Earthen pots of 20 cm diameter with a drainage hole were filled with approximately 4 kg of soil.

Experimental design and treatments

The pot experiments was conducted in the net house in the botanical garden, Department of Botany, Karnatak University, Dharwad. The CRB (completely randomized block) design was used for the experiment. There were 16 soil treatments with control, *Glomus* only, *Azotobacter chroococcum* only, and *Glomus macrocarpum* + *Azotobacter chroococcum* only, at four different levels of nitrogenous fertilizer; that is, zero kg/ha, 24 kg/ha, 48 kg/ha and 144 kg/ha. These 16 soil treatments were replicated three times. Nitrogen was applied in three splitted doses. The first dose, consisting of 1/3 the normal dose, was applied before transplantation; the second 1/3 at the time of tillering; arid the last 1/3 at the panicle initiation phase.

Collection of seeds and raising seedlings

Solanum tuberosum (Potato) seeds were collected from the Department of Agriculture, University Dharwad. Seeds were surface sterilized with 2% Sodium hypochlorite. They were sown on sterilized sand on trays after germination. When the seedlings were three weeks old, three seedlings were transplanted into pots with different treatments.

Bioinoculants used

The microbial inocula used were *Glomus macrocarpum* and *Azotobacter chroococcum. Glomus macrocarpum* was isolated from different rice rhizospheres. The mycorrhizal inoculum consisting of spores, soil, and infected root fragments were obtained from the pot cultures of *Chloris gayana* Kunth. as the host plant, each pot received an inoculum of 10 g at 2 cm below the soil surface near the root system. The non-mycorrhizal pots received the same quantity of autoclaved inoculum.

Azotobacter chroococcum, was isolated from the rice rhizosphere using Ashby's medium and maintained on agar medium as *Azotobacter chroococcum* was enriched on nutrient media. After 24 hours, the cells were collected by centrifugation at 2000 rpm for 20 minutes and suspended in % strength Ringer's solution so as to get 5 × 105 cells ml"1 by using haemocytometer. Each *Azotobacter chroococcum* treated pot obtained 3 ml of this cell suspension.

Measurement

Plant height and tiller number were recorded thrice at an interval of 30 days after 1 month of transplantation. Ear number and 1000-grain weight were recorded after harvesting. Shoot and root dry weights were recorded after drying the plants in an oven at 70°C for 48 hours and then cooling them in a dessicator. Spore count (50 g/ads [amended soil]) was estimated by the wet sieving and decanting method (Gerdemann and Nicolson, 1963) and the percentage of root length colonized by AMF was estimated by examining stained samples (Koske and Gemma, 1989) microscopically (Brundrett, Piche and Peterson, 1984).

Determination of plant nutrient concentration

The shoot phosphorus (P%) and nitrogen content (N%) of the plants were determined by using the ascorbic acid procedure as described in the *Laboratory methods of soil and plant analysis: a working manual* (Okalebo, 1993) and Indophenol Blue Method (Allen, 1974), respectively, after an acid digestion treatment.

Statistical analysis

Results were subjected to two-way analysis of variance and the significance was determined according to Duncan's Multiple Range Test (Gomez and Gomez, 1984).

RESULTS AND DISCUSSION

Solanum tuberosum plants varied in their response to inoculation with *Glomus macrocarpum, Azotobacter chroococcum,* and their combinations. Generally, inoculated *Solanum tuberosum* plants had greater growth compared to uninoculated controls (Table 20.1). *Solanum tuberosum* plants inoculated with *Glomus macrocarpum* attained maximum height when nitrogen was added at 48 kg/ha. The *Azotobacter chroococcum* inoculated plants, however, required less nitrogen, that is, nitrogen at 24 kg/ha to attain height at par with *Glomus macrocarpum* inoculated plants. The dually inoculated plants attained maximum height when soil was fertilized with nitrogen at 48 kg/ha (Table 20.1).

Dual inoculated plants irrespective of rate of nitrogen application produced maximum shoot biomass. Inoculation of *Solanum tuberosum* plants with *Azotobacter chroococcum* either alone or in combination with *Glomus macrocarpum* produced maximum shoot biomass. Increasing soil nitrogen levels had a negative impact on shoot biomass in both inoculated and uninoculated plants. Efficiency of mycorrhizal fungi was significantly improved when they were used along with *Azotobacter chroococcum.*

The positive effect of microbial inoculants on tiller number, ear number, and grain yield were observed at all the levels of nitrogen-fertilization. The maximum number of tillers and ears was observed at a low level of nitrogen addition (24 kg/ha) in all microbial treatments (Table 20.1). More grain yield was also recorded from plants growing on low levels of nitrogen addition. Rhizobacterization of plants with *Azotobacter chroococcum* resulted in a significant increase

Table 20.1.: Growth of *Solanum tuberosum* L. Plants as Influenced by *Glomus macrocarpum* and *Azotobacter chroococcum* for 60 Days

Microbial inoculums	N-level (kg/ha)	Height (cm)	Biomass (g)	Tuber No.	1000 grain weight (g)	Shoot N (%)	Shoot P (%)
Control	0	74.00	10.00	4.00	17.64	2.08	0.073
	(e)	(f)	(i)	(h, i)	(e, f)	(j)	(j)
	24	82.00	15.45	5.00	19.07	2.70	0.081
		(c, d, f)	(e)	(g, h, i)	(c, d, e)	(g, h)	(g, h)
	48	87.33	11.00	4.66	18.91	3.19	0.090
		(c, d, e)	(f)	(h, i)	(c, d, e)	(c, d, e)	(c, d)
	144	76.00	8.50	3.00	13.25	3.13	0.076
		(d, e)	(g)	(i)	(g)	(d, e)	(i)
Glomus macrocarpum	24	87.66	25.34	12.00	20.89	2.93	0.093
		(a, b, c)	(b)	(b, c)	(b, c, d)	(f)	(b, c)
	48	96.33	21.64	9.00	19.80	3.28	0.089
		(a)	(c)	(d, e)	(b, c, d, e)	(b, c, d)	(d)
	144	87.00	11.35	5.00	15.66	3.36	0.07
		(b, c)	(f)	(g, h, i)	(f)	(a, b, c)	(h)
Azotobacter chroococcum	0	77.00	15.00	6.00	20.68	2.25	0.079
		(d, e)	(e)	(f, g, h)	(b, c, d)	(i)	(h)
	24	91.66	32.67	12.00	21.19	3.08	0.085
		(a, b)	(a)	(b, c)	(a, b, c)	(e, f)	(e, f)
	48	87.66	20.00	8.00	20.74	3.35	0.094
		(a, b, c)	(c, d)	(d, e, f)	(b, c, d)	(a, b, c)	(b)
	144	78.00	16.00	7.00	17.20	3.40	0.082
		(d, e)	(e)	(e, f, g)	(e, f)	(a, b)	(f, g)
Glomus macrocarpum+ Azotobacter chroococcum	0	78.00	18.50	7.00	22.38	2.62	0.092
		(d, e)	(d)	(e, f, g)	(a, b)	(h)	(b, c, d)
	24	84.00	34.15	18.00	23.82	3.28	0.10
		(b, c, d)	(a)	(a)	(a)	(b, c, d)	(a)
	48	90.33	27.00	14.00	22.50	3.44	0.103
		(a, b, c)	(b)	(b)	(a, b)	(a, b)	(a)
	144	76.33	19.50	10.00	18.40	3.52	0.085
		(d, e)	(d)	(c, d)	(d, e)	(a)	(e, f)

Means with the same letter in a column are not significantly according to DMRP, P<0.05

in tiller number irrespective of whether they were mycorrhizal or non-mycorrhizal. Higher levels of soil nitrogen were inhibitory for growth and yield of rice, irrespective of the nature of microbial inoculants used (Table 20.1). No significant difference was observed between *Glomus macrocarpum* and *Azotobacter chroococcum* for, their ability to induce tiller production at the same level of soil nitrogen. However, *Azotobacter chroococcum* could produce more ears than *Glomus* sp. at the highest level of nitrogen sources.

There is significant increase in shoot phosphorus content in the inoculated plants compared to uninoculated control at all the levels of nitrogen fertilization (Table 20.1). Plants inoculated dually with *Glomus macrocarpum* and *Azotobacter chroococcum* possessed higher shoot phosphorus content than the singly inoculated ones. The highest shoot nitrogen content was also recorded from dual inoculated plants growing on highest soil nitrogen sources.

The endogonaceous spore population was more in inoculated plants than uninoculated control plants, irrespective of the soil nitrogen concentrations (Figure 20.1). *Solanum tuberosum* plants inoculated with *Glomus macrocarpum* harboured more spore numbers than *Azotobacter chroococcum* inoculated ones. Spore numbers in the rhizosphere of dually inoculated rice plants were significantly higher than those of the singly inoculated ones. The maximum spore number was recorded from the rhizosphere of dual inoculated plants growing on 24 kg/ha of nitrogen application and the minimum spore number was recorded from uninoculated unfertilized plants.

The root colonization of inoculated rice plants increased significantly from the uninoculated control (Figure 20.2). *Glomus macrocarpum* only inoculated plants had more roots infected as compared to *Azotobacter chroococcum* only inoculated plants growing on all the levels of soil nitrogen concentrations. More roots were colonized when plants were

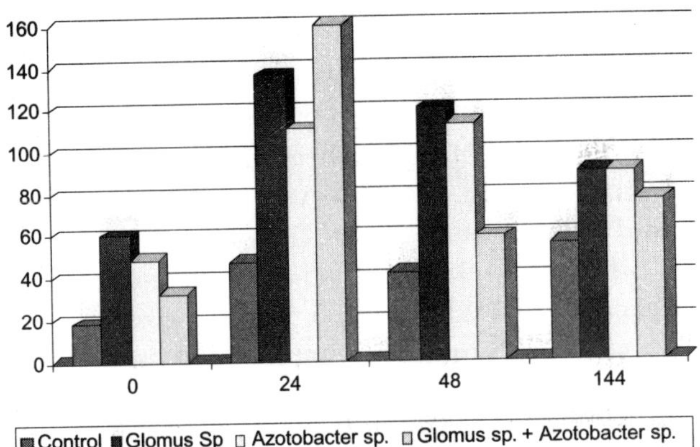

Fig. 20.1.: Effect of *Glmous* Sp. and *Azotobacter* Sp. on Mycorrhizal Spore Population on *Solanum tuberosum* Plants. Bars with Same Letter(s) do not Dif ferent Significantly According to Duncan's Multiple Test at P<0.05

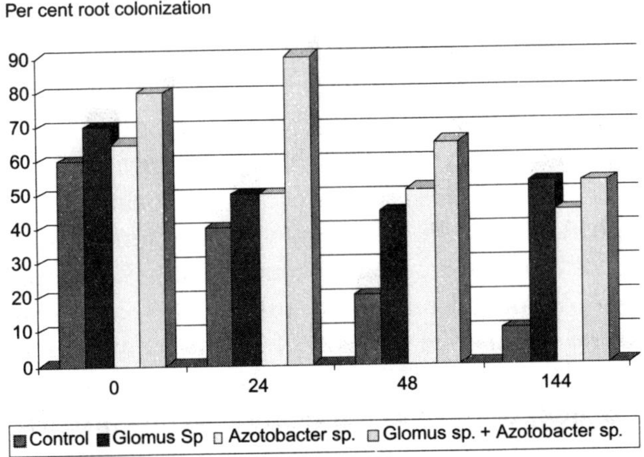

Fig. 20.2.: Effect of *Glmous* sp. and Azotobacter sp. on Percentage Root Infection on *Solanum tuberosum* Plants. Bars with Same Letter(s) Do Not Different Signifi cantly According to Duncan's Multiple Test at P<0.05

inoculated with both the inoculants than with only *Glomus macrocarpum*.

It has been observed that sets inoculated with *Glomus macrocarpum* and *Azotobacter chroococcum*, either alone or in combination with different levels of soil nitrogen application, positively affected crop growth over the treatments that received nitrogen only. As observed earlier (Azcon, Ruiz-Rozano and Rodriguez, 2001; Giovannetti *et al.*, 2002), increasing nitrogen application had a negative effect on plant height, irrespective of whether the plants were inoculated or not. The present result indicates that inoculation of rice plants with *Glomus* sp. and, the dual inoculation of *Glomus* sp. and *Azotobacter chroococcum* enhanced the plant growth when urea was incorporated up to the recommended dose, that is, 48 kg/ha.

Solanum tuberosum plants inoculated with *Glomus macrocarpum* either alone or in combination with *Azotobacter chroococcum*, grew taller than the uninoculated control plants. This is in agreement with earlier works on cotton, maize, and soybean grown on unsterilized soil (Mohan, Bagyaraj and Manjunath, 1984; Click, 1995). Rice plants for tiller production responded more to dual inoculation of AMF and only PGPRs than to either only mycorrhizal fungi or only PGPR. The present results, together with previous reports (Zambre, Konde and Sonar, 1984), confirm that association of crop yields with AMF and PGPRs enhanced the number of tillers per pot. Dhillion, Kler, and Chahal (1980); Rai and Gaur (1982); and Zambre, Konde and Sonar (1984) also reported that a greater number of tillers was produced when the crop was inoculated with PGPR. Rice plants inoculated with mycorrhiza and/or rhizobacteria insignificantly increased the 1000-grain weight over uninoculated control. (Raut and Ingle, 1980; Reyndars and Vlassac, 1982); Kundu and Gaur (1980); and Zambre, Konde and Sonar (1984) have also reported that inoculation of wheat with *Azospirillum* and *Azotobacter* increased grain yield. Such growth responses are variable and depend upon the initial fertility status of soil and the type of crop planted (Subba Rao, Tilak, Lakshmikumari *et al.*, 1980). The increase in growth and yield *of Azotobacter chroococcum* inoculated plants is not necessarily due to the nitrogen fixation by the added rhizobacteria (Lethbridge and Davidson, 1983), but might also be due to the growth hormones secreted by the rhizobacteria (Kucey *et al.*, 1989) and tropical growth conditions (Wani, Chandrapalaiah and Dart, 1982).

In unsterilized soil, inoculated plants showed higher spore population and higher percentage of root infection compared to the uninoculated control plants. This indicates that inoculation with mycorrhizal fungi stimulated *Solanum tuberosum* growth beyond indigenous AMF. This might be because of low indigenous AMF, which allowed effectiveness, introduced AMF to be adequately tested.

It was hypothesized that AMF would be effective only for the acquisition of slowly diffusing nutrients by plants (Lakshman and Ratageri, 2005). However, the present work demonstrates the effectiveness of mycorrhizal roots to promote nitrogen uptake, even when nitrogen, in the form of urea, is present in the soil in non-limiting amounts. This is in agreement with the recent findings of (Medina *et al.*, 1988; Azcon, Ruiz-Rozano and Rodriguez, 2001) that mycorrhizal roots promote nitrogen uptake when present in non-limiting amounts.

The addition of nitrogen at 24 kg ha^{-1} improved the growth parameters in inoculated plants. The statistically significant interaction between nitrogen addition and microbial treatments suggests that the differences in growth parameters (Moyed *et al.*, 1997) owing to different microbial treatments were due to different rates of nitrogen application, and the nitrogen response differed between inoculated and non-inoculated plants. The results of the present work strongly suggest that application of bioinoculants like AMF plus PGPR would enable farmers for optimizing *Solanum tuberosum* production with minimum input of inorganic fertilizer.

REFERENCES

Allen, S. E. (1974). Chemical analysis of ecological materials Blackwell scientific publication, New Delhi. Pp. 1921–193.

Azcon, R., Ruiz-Rozano, J. M. and Rodriguez R. (2001). Different Contribution of arbusctilar mycorrhizal fungi to plant nitrate uptake (15 N) under increasing N supply to the soil. Canadian Journal of Botany. 97: 1175–1180.

Berea, J. M. (2000). Rhizosphere and mycorrhiza of field crops. In biological resources management: connecting science and Policy edited by J.P. Toutant, E. Galante, J. M. Lynch, J. S. Schepers, D. Warnerand and P. A. Werry. (OECD) INRA Editions and Springer, Berlin, Heilberg New York, pp. 110–125.

Borea, J. M. (2000). Rhizosphere and mycorrhiza of field crops In: *Biological Resource Management: Connecting Science and Policy* edited by J. P. Toutant, E. Balazs E. Galante, J. M. Lynch, J. S. Schepers, D. Werner and P. A. Werry (OECD) INRA Editions and Springer, Berlin, Heidelberg New York, pp. 110–125.

Brundrett, M. C., Piche, Y. and Peterson, R. L. (1984). A new method for observing the morphology of Vesicular Arbuscular mycorrhizae. *Indian Journal of Botany* 62: 2128–2134.

Dhillion, G. JS., Kler, D. S. and Chahal, V. P. S. (1980). Effect *of Azotobacter* inoculation along with different doses of nitrogen on growth and yield of wheat *Indian Journal of Agronomy*. 25: 533–535.

Gerdemann, J. W. and Nicolson, T. H. (1963). Spores of mycorrhizal Endogone extracted from soil by wet sieving and decanting. *Transaction of British Mycological Society*. 46: 235–244.

Goldstein, A. H. (1995). Recent progress in understanding the molecular genetics and biochemistry of calcium phosphate solublization by gram negative bacteria. Biol.Agri.Hort. 12:185–193.

Giovannetti, M. S., Brana, C. and Avio, L. (2002). Arbuscular mycorrhizal fungal mycelium: from germilings to hyphal networks In: Mycorrhyzal *Technology in Agriculture from*

Genes to Bioproducts, edited by S. Gianinazzi, H Schuepp, J. M. Barea and K. Haselwandter Birkhauser, Basel, pp. 49–58.

Glick, B. R. (1995). The enhancement of plant growth by free-living bacteria *Canadian Journal of Microbiology*. 41:109–117

Gomez, K. A. and Gomez, A. A. (1984). Statistical producers for agricultural research (2nd edn). An international Rice Research Institute book. John Wiley and Sons, New York.

Kacey, R. M. N., Janzen, H. H. and Legett, M. E. (1989). Microbioloy mediated increases in plant available phosphorus. Adv. Agron. 42:198–228.

Koske, R. E. and Gremma J. N. (1989). A modified procedure for staining roots to detect VA Mycorrhizas. *Mycological Research*. 92: 486–505.

Kundu, B. S. and Gaur, A. C. (1980). Establishment of nitrogen-fixing and phosphate solublizing bacteria in rhizosphere and their effect on yield and nutrient uptake of wheat crop. *Plant and soil*. 57:223–230.

Lakshman, H. C. and Ratageri, R. H. (2005). Dual inoculation of VA-mycorrhiza and Rhizobium beneficial to *Pithecblobium dulce* Roxb. *J. Microbial. Word*. 7(2): 163–169.

Lethbridge, G. and Davidson, M. S. (1983). Root associated nitrogen fixing bacteria and their role in the nitrogen nutrition of wheat estimated by ^{15}N isotope dilution. *Soil Biology and Biochemistry*. 15:365–374.

Lloyd, A., Webb, J., Archer, J. R. and Sylvester-Bradley, R. (1997). Urea as a nitrogen fertilizer for cereals. *Journal of Agricultural Science*. 128: 263–271

Martinez-Toledo, M. V., De la Rubia, T., Moreno, J. and Gqjazalez-Lopez, J. (1988). Root exudates of *Zea mays* and production of auxins, gibberelliris and cytokinins *Azotobacter Chroococcum*. *Plant and Soil*. 110: 149–152.

Me Gonigle, T. P. (1998). A numerical analysis of published field trials with vesicular arbuscular mycorrhizal fungi. Functional Ecology. 2: 473–478.

Medina, O. A., Sylvia, D M. and Keretschmer Jr. A. E. (1988). Response of sirato to vesicular arbuscular fungi of different phosphorus levels: *Soil Science Society of American Journal*. 52: 420–423.

Mohan R., Bagyaraj D. J. and Manjunath A. (1984). Response of crop plants to Vesicular-Arbuscular Mycorrhizal inoculation in five unsterile soils Karnataka. *Journal of Soil Biology and Ecology* 4(1):6–12.

Okalebo, J. R., Gathua, K. W. and Woomer, P. L. (1993). Laboratory Methods of Plant and Soil Analysis: A Working Manual. *Technical Bulletin No. 1, Soil Science Society, East Africa.*

Rai, S. N. and Gaur, A. C. (1982). Nitrogen fixation by *Azospirillum* spp and effect of *Azospirillum lipoferum* on the yield and N-uptake of wheat crop. *Plant and Soil*. 69:233–238.

Raut, R. S. and Ingle, U. M. (1980). *Azospirillum brasilense* as a biofertilizer to wheat (HD1593). *Research Bulletin Marathwada Agricultural University* 4: 8–9.

Reynders, L. and Vlassak K. (1982). Use of *Azospirillum brasilence* as biofertizer in intensive wheat cropping. *Plan and Soil*, 66: 217–223.

Schreiner, R. P. and Bethenfalvay, G. J. (1995). Mycorrhizal interactions in sustainable agriculture. *Critical Review biotechnology*. 15:271–287.

Smith S. E. and Read, D. J. (1997). Mycorrhizal Symbiosis Academic Press, San Diego. pp. 605.

Subba Rao, N. S., Tilak, K.V.B.R., Lakshmikumari, M. and Singh, C. S. (1980). Azospirillum a new bacterial fertilizer. *Indian Farming*. 30: 3–5.

Wani, S. P., Chandrapalaiah, S. and Dart, P. J. (1985). Response of pearl-millet cultivars to inoculation with nitrogen fixing bacteria. *Expl Agric* 21: 175–182.

Wani, S. P., Dart, P. L. and Subba Rao, R. V. (1982). Factors affecting the nitrogenase activity of sorghum and millet estimated by soil-root core assay Method *National Symposium on Biological Nitrogen Fixation*. IARI. New Delhi pp. 57.

Zambre, M. A., Konde, B. K. and Sonar, K. R. (1984). Effect of *Azotobacter chroococcum* and *Azospirillum brasilense* inoculation under graded levels of Nitrogen on growth and yield of wheat. *Plant and Soil*. 79: 61–67.

21

Plant Growth Promoting Rhizobacteria as Bioinoculants

Avita K Marihal, S M Pradeep and K S Jagadeesh

INTRODUCTION

Preparations of live microorganisms (bacteria, fungi) utilized for improving plant growth and crop productivity are generally referred to as **microbial inoculants** (Subba Rao and Dommergues, 1998). During the late 19th and early 20th centuries, inorganic compounds containing nitrogen, potassium and phosphorus (NPK) were synthesized and used as fertilizers. Due to the growth in human populations, fertilizers were used to increase crop production and meet the rising demands for food. Increases in the production cost, and the hazardous nature of chemical fertilizers for the environment has led to a resurgence of interest in the use of biofertilizers for enhanced environmental sustainability, lower cost production and good crop yields. Research in the field of biofertilizers has resulted in the development of different kinds of microbial inoculants or biofertilizers including nitrogen fixing bacteria, phosphate solubilizing microorganisms, vesicular-arbuscular mycorrhizae (VAM) and plant growth promoting rhizobacteria (PGPR).

Dept. of Agricultural Microbiology, University of Agricultural Sciences, Dharwad 580 005, Karnataka, India.

Plant growth-promoting rhizobacteria (PGPR) are free-living soil-borne bacteria that colonize the rhizosphere, and when applied to seed or crops enhance the growth of plants (Kloepper *et al.*, 1980). The rhizosphere is the soil found around the root and is under the influence of roots. It is a site with complex interactions between the root and associated microorganisms (Sylvia *et al.*, 1998). In the past 10–15 years, close to 4,000 publications have appeared in the field of plant growth-promoting bacteria (Bashan and Holguin, 1998). Plant growth-promoting rhizobacteria enhance plant growth either by direct or indirect mechanisms (Glick, 1995). The enhancement of plant growth by PGPR indicates their potential as biofertilizers in the field of agriculture. A variety of plant associated bacteria improve plant growth and it is estimated that about 2–5% of the rhizosphere bacteria are plant growth promoting rhizobacteria (Antoun and Kloepper, 2001). These bacteria belong to diverse genera, including *Acetobacter, Achromobacter, Anabaena, Arthrobacter, Azoarcos, Azospirillum, Azotobacter, Bacillus, Burkholderia, Clostridium, Enterobacter, Flavobacterium, Frankia, Hydrogenophaga, Kluyvera, Microcoleus, Phyllobacterium, Pseudomonas, Serratia, Staphylococcus, Streptomyces,* and *Vibrio*, as well as the well-known legume symbiont *Rhizobium.*

MECHANISMS OF PGPR PLANT GROWTH PROMOTION

PGPR enhance plant growth by direct and indirect means, but the specific mechanisms involved have not all been well-characterized. Direct mechanisms of plant growth promotion by PGPR can be demonstrated in the absence of plant pathogens or other rhizosphere microorganisms, while indirect mechanisms involve the ability of PGPR to reduce the deleterious effects of plant pathogens on crop yield. PGPR have been reported to directly enhance plant growth by a variety of mechanisms like:

(a) **Phosphorus solubilization**: Some rhizosphere bacteria solubilize organic and/or inorganic phosphorus by means of acid phosphatases production in the first case and organic acids production in the second case (Goldstein, 1986; Rodriguez and Fraga, 1999). Phosphorus solubilisation by bacteria increases the availability of this element to plants.

(b) **Production of phytohormones**: One of the direct mechanisms by which PGPR promote plant growth is by production of plant growth regulators or phytohormones (Glick, 1995). Frankenberger and Arshad (1995) have discussed in detail the role of auxins, cytokinins, gibberellins, ethylene and absicisic acids (ABA) which, when applied to plants, help in increasing plant yield and growth. Microbial production of individual phytohormones such as auxins and cytokinins has been reviewed by various authors over the last 20 years (Fallik and Okon, 1989, Barbieri and Galli, 1993; Patten and Glick, 1996; Patten and Glick, 2002). A great number of rhizobacteria synthesise auxins; about 80% of the bacteria isolated from plant rhizosphere produce indole-3-acetic acid, one of the most physiologically active auxins (Cherryl and Glick, 1996; Arshad and Frankenberger, 1991). Cytokinins have been found in bacterial cultures of *Azotobacter chroococcum, Azospirillum brasilense, Bacillus cereus mycoides, Agrobacterium tumefaciens* and others (Arshad and Frankenberger, 1991; Tien *et al.*, 1979), and gibberellins have been detected in cultures of *Azotobacter vinelandii, A. beijerinkii* and *Azospirillum brasilense* (Azcon and Barea, 1975; Tien *et al.*, 1979). Salamone (2000) reported the growth-promoting effect of *P. fluorescens* strain G20–18 on wheat and radish plants by production of cytokinin phytohormones. Bent *et al.* (2001) reported that the production of indole compounds by three different strains, *Paenibacillus polymyxa* L6, *P. poly*myxa Pw-2, and *Pseudomonas fluorescens* M20 increased in concentration with increasing concentrations of tryptophan (0–200 mg/ml) at different times. Several observations indicated that production of phytohormones enhanced plant growth.

(c) **Production of siderophores**: The production and secretion of siderophores with a very high affinity to ferric iron under conditions of limited availability of this element allow PGPR to prevent the proliferation of phytopathogens. Pseudobactins are siderophores produced by fluorescent pseudomonads. Pseudobactins immobilize ferric iron present in the rhizosphere, thus, depriving pathogenic microorganisms of this element and inhibiting their growth (Leeman *et al.*, 1996). Under these conditions, plants do not suffer from iron deprivation because they have the capacity to bind the bacterial iron-siderophore complex and to assimilate it.

(d) **Antibiosis**: Antibiosis perhaps is the most investigated mechanism of biocontrol. Usually, it consists of biochemical warfare among organisms. Several PGPR inhibit or kill phytopathogenic microorganisms by secreting metabolic products such as antibiotics, enzymes, lytic agents, volatile compounds and other toxic substances (Berg *et al.*, 2000; Maurhoffer *et al.*, 1992; Graham and Mitchell, 1998). Antibiotics are the substances produced by microorganisms, which in very dilute concentrations, have the capacity to inhibit growth of other microorganisms or kill them. Many phytopathogens are sensitive to *Bacillus* species which produce antibiotics (Pusey, 1989).

SELECTION AND CHARACTERIZATION OF PGPR

One of the challenges in developing PGPR for commercial application is ensuring that an effective selection and screening procedure is in place, so that the most promising organisms are identified and brought forward. In the agricultural chemical industry, thousands of prospective compounds are screened annually in efficient high-throughput assays to select the best one or two compounds for further development. Similar approaches are not yet in place for PGPR. Effective strategies for initial selection and screening of rhizobacterial isolates are required. It may be important to consider host plant specificity or adaptation to a particular soil, climatic conditions or pathogen in selecting the isolation conditions, and screening assays (Bowen *et al.*, 1999). The spermosphere model, an enrichment technique that relies on seed exudates as the nutrient source, has been used for selection and isolation of promising N_2–fixing rhizosphere bacteria from rice (Thomas-Bauzon, 1982). One approach for selection of organisms with the potential to control soil-borne phytopathogens is to isolate from soils that are suppressive to that pathogen (Weller, 2002). Other approaches involve selection based on traits known to be associated with PGPR such as root colonization (Silva, 2003), 1-aminocyclopropane-1-carboxylate (ACC) deaminase activity (Cattelan *et al.*, 1999; Glick, 1995), and antibiotic (Giacomodonato *et al.*, 2001) and siderophore production (Cattelan *et al.*, 1999). The development of high throughput assay systems and effective bioassays will facilitate selection of superior strains (Mathre, 1999; McSpadden, 2002).

DELIVERY SYSTEMS

Plant growth promoting rhizobacteria are delivered through several means, based on survival nature and mode of infection of the pathogen. They are delivered through seed, soil, foliage, rhizomes, sets, or through combination of several methods of delivery.

INOCULATION TO SEEDS

Application of PGPR has given positive results in a number of crops. Organic farmers are familiar with the practice of inoculating legume seed with specific *Rhizobium* strains to ensure adequate nitrogen fixation. However, it is also possible to inoculate seeds with other bacteria that increase the availability of other nutrients, including solubilizing phosphate, oxidizing sulphur, chelating iron and copper.

Seed bacterization with PGPR has emerged as a powerful technology to enhance plant growth and yield, besides providing protection against diseases. The direct promotion of plant growth by PGPR generally entails providing the plant with a compound that is synthesized by the bacterium or facilitating the uptake of nutrients from the environment (Glick, 1995 and Glick *et al.*, 1999). Plant growth benefits due to the addition of PGPR include increase in germination rates, root growth and yield including grain, increase in leaf area, chlorophyll content, magnesium, nitrogen and protein content, hydraulic activity, tolerance to drought and salt stress, shoot and root weights and delayed leaf senescence (Lucy *et al.*, 2004). Earnapalli *et al.* (2005) isolated several rhizobacteria, screened and selected promising strains effective against *A. solani*. Results revealed that seven strains possessed high biocontrol potential against early blight of tomato in a glasshouse. Out of these, *Pseudomonas* strain B–25 was found to be the most promising strain with a disease control of 57.78 per cent and good plant growth promotion. Zehnder *et al.* (2001) reported that delivering of *Serratia marcescens* strain 9–166 as seed dip before planting, and application of 100 ml of the same @ 10^8 cfu/ml to the sterilized soil less planting mix, reduced bacterial wilt of cucumber and controlled cucumber beetles besides increasing the fruit weight. Many researchers have reported increased seed P content by phosphate solubilizing microorganisms (Kucey, 1987; Zaidi *et al.*, 2004)

Talc based formulation of *P. fluorescens* Pf1 was coated on to seeds @ 4g/kg (10^7cfu/g) of chickpea seeds for the management of chickpea wilt (Vidhyasekaran and Muthamilan, 1995). Treatment of cucumber seeds with strain mixtures comprising of *Bacillus pumilus*- 1NR7, *B. subtilis*-GB 03 and *Curtobacterium flaccum faciens*- ME1 with a mean bacterial density of 5×10^9 cfu/seed reduced the intensity of angular leaf spot and anthracnose, equivalent to the synthetic elicitor Actigard and was better than seed treatment with individual strains (Raupach and Kloepper, 1998).

In a series of experiments on cucumber, plants were seed inoculated with the PGPR *Pseudomonas fluorescens* 89B–27 and *Serratia marcescens* 90–166 and challenged with Cucumber Mosaic Virus (CMV). The bacterial treatments reduced the number of plants showing mosaic symptoms in 28 days after virus inoculation (Raupach *et al.*, 1996). Soaking of rice seeds in water containing 10 g of talc based formulation of *P. fluorescence* consisting of PF1 and PF2 (10^8 cfu/g) for 24 h controlled rice sheath blight under field condition (Nandakumar *et al.*, 2001). Seed treatment of lettuce with either vermiculite or kaolin based carrier of *B. subtilis* (BACT–0) significantly reduced root rot caused by *Pythium aphanidermatum* and it also increased the fresh weight of lettuce under greenhouse conditions. Seed treatment with vermiculite based *P. putida* reduced *fusarium* root rot of cucumber and increased the yield and growth of cucumber (Amer and Utkede, 2000).

SOIL APPLICATION

Vidhyasekaran and Muthamilan (1995) stated that soil application of peat based formulation of *P. fluorescens* (Pf_1) at the rate of 2.5 kg of formulation mixed with 25 kg of well decomposed farmyard manure; in combination with seed treatment increased rhizosphere colonization of Pf_1 and suppressed chickpea wilt caused by *Fusarium oxysporum*.

Broadcasting of talc based formulation of strain mixtures (Pf_1 and FP7) by blending 2.5 kg of formulation with 50 kg of sand after 30 days of transplanting paddy seedlings to main field significantly reduced sheath blight and increased yield under field, conditions (Nandakumar *et al.*, 2001).

Vasudevan *et al.* (2002) revealed that incorporation of commercial chitosan based formulations LS 254 (comprising of *Paenobacillus macerans* + *B. pumilus*) and LS 255 (comprising of *P. macerans* + *B. subtills*) in to soil at the ratio of 1:40 (formulation : soil) increased bio-matter production by increasing both root and shoot length and yield. Soil application of the strain mixture formulations LS 256 and LS 257 comprising of two different *Bacillus* spp., was better than seed treatment and suppressed downy mildew under greenhouse and field conditions (Niranjan *et al.*, 2003).

FOLIAR SPRAYING METHOD

The efficacy of biocontrol agents for foliar diseases is greatly affected by fluctuation of microclimate.

Application of *Bacillus subtilis* to bean leaves decreased incidence of bean rust (*Uromyces phaseoli*) by 75 per cent which was equivalent to weekly treatments with the fungicide mancozeb (Baker et al., 1983). Meena *et al.* (2002) found that application of *P. fluorescens* on to foliage (@ 1 kg talc based formulation/ha) on 30, 45, 60, 75 and 90 days after sowing, reduced leaf spot and rust of groundnut under field conditions.

Pre-harvest foliar application of talc based fluorescent pseudomonad strain Fp7 supplemented with chitin at fortnightly intervals (5g/plant spray volume 20 L/tree) on to mango trees from per flowering to fruit maturity stage induced flowering to the maximum, reduced the latent infection by *Colletotrichum gleosporioides* besides increasing the fruit quality and yield (Vivekananthan *et al.*, 2004).

COMBINATION OF DELIVERY SYSTEMS

Seed treatment of pigeonpea with talc based formulation of fluorescent *Pseudomonas* @ 4 g per kg of seed followed by soil application at the rate of 2.5 kg/ha at 0, 30 and 60 days after sowing controlled pigeonpea wilt incidence under field conditions. The additional soil application of talc based formulation improved disease control and increased yield compared to seed treatment alone (Vidhyasekaran *et al.*, 1997).

Delivering of *Pseudomonas fluorescens* as seed treatment followed by three foliar applications suppressed rice blast under filed conditions (Krishnamurthy and Gnanamanickam, 1998). Benhamou *et al.* (1996) reported that compatible mixing PGPR strains may enhance several plant defense mechanisms with synergistic activity culminating to have a broad spectrum of resistant activity.

GENETIC ENGINEERING OF PGPR STRAINS

Genomic tinkering of naturally occurring PGPR strains with genes that are beneficial to plants will lead to the accentuated expression of the genomic products which could alleviate the attack of both pests and diseases. This will facilitate for the introduction of a single bacterium with multiple modes of action to benefit the growers to save their crop with increased returns by reducing the inputs invested for plant protection measures. However, the release of genetically modified organisms is a policy decision to be made by the policy makers. Hence, appraising of policy makers about the safe usage of beneficial bacteria will be a boon to the farming community and environment.

STRATEGIES TO PROMOTE COMMERCIALIZATION

Commercialization of biocontrol could be promoted by :

- Popularization of biocontrol agents
- Industrial linkages

1. Popularization of biocontrol agents

Motivating the growers through

(a) Publicity

(b) Field demonstrations

(c) Farmers days

(d) Biovillage adoption

(e) Conducting periodical trainings for commercial producers and farmers to increase / improve the supply.

2. Industrial linkages

(a) Technical support should be made available to entrepreneurs on quality control and registration.

(b) Regular monitoring is essential to maintain the quality.

(c) Constant research support should be extended to standardize the dosage, storage, and delivery systems. Positive policy support from Government to use more of biocontrol agents in crop protection.

FUTURE PROSPECTS

As our understanding of the complex environment of the rhizosphere, of the mechanisms of action of PGPR, and of the practical aspects of inoculant formulation and delivery increases, we can expect to see new PGPR products becoming available. The success of these products will depend on our ability to manage the rhizosphere to enhance survival and competitiveness of these beneficial microorganisms (Bowen *et al.*, 1999). Rhizosphere management will require consideration of soil and crop cultural practices as well as inoculant formulation and delivery (Bowen *et al.*, 1999). Genetic enhancement of PGPR strains to enhance colonization and effectiveness may involve addition of one or more traits associated with plant growth promotion (Glick, 1995). Genetic manipulation of host crops for root-associated traits to enhance establishment and proliferation of beneficial microorganisms (Mansouri, 2002) is being pursued. However, regulatory issues and public acceptance of genetically engineered organisms may delay their commercialization. The use of multi-strain inocula of PGPR with known functions is of interest as these formulations may increase consistency in the field (Jetiyanon, 2002). They offer the potential to address multiple modes of action, multiple pathogens, and temporal or spatial variability. Thus, PGPR offer an environmentally sustainable approach to increase crop production and health.

REFERENCES

Amer, G. A. and Utkhede, R. S. (2000). Development of formulations of biological agents for management of root rot of lettuce and cucumber. *Can. J. Microbiol.* 46: 809–816.

Antoun, H. and Kloepper, J.W. (2001). Plant Growth Promoting Rhizobacteria. *Encyclopedia of Genetics*. London: Academic Press.

Arshad, M. and Frankenberger Jr, W. T. (1991). Microbial production of plant hormones. *Plant Soil.* 133: 1–8.

Azcon, R. and Barea, J. M. (1975). Synthesis of auxins, gibberellins and cytokinins by *Azotobacter vinelandii* and *Azotobacter beijerinckii* related to effects produced on tomato plants. *Plant Soil.* 43: 609–619.

Baker, S. C., Stavely, J. R., Thomas, C. A., Sasser, M. and Mac. Fall, S. J. (1983). Inhibitory effect of *Bacillus subtilis* on *Uromyces phaseoli* and on development of rust pustules on bean leaves. *Phytopathol.* 73:1148 –1152.

Barbieri, P. and Galli, E. (1993). Effect on wheat root development of inoculation with an *Azospirillum brasilense* mutant with altered indole-3-acetic acid production. *Research Microbiol.* 144: 69–75.

Bashan, Y. and Holguin, G. (1998). Proposal for the division of plant growth promoting rhizobacteria into two classifications. Biocontrol PGPB and PGPB. *Soil Biol. Biochem.* 20: 1225–1228.

Benhamou, N., Kloepper, J. W. and Tuzun, S. (1998). Induction of resistance against *Fusarium* wilt of tomato by combination of chitosan with an endophytic bacterial strain: ultrastructural and cytochemistry of the host response. *Planta.* 204: 153–168.

Bent, E., Tuzun, S., Chanway, C.P. and Enebak, S. (2001). Alterations in plant growth and in root hormone levels of lodgepole pines inoculated with rhizobacteria. *Can. J. Microbiol.* 47: 793–800.

Berg, G., Kurze, S., Buchner, A., Wellington, E. M. and Smalla, K. (2000). Successful strategy for the selection of new strawberry-associated rhizobacteria antagonistic to *Verticillium* wilt. *Can. J. Microbiol.* 46: 1128–1137.

Bowen, G. D. and Rovira, A. D. (1999). The rhizosphere and its management to improve plant growth. *Adv. Agron.* 66:1–102.

Cattelan, A. J., Hartel, P. G. and Fuhrmann, J. J. (1999). Screening for plant growth-promoting rhizobacteria to promote early soybean growth. *Soil Sci. Soc. Am. J.* 63:1670–1680.

Cherryl, P. and Glick, B. (1996). Bacterial biosynthesis of indole-3-acetic acid. *Can. J. Microbiol.* 42: 207–220.

Earnapalli, V.N. (2005). *Screening of antagonistic microorganisms for biological control of early blight of tomato caused by Alternaria solani (Ell. & Martin). Thesis (M.Sc Agri)*, Department of Agril. Microbiology, University of Agricultural Sciences, Dharwad, Karnataka.

Fallik, E. and Okon, Y. (1989). Identification and quantification of IAA and IBA in *Azospirillum brasilense* inoculated maize roots. *Soil Biol. Biochem.* 21:147–153

Frankenberger, Jr. W.T. and Arshad, M. (1995). *Phytohormones in soils: Microbial production and function.* Marcel Dekker, Inc. New York, NY, p. 503.

Giacomodonato, M. N., Pettinari, M. J., Souto, G. I., Mendez, B. S. and Lopez, N. I. (2001). A PCR-based method for the screening of bacterial strains with antifungal activity in suppressive soybean rhizosphere. *World J. Microbiol. Biotechnol.* 17:51–55.

Glick, B.R. (1995). The enhancement of plant growth by free-living bacteria. *Can. J. Microbiol.* 41: 109–117.

Glick, B.R., Patten, C.L., Holguin, G. and Penrose, D.M. (1999). *Biochemical and genetic mechanisms used by plant growth promoting bacteria*. Imperial College Press, London, United Kingdom, p. 267.

Goldstein, A. H. (1986). Bacterial solubilization of mineral phosphates: historical perspective and future prospects. *American J. Alter. Agri.* 1: 51–57.

Graham, J. H. and Mitchell, D. J. (1998). *Biological control of soilborne plant pathogens and nematodes. In: Principles and Applications of Soil Microbiology* (eds. D. M. Sylvia, J. J. Fuhrmann, P. G. Hartel and D. A. Zuberer). Prentice Hall, New Jersey, USA, pp. 427–446.

Jetiyanon, J. and Kloepper, J. W. (2002). Mixtures of plant growth-promoting rhizobacteria for induction of systemic resistance against multiple plant diseases. *Biol. Control.* 24:285–291.

Kloepper, J.W., Leong, J., Teintze, M. and Schroth, M.N, (1980). Enhanced plant growth by siderophores produced by plant growth promoting rhizobacteria. *Nature.* 286: 885–886.

Krishnamurthy, K. and Gnanamanickam, S. S. (1998). Biological control of rice blast by *Pseudomonas fluorescens* strain Pf7–14: Evaluation of a marker gene and formulations. *Biol. Control.* *13*: 158–165.

Kucey, R.M.N. (1987). Increased phosphorus uptake by wheat and field beans inoculated with a phosphorus-solubilizing *Penicillium bilaji* strain and with vesicular-arbuscular mycorrhizal fungi. *Appl. Environ. Microbiol.* 53: 2699–703.

Leeman, M., Den Ouden, F. M., Van Pelt, J. A., Dirkx, F. P. M., Steijl, H., Bakker, P.A. H. M. and Schippers, B. (1996). Iron availability affects induction of systemic résistance to *Fusarium* wilt of radish by *Pseudomonas fluorescens*. *Phytopathol.* 86: 149–155.

Lucy, M., Reed, E. and Glick, R. B. (2004). Applications of free living plant growth promoting rhizobacteria. *Antonie van Leeuwenhoek.* 86: 1–25.

Mansouri, H., Petit, A., Oger, P. and Dessaux, Y. (2002). Engineered rhizosphere: the trophic bias generated by opine-producing plants is independent of the opine type, the soil origin, and the plant species. *Appl. Environ. Microbiol.* 68:2562–2566.

Mathre, D. E., Cook, R. J. and Callan, N. W. (1999). From discovery to use. Traversing the world of commercializing biocontrol agents for plant disease control. *Plant Dis.* 83: 972–983.

Maurhoffer, M., Keel, M. C., Schnider, U., Voisard, C., Haas, D. and Defago, G. (1992). Influence of enhanced antibiotic production in *Pseudomonas fluorescens* strain CHAO on its disease suppressive capacity. *Phytopathol.* 82: 190–195.

McSpadden Gardener, B. B. and Fravel, D. R. (2002). *Biological control of plant pathogens: Research, commercialization, and application in the USA*. Online. Plant Health Progress doi: 10.1094/PHP-2002-0510-01-RV.

Meena, B., Radhajeyalakshmi, R., Marimuthu, T., Vidhyasekaran, P. and Velazhahan, R. (2002). Biological control of groundnut late leaf spot and rust by seed and foliar applications of a powder formulation of *Pseudomonas fluorescens*. *Biocont. Sci. Technol.* 12: 195–204.

Nandakumar, R., Babu, S., Viswanathan, R., Sheela, J., Raguchander, T. and Samiyappan, R. (2001). A new bio-formulation containing plant growth promoting rhizobacterial mixture for the management of sheath blight and enhanced grain yield in rice. *Biocontrol.* 46: 493–510.

Niranjan Raj, S., Deepak, S. A., Basavaraju, P., Shetty, H. S., Reddy, M. S. and Kloepper, J. W. (2003). Comparative performance of formulations of plant growth promoting rhizobacteria in growth promotion and suppression of downy mildew in pearl millet. *Crop Protect.* 22: 579–588.

Patten, C. and Glick, B.R. (1996). Bacterial biosynthesis of indole-3-acetic acid. *Can. J. Microbiol.* 42: 207–220.

Patten, C.L. and Glick, B.R. (2002). Regulation of indoleacetic acid production in *Pseudomonas putida* GR12-2 by tryptophan and stationary-phase sigma factor RpoS. *Can. J. Microbiol.* 48: 635–642.

Pusey, P.L. (1989). Use of *Bacillus subtilis* and related organisms as biofungicides. *Pest. Sci.* 27:113–140.

Raupach, G. S. and Kloepper, J. W. (1998). Mixtures of plant growth promoting rhizobacteria enhance biological control of multiple cucumber pathogens. *Phytopathol.* 88: 1158–1164.

Rodriguez, H. and Fraga, R. (1999). Phosphate solubilizing bacteria and their role in plant growth promotion. *Biotech. Adv.* 17: 319–349.

Salamone, I. E. G. (2000). *Direct beneficial effects of cytokinin producing rhizobacteria on plant growth. Ph.D. Thesis,* University of Saskatchewan, Saskatoon, SK, Canada.

Silva, H. S. A., Romeiro, R. S. and Mounteer, A. (2003). Development of a root colonization bioassay for rapid screening of rhizobacteria for potential biocontrol agents. *J. Phytopathol.* 151:42–46

Subba Rao, N.S. and Dommergues, Y.R. (1998). *Microbial interactions in Agriculture and Forestry*, Science Publishers, Inc, U.S.A. pp. 278.

Sylvia, D.M., Fuhrmann, J.J., Hartel, P.G. and Zuberer, D.A. (1998). *Principles and applications of Soil Microbiology.* Prentice-Hall, Inc. Upper Saddle River, NJ. pp. 550.

Thomas-Bauzon, D., Weinhard, P., Villecourt, P. and Balandreau, J. (1982). The spermosphere model. I. Its use in growing, counting and isolating N_2-fixing bacteria from the rhizosphere of rice. *Can. J. Microbiol.* 28:922–928.

Tien, T. M., Gaskins, M. H and Hubbell, D. H. (1979). Plant growth substances produced by *Azospirillum brasilense* and their effect on the growth of pearl millet *(Pennisetum americanum* L.). *Appl. Environ. Microbiol.* 37: 1016–1025.

Vasudevan, P., Reddy, M. S., Kavitha, S., Velusamy, P., avid PaulRaj, R. S., Purushothaman, S. M., Brindha Priyadarsini, V., Bharathkumar, S., Kloepper, J. W. and Gnanamanickam, S. S. (2002). Role of biological preparations in enhancement of rice seedling growth and grain yield. *Curr. Sci.* 83: 1140–1144.

Vidhyasekaran, P. and Muthamilan, M. (1995). Development of formulations of *Pseudomonas fluorescens* for control of chickpea wilt. *Plant. Dis.* 79: 782–786.

Vidhyasekaran, P., Sethuraman, K., Rajappan, K and Vasumathi, K. (1997). Powder formulation of *Pseudomonas fluorescens* to control pigeonpea wilt. *Biol. Contro.* 8: 166–171.

Vivekananthan, R., Ravi, M., Ramanathan, A. and Samiyappan, R. (2004). Lytic enzymes induced by *Pseudomonas fluorescens* and other biocontrol organisms mediate defence against the anthracnose pathogen in mango. *World J. Microbiol. Biotechnol.* 20: 235–244.

Zaidi, A., Khan, M.S. and Aamil, M. (2004). Bioassociative effect of rhizospheric microorganisms on growth, yield, and nutrient uptake of greengram. *J. Plant Nurition.* 27: 601–12.

Zhender, G. W., Murphy, J. F., Sikora, J. E. and Kloepper, J.W. (2001). Application of rhizobacteria for induced resistance. *Eur. J. Plant Pathol.* 107: 39–50.

MYCORRHIZAL TECHNOLOGY IN ADVERSE ENVIRONMENT

22

Application of Microbial Biotechnology for Sustainable Legume Production in Desert Conditions

*Nishi Mathur**, *Joginder Singh, Sachendra Bohra, Avinash Bohra** *and Anil Vyas*

INTRODUCTION

There is a strong awareness that in the next 30 years the world population will grow to 8 billion people. That growth will require an increase of agricultural production of 75–100% to provide the needed food. If such an increase in production should be realized with current agricultural management that would similarly require a doubling of the use of fossil fuel energy for fertilizer production and cause economic hardship and surpassing damage to the environment (Graham and Vance, 2003; Norse, 2003). The availability of a useful nitrogen (N) source is, except for water, the major limiting factor in agricultural productivity. That has commonly resulted in the heavy use of chemical nitrogen fertilizer to replenish soil

* Department of Biotechnology, Mahila P. G. Mahavidhyalaya, Jodhpur- 342 001 Microbial Biotechnology and Biofertilizer Laboratory, Department of Botany, J. N. V. University, Jodhpur- 342 003

nitrogen, an approach that suffers from high costs and severe, environmental effects. In the late 1870's, Dr. Martinus Willem Beijerinck discovered the bacteria *rhizobium* and identified that it created nodules on legumes (species of Leguminosae, like soybeans), that "fixed" nitrogen from the air. It became known as biological nitrogen fixation (BNF). Biological nitrogen fixation (BNF) is the process whereby atmospheric nitrogen (N=N) is reduced to ammonia in the presence of nitrogenase. Nitrogenase is a biological catalyst found naturally only in certain microorganisms such as the symbiotic *Rhizobium* and *Frankia*, or the free-living *Azospirillum* and *Azotobacter*. Biological nitrogen fixation is brought about by free-living soil microorganisms and by symbiotic associations of microorganisms with higher plants. Leguminous plants, like soybeans, fix atmospheric nitrogen by working symbiotically with special bacteria, rhizobia, which live in the root nodules. Rhizobia infect root hairs of the leguminous plants and produce the nodules. The nodules become the home for bacteria where they obtain energy from the host plant and take free nitrogen from the soil air and process it into combined nitrogen. In return, the plant receives the fixed N from nodules and produces food and forage protein. Parts of the fixed nitrogen can then can be used for the soybean plant itself or carried-over in the soil for the next year's crop; the major reason why soybeans are often cropped in rotation with very N-demanding crops (such as corn for example). Inoculation with very efficient rhizobia bacteria strains provides the opportunity for legumes to fix N in the soils biologically beyond their natural capability or in cases where the rhizobia bacteria naturally present in the soil have become low in number or ineffective. Inoculation is a means of providing nitrogen to legumes (and succeeding crops) in addition to the amounts that could have been naturally fixed and therefore is an environmentally sound alternative for chemical nitrogen fertilizer for soybeans and crops grown in rotation with soybeans.

Throughout history, agriculture has depended on cropping systems that combined a nitrogen-consuming cereals with nitrogen-fixing legumes. In recent history cereals productivity has dramatically increased under high input systems whilst legume yields have neared a plateau, stagnated or even reduced. This has resulted in unbalanced cereal-legume global production, and thus in higher and unsustainable dependence on N chemical fertilizer inputs which reached $ 20 billion annually (Crouch *et al.*, 2004). Increasing cultivation of legumes will be required to ameliorate environmental degradation, reduce depletion of non-renewable resources and provide adequate nitrogen for sustainable agriculture. Most of the new lands to be opened for legume cultivation in the developing countries are located in the dry desert arrears. Symbiotic nitrogen fixation (SNF) by legumes plays major role in sustaining crop productivity of marginal lands and in small holders systems. Farmers in dry areas depend on legumes as an important crop in their cropping systems due to the capacity of these plants to fix nitrogen from air by the interaction with nitrogen fixing rhizobia. It is well known that nitrogen is abundant in the atmosphere, but plants cannot directly utilize the elemental nitrogen from the air. Symbiotic nitrogen fixation occurs mainly through symbiotic association of legumes with N_2-fixing rhizobia that convert elemental nitrogen into ammonia. This type of biological nitrogen fixation (BNF) is therefore less costly and more sustainable as compared with nitrogen fertilizes for production of plant proteins. Scientific and technological progress has opened tremendous opportunities for the befit of small farmers. Most of the lands for future agricultural expansion in different countries are located in deserts, which are prevalent in these dry areas of African and Asian regions. These areas are characterized by nutrient deficient sandy soil with poor water holding capacity, scarcity of water and high temperature (Bohra *et al.*, 2005). The experience gained by agricultural institutions in these regions for the production of food, feed and fibers overcame the challenges posed by the conditions of dry environment to turn the desert soils into productive agricultural land. Legumes played an important role in this type of agriculture. It is well known that legumes are among the key plants necessary for building the sustainable agricultural production systems particularly in the newly reclaimed soils. The integration of legume-rhizobia biological nitrogen fixation (BNF) technology in desert agriculture reduces the cost of application of microbial biotechnology for sustainable legume production in desert conditions production and preserves underground water from contamination with the excess nitrogen fertilizers required to insure the high productivity. High proportion of soluble nitrogen fertilizers could be leached with irrigation water and reaches the ground water. The employment of BNF technology to supply host legumes with the nitrogen requirements essential for plant growth, attracted the attention of scientists since early studies of biology and microbiology. However, the dependence on legume-rhizobia symbiotic systems for agricultural expansion in the desert areas requires careful assessment of the performance of these biological systems under the harsh conditions prevailing in the desert environment. Substantial information is available on the net benefits of BNF by legumerhizobia symbiosis. However, this biological system is particularly sensitive to various environmental stresses. In desert areas the major stress factors is the drought associated with the scarcity of water availability. It is reported that most of the stress factors including drought influence all physiological process in plants (VanHaroon *et al.*, 2001; Mathur *et al.*, 2005; Bouhmouch *et al.*, 2005; Sinclair *et al.*, 2007; Farooq *et al.*, 2009). Abdel Al (2004) also showed that legume-*Rhizobium* symbiosis is particularly sensitive to drought. This stress factor may impair the development of root hairs and the site of entry of rhizobia to the host. It is generally agreed that rhizobial strains are relatively more tolerant to the drought stress than the plant host (Williams and Mallorca, 1984). A deeper root system with enhanced water uptake capacity is considered the major plant developmental trait for drought avoidance in legume crop native to desert and arid conditions (King *et al.*, 1994).

The characteristics of a sustainable agriculture are decreased fossil energy consumption, increased use of renewable resources, introduction of more diversity in crops, decreased transportation, and to ensure soil fertility and food quality. We will briefly examine the contribution of legumes to these different aspects. Legumes minimise agricultural consumption of fossil energy and so lower its contribution to the greenhouse effect Legumes have the unique ability to fix nitrogen, which is a renewable resource, thanks to the formation of root nodules in which they house symbiotic bacteria, the Rhizobia. This symbiotic nitrogen fixation is extremely efficient and worldwide the production of combined nitrogen (ammonia) by the *Rhizobium*-legume symbiosis is higher than the production of nitrogen fertiliser by the chemical industry. Using nitrogen fertilisers consumes fossil energy: the equivalent of two tons of fuel is required to produce and spread one ton of nitrogen fertiliser. The use of nitrogen fixing plants i.e. legumes to produce plant proteins results in a decrease in the consumption of fossil energy and therefore decreases the agricultural contribution to global warming. The cultivation of a legume saves the consumption of approximately 0.2 tons of fuel per hectare, corresponding to the production of 600 kg of CO_2 per hectare. Legumes are thus a pillar in the development of sustainable agricultural systems. It has been estimated that more than 50% of the energy costs of non-legume crops are associated with the generation and application of nitrogen fertilizer.

The present investigation illustrates our research efforts on the integration of BNF by legume-rhizobia symbiotic systems in several legumes adapted for growth in arid and desert environments. The analysis of the factors involved in the success or failure of SNF systems provide the knowledge that facilitates development of appropriate management options for sustainable legume production in desert conditions.

MATERIALS AND METHODS

The present study covers a wide range of greenhouse and field-testing of rhizobial strains performance in desert soils. The performance was assessed by the application of various techniques. For the quantification of the amount of nitrogen fixed under field condition the ^{15}N isotope dilution technique was used (Burris and Miller, 1941, Fried and Middelboe, 1977, Amarger *et al.*, 1979). The competition for nodulation between native and inoculants rhizobial strains was assessed by fluorescent antibody (FA) technique (Somasegaran and Hoben, 1994). For studying the factors limiting the response to inoculation in several legumes in sandy soils, the advanced molecular biology techniques, such as Rep-PCR of rhizobial DNA was applied (Louws *et al.*, 1994) in addition to FA analysis of nodule occupancy.

RESULTS AND DISCUSSION

Sustainable agricultural production is attractive system because it is supposed to cause least damage to the environment while ensuring high productivity. This is where SNF becomes an indispensable component of sustainable farming systems. Improvement in SNF would come from selection and ability to use successfully better strains of N_2-rhizobia. Where the naturally occurring strains in the soil are proven to be highly effective, inoculation likely to be not necessary, which may be the case in many areas where particular legumes have been cultivated successfully over many years. In other words, the benefit from inoculation is not easy in soils that harbor high numbers of very competitive but ineffective N_2-fixing strains, but should not be regarded as impossible. The worldwide *Rhizobium* Ecology Network (WREN) supported by NSF to establish prediction model for inoculation success for economically important legumes

(Moawad *et al.,* 1994). It is recognized that native soil population of rhizobia can be characterized functionally for nitrogen fixation potential by determining their numbers and effectiveness (Singleton and Travers, 1986). The MPN counts of rhizobia nodulating the tested legumes and the effectiveness of the native rhizobia for these legumes are presented in Table 22.1.

Table 22.1. : MPN of Rhizobia in Soil and Its Effectiveness

Host plants	Counts g^{-1} soil	Effectiveness*
Glycine max	0.0	0.0
Vigna aconitifolia	0.0	0.0
Trifolium alexandrenum	1×10^2	55.0
Vicia faba	5×10^2	70.0
Vigna unguiculata	4.2×10^2	60.0

*Percentage of pink nodules formed by native population in untreated plots

The table shows that the tested soils were void of rhizobia for soybean and beans. However, rhizobia modulating faba bean and clover occurred in densities of 100 cell/g and for cowpea 10,000 cells/g. The date show that the effectiveness of rhizobia was high within the native population of faba bean rhizobia (70%) and less for cowpea and clover rhizobia being 60% and 55% respectively. The response to inoculation in these legumes in pot experiments (Table 22.2) show significant yield responses to rhizobial inoculation with selected strains of rhizobia with soybean, beans, cowpea, lentil and clover. No significant differences were found between inoculated and N-fertilized treatments. No yield response was found in faba bean either inoculated or fertilized with nitrogen. The data show the application of microbial biotechnology for sustainable legume production in desert conditions importance of rhizobial inoculation in the studied legumes with the exception of faba bean which showed no response to inoculation.

Table 22.2. Response of Six Legume Crops to Rhizobial Inoculation (Yield tons/ha)

Treatments	Soybean	Beans	Cowpea	Faba bean	Lentil	Clover
Uninoculated	2.16 a	0.76 a	1.20 a	3.02 a	1.27 a	24.87 a
Inoculated	3.25 b	0.99 b	1.99 b	3.06 a	1.78 b	27.47 b
900 kg N/ha	3.22 b	1.07 b	2.08	3.15 a	1.82 b	27.65 b

The competitive ability of inocula strains (Table 22.3) shows that the inoculants trains dominated nodules of soybean, beans, cowpea, clover and lentil. However, most of faba bean nodules were formed from the indigenous rhizobia, which likely are very efficient in N_2 fixation.

Table 22.3: Competition for Nodulation Between Inoculants Strains and Native Rhizobia in Several Legumes

Host Legume	Strain nodule occupancy (%)	
	Inoculants	Native
Soybean	100	0
Beans	100	0
Cowpea	64	36
Clover	56	44
Faba bean	4	96
Lentil	68	32

The quantification of SNF in the field is of great importance to establish nitrogen fertilization requirements. Nonetheless, the estimation of nitrogen fixed by various legume-rhizobia symbiotic systems under field conditions is not an easy task. The problem becomes more complicated if one wants to assess the N_2 fixing capacity of particular potential inoculants strain in a field with native background rhizobia. In this case the N_2 fixed may come from two sources; the inoculants strain and/or the native strains existing in the soil. The ^{15}N isotope dilution technique (Fried and Middleboe, 1977) is the only available approach for the quantification of N_2 Fixation in the field. We used this approach in several field studies (Moawad *et al.*, 1989 and 1998). The opening of new desert areas for legume cultivation usually creates unique opportunities for accurate measurement of N_2 fixation with the first field cultivation with legume crop with no need for isotope technique. In such soils the success of nodulation depends primarily on the introduction of inoculant strain to the host legume. The quantification of N_2 fixed can be calculated by comparing the total nitrogen in both inoculated and uninoculated fields (difference method). This approach was applied in one cowpea field trial in newly reclaimed soil at Mandore, Jodhpur, Rajasthan. The rhizobial field application was done using two separate approaches; peat-based inoculants and liquid inoculants injection with pivot irrigation system. The cells of the three recommended rhizobial strains were equally mixed for inoculant preparation. Both inoculation techniques were capable of supplying peanut under field conditions with enough nitrogen to support good growth and yield with high net N_2 gains (Table 22.4). Dart and Krantz (1977) reported that peanut can fix appreciable amounts of nitrogen and estimated as much as 240 kg N/ha.

Table 22.4. : Yield and N_2 Fixation in Field Grown Peanut Using Two Inoculation Techniques

Inoculation	Yield kg ha^{-1}	N_2 fixed kg N ha^{-1}
Uninoculated	1855 a*	23 a
Peat-based	5398 b	192 b
Liquied	4425 b	175 b

*Values with different letters within the same vertical column are significantly different.

In soils with history of cultivation of certain legume particularly with repeated inoculation the competition for nodulation play significant role in SNF. It is reported that the background native rhizobia in the field may adversely affect the chances of nodulation by the desired inoculant strains (Moawad and Schmidt, 1984, Moawad and Bohlool, 1984, Moawad *et al.*, 1988 and 1991).

Table 22.5.: Nodule Occupancy by Native Rhizobial Strains in Several Legume Cultivars Using FA Analysis.

Legume	% of nodules	Reference
Lentil		Moawad et al. 1994
Giza 9	24	
Giza 370	28	
Beans		Moawad and Wafaa, 2002
Dual	50	
Paultista	20	
Smamtha	30	

Table 22.5 shows that the comprehensive screening programs to identify the highly fixing strains of rhizobia for specific host faces the challenge of native strains which limits the benefits of SNF capacity of the selected strains when inters into competition with these native soil rhizobia. In the two legume crops listed in this table the native rhizobial strains occupied substantial proportion of nodules indicating that inoculant strains were not capable of nodulating the host legume. *Vigna aconitifolia* is among the legume crops which grow very well in sandy and loam soils. The symbiotic and competitive performance of two highly effectiverhizobia was tested with two cultivars; Jwala and Maru moth. Data in Table 22.6 show that strain RMO-40 was the best nitrogen fixer with Jwala, whereas strain CZM-99 was better with Maru moth.

Table 22.6. Response of Two Bean Cultivars to Rhizobial Inoculation

Treatments	Jwala		Maru moth	
	Shoot dry weight g/plant	N uptake mg/plant	Shoot dry weight G/plant	N uptake mg/plant
Uninoculated	4.83 b*	135.2	4.16 c	158.9
+N**	9.03 a	307.0	8.08 a	315.1
CZM-99	2.96 b	109.5	5.55 b	199.8
RMO-40	9.54 a	282.2	4.32 c	113.4

* Values labeled with the same letters are not significantly different using Duncan's multiple-range test.

** Fertilized with nitrogen.

Two techniques were applied to assess the success of nodulation by inoculant trains with the two bean cultivars. These techniques are FA and REP-PCR fingerprinting. The results from both techniques were close (Table 7).

Table 22.7.: Nodule occupancy (%) by Inoculant Strains with Two Bean Cultivars

Treatments	Jwala		Maru moth	
	CZM-99	RMO-40	CZM-99	RMO-40
Uninoculated	-	-	-	-
+N**	0	0	0	0
Ph #163	30 (40)*	0 (0)	**30 (40)	0
CE3	0 (0)	50 (40)	0 (0)	20 (10)

* Numbers in side parenthesis are % nodule occupancy using FA.

** Numbers without parenthesis are % nodule occupancy using REP-PCR Fingerprinting.

None of the strains was able to take all the nodule sites on both cultivars. Strain RMO-40 gave 40–50% nodulation on Jwala, whereas strain CZM-99 gave 30–40% on Maru moth. Same nodule occupancy was found with strain CZM-99 on cultivar Jwala (30–40%). The native rhizobia occupied 50% of nodules on Jwala and much more on Maru moth. These data the importance of selection of the proper rhizobial strains for use as inocula in different bean cultivars, to insure the nodulation by desired strain. The benefit from SNF system can be increased with the in depth studies of the performance of selected rhizobial strains under the prevailing environmental conditions. The more we study these biological systems the better we can sustain legume production in arid and semi arid conditions to open new agricultural land in the marginal arid soils particularly desert which dominate the future potential land reserve for production of food and feed for ever increasing population in these regions.

REFERENCES

Abdel, Al H. M. (2004). Application of Microbial Biotechnology for Sustainable Legume Production in Desert Conditions. *International Conf. on Water Resources and Arid Environment.* Pp. 1–10.

Amarger, N., Mariotti A., Mariotti F., Durr J.C., Bourguignon, C. and Lagachere, B. (1979). Estimate of symbiotically fixed nitrogen in field grown soybeans using variation in ^{15}N natural abundance. *Plant and Soil.* 52: 269–280.

Bouhmouch, I., Souad-Mouhsine, B., Brhada F. and Aurag, J. (2005). Influence of host cultivars and *Rhizobium* species on the growth and symbiotic performance of *Phaseolus vulgaris* under stress. *Journal of Plant Physiology.* 162(10): 1103–1113.

Bohra, S., Singh, J., Mathur, N. and Vyas, A. (2005). Increased nutrient uptake and productivity of *Dolichos biflorus* by mycorrhizal fungi. *J. Arid legumes.* 2(1): 187–189.

Burris, R.H. and Miller, C.E. (1941). Application of 15N to the study of biological nitrogen fixation *Science.* 93: 114–115.

Crouch, J.H., Buhariwalla H.K., Blair M., Mace E., Jayashree B. and Serraj R. (2004). Biotechnology-based Contributions to Enhancing Legume Productivity in Resource-poor Areas. Pp. 47–66. *In Symbiotic Nitrogen Fixation, Prospects for Enhanced Application in Tropical Agriculture.* Editor: Rachid Serraj, Oxford and IBH. Publishing Co. Pvt. Ltd. New Delhi.

Dart, P.J. and Krantz, B. (1977). A. Legumes in the semiarid tropics. In: Exploiting the legume-*Rhizobium* symbiosis in tropical agriculture, J.M. Vincentm, A.S.Whinery & J. Bose (Eds.). Univ. Hawaii, Coll. *Trop. Agric. Misc. Publ.* 145:119–154.

Farooq, M., Wahid, A., Kobayashi, N., Fujita, D. and Basra, S.M.A. (2009). Plant drought stress: effects, mechanisms and management, *Agron. Sustain. Dev.* 29: 185–212.

Fried, M. and Middelboe, V. (1977). Measurement of amount of nitrogen fixed by a legume crop. *Plant and Soil.* 47: 713–715.

Graham, P.H. and Vance, C.P. (2003). Legumes: Importance and constraints to greater utilization. *Plant Physiol.* 131:872–877.

King, J., Gay, A., Sylvester-Bradley, R., Bingham, I., Foulkes, J., Gregory, P. and Robinson, D. (2003). Modelling Cereals Root

Systems for Water and Nitrogen Capture: Towards Economic Optimum. *Annals of Botany.* 91: 383–390.

Louws F.J., Fulbright D.W., Taylor S.C. and de Bruijn F.J. (1994). Specific genomic fingerprints of *phytopathogenic Xanthomonas* and *Pseudomonas pathovars* and strains generated with repet-protein-due binding, *Anal. Biochem.* 72: 248–254.

Mathur, R., Singh, J., Mathur, N., Bohra, S. and Vyas, A. (2005). Mycorrhizal relationship of Mung bean: Influence of Abiotic factors. *Journal of Arid Legumes.* 2 (1): 193–195.

Moawad H. and Wafaa M, Abd El-Rahim. (2002). Assessment of symbiotic performance of several bean cultivars inoculated with their specific rhizobia. *Egyptian Journal of Microbiology.* 37(3): 247–262.

Moawad H. and Wafaa M., Abd El-Rahim, (2004). Dossoki Abd El-Aleem. Performance of phaseolus bean rhizobia in soils from the major production sites in Nile Delta. *Journal of Comptes rendus - Biologies.* 327(5): 445–453.

Moawad, H. and Badr El-Din, S.M.S. (1989). Assessment of rhizobial inoculation response in field-grown lentil using ^{15}N dilution technique. In Proceedings of 12th North American Symbiotic Nitrogen Fixation Conference, Ames, Iowa, pp. 80.

Moawad, H. Badr, El-Din, S.M.S. and Radwan, S.M.A. (1994). Assessment of need to inoculate for widely cultivated Legumes crop in Egypt. In: Recent Development in Biological Nitrogen Fixation in Africa, M. Sidiki and A. Hilali (eds.), IAV Hassan II, Rabat, Morroco, pp.16–23.

Moawad, H. and Bohlool, B. (1984). Competition among *Rhizobium* spp. strains for nodulation of *Leucaena leucocephala* in two tropical soils. Appl. *Environm. Microbiol.* 48:5–9.

Moawad, H., Badr el-Din, S.M.S. and Khalafallah, M.A. (1988). Persistence and competitiveness of three *Bradyrhizobium japonicum* inoculant strains in clayloam Nile Valley soil. *Plant and Soil. 108*: 137–141.

Moawad, H., Ellis, W.R. and Schmidt. (1984). E.L. Rhizosphere response as a factor in competition among three serogroups of indigenous *Rhizobium japonicum* fornodulation of field-grown soybeans. *App. Environm. Microbiol.* 47:607–612

Moawad, H., Khalafallah, M.A. and Badr El-Din. S.M.S. (1991). Competition between three *Bradyrhizobium japonicum* as affected by strain proportion in inocula. *Egypt. J. Microbiol.* 26:230–277.

Moawad, H., Badr El-Din, S.M.S. and Abdel-Aziz, R.A. (1998). Improvement of biological nitrogen fixation in Egyptian winter legumes through better management of *Rhizobium*. *Plant and Soil*. 204: 95–106.

Norse, D. (2003). Fertilizers and world food demand implications for environmental stresses. In: IFA-FAO Agriculture Conference, Global Food Security and the Role of Sustainable Fertilization.

Rachid Serraj. (2004). Symbiotic Nitrogen Fixation, Prospects for Enhanced Application in Tropical Agriculture. Oxford and IBH Publishing Co. Pvt. Ltd. New Delhi.

Sinclair, T.R., Purcell, L.C., King, C.A., Sneller, C.H., Chen, P. and Vadez, V. (2007). Drought tolerance and yield increase of soybean resulting from improved symbiotic N_2 fixation. *Field Crop Research*. 101(1): 68–71.

Singleton, P.W. and Tavares, J.W. (1986). Inoculation response of legumes in relation to number and effectiveness of indigenous *Rhizobium* populations. *Appl. Environ. Microbiol.* 51: 1013–1018.

Somasegaran, P. and Hoben, H. J. (1994). Handbook for Rhizobia (Methods in legume-*Rhizobium* Technology). Springer-Verlag New York, Inc., 175 Fifth Avenue, New York, USA.

VanHoorn, J.W, Katerji N, Hamdy, A. and Mastrorffli, M. (2001). Effect of salinity on yield nitrogen uptake of four grain legumes and on biological nitrogen contribution from the soil. *Agricultural Water Management*. 51:87–98.

Williams, P.M. and deMallorca S.M. (1984). Effects of osmotic ally induced leaf moisture stress on nodulation and nitrogenase activity of *Glycine max*. *Plant Soil*. 80: 267–283.

23

Application of Mycorrhiza on Some Important Oil Yielding Plants in Alleviating High Concentration of Zinc

Roopa K Jambotkar[1], *H C Lakshman*[2] *and Shweta J Sabannavar*[2]

INTRODUCTION

Living organisms in the biosphere exhibit a number of interactions which either alter their environment and or the size and composition of each other populations. Perhaps the most striking relationship is "symbiosis" in which both partners derive benefit from each other. Mutualistic symbiosis established between photosynthetic plants and soil microbes play an important role in the ecosystem. The well known example of such symbiosis between non pathogenic soil fungi and roots of flowering plants is "Mycorrhizae".

Arbuscular mycorrhizal fungi are geographically ubiquitous and occur over a broad ecological range. They are

1. J.S.S's Shri Manjunatheshwar P.U. Science College, Vidyagiri, Dharwad-580005, India.
2. Post Graduate Department and Research in Botany (Microbiology Lab) Karnatak University, Dharwad-580003, India.

commonly associated with agriculture, horticulture pastures and tropical forests. (Lakshman, 1992)

The advantages of AM fungi in crop productivity raise concern for their survival and preservation in soils. The conservation in soils covers many aspects of agricultural crop production and management programmes. (Hayman and Mosse, 1973). The beneficial functions of AM fungi are concerned with the uptake, translocation and transfer of nutrients, mainly phosphate from soil to host cells thereby improving nutrition and growth of host plants particularly the plant nutrients which are available at low concentration in soil.

Although, these fungi have extremely wide host range, they are not specific in their effect on host plant species. Their effect can vary on cultivars within a single plant species and also on plants in different ecosystems, soil types and similar soils differing in their physico-chemical constituents.

Phytoremediation is a sustainable and inexpensive technology based on the removal of pollutants from the environment by plants, is becoming an increasingly important objective in plant research. However, Phytoremediation is a slow process, improvement of efficiency and thus increased stabilization or removal of heavy metals from soils is an important goal. Arbuscular mycorrhizal fungi provide an attractive system to advance plant based environmental clean-up. During symbiotic interactions, the hyphal network functionally extends the root system of their hosts. Thus plants in symbiosis with AM fungi have the potential to take up heavy metal from an enlarged soil volume.

Montilla, M.A., Parra *et al.*, studied the effect of Zinc Phytotoxicity in oilseed such as rapeseed grown on zinc-loaded substrates consisting of Fe oxide coated and calcite sand in the year (2003). Borbala Biro and his coworkers in the year (2005) worked on the mycorrhiza. That they have shown mycorrhiza functioning as a part of the survival mechanism of barley (*Hardeum Vulgare* L) at a long term heavy metal stress. In the year 2006 Vera Gohre and Uta Paszkowski studied the contribution of arbuscular mycorrhizal symbiosis to heavy metal (Phytoremediation). Roopa and Lakshman (2007) studied the phytoremediation on oil yielding plants inoculated with mycorrhiza and treated with Zinc.

Zinc toxicity in plants results in reduced growth of both root and shoot (Choi *et al.*, 1996; Ebbs and Kochian, 1997; Fontes and Cox, 1998) and chlorosis in the younger leaves. This can extend to older leaves after prolonged exposure to high soil Zn levels (Ebbs and Kochian, 1997). The chlorosis may arise partly from and induced iron (Fe) deficiency as hydrated Zn^{+2} and Fe^{+2} ions have similar radii (Marschner, 1998). Excess Zn can also give rise to manganese (Mn) and copper (Cu) deficiencies in shoot: such deficiencies have been ascribed to hinder transfer of these micronutrients from root to shoot based on the fact that the Fe ad Mn concentrations in plants grown in Zn-rich media are greater in the root than the shoot (Ebbs and Kochian, 1997). One other typical effect of Zn toxicity is the appearance of a purpulish-red color in leaves, which is ascribed to phosporus (P) deficiency (Li *et al.*, 2004).

The concentration of Zinc (Zn) in soil has increased in various regions over the last few decades as a consequence of human activities. Zinc concentration in soil usually ranges from 10 to 300 mg/kg. The average being 50 mg/kg (Mortvedt, 2000), however it can reach thousands of mg/kg in severely Zinc contaminated soils. Anthropogenic sources of Zinc in soil include sewage sludge, compost, agrochemicals and mine tailings (Kiekens, 1990). The critical soil zinc toxicity level for plants varies as a function of plant genotype and various soil and climatic factors.

MATERIALS AND METHODS

The experiment was conducted under polyhouse condition using sterile red sandy loamy soil. The soil is mixed with four different concentrations of Zinc (Zn) to determine the effect of colonization by the arbuscular mycorrhizal fungus on the three different varieties of oil yielding plants such as Sunflower, Safflower and Mustard, under soil-zinc conditions. Plants of Sunflower (*Helianthus annuus L.*) Var. Morden, Safflower (*Carthamus tinctorius L.*) Var. Annigeri and Mustard (*Brassica juncea L.*) local variety were grown in inoculated or uninoculated substrate and subjected to four soil - (Zn) concentrations such as (0, 50, 100 and 150mg/kg of soil).

The seeds of the said varieties were procurred from oil seed research center of Agricultural University Dharwad. The experimental soil was sterilized with 5% methyl bromide and its physico-chemical characteristics were analysed as outlined by Jackson (1973). The seedlings of Sunflower, Safflower and Mustard were raised in earthen pots measuring in 15 × 8cm diameter. Pots containing 3 kg of sterilized red sandy loamy soil. The soil and pure sand is mixed in the ratio of 3:1. Arbuscular mycorrhizal fungi *Glomus fasciculatum* was multiplied by using onion as host plant. 15g of air dried AM inoculum of *Glomus fasciculatum* was given for soil, prior to sowing of seeds. The inoculum was added to the soil as a thin layer 2 cm below the soil surface. The inoculum consists of 3 g roots bits plus 12 g of rhizospheric soil of host plant with hyphae and sporocarps. After adding inoculum to the soil, it is mixed with different concentrations of zinc as mentioned below :

1. Control Soil (Non-AM) + 0 conc. of Zn.
2. Gf soil (AM) + 0 conc. of Zn.
3. Control soil + 50 mg of Zn conc.
4. Gf soil + 50 mg of Zn conc.
5. Control soil + 100 mg of Zn conc.
6. Gf soil + 100 mg of Zn conc.
7. Control soil + 150 mg of Zn conc.
8. Gf soil + 150 mg of Zn conc.

Experimental pots were maintained in triplicates in polyhouse. Plants were watered on alternate days to maintain moisture level. 2 ml of Hoagland solution without P was treated for each plant at interval of 15 days. They are kept free of weeds and irrigated properly. The observations were recorded at a period of 30, 60 and 90 days after sowing. Plants were examined for their growth parameters such as Shoot length, Fresh weight of shoot, Dry weight of shoot, root length, fresh weight of root, dry weight of root, number of leaves, number of inflorescence, number of fruits, percent of root colonization.

Percent of Mycorrhizal colonization in the roots of the plants was estimated according to the method following (Phillips and Hayman, 1970) and Giovanetti and Mosse (1980). The spore count was carried out by wet sieving and decanting technique (Gerdemann and Nicolson, 1963).

The percentage of colonization was calculated by the following formula :

$$\text{Percent of root colonization} = \frac{\text{No. of root bits showing colonization}}{\text{Total no. of root bits observed}} \times 100$$

ANALYSIS OF HEAVY METALS IN EXPERIMENTAL PLANTS

Zinc content was determined using atomic absorption spectrophotometer according to (Koide, 1991). It was done using cathode tube specific for zinc and wavelength adjusted to 215 nm. Zinc chloride solution was used for preparing the standard curve. The digested and diluted samples of phosphorus estimation were fed to the atomic absorption spectrophotometer and the readings were recorded. Zinc concentration was computed using the standard curve.

PLANT DRY WEIGHT

Shoot and root samples were dried in an oven at 70°C to a constant weight. The weight of both shoot and roots were recorded and expressed per plant basis. The dried root and shoot samples were powdered in a grinder and used for nutrient analysis.

PHOSPHORUS ESTIMATION IN SHOOTS OF EXPERIMENTAL PLANT

Phosphorus was analysed colorimetrically followed by the Vanadomolybdate Yellow colour method (Jackson, 1973). The finely ground sample (0.5 g) was digested in triacid mixture containing conc. of nitric acid, perchloric acid and sulphuric acid (7:3:1 v/v). The residue was dissolved in distilled water and the volume made upto 100 ml. From this 10 ml aliquot was taken in a 50 ml volumetric flask, 10 ml vanadomolybedate reagent added and the volume made upto 50 ml. The intensity of yellow colour due to phosphovanadomolybdate complex was read at 420 nm in a spectronic 20 colorimeter. The phosphorus concentration was determined by comparing with a standard curve.

RESULTS AND DISCUSSION

The three different oil yielding plants when treated with 4 different Zinc concentrations (0, 50, 100. 150 mg/kg soil) showed that the growth of non-mycorrhizal and mycorrhizal plants at a '0' concentration of Zinc there was an increase in the shoot length, Fresh weight, dry weight, root length, number of leaves, number of inflorescence, number of fruits, percent colonization, spore number per 50 g of soil significantly increased growth parameter in mycorrhizal plants than non-mycorrhizal plants. When these three plants treated with 50 mg of Zinc concentration to both AM and non-AM plants, it is evident from Tables (23.1, 23.2 and 23.3) that there is an increase in growth parameters in mycorrhizal plants compared to non-mycorrhizal plants.

Further when these plants treated with 100 mg of Zinc concentration in inoculated and uninoculated plants. There was an increase in growth parameters in inoculated plants than compared to uninoculated plants. The plants treated with 50 mg of Zinc when compared with the plants treated with 100mg of Zinc. It is evident from the Tables (23.1, 23.2 and 23.3) that there was an increase in growth parameters of inoculated plants treated with 50 mg of Zinc than compared to 100 mg of Zinc of inoculated plants.

Further these three plants treated with 150 mg of Zinc concentration to both inoculated and uninoculated, there was an increase in growth parameters of inoculated plants than compared to uninoculated. But when this treatment is compared with 50 mg of Zinc, 100 mg of Zinc of inoculated and uninoculated plants there is reduced in growth parameters. All the data was collected at 30, 60 and 90 days of plant harvest period from inoculated plants treated with different Zinc concentrations in relation to growth parameters over uninoculated plants as shown in Table (23.1, 23.2 and 23.3).

From available data it can be furnished that plants treated with *Glomus fasciculatum* had better tolerance to Zinc toxicity compared to uninoculated control plants. Further it can be recorded that plants treated with low quantity of Zinc (50 mg/kg soil) and *Glomus fasciculatum* performed well compared to uninoculated Zinc treated plants. However growth is hindered by increasing concentration of Zinc. No. of flowers and inflorescence were observed in 30 days.

The total Zinc content significantly ($P<0.05$) increased as soil (Zn) concentration increases in shoot of both inoculated and uninoculated plants. (Table Nos. 23.1, 23.2 and 23.3) and Fig. 23.2 more specifically the Zn concentration were significantly lower in shoot of uninoculated plants than compared to inoculated plants. In the shoots, the P concentrations significantly decreased as soil (Zn) concentration increased (Table No. 23.1, 23.2 and 23.3) and Fig. 23.1.

Table 23.1. Effect of Zinc on Shoot Length, Fresh Weight of Shoot, Dry Weight of Shoot, Root Length, Fresh Weight of Root, Dry Weight of Root, Number of Leaves, Number of Inflorescence, Number of Fruits, per cent Root Colonization, Spore Number/50 gm of Soil, Phosphorous Uptake and Zinc Uptake with Respect to *Glomus fasciculatum*.

Safflower													
	SL (cm)	FWS (gm)	DWS (gm)	RL (cm)	FWR (gm)	DWR(gm)	NL	NI	NFR	PC%	SN	P%	Zn (ppm)
30 Days													
CN	12.100h	0.36h	0.122h	5.400h	0.193h	0.092h	7.000d	0.000h	0.000h	0.000e	0.000e	0.120g	32.930f
Gf	24.233c	1.195g	0.142g	6.100g	0.363g	0.118g	12.000e	0.000g	0.000g	72.166b	38.000c	0.220bc	43.533e
CN+50mg of Zn	18.166e	2.967d	0.565d	9.300d	0.466d	0.293d	14.00b	0.000f	0.000f	0.000e	0.000e	0.150f	48.320d
Gf+50mg of Zn	28.300a	5.624a	0.862a	13.100a	0.626a	0.483a	17.000a	0.000e	0.000e	86.100a	56.000a	0.240bc	57.710c
CN+100mg of Zn	16.566	2.303e	0.482e	8.100e	0.452e	0.264e	12.000c	0.000d	0.000d	0.000e	0.000e	0.180de	53.320c
Gf+100mg of Zn	25.200b	4.868b	0.772b	11.233b	0.522b	0.392b	15.00b	0.000c	0.000c	70.100c	41.00b	0.260b	63.440b
GN+150mg of Zn	14.100g	1.533f	0.373f	7.433f	0.392f	0.213f	12.000c	0.000b	0.000b	0.000e	0.0009e	0.180de	65.710a
Gf+150mg of Zn	23.100d	4.113c	0.626c	10.566c	0.482c	0.363c	15.000b	0.00a	0.000a	46.166d	31.000d	0.800a	66.180a
60 Days													
CN	26.100h	0.75h	0.213h	7.100h	0.393h	0.193h	7.000f	0.333a	0.000h	0.000e	0.000e	0.150f	52.150h
Gf	35.233g	1.282g	0.958g	9.100f	0.426g	0.233g	13.000e	0.666a	0.000g	75.100c	41.000d	0.230d	64.940g
CN+50mg of Zn	56.166c	4.954d	1.828c	11.333d	0.522d	0.393d	15.000cd	0.666a	0.000f	0.000e	0.000e	0.180e	86.040f
Gf+50mg of Zn	61.233a	8.186a	2.626a	16.233a	0.862a	0.492a	18.000a	1.000a	0.000e	95.233a	57.000a	0.280b	95.620e
CN+100mg of Zn	55.233d	3.323e	1.226e	10.166e	0.513e	0.336e	14.000de	0.333a	0.000d	0.000e	0.000e	0.2250c	107.180d
Gf+100mg of Zn	58.100b	7.226b	1.982b	14.100b	0.823b	0.466b	17.000ab	0.666a	0.000c	82.166b	54.000b	0.350a	116.660c
CN+150mg of Zn	48.566f	2.127f	0.994f	8.100g	0.442f	0.288f	14.000de	0.333a	0.000b	0.000e	0.000e	0.280b	124.850b
Gf+150mg of Zn	51.100e	6.287c	1.626d	12.133c	0.653c	0.412c	16.000bc	0.666a	0.000a	73.166d	48.000c	0.360a	132.380a
90 Days													
CN	44.100h	1.168h	0.553g	9.300g	0.613h	0.322h	8.000f	0.666b	0.000h	0.000e	0.000e	0.270g	162.080g
Gf	51.166g	2.353g	1.922f	10.433f	0.693f	0.394g	13.000e	1.000ab	0.000g	86.566c	45.000d	0.680b	194.260e
CN+50mg of Zn	57.300d	5.421d	2.626d	14.100d	0.783d	0.583d	25.000c	1.333ab	0.000f	0.000e	0.0009e	0.350f	126.380h
Gf+50mg of Zn	74.100a	9.953a	3.982a	19.166a	0.988a	0.693a	31.000a	1.666a	0.000e	99.100a	71.000a	0.580c	211.350d
CN+100 mg of Zn	56.166e	5.213e	2.982c	12.166e	0.733e	0.532e	24.000c	0.666b	0.000d	0.000e	0.000e	0.450e	192.350f
Gf+100mg of Zn	70.166b	9.905b	3.622b	17.100b	0.932b	0.682b	27.000b	1.333ab	0.000c	93.366b	61.000b	0.680b	248.730c
CN+150mg of Zn	53.366f	4.822f	2.222e	10.366f	0.653g	0.492f	22.000d	0.666b	0.000b	0.000e	0.000e	0.540d	265.776b
Gf+150mg of Zn	65.100c	8.113c	2.982c	15.400c	0.822c	0.593c	24.000c	1.000ab	0.000a	80.100d	54.000c	0.760a	356.280a

SL—Shoot length, FWS—Fresh weight of shoot, DWS—Dry weight of shoot, RL—Root length, FWR—Fresh weight of root, DWR—Dry weight of root, NL—Number of leaves, NI—Number of inflorescence, NFR—Number of fruits, PC—Percentage Colonization, SN—Spore Number, CN—Control, Gf—Glomus fasciculatum, CN+50mg of Zn—Control + 50mg of Zn (uninoculated soil), Gf + 50mg of Zn—Glomus fasciculatum + 50mg of Zn (AMF)
Means sharing a letter in columns are not significantly different according to Ducan's test $P<0.05$.

Table 23.2. Effect of Zinc on Shoot Length, Fresh Weight of Shoot, Dry Weight of Shoot, Root Length, Fresh Weight of Root, Dry Weight of Root, Number of Leaves, Number of Inflorescence, Number of Fruits, per cent Root Colonization, Spore Number/50 gm of Soil, Phosphorous Uptake and Zinc Uptake with Respect to *Glomus fasciculatum*.

Sunflower													
	SL (cm)	FWS (gm)	DWS (gm)	RL (cm)	FWR (gm)	DWR(gm)	NL	NI	NF	PC%	SN	P%	Zn (ppm)
30 Days													
CN	18.366h	0.754g	0.158h	5.100g	0.252h	0.038h	7.000d	0.000h	0.000h	0.000e	0.000e	0.140g	27.320g
Gf	21.233f	1.415f	0.276g	11.233f	0.368f	0.184g	10.000c	0.000g	0.000g	52.500c	39.000d	0.220e	57.760e
CN+50mg of Zn	23.166d	2.982c	1.322c	15.166d	1.392c	0.912c	11/000bc	0.000f	0.000f	0.000e	0.000e	0.190f	53.250f
Gf+50mg of Zn	31.166a	3.422a	1.822a	18.100a	1.608a	0.996a	14.000a	0.000e	0.000e	81.166a	66.000a	0.250d	62.330d
CN+100mg of Zn	22.366e	2.724d	0.956e	13.366e	0.957d	0.722d	8.000d	0.000d	0.000d	0.000e	0.000e	0.270c	74.443c
Gf+100mg of Zn	25.566b	3.222b	1.523b	17.233b	1.512b	0.975b	12.000b	0.000c	0.000c	76.366b	57.000b	0.280c	77.330b
GN+150mg of Zn	21.100g	2.213d	0.924f	11.366f	0.346g	0.192f	7.000d	0.000b	0.000b	0.000e	0.000e	0.300b	75.180c
Gf+150mg of Zn	24.366c	2.982c	0.988d	16.100c	0.376e	0.198e	10.000c	0.000a	0.000a	33.166d	51.000c	0.340a	84.660a
60 Days													
CN	25.233h	1.492h	0.322h	8.100g	0.322h	0.192h	7.000d	0.333a	0.000h	0.000e	0.000e	0.270g	53.640f
Gf	31.366f	2.953g	0.372g	14.100ed	0.412g	0.295g	10.000c	0.666a	0.000g	75.500c	41.000d	0.440d	112.320e
CN+50mg of Zn	35.166d	4.902c	2.982c	17.100c	2.986d	1.966c	12.000b	0.666a	0.000f	0.000e	0.000e	0.380f	123.660d
Gf+50mg of Zn	45.100a	6.917a	4.982a	21.166a	3.896a	2.522a	14.000a	1.000a	0.000e	86.100a	78.000a	0.480dc	132.276c
CN+100mg of Zn	32.100e	4.438d	2.626d	15.3656d	2.396e	1.322e	11.000bc	0.333a	0.000d	0.000e	0.000e	0.420e	131.480c
Gf+100mg of Zn	44.500b	5.475b	3.726b	19.300bh	3.275b	2.126b	12.000b	0.666a	0.000c	79.300b	58.000b	0.560ab	147.870b
CN+150mg of Zn	30.233g	3.428f	2.226e	13.166f	2.113f	1.118f	10.000c	0.333a	0.000b	0.000e	0.000e	0.550b	148.370b
Gf+150mg of Zn	41.166c	4.295e	2.126f	17.166c	3.122c	1.933d	11.000bc	0.333a	0.000a	69.100d	52.000c	0.570a	153.326a
90 Days													
CN	33.100h	1.982h	0.986h	9.100f	0.993h	0.922g	8.000e	0.666a	0.000h	0.000e	0.000e	0.520g	106.380g
Gf	38.300d	2.985g	0.996g	16.233e	1.922g	0.626hy	11.000d	0.666a	0.000g	81.300c	51.000d	0.720f	226.040f
CN+50mg of Zn	37.100e	5.982d	3.953c	19.233c	2.998d	1.982d	13.000bc	0.666a	0.000f	0.000e	0.000e	0.730f	246.310e
Gf+50mg of Zn	47.300a	9.725a	5.953a	23.1009a	3.992a	2.726a	17.000a	1.333a	0.000e	90.100a	81.000a	0.860d	263.550d
CN+100 mg of Zn	36.100f	5.627e	3.8921d	18.166d	2.866e	1.668e	12.000cd	0.666a	0.000d	0.000e	0.000e	0.810ed	261.980d
Gf+100mg of Zn	46.500b	8.826b	4.964b	21.366b	3.826b	2.594b	14.000b	1.000a	0.000c	85.366b	61.000b	0.980c	294.760b
CN+150mg of Zn	34.166g	4.984f	2.982f	16.100e	2.622c	1.393f	11.000d	0.666a	0.000bh	0.000e	0.000e	1.070b	283.330c
Gf+150mg of Zn	43.100c	7.926c	3.664e	19.233c	3.622f	1.993c	12.000cd	1.000a	0.000a	75.233d	58.000c	1.110	297.770a

SL—Shoot length, FWS—Fresh weight of schoot, DWS—Dry weight of shoot, RL—Root length, FWR—Fresh weight of root, DWR—Dry weight of root, NL—Number of leaves, NI—Number of inflorescence, NFR—Number of fruits, PC—Percentage Colonization, SN—Spore Number, CN—Control, Gf—Glomus fasciculatum, CN+50mg of Zn—Control + 50mg of Zn (uninoculated soil), Gf + 50mg of Zn—Glomus fasciculatum + 50mg of Zn (AMF)
Means sharing a letter in columns are not significantly different according to Ducan's test P<0.05.

Table 23.3. Effect of Zinc on Shoot Length, Fresh Weight of Shoot, Dry Weight of Shoot, Root Length, Fresh Weight of Root, Dry Weight of Root, Number of Leaves, Number of Inflorescence, Number of Fruits, per cent Root Colonization, Spore Number/50 gm of Soil, Phosphorous Uptake and Zinc Uptake with Respect to *Glomus fasciculatum*.

Mustard													
	SL (cm)	FWS (gm)	DWS (gm)	RL (cm)	FWR (gm)	DWR(gm)	NL	NI	NF	PC%	SN	P%	Zn (ppm)
30 Days													
CN	5.100g	0.297h	0.047h	3.300h	0.218h	0.112h	5.000d	0.000h	0.000h	0.000e	0.000e	0.160g	12.180f
Gf	7.300e	0.317g	0.092g	5.333g	0.?12g	0.192g	8.000bc	0.000g	0.000g	40.166c	31.000d	0.260d	22.360e
CN+50mg of Zn	8.166d	0.826d	0.493d	8.166d	0.468d	0.293d	9.000b	0.000f	0.000f	0.000e	0.000e	0.220f	13.460f
Gf+50mg of Zn	10.333a	1.922a	0.986a	14.166a	0.682a	0.396a	11.000a	0.000e	0.000e	60.500a	51.000a	0.280c	23.880e
CN+100mg of Zn	7.566e	0.812e	0.413e	7.200e	0.413e	0.236e	8.000bc	0.000d	0.000d	0.000e	0.000e	0.240e	28.850d
Gf+100mg of Zn	9.233b	1.626b	0.887b	10.100b	0.636b	0.366b	11.000a	0.000c	0.000c	45.233b	44.000b	0.320b	38.720b
GN+150mg of Zn	6.233f	0.792f	0.292f	6.100f	0.382f	0.123f	7.000c	0.000b	0.000b	0.000e	0.000b	0.270cd	36.180c
Gf+150mg of Zn	8.100c	1.222c	0.813c	9.100c	0.613c	0.318c	9.000b	0.000a	0.000a	32.100d	32.000c	0.360a	66.250a
60 Days													
CN	13.100h	0.313h	0.092h	5.100h	0.292h	0.192h	7.000d	0.000b	2.000e	0.000e	0.000d	0.320g	23.380g
Gf	30.366g	0.398g	0.162g	7.166g	0.393g	0.216g	8.000cd	0.333ab	3.666de	42.133c	38.000c	0.370f	43.710e
CN+50mg of Zn	39.100d	1.262d	0.986d	9.200d	0.562d	0.366d	10.000ab	0.333ab	7.000b	0.000e	0.000d	0.420ed	35.820f
Gf+50mg of Zn	55.166a	2.892a	1.986a	15.233a	0.826a	0.482a	11.000a	1.000a	9.000a	70.300a	57.000a	0.470c	46.770d
CN+100mg of Zn	35.233e	1.213e	0.962e	8.366e	0.513e	0.322e	9.000bc	0.333ab	5.666bc	0.000e	0.000d	0.450d	55.150c
Gf+100mg of Zn	45.100b	2.336b	1.326b	12.133b	0.813b	0.436b	10.000ab	0.666abn	7.333ab	50.100b	51.000b	0.580b	64.730b
CN+150mg of Zn	32.366f	1.192fr	0.892f	7.766f	0.492f	0.312f	8.000cd	0.333ab	4.666cd	0.000e	0.000d	0.480c	65.480b
Gf+150mg of Zn	42.100c	2.122c	1.113c	10.,100c	0.766c	0.419c	9.000bc	0.333ab	6.666b	39.233d	37.000c	0.660a	122.380a
90 Days													
CN	15.166h	0.786h	0.216h	7.100h	0.322h	0.216h	8.000d	0.666a	4.666d	0.000e	0.000d	0.630f	45.780g
Gf	34.166g	1.098g	0.713g	9.500g	0.413g	0.266g	9.000cd	1.000a	8.666bc	45.100c	41.000c	1.320a	82.150e
CN+50mg of Zn	48.30-0d	2.982d	1.982d	12.233d	0.692d	0.388d	10.000bc	1.000a	9.666b	0.000e	0.000d	0.710e	68.640f
Gf+50mg of Zn	59.100a	6.132a	3.626a	18.100a	0.966a	0.663a	12.000a	1.333a	11.666a	75.233a	61.000a	0.820d	83.560e
CN+100 mg of Zn	42.500e	2.326e	1.822e	11.600e	0.626e	0.3633	8.000d	0.666a	8.333bc	0.000e	0.000d	0.830d	108.460d
Gf+100mg of Zn	51.300b	6.122b	3.392b	17.166b	0.933b	0.622b	11.000ab	0.666a	1.333a	54.100b	54.000b	1.180b	125.793c
CN+150mg of Zn	40.,433f	2.122f	1.792f	10.566f	0.526f	0.332f	9.000cd	0.666a	7.333c	0.000e	0.000d	0.950c	128.970b
Gf+150mg of Zn	49.166c	5.982c	3.1113c	16.100c	0.892c	0.525c	10.000bc	0.666a	11.666a	42.233d	42.000c	1.310a	243.760a

SL—Shoot length, FWS—Fresh weight of shoot, DWS—Dry weight of shoot, RL—Root length, FWR—Fresh weight of root, DWR—Dry weight of root, NL—Number of leaves, NI—Number of inflorescence, NFR—Number of fruits, PC—Percentage Colonization, SN—Spore Number, CN—Control, Gf—Glomus fasciculatum, CN+50mg of Zn—Control + 50mg of Zn (uninoculated soil), Gf + 50mg of Zn—Glomus fasciculatum + 50mg of Zn (AMF)

Means sharing a letter in columns are not significantly different according to Ducan's test $P<0.05$.

(*a*) Sufflower

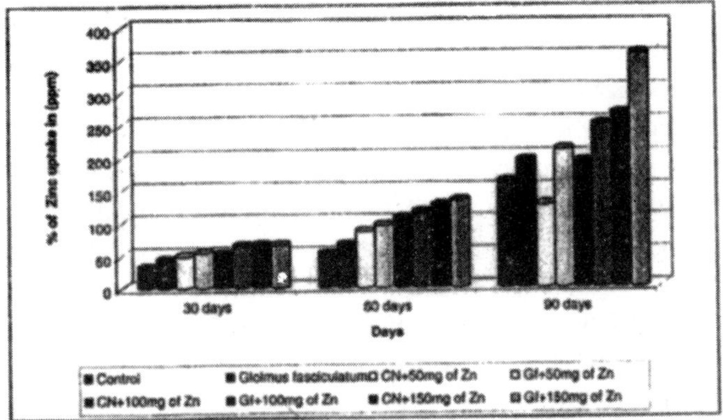

(*b*) Sunflower

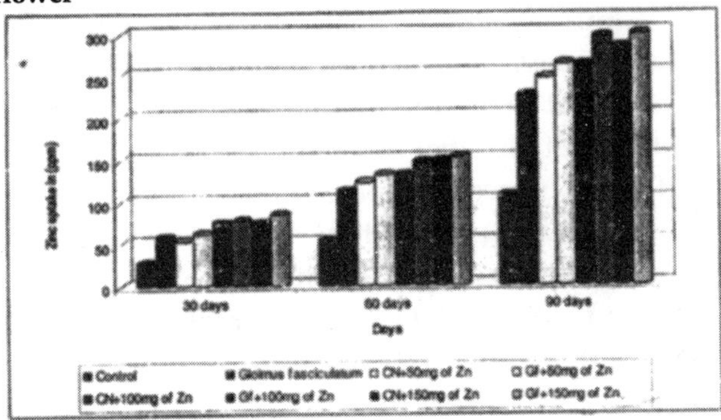

(*c*) Mustard

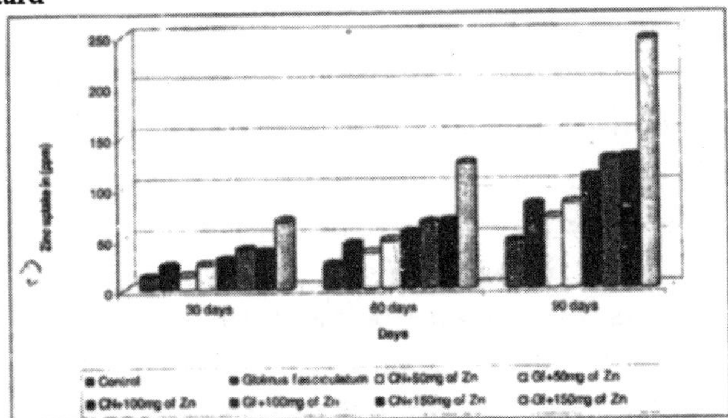

Fig. 23.1. Effect of Arbuscular Mycorrhizal Fungi on Phosphorous Uptake in Shoots of Safflower, Sunflower and Mustard Plants Treated with Zinc.

Pot experiments of inoculated plants treated with four different Zinc concentrations (0, 50, 100,150 mg/kg soil) to three different soil yielding plants Sunflower, Safflower and Mustard have showed that *Glomus fasciculatum* had better tolerance to Zinc toxicity compared to uninoculated control plants. In 0 concentration of Zinc in inoculated and uninoculated plants there was an increase in shoot length, fresh weight of shoot, dry weight of shoot, root length, fresh weight of root, dry weight of root number of leaves, number of inflorescence, number of fruits, percent colonization and spore no. per 50 g of soil (growth parameter) in inoculated plants than compared to uninoculated plants. In 50 mg of Zinc concentration in three different oil yielding plants such as Safflower, Sunflower and Mustard there was a increase in growth parameters such as shoot length, fresh weight of shoot, dry weight of shoot, root length, fresh weight of root, dry weight of root, number of leaves, number of inflorescence, number of fruits, percent colonization and spore no. per 50 g of soil in inoculated plants than compared to uninoculated plants. But further increase in Zinc concentration i.e. when inoculated and uninoculated plants treated with 100 mg of Zinc there is a reduction in the growth parameters, but there is an increase in growth parameters in inoculated plants than compared to uninoculated plants.

Application of Mycorihizae on Oil Yielding Plants in Alleviating Soil Polluted with high concentration of Zinc

1. Safflower (*Carthamus tinctorius* L)

A. Uninoculated (Control)
B. Inoculated (*Glomus farciculatum*)
C. Control + 50 mg of Zinc
D. *Glomus farciculatum* + 50mg of Zinc
E. Control + 100 mg of Zinc
F. *Glomus farciculatum* + 100 mg of Zinc
G. Control + 150 mg of Zinc
H. *Glomus farciculatum* + 150 mg of Zinc

2. Sunflower (*Helianthus Annuus* L)

A. Uninoculate (Control)
B. Inoculated (Glomus farciculatum)
C. Control + 50 mg of Zinc
D. *Glomus farciculatum* + 50 mg of Zinc
E. Control + 100 mg of Zinc
F. *Glomus farciculatum* + 100 mg of Zinc
G. Control + 150 mg of Zinc
H. *Glomus farciculatum* + 150 mg of Zinc

3. Mustard (*Brassica juncea* L)

A. Uninoculate (Control)
B. Inoculated (*Glomus farciculatum*)
C. Control + 50 mg of Zinc
D. *Glomus farciculatum* + 50 mg of Zinc
E. Control + 100 mg of Zinc
F. *Glomus farciculatum* + 100 mg of Zinc
G. Control + 150 mg of Zinc
H. *Glomus farciculatum* + 150 mg of Zinc

According to (Choi J.M., *et al.*, 1996; Ebbs and Kochian 1997; Fontes and Cox, 1998) have showed that Zinc toxicity in plants results in reduced growth of both root and shoot. In 150 mg of Zinc concentration there is a reduction in the growth of all three plants. But there was an increase in growth parameters in inoculated plants than compared to uninoculated plants. Table Nos. (23.1, 23.2 and 23.3).

On the other hand there are some reports on arbuscular mycorrhiza from metal-contaminated soils (Arnold and Kapustka, 1987; Bajwa *et al.*, 1991; Gildon and Tinker, 1983 Koomen *et al.*, 1990 ; Leyval *et al.*, 1991) which suggest that selection of metal tolerant strains has occurred. There is also mounting evidence that AM fungal infection may decreases metal accumulation in plants growing in polluted soil and

Fig. 23.2

thus protect the host against phytotoxic metal effects (Arines *et al.*, 1989; Bethlenfalvay and Franson, 1989; Dueck *et al.*, 1986; Kerbawy *et al.*, 1989 Leyval *et al.*, 1991, Mc Gee, 1987; Pacovsky, 1986; Schuepp *et al.*, 1987; Lakshman and Srinivasalu, 2002). Nutrient uptake in the inoculated and unicoculated soil yielding plants such as Safflower, Sunflower and Mustard with four different Zinc concentrations have showed that there was an increase in the Zinc concentration in shoot as soil (Zinc) concentration increases but increased in the Zinc uptake there is decreased in the phosphorus uptake. As a rule, mycorrhizal infection enhances plant growth by increasing nutrient uptake via increases in the absorbing surface area, by mobilizing sparingly available nutrient sources or ectoenzymes. Mycorrhizal infection may also protect roots from soil pathogens (Perrin, 1990) and there by increase in the nutrient uptake.

According to Singh *et al.*, (2008) the P concentrations in Wild tobacco shoots decreased as soil (zinc) concentration increased, where as Li *et al.*, (2004) have showed that Zinc concentrations in barley decreased as soil-(P) increased, this suggested a reciprocal antagonism between Zn and P (Shetty *et al.*, 1994).

CONCLUSION

AM fungi inoculated plants showed an increase growth and nutrient uptake when oil yielding plants colonized with AM fungi. Similarly, the lower to moderate Zinc concentration of the soil with AM inoculation showed significant increased plant biomass production. The present work clearly brought that AM fungi inoculated Mustrad plants with Zinc 50 mg/kg soil was most efficiently balanced than lower Zinc concentration or higher than Zinc 50 mg of concentration. So that soil yielding Mustard, Sunflower and Safflower plants with Zinc 50 mg/kg soil with AM fungal inoculation is recommended.

REFERENCES

Arines, J., Vilarino, A. and Sainz M. (1989). Effect of different inocula of vesicular Arbuscular mycorrhizal fungi on manganese content and concentration in red clover (*Trifolium pretense* L.) plants. *New Phycol.* 112:215–219.

Arnold and Kapustka (1987). VA Mycorrhizal colonization and spore populations in an abandoned agricultural field after five years of sludge application. *Ohio J. Sci.* 87:112–114.

Bajwa, R., Khan, F. and Mahmood, S. (1991). Vesicular - arbuscular mycorrhiza in industrially polluted soils. Sci. Int. (Lahore), Monograph Contemporary studies in mycorrhiza and Biomagnetism, pp.31–37.

Bethlenfalvay G.J. and Franson R.L. (1989). Manganese toxicity alleviated by mycorrhizae in Soybean. *J. Plant Nutr.* 12:953–970.

Borbala Biro, Katalin Posta, Anna Fuzy, Imre Kadar, Tamas, Nemeth. (2005). Mycorrhizal functioning as a part of the survival mechanisms of barley (*Hordeum Vulgare* L) at long term heavy metal stress. Proceedings of the 8th Hugarian congress on plant physiology and the 6th Hungarian congres on plant physiology and the 6th Hungarian conference on photosynthesis, 49:65–67.

Choi, J.M., Pak, C. H. and Lee, C. W. (1996). Micronutrient toxicity in French Marigold. *J. Plant Nutr.* 19:901–916.

Dueck, T. A., Visser, P., Ernst, W.H.O. and Schat, H. (1986). Vesicular -arbuscular mycorrhizae decrease Zinc-toxicity to grasses growing in Zinc polluted soil. *Soil Biology and Biochemistry.* 18:331–337.

Ebbs, S. D. and Kochian L. V. (1997). Toxicity of zinc and copper to Brassica species: Implications for phytoremediation. *J. Environ Qual* 26, 776–781.

El-kherbawy, M., Angle, J. S., Heggo, A. and Chaney, R. L. (1989). Soil PH, rhizobia and vesicular arbuscular mycorrhizae inoculation effects on growth and heavy metal uptake of alfalfa (*Medicago Sativa* L). *Biology and Fertility of Soil.* 8 (1): 61–65.

Fontes, R.L.S. and Cox, F. R. (1998). Zinc toxicity in Soyabean grown at high iron concentration in nutrient solution. *J. Plant Nutr.* 21; 1723–1730.

Gerdemann, J. W. and Nicolson, T.H. (1963). Spores of mycorrhizal endogone species extracted from the soil by wet sieving and decanting. *Transaction of British mycological society.* 46:235-244.

Gildon, A. and Tinker, P.B. (1983). Interactions of vesicular arbuscular myccorrhizal infection and heavy metal in plants. The effect of heavy metal on the development of vesicular-arbuscular mycorrhizae. *New Phytologist.* 95:247-261.

Giovannetti, M. and Mosse, B. (1980). An evaluation of techniques for measuring VAM infection in roots. *New phytologist.* 84:489-500.

Hayman, D. S. and Mosse, B. (1973). Advances in the study of vesicular arbuscular mycorrhiza *Ann. Rev phytopathol.* 11:171-196.

Jackson, M. L. (1973). Soil chemical analysis. Prantice Hall of India Pvt Ltd., New Delhi, pp. 10-330.

Kiekens, L. (1990). Zinc in heavy metals in soils Ed. B. J. Alloway, 2nd ed, Blackie Academic and professional. Glasgow, pp.284-305.

Koide, R. T. (1991). Nutrient supply. Nutrient demand to mycorrhizal infection. *New phytol.* 117:355-385.

Koomen, I., McGrath, S. P. and Giller, I. (1990). Mycorrhizal infection of clover is delayed in soils contaminated with heavy metals from part Sewage sludge applications, *Soil Biol. Biochem.* 22:871-873.

Lakshman, H. C. and Srinivasalu, Y. (2002). Importance of AM fungal association an aquatic plants growing in a polluted fresh water ponds. In: Ecology polluted water Ed. Arvindkumar APH Publishers, New Delhi, pp. 170-200.

Lakshman, H. C. (1992). AM Fungal association in some hydrophytic and xerophytic tropical plants Nat. Con. 73:107-112.

Leyval, C., Berthelin, J., Schontz, D., Weissenhorn, I. and Morel, J. L. (1991). Influence of endomycorrhizas on maize uptake of Pb, Cu, Zn and Cd applied as mineral salts or sewage sludges. *In heavy metal in the Environment.* Ed J. G. Farmer. Pp. 204-207.

Li Hy, Zhu, Y.G., Smith, S.E. and Smith, F.A. (2004). Phosphorus-zinc interactions in two barley cultivers differing in phosphorus and zinc efficiencies. *J. Plant Nutr.* 26:1085-1099.

Marschner, H. (1998). Mineral nutrition of higher plants. Academic press. London. pp. 498.

McGee P. A. (1987). Alteration of growth of *Solanum opacum* and *Plantago drummondii* and inhibition of regrowth of hyphae of vesicular-arbuscular mycorrhizal fungi from dried root pieces by manganese. *Plant and Soil*. 101:227–233.

Montilla, M. A. P. and Torrent, J. (2003). Zinc phytotoxicity to oilseed rape grown on zinc loaded substrates consisting of Fe-oxide coated and calcite sand. *Plant and soil*. 257–236.

Mortvedt, J. J. (2000). Bioavailability of micronutrients in Handbook of Soil Science. Ed. In chief M.E. Summer, pp. D 71–D86.

Pacovsky, R. S. (1986) Micronutrient uptake and distribution in mycorrhizal or Phosphorus fertilized Soyabeen. *Plant and Soil*. 95:379–388.

Phillips, J. M. and Hayman, D. S. (1970). Improved procedures for clearing roots and staining parasite and vesicular arbuscular mycorrhizal fungi for rapid assessment of infection. *Transaction of British mycological society*. 55:158–160.

Roopa, K. Jambotkar. and Lakshman, H.C. (2007). Effect of Zinc concentration soil with AM fungus on Brassica juncea L. an oil yielding plant. *I.J.Plant.Sc.* 4(1): 52–54.

Schuepp, H. (1987). Interaktionen zwischen VA-mykorrhizenand schwermetalbelastungen in mycorrhiza and plant stress. *Angew. Bott.* 61:85–95.

Shetty, K. G., Hetrick, B.A.D. and Schwab, A.P. (1994). Effect of mycorrhize and other soil microbes on revegetation of heavy metal contaminated mine spoil. *Environ Pollut.* 86:181–188.

Singh, P.K., Mishra, M. and Vyas, D. (2008). Phytoprotection role of AM fungi in the rhizosphere of tomato. In Proc. Nat. Seminar vision 2008. Current challenges in Biological Sciences. Eds. Kandya, D.C., Khare, A.K., Vyas, P.K., Chourasia, D.B.K and Singh, P.K. New Delhi. pp. 51.

Vera Gohre., Uta Paszkowski. (2006). Contribution of the arbuscular mycorrhizal symbiosis to heavy metal phytoremediation. *Planta*. 223:115–122.

24

Phytoprotection by Arbuscular Mycorrhizae

Deepak Vyas, Pradeep Kumar Singh, Meenakshi Mishra,
** R K Gupta*

INTRODUCTION

In 1885 Frank was the first to describe the symbiosis between a fungus and the roots of trees and he coined the word mycorrhiza (fungal root). He himself already recognized the enormous potential and importance of his findings, calling the fungus the "wet nurse" of the tree. Today we know that mycorrhizas are the most widespread associations between microorganisms and higher plants, occurring on roots of more than 80% of all terrestrial plants (Sieverding, 1991). According to fossil records, plants colonizing terrestrial ecosystems four hundred million years ago were mycorrhizal (Remy *et al.*, 1994). Hence mycorrhizal fungi have been involved in the adaptation of plants to unfavorable conditions from the very beginning (Penninsi, 2004).

The most prominent effect of the fungus is improved phosphorus nutrition of the host plant in soils with low

---

Lab of Microbial Technology & Plant Pathology Department of Botany, Dr. H.S. Gour University, Sagar (M.P.) 470 003

* Department of Botany, Govt. P.G. College, Rishikesh, Uttrakhand

phosphorus levels (Koide, 1991). However uptake of nitrogen, zinc, copper and minor nutrients are enhanced as well. Improved plant nutrition is due to (i) increased root surface through extraradical hyphae, (ii) degradation of organic material and (iii) alteration of the microbial composition in the rhizosphere (Marschner, 1998; Hodge and Campbell, 2001). In exchange for the nutrients, the fungus is supplied with photosynthates (about 20% of the plant assimilation capacity) and these assimilates are either respired or released into the soil. As a result of higher carbon drain, microbial rhizosphere composition is altered and therefore the term "mycorrhizosphere" was coined (Rambelli, 1973). As the majority of all plants are mycorrhizal, mycorrhizosphere versus rhizosphere might be the rule rather than the exception (Linderman, 1988). In addition, mycorrhizas increase drought resistance (Augé and Stodola, 1990) and heavy metal tolerance (Brundrett, 1991), improve soil aggregation (Andrade *et al.*, 1998) and promote overall plant health (Linderman, 1992).

Arbuscular mycorrhizal fungi (AMF) are by far the most abundant of all mycorrhizas. They are characterized by the formation of an extraradical mycelium and branched haustorial structures within the cortical cells termed arbuscules. The latter are the main sites of exchange between the two symbiotic partners (Hock and Varma, 1995). In contrast to ectomycorrhizal fungi, AMF are considered to be obligate biotrophic microorganisms and cannot be cultured axenically. The general differentiation between symbiosis and pathogenicity is not easy and this is particularly true formycorrhizas. Depending on environmental conditions, root colonization with arbuscular mycorrhizal can either stimulate or inhibit plant growth (Sanders, 1993) and during the formation of the arbuscular mycorrhizal symbiosis several defence responses have been observed in the plant (Gianinazzi-Pearson *et al.*, 1996; García-Garrido and Ocampo, 2002).

The ability of AMF to suppress root diseases caused by soil-borne pathogens has been intensively studied in the last thirty years. Fungal root pathogens are inhibited by mycorrhiza inoculation in the cases of *Fusarium oxysporum* on different crops (Dehne and Schönbeck, 1979; Caron *et al.*, 1986; St. Arnaud *et al.*, 1997; Filion *et al.*, 1999), various Phytophthora species (Davis and Menge, 1980; Cordier *et al.*, 1996; Trotta *et al.*, 1996), *Rhizoctonia solani* (Yao *et al.*, 2002) and *Pythium ultimum* (Calvet *et al.*, 1993). Bacterial diseases are also reduced by mycorrhiza establishment on the root (Dehne, 1982). Numerous reports were published on the suppression of nematode penetration and development following AMF inoculation (Saleh and Sikora, 1984; Habte *et al.*, 1999; Talavera *et al.*, 2001; Elsen *et al.*, 2001; Waceke *et al.*, 2001; Diedhiou *et al.*, 2003). Contradictory results also have been reported, which is believed to be due to the heterogeneity of experimental design and varying environmental factors (Trotta *et al.*, 1996).

Though nutrient uptake has been the focus of much research on the AM association there is evidence that AMF also play a role in the suppression of crop pests and diseases, particularly soil-borne fungal diseases. (Schöenbeck, 1979; Paulitz and Linderman 1991; Linderman, 1994; Borowicz, 2001) and there have been a number of reviews of the subject, for example, Azocon-Aguilar and Barea (1996), Harrier and Watson (2004) and Whipps (2004). Table (24.1) shows some of the soil-borne diseases, which have been experimentally suppressed using AMF, a more complete list is given in Whipps (2004). Reductions in the severity of disease occur rather than total inhibition, but this can still result in a significant increase in yield over plants not inoculated with AMF, and occurs despite the fact that pathogen infection generally reduces AM colonisation (Torres-Barragan *et al.*, 1996; Dar *et al.*, 1997; Karagiannidis *et al.*, 2002). In some cases, the apparent resistance of a plant to a pest or disease may be simply the result of improved nutrition (Cordier *et al.*, 1996;

Karagiannidis *et al.*, 2002 *et al.*,), though there is evidence for multiple mechanisms of resistance, probably operating simultaneously (Whipps, 2004). Perhaps the most important of these is exclusion, which seems to be a simple case of competition for space (Azcon-Aguilar and Barea, 1996). Once colonisation of a root cell by AMF occurs the pathogen is excluded from that cell. As a result, the most effective control is achieved when colonisation by AMF takes place before attack by the pathogen (Slezac *et al.* 1999; Matsubara *et al.*, 2001; Sylvia and Chellemi, 2001). Other factors involved may be related to changes in root exudates (Filion *et al.*, 1999; Norman and Hooker, 2000) which can cause changes in the rhizosphere microbial community (Dar *et al.*, 1997) changes to set to the host root architecture (Yano *et al.*, 1996; Vigo *et al.*, 2000) or changes to plants defense mechanism (Azcon-Aguilar and Barea, 1996; Ganinazzi-Pearson *et al.*, 1996), changes to plant defence mechanisms or so-called induced resistance result from a priming effect of the AM colonisation, which does not in itself cause a significant defensive response by the plant but induces the plant to respond faster to infection by pathogenic fungi (Whipps, 2004).

PHYTOPROTECTION BY AM FUNGI AGAINST PLANT DISEASES

Control of fungal diseases

The interactions between AM and plant diseases in crops have received only minimal experimental consideration. To evaluate the influence of AM fungi on disease control, plant pathogen, symbiotic fungus and environmental conditions are to be considered simultaneously. As AM fungi could not be cultured, the effect could be observed only on the host plant. Most of the experiments done differ in findings in this field. In general, mycorrhizal plants are less prone to disease and suffer less damage in comparison to non-mycorrhizal ones. Pathogen development is slow, suppressed or inhibited in mycorrhizal plants. However, increased disease severity has also been reported in mycorrhizal plants (Gerdemann, 1968; Mosse, 1973; Schöenbeck, 1979; Rhodes, 1980; Bali and Mukerji, 1989; Duchesne, 1994; Traquair, 1995, Raman, 1996).

More number of chlamydospores of AM fungi were observed in the rhizosphere of healthy plants than in the diseased ones (Zaidi and Mukerji, 1983).

Control of soil and root borne pathogens

Since AM are formed in the roots of plants, maximum attention has been paid towards mycorrhizae disease incidence interactions in relation to soil and root borne pathogens (Table 24.1). AM fungi impart resistance to host plant against root pathogens. Root pathogens are major limiting factor for plant production. Most widespread root borne pathogens belong to genera, *Phytophthora, Fusarium, Pythium* and *Rhizoctonia*. They kill roots or reduce their ability

TABLE 24.1. : Show the Interaction Between the Host-Pathogen and AM Fungi

Host	Disease	Pathogen	Average number of per 10 g soil	
			Diseased	Healthy
Melilotus indicus	Downy mildew	*Perenospora trifoliarum*	34	46
Melilotus indicus	Powdery mildew	*Erysiphe trifolii*	88	156
Brassica napa	Powdery mildew	*Erysiphe chicoracearum*	32	48
Euphorbia pulcherrima	Rust	*Melampsora euphorbii*	64	86
Brassica campestris	White rust	*Albugo candida*	92	166
Triticum aestivum	Loose smut	*Ustilago tritici*	32	46
Triticum aestivum	Leaf spot	*Drechslera graminearum*	54	71
Coriandrum sativum	Stem gall	*Protomyces macrosporus*	46	64

(K.G. Mukerji, 1997)

Table 24.2. : Selected Soil-Borne Fungal Diseases Controlled by AMF

Pathogen	Disease	Crop	Reference
Sclerotium cepivorum	White rot	Onions (Allium cepa)	Torres-Barragan *et al.* (1996)
Fusarium oxysporum	*Fusarium* root rot	Asparagus (Asparagus officinalis) French bean (phaseolus vulgaris)	Matsubara *et al.* (2001) Dar *et al.* (1997)
Verticillium dahliae	*Verticillium* wilt	Tomatoes (Lycopersicon esculentum) Aubergines (Solanum melongena)	Karagiannidis *et al.*, (2002) Matsubara *et al.* (2000)
Helicobasidium-mompa *Rhizoctonia solani* *Aphanomyces euteiches*	Violet root root and stem rots root rot	Asparagus Mung bean vigna radiate) Pea (Pisium sativum)	Kasiamdari *et al.* (2002) Kioller and Rosendahl (1996) Bodker *et al.* (2002)

(Gossling, 2005)

to absorb water and nutrients by penetrating root tissues and producing toxins. They share their mode of nutrition and habitat with mycorrhizal fungi. Hence, two categories of fungi strongly compete with one another for nutrition and space.

When both the AM and pathogen are free living in the soil before penetration, pathogens often have competitive advantage because or their saprophytic ability, many of them are less strictly biotrophic than the mycorrhizal fungi. But situation is different when mycorrhizal fungus is already established, the mycorrhizal fungus does not have to compete for substrate and may prevent penetration by pathogens. According to Caron *et al.* (1986 a,b) studied the effect of inoculation sequence of *Glomus intraradices* reduces the damage caused by *Fusarium oxysporum f.* sp. *radicis lycopersici* on the growth of tomato.

Increased phosphorus uptake due to AM inoculation results in the reduction of damage caused by pathogens. Citrus inoculated with *Glomus fasciculatum* alone showed more plant height, top and root weight than in non-mycorrhizal seedlings inoculated with *Phytophthora parasitica* or seedlings inoculated with both *G. fasciculatum* and *P. parasitica* (Davis *et al.*, 1978). At high soil phosphorus levels, *phytophthora* produces fewer zoospores leading to lowering of inoculum potential of pathogen. Mycorrhizal onion roots were shown to be less susceptible to pink rot caused by *Pyrenochaeta terrestris* (Safir, 1968). This was related to large amounts of reducing sugars in mycorrhizal roots. Less stunting of strawberry was observed when inoculated with *Glomus fasciculatum* and pathogen *Cylindrocarpon destructans* than when inoculated by pathogen alone.

Schöenbeck and Dehne (1977) also proved that damage caused to cotton seedlings was less even when VAM fungi were inoculated after the pathogen *Thielaviopsis basicola*. It was later on reported that chlamydospore production by *T. basicola* was negatively correlated to percent mycorrhizal colonization of tobacco and alfalfa (Baltruschat and Schöenbeck, 1972). Chlamydospore reduction is attributed to high levels of free amino acids arginine and citrulline.

Different AM fungi are known to confer variable tolerance to *Phytophthora parasitica*. Davis and Menge (1980, 1981) used five different AM fungal species on three different plants. They observed that maximum resistance to the pathogens was due to two AM fungi namely. *G. fasciculatum* and *G. constrictum* in *Citrus* sp. AM *mycorrhizal* peanut plants become resistant to root rot pathogen *Sclerotium rolfsii* (Krishna and Bagyaraj, 1983). Number of sclerotia produced by the pathogen were reduced on the mycorrhizal roots. Biomass production and phosphorus content were maximum in mycorrhizal plants and minimum in pathogen inoculated plants.

Take all disease of wheat, caused by *Guamannomyces graminis* var. tritici was reduced when mycorrhizal fungus was inoculated in plants. The disease is favoured by inadequate plant nutrition leading to mineral deficiency especially phosphorus (Graham and Menge, 1982). AM fungi are able to increase nutritional status of the host plant thereby improving resistance against the pathogens. Recenlty Singh (2008) reported prior inoculation of AM fungi pronounced greater reduction in wilt disease caused by Foc in tomato plant. AM fungi and *Trichoderma* sp. produced more significant biocontrol effect.

Table 24.3. : Interaction Between VAM Fungi and Soil and Root Borne Plant Pathogens (Fungi)

Pathogen	Host	Effect of mycorrhizal plants		Reference
		Damage	Interaction	
Cylindrocarpon destructans	Strawberry	-	-	Paget, 1975; Traquair, 1995
Cylindrocladium scoparium	Yellow poplar	-	-	Barnard 1977, Chakravarty and Unestam 1987
Fusarium oxysporum	Cucumber	-	-	Schoenbeck, 1979
F. oxysporum	Easter lilly	+	-	Ames and Linderman, 1978
F. avenaceum	Clover	-	-	Dehne, 1982
F. solani	Soybean	-	-	Zambolim and schenck, 1983
F. oxysporum	Onion	-	-	Dehne, 1982
Pathogen	**Host**	**Effect of mycorrhizal plants**		**Reference**
		Damage	**Interaction**	
F. oxysporum lycopersici	Cucumber	-	-	Dehne and Schoenbeck, 1979
F. oxysporum lycopersici	Tomato	+	-	McGraw and Schenck, 1981
Fusarium spp.	Tomato	-	-	Caron *et al.*, 1986a, b
Gaeumannomyces graminis	Wheat	-	-	Graham and Menge, 1982

Macrophomina phaseolina	Soyabean	No effect		Stewart and Pfleger, 1977
Olpidium brassicae	Tobacco	-	-	Schoenbeck and Dehne, 1977
	Lettuce	-	-	Schoenbeck and Schinzer, 1972
P. palmivora	Papaya	-	No	Ramirez 1974
P. parasitica	Citrus	-	-	Schenck and Nicolson, 1977
	Citrus	-	-	Davis and Menge, 1981
Phoma terrestris	Onion	-	-	Becker, 1976
Pythium ultimum	Soyabean	-	-	Christie *et al.*, 1978
P. ultimum	Poinsettia	-	-	Harley and Wilson, 1959
P. ultimum	Soyabean	No effect		Kaye *et al.* 1984 Chou and Schmitthenner 1974
Pyrenochaeta terrestris	Onion	-	-	Backer, 1876
	Onion	-	-	Schoenbeck and Dehne, 1981
P. terrestris	Onion	-	-	Schoenbeck and Dehne, 1981
Rhizoctonia solani	Poinsettia	-	-	
	Cotton	-	-	Mathre, 1968
	Citrus	-	-	Menge *et al.*, 1977
Verticillium dahliae	Chrysanth emum	-	-	Pegg and Joughlaekha, 1981
V. dehliae	Cotton	+	+	Davis *et al.*, 1979
V. spp.	Tomato	-	-	Baath and Hayman, 1983
Fusarium oxysporum.	Tomato	-	+	Singh, 2008

C. Control of shoot and leaf diseases

AM fungi in the roots also alter the plants resistance to diseases in shoot and leaves. The non mycorrhizal plants growing under the same conditions as mycorrhizal ones show more susceptibility towards leaf disease pathogens, thereby affecting the whole plant (Schoenbeck and Dehne, 1981).

When the young leaves of cucumber were infected with the powdery mildew fungus then only comparable differences were observed between mycorrhizal and non-mycorrhizal plants. The increase in susceptibility of the shoot was due to an enhanced development of the pathogen than to an increased frequency of infection. This influence seems to be correlated with nutritional aspect of mycorrhiza disease interactions and higher physiological activities in the whole host plant. Genetically resistant plants remain resistant, whereby susceptibility may be modified by the symbiosis (Schoenbeck and Dehne, 1981).

D. Control of viral pathogens

Generally, the interaction between AM and pathogens could be observed on the roots. In contrast to restriction of soil borne plant pathogens to site of symbiosis between the plant and the endophyte, virus diseases are less restricted because they can spread systemically throughout the whole plant. In all cases, whether the infection by different mycorrhizal host plant always exhibited multiplication of the virus and also its sensitivity was enanced.

Vigorously growing plants are more prone to viral diseases. These effect may be due to better availability of nutrition in mycorrhizal plants.

Schoenbeck and Schinzer (1972) showed increase in the viral disease and TMV concentration in mycorrhizal plants than non-mycorrhizal ones. Immunofluorescent techniques were also used to confirm the above observation (Schoenbeck and Dehne, 1981; Dehne, 1982). They observed increased viral concentration in mycorrhizal roots at arbuscular stage of the endophyte. This indicates that the high metabolic activity in the host cytoplasm of arbuscular cells are favourable for the accumulation of viral particles. The increased exchange of substance in the host cell characterised by high phosphate metabolism and high concentration of nucleic acids and proteins, probably lead to increased viral infection and multiplications (Schoenbeck, 1979; Schoenbeck and Dehne, 1981).

Table 24.4.: Effects of *VA-mycorrhiza* on Some Virus Diseases

Pathogen	Host	Effects in mycorrhizal plants	Author
Tobacco mosaic virus	tobacco	more local lesions	Schonbeck and Schinzer, 1972
Tomato mosaic virus	tomato	increase in virus titre	Daft and Okusqanya, 1973a
Potato virus X	tomato	increase in virus titre	Daft and Okusqanya, 1973a
Arabis mosaic virus	petunia	increase in virus titre	Daft and Okusqanya, 1973a
Arabis mosaic virus	strawberry	increase in virus titre	Daft and Okusqanya, 1973a

It seems that mycorrhizal fungi become better than control agents.

E. Control of Bacterial Disease

There are only two reports in which plant susceptibility to bacterial pathogens was studied in relation to the incidence of VA fungi (Bagyaraj, 1984), and different results have been observed. The susceptibility of peach plants to the bacterial canker caused by *Pseudomons syringae* was not affected by VA myorrhizal infection, but a reduction in the severity of bacterial with of VA mycorrhizal infected tomatoes caused by *P. solanacearum* was found.

VA mycorrhizal inoculation of lettuce offers a case in point. Lettuce production in California and Florida is hampered by corky root disease caused by a gram-negative bacterium (Van Bruggen *et al.*, 1988). Preliminary evidence indicates that VA mycorrhizal colonization of lettuce roots may suppress disease development. This in turn suggests that inoculation of transplants prior to installation in the field may offer some disease control. Lettuce transplants, however, are grown in 4-week rotations and establishing the high levels of colonization necessary to promote disease protection remains problematic. Low levels of VA mycorrhizal colonization on lettuce seedlings at levels of fertilization necessary to give rapid seedling growth.

F. Control of nematodes

Plant parasitic nematodes and vesicular arbuscular mycorrhiza occur together in roots and rhizosphere, each having characteristic but opposite effect on plant growth and vigour. It has been observed that AM fungi have potential as biocontrol agent when both group of organisms occur simulataneously in the roots or rhizosphere of the same plant. Antagonistic effect of AM on plant parasitic nematodes could have either a physical or physiological basis. Mycorrhizal fungi render susceptible plants tolerant to plant parasitic nematodes, when applied, may prove to be very valuable in limiting crop losses due to these pathogens (Mittal *et al.*, 1991). Cotton and soybean are commonly used host and *Meloidogyne incognita* most frequently used nematode pest. These experiments were conducted under greenhouse conditions (Hussey and Roncadori, 1982).

Mycorrhizal fungi are known to offset damage due to nematodes in the susceptible cultivar. The effect is due to increased host vigour rather than to antagonistic physiological effect (Roncadori and Hussey, 1977). In grapes interaction between *Meloidogyne arenaria* and *G. fasciculatum* stimulated growth of host plants in dual inoculated plants than in nematode infested plants (Atilano *et al.*, 1976). Similar results were reported in oats (Sikora and Schoenbeck, 1975), peach (Strobel *et al.*, 1982) peanut (Hussey and Roncadori, 1982), soybean (Kellam and Schenck, 1980), tobacco (Fox and Spasoff, 1972) and tomato (Sikora and Schoenbeck,1975).

In a dual inoculation study, it was observed that nematode was not antagonistic to mycorrhizal synthesis but AM fungi *G. etunicatus* and *G. mosseae* influenced reporduction of *M. incognita* in many ways. Number of eggs of nematodes per gram of root was lower in mycorrhizal plants than in controls (Hussey and Roncadori, 1982). Although mycorrhizal symbionts are known to provide resistance to host plants, degree of resistance provide vary with the species of fungus forming mycorrhiza. Schenck *et al.* (1975) inoculated soybean with nematode *M. incognita* and three AM fungal species, namely, *Gigaspora heterogama, Glomus macrocarpum* and *Scutellospora calospora* (=Endogone calospora). Among mycorrhizal plants, nematode densities (juveniles per volume of soil) were highest on a nematode susceptible soybean colonized by *S. calospora* and lowest on plants colonized by *G. macrocarpum.* In another greenhouse experiments soybeans were inoculated with *M. incognita* and either *G. margarita* or *G. etunicatus.* Both AM fungi stimulated plant growth but nematode reproduction was unaffected by *G. margarita* and suppressed by *G. etunicatus.*

2. Role of Nutrient in disease control

VA mycorrhizal hyphae may extend several centimetres from the root surface and thereby enlarge the depletion zone for P accordingly. The external hyphae of AM have the capacity to deliver a high proportion of the host plant demand not only of P but also Zn and Cu (George *et al.*, 1995). In contrast, the capacity for uptake and delivery of K, Ca and Mn seems to be very low or negligible. This may lead to a decrease in concentration of K and Mn in the host plant when VA mycorrhizal infection decreases root surface area (and thus K-acquisition) or root exudation (and thus activity of Mn-reducing rhizosphere microorganisms) (George *et al.*, 1995). There is good evidence for a substantial capacity of external hyphae to absorb also NH_4-N and deliver it to the host plant. In the hyphosphere as in the rhizosphere, uptake of NH_4.N is associated with substrate acidification. Although hyphae interconnect roots of different species, including both legumes and non legumes, hyphal transfer of fixed N from legumes to non legumes seems to be very small.

Table 24.5. : Interaction Between AM Fungi and Plant Parasitic Nematodes and Their Effect on Host Plants

Host	Mycorrhizal fungus	Plant parasitic nematode	Effect on host plant	Reference
Peanuts	*Gigaspora margarita*	*Meloidogyne hapla*	Decrease in reproduction of nematodes Mycorrhizal plants show increased plant growth Does not effect mycorrhizal colonization	Gradison and Cooper, 1986
Tomato	-	*M. hapla*	Decreased reproduction & disease damage due to nematode	Cooper and Gradison, 1986
White clover	"	*M. hapla*	"	
Cotton	"	*M. incognita*	"	Smith *et al.*, 1986
Beans	*Glomus etunicatum*	*M. javanica*	Less disease damage and reproduction of nematodes	Olivera and Zambolim, 1988
Tobacco	"	*Heterodera solanacearum*	Decreased reproduction in mycorrhizal plants	Fox and Spasoff, 1972
Soybean	*Gigaspora margarita*	*M. incognita*	Increased nematode reproduction	
Soybean	"	"	Decreased damage to mycorrhizal plants	Kellam and Schenck, 1980
Cucumber	"	"	Decreased damage to plant and decreased infection & reproduction due to nematodes	Priestel, 1980
Cotton	"	"	Decreased damaged to plant increased reproduction	Hussey and Rancadori, 1982
Tomato	"	"	Decreased infection nematodes	
Cotton	"	*Pratylenchus brachyurus*	Decreased damage to mycorrhizal plants	Roncadori, 1977

Host	Mycorrhizal fungus	Plant parasitic nematode	Effect on host plant	Reference
Cucumber	*P. Penetraus*	"	Decreased nematode reproduction	Priestel, 1980
Carrot	*Glomus mosseae*	*Meloidogyne hapla*	Not known	Sikora and Schoenbeck, 1975
Citrus	*Glomus fasciculatum* *G. mosseae* *G. etunicatus*	*T. semipenetrans* *Tylenchulus semipenetrans* *Rhodopholus similis*	Growth enhanced Decrease in nematode number	O'Banno *et al.*, 1979 O Bannon and Nemec, 1979 O Bannon and Nemec, 1979
Cotton	*Gigaspora margarita*	*Meloidogyne hapla*	No effect	Hussey and Roncadori, 1977
		Robychulus renfformis	Decrease in infection and damage to plant	Sitaramaiah and Sikora, 1982 a, b
	Glomus etunicatus	*Pratylenchus*	"	Hussey and Ronacadori, 1982
	G.etunicatus	*Meloidogyne*	"	Roncadori and Hussey, 1977
	G. mosseae	*Meloidogyne incognita*	"	Smith *et al.*, 1986
	Gigaspora margarita	*Aphelenchus avenae*	Growth stimulated	Hussey and Roncadori, 1982
Grapes	*Glomus fasciculatum*	*Meloidogyne arenaria*	Growth stimulated	Atilano *et al.*, 1976
Oats	*G. mosseae*	*M. incognita*	Not determined	Sikora and Schoenbeck, 1975
Peach	*Gigaspora margarita*	*Meloidogyne incognita*	Growth enhanced	Storbel *et al.*, 1982
	Glomus etunicatus	*M. incognita*	Growth enhanced	Strobel *et al.*, 1982

(Mukerji, 1996)

3. Rhizospheric intraction and disease control

Singh, and Vyas (2007), recently we have found reduction in the growth of *Fusarium oxysporum* on *Cicer arietinum*. Addition of different AM fungi produces different results. Among the four test AM fungal species three species *Glomus mosseae, Acaulospora spinosa* and *Gigaspora albida* produces phytoprotection effect and *G. fasciculatum* was found uneffective.

The result depicted in table (24.6-24.7) shows percent germination of chickpea, pre-inoculated with native *Trichocderma* isolates and interaction with AM fungi, *Rhizobium* sp. and pathogen *Foc*. It was observed that chickpea plants pre-inoculated by different *Tricoderma* isolates when further treated with AM fungi, *Rhizobium* sp. or *Foc* shows differential response for their percent germination. TvSV-11 pre-inoculated chickpea plant shows greater percent germination with AM fungi followed by TvSV-21, ThSV-235, ThSV-215 and ThSV-221. Interestingly, when dual inoculum (AMF+*Rhizobium)* combination was used performance of ThSV-221, pre-inoclated plant rises about 88%.

Table 24.6.: Phytoprotection Role of AM Fungal Species Against Foc in Chickpea Variety ICC 4951 on % Germination with or without Trichoderma Isolates

Treatment	Germination (%)										
	Control	foc*	Gf	Gf + foc	Gm	Gm + foc	As	As + foc	Ga	Ga + Foc	Mean
Control[x]	75.2	49.0	76.4	46.3	77.3	50.1	75.3	50.2	77.3	44.0	62.1
TvSV-11	85.6	56.2	82.2	70.5	87.2	73.5	86.6	86.6	86.0	56.2	75.6
TvSV-21	80.5	50.2	78.9	65.2	82.3	68.1	81.0	65.1	82.0	60.2	71.7
TvSV-215	78.2	52.1	77.4	62.0	82.2	70.1	82.0	64.6	82.0	59.1	71.4
TvSV-221	78.5	51.0	80.2	66.0	80.6	71.0	81.1	63.2	81.1	61.0	71.4
TvSV-235	80.5	50.0	80.0	67.0	84.6	65.0	83.0	65.0	83.0	60.3	72.1
Mean	79.8	51.4	81.7	62.8	82.4	66.3	81.3	62.8	81.9	65.8	

CD (P=0.05) = 45.9 (Singh, 2008)

Table-7: Phytoprotection role of AM fungal species and *Rhizobium* against Foc in chickpea variety ICC 4951 on % germination with or without *Trichoderma* isolates

Treatment	Germination (%)								
	Control	AMF	Rhizobium	AMF + Rhizobium	Foc	AMF + Foc	Rhizobium +Foc	AMF + Rhizobium +Foc	Mean
Control[X]	75.2	77.3	78.4	81.2	49.0	80.1	52.3	70.2	70.5
TvSV-11	85.6	87.2	84.9	87.5	56.2	73.5	64.6	78.9	77.3
TvSV-21	80.5	82.3	79.0	83.2	50.2	68.1	61.0	75.0	72.4
TvSV-215	78.2	82.2	78.6	84.0	52.1	70.1	62.0	74.2	72.7
TvSV-221	78.5	80.6	80.2	88.0	51.0	71.0	61.1	76.0	73.3
TvSV-235	80.5	84.5	82.6	86.0	50.0	65.0	63.0	78.0	73.7
Mean	79.7	82.4	80.6	84.9	51.4	71.3	60.7	75.4	

CD (P=0.05) =45.9
(Singh, 2008)
Foc= *Fusarium oxysporum f.sp. ciceris*
Gf= *Glomus Faciculatum*
Gm= *Glomus Mosseae*
As= *Acaulospora spinosa*
Ga= *Gigasspora albida*
X= without *Tricoderma* isolates
Y= without *Foc*, AMF and *Rhizobium*

Greater protection of AMF was obtain with TvSV-11 and ThSV-235 followed by ThSV-215 and TvSV-21 and lesser germination obtained with ThSV-221, when the same set of experiment conducted with AMF + Rhizobium + Foc there was no significant difference for bioprotection mechanism against Foc.

B. Rhizospheric interaction and disease control

Microbial populations in the rhizosphere are known either to interfere with or to benefit the establishment of mycorrhizal symbiosis (Gryndler, 2000). A typical beneficial effect is that exerted by 'mycorrhiza-helper-bacteria' (MHB), a term that was coined by Garbaye (1994) for those bacteria known to stimulate mycelial growth of mycorrhizal fungi and/or enhance mycorrhizal formation. This applies both to *Ectomycorriza* (Garbaye, 1994; Founoune *et al.*, 2002; Frey-Klett *et al.*, 2005) and to AM associations (Azcon-Aguilar and Barea, 1995; Gryndler, 2000; Barea *et al.*, 2004; Johansson *et al.*, 2004). Soil microorganisms are known to produce compounds that increase the rates of root exudation. This, in turn, stimulates AM fungal mycelia in the rhizosphere or facilitates root penetration by the fungus. Plant hormones, as produced by soil micro-organisms, are known to affect AM establishment (Azcon-Aguilar and Barea, 1992). Rhizosphere microorganisms are also known to affect the presymbiotic stages of AM development, such as spore germination rate and mycelial growth (Azcon-Aguilar and Barea, 1992, 1995).

AM-induced changes in plant physiology affect the microbial populations, both quantitatively and qualitatively, in either the rhizosphere and/or the rhizoplane. Therefore, the rhizosphere of a mycorrhizal plant can have features that differ from those of a non-mycorrhizal plant (Barea *et al.*, 2002a, b; Johansson *et al.*, 2004). However, there are specific modifications in the environment surrounding the AM mycelium itself, the mycorrhizosphere (Linderman, 1988; Gryndler, 2000). In addition to this term, the soil space affected by extraradical hyphae is also called the mycosphere (Linderman, 1988) or hyphosphere as an analogy with the term rhizosphere (Gryndler, 2000). Large number of bacteria (including actinomycetes) and fungi can be associated with both AM fungal structures. Since the AM mycelium releases energy-rich organic compounds, an increased growth and activity of microbial saprophytes can be expected to occur in the mycorrhizosphere. However, the enrichment of this particular environment by organic compounds is much lower than that of the rhizosphere, corresponding to lower counts of bacteria in mycorrhizoshere soil, compared with those in the rhizosphere (Andrade *et al.*, 1997).

The establishment of PGPR inoculates in the rhizosphere can be affected by AM fungal co-inoculation. In particular, AM inoculation improves the establishment of both inoculated and indigenous phosphate-solubilizing rhizobacteria acting as MHB (Toro *et al.*, 1997; Barea *et al.*, 2002c).

C. Other interaction of AM Fungi

This is the key to exploiting the possibilities of dual (AM fungi and PGPR) inoculation to aid plant defence against root pathogens. Barea *et al.* (1998) carried out a series of experiments to test the effect of *Pseudomonas* strains producing 2,4-diacetylphloroglucinol (DAPG) on AM formation and functioning. Three *Pseudomonas* strains were tested for their effects on AM fungi: a wild type (F113) producing the antifungal compound DAPG; the genetically modified strain (F113 G22), a DAPG-negative mutant of F113; and another genetically modified strain [F113 (pCU203)], a DAPG-over-producer. The results from *in vitro* and *in situ* experiments under controlled conditions demonstrated no negative effects of these *Pseudomonas* strains on spore germination. There was, however, a stimulation of hyphal growth of *G. mosseae*. A field aexperiment was designed to validate these results. None of these *Pseudomonas* strains affected : (i) the numbers or diversity of the native AM fungal population; (ii) the percentage of root length that became mycorrhizal; or (iii) AM performance. Furthermore, the antifungal Pseudomonas improved plant growth and nutrient (N and P) acquisition by the mycorrhizal plants (Barea *et al.*, 1998).

4. Mechanism of plant defence

Different mechanisms have been shown to play a role in plant protection by AMF, namely, improved plant nutrition, damage compensation, competition for colonization sites or photosynthates, changes in the root system, changes in rhizosphere microbial populations, and activation of plant defense mechanisms. Several mechanisms can be operative simultaneously, with contributions depending on environmental conditions, timing of the interaction and partners involved (Azcon-Aguilar and Barea ,1996).

There is evidence for the accumulation of defensive plant compounds related to mycorrhization, although to a much lower extent than in plant-pathogen interactions. Accumulation of reactive oxygen species, activation of phenylpropanoid metabolism, and accumulation of specific isoforms of hydrolytic enzymes such as chitinases and glucanases has been reported in mycorrhizal roots. There reactions, however, are generally localized, suggesting. Concerning above ground effects, accumulation of insect anti feedant compounds and transcriptional regulation of defense-related genes (Pozo *et al.*, unpublished data) have been described in the shoots of mycorrhizal plants. Furthermore, the volatile blends released by AM plants can be more attractive to aphid paraisotoids than those from non-mycorrhizas. Nevertheless, accumulation of PR proteins, salicyclic acid, or expression of market genes associated with systemic acquired resistance has not been reported in systemic tissues.

A. Mechanism of mycorrhiza-induced resistance

Priming seems to be the main mechanism operating in MIR. The lack of systemic activation of cellular or biochemical defense mechanisms in mycorrhizas and the stronger defense reactions observed upon pathogen challenge support this hypothesis. Root colonization by AMF systemically protected tomato plants against *Phytophthora parasitica* infection. Only mycorrhizal plants formed papilla-like structure around the sites of pathogen infection through decomposition of non-sterified pectins and callose, preventing the pathogen from spreading further, and they accumulated significantly more PR-1a and basic β-1,3 glucanases than non-mycorrhizal plants upon *Phytophthora* infection. (Cordier *et al.*, 2002; Pozo *et al.*, 1999).

B. Signaling pathways involved

The establishment of mutualistic associations involves mutual recognition and a high degree of co-ordination based on a continuous molecular dialogue between both symbionts. Advances in our knowledge about plant microbe signaling are unraveling mechanisms regulating biotic interactions. As host recognition compounds for AMF has uncovered a possible mechanism mediating the protector effect of AM against parasitic plants.

Fig. 24.1.:Spectrum of Efficiency of MIR, AM Symbioses Generally Reduce Incidence and/or Damage Caused by Soil-Borne Pathogens, Nematodes, and Root-Chewing Insects. (Pozo, 2007)

Upon perception of any 'foreign' organism, different signaling pathways operate within the plant to co-ordinate the appropriate response. These pathways are orchestrated by 'alarm signals' that regulate differential sets of defense related genes. Salicylic acid (SA) co-ordinate defense mechanisms that are generally effective against biotrophic pathogens, whereas jasmonates (JA) regulate wounding responses and resistance against necrotrophs.

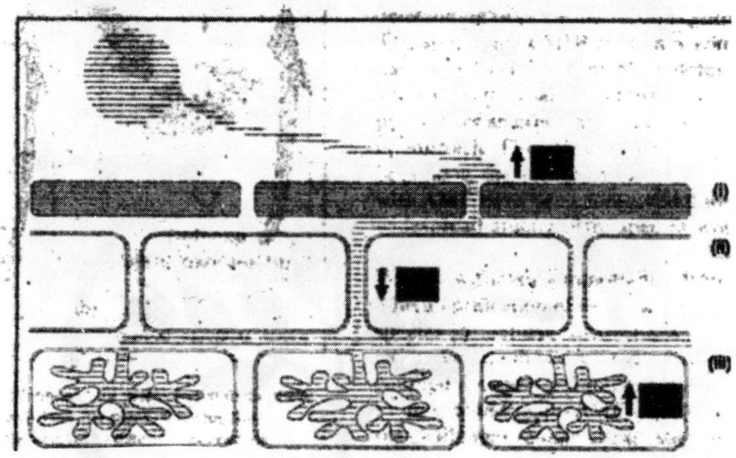

Fig. 24.2.: Modulation of Plant Defense Signaling During Mycorrhiza Establishment.
(i) Upon germination, AM fungi grow toward the root and form appresoria at the root surface. At this stage, the plant reacts with an increase in SA levels. (ii) In a compatible interaction, SA levels decrease as the fungus colonizes the cortex. (iii) JA synthesis occurs in arbuscule containing cells. (Pozo, 2007)

Nevertheless, there is some overlap in their spectrum of efficiency, especially on pathogens with intermediate life styles. Insect feeding guilds also determine the response they trigger in the plant. Generalist chewing insects, but not phloem-feeding ones, cause wounding and trigger JA-regulated responses. AMF repress SA-dependent defense responses in the host in order to achieve a compatible interaction. In the mycorrhizal plants, supression of SA responses such attenuation could explain the delay in systemic accumulation of PR proteins upon treatment with SA or analogs. (Shaul O *et al.*, 1999; Dumas-Gaudot *et al.*, 2000).

C. Siderophore Formation

There is some evidence that AM fungi may produce siderophores. The AM grass Hilaria jamesii, which showed greater iron uptake than a non-mycorrhizal control, tested positive for siderophores when bioassayed (Cress *et al.*, 1986). Arbuscular mycorrhizal fungi are shown to suppress the plant diseases due to increased uptake of macro- and micro-nutrients or drought tolerance of the AM fungal plant. Alleviation of abiotic stress, such as decreased toxicity to salt and heavy metals by AM fungal colonised plants have shown to decrease disease in some cases (Hooker *et al.*, 1994; Linderman, 1994; Karagiannidis *et al.* 2002). Altered root branching or root morphology due to AM fungal colonisation may also decrease the negative effect of plant pathogens (Norman *et al.*, 1996; Fusconi *et al.*, 1999).

D. Phytoalexins

The phytoalexins is considered to be an important mechanism of disease resistance in plants (Morandi *et al.*, 1984). Only a few reports are available on phytoalexin production in AM plants. Morandi *et al.* (1984) observed accumulation of isoflavonoid compounds during VAM symbiosis. To understand the role of AM fungi in disease resistance, we examined the interaction between AMF and a wilt causing fungus, *Fusarium oxysporum*, in cowpea. Further, the elicitation and accumulation of phytoalexin compounds in response to fungal infection and their antifungal activity were examined *in vitro*.

Accumulation of phytoalexins in roots (Sundaresan *et al.*, 1989) cowpea plants were treated with $CuSO_4$, phytoalexin production was enhanced with about 9.7-, 10.3- and 13-fold increase in the concentration of compounds I, II and III respectively over the untreated plants (Table 24.8). AM infection in plants greatly induced the synthesis of phytoalexin compounds; the extent of phytoalexin production was positively correlated with the degree of VAM colonization (Table 24.8). At the final harvest, AM plants had 3.0, 5.0 and 4.0 times more concentration of daidzein, compounds II and III respectively than the non-mycorrhizal plants.

Table 24.8. : Synthesis of Phytoalexin Compounds in Cowpea Roots Elicited with $CuSO_4$

Treatment	Control	$CuSO_4$
Daidzein*	0.9	8.7
Compound II**	0.006	0.06
Compound III***	0.003	0.04

Values are the means of three replicates.

* Expressed in µg/g fresh weight.

** The concentration of compounds II and III were expressed in relative absorbance at 283 and 340 nm respectively since the compound were not identified. Values are the absorbance of compounds separated from 1 g of fresh roots.

(Sundaresan *et al.*, 1992)

Table 24.9.:Phytoalexin Accumulation in Cowpea Roots in Relation to the Development of AM Infection

Treatment	Days	AM infection (%)	Deidzein*	Compound II**	Compound III***
Control	15	-	5.7	0.3	0.01
	30	-	9.7	0.9	0.03
	45	-	11.9	1.3	0.37
	60	-	12.7	1.6	0.78
AM	15	18	17.8	2.6	0.11
	30	42	30.4	4.3	0.92
	45	64	33.5	5.9	1.91
	60	75	42.1	8.2	3.22

** The concentration of compounds II and III were expressed in relative absorbance at 283 and 340 nm respectively since the compounds were not identified. Values are the absorbance of compounds separate from 10 g of fresh roots.

(Sundaresan *et al.*, 1992)

Table 24.10 Inhibitory Concentration of Phytoalexins on Germination of *F. oxysporum* Spores

Treatment	Minimum inhibitory concentration µg/ml
Daidzein	>100
Compound II	50
Compound III	10

Values are the means of 3 replicates
(Sundaresan *et al.*, 1992)

According to Sundaresan *et al.* (1992) mycorrhizal infection increased the production of phytoalexins in cowpea. Further, the extent of accumulation of phytoalexin compounds was positively associated with colonization of *G. fasciculatum*. The production of phytoalexins was also found to be elicited by $CuSO_4$. Earlier Morandi *et al.*, (1984) reported an increase in the accumulation of phytoalexin compounds in soybean with increase in infection by *G. mosse* and *G. fasciculatum*; they also reported elicitation of phytoalexins by $CuSO_4$.

E. Enzyme

Many plants enzymes are involved in defense reactions against plants pathogens. These include oxidative enzymes such as POD and PPO which catalyse the formation of lignin and other oxidative phenols that contribute to the formation of defense barriers that contribute to the formation of defense barriers for reinforcing the cell structure (Avdiushko *et al.*, 1993). Other enzymes such as PAL are involved in phytoalexin or phenolic compound biosynthesis. These enzymes have been correlated with defense against pathogens in several plants, including tomato (Borden and Higgins, 2002), pepper (Jung *et al.*, 2004), on wheat. 1`POD (peroxidase), a member of the PR-9 family, is of the lignin-formation type of plant defense response (Ray *et al.*, 1998). POD activity is associated with disease resistance in plant (Lin and Kao, 2001), and increases in host plants following pathogen infection (Bordena and Higgins, 2002; Koike *et al.*, 2001; Zheng *et al.*, 2004). Plant POD has been reported to catalyse the last step in the biosynthesis of lignin and hydrogen peroxide (Breusegem *et al.*, 2001). Mohammadi and Kazemi (2002) reported that POD was induced wheat head by infection with pathogenic strain of *F. graminearum.*

PPO (Polyphenol oxidase) usually accumulates during pathogenic attack, in wounds, or tissue senescence in plants. Plant calls membrane disruption occurrence might initiate formation of quinines following an increase in accessibility of PPO to its substrate. The increase activity of PPO could be detected in the cucumber leaf in the vicinity of lesions caused by some foliar pathogens or phosphate application (Avdiushko *et al.*, 1993). PPO activity in Pc-treated roots gradually increased in the first 3 days after the pathogen infection and then rapidly decreased, while the activity in Gi+Pc-treated roots was unchanged till the third day, and slight decreased thereafter. There as no significant difference in POD activity between control and Gi+Pc-treated plants during the experimental period in root. This might suggest that the induced PPO activity in Pc-treated roots could be a defense response against *P. capsici* infection at the early stage. Similar results obtain in plants-pathogens fungal interactions such as wheat. I *graminearum* (Mohammadi and Kazemi, 2002). In Gi+ Pc-treated roots, the PPO activity remained similar to that of the control, suggesting the antagonistic effect of *G. intraradices*. The capability of phytopathogenic stress was related to the increase or sustained ability to from defense barriers for reinforcing the cell structure by lignification-related enzymes. However, decrease in the activities if lignification-related enzymes, particularly PPO activity, might be the result from prolonged pathogens stress.

PAL (Phenylalanine ammonia lyas), one of the key enzymes in the phenylpropanoid pathway and the flavonoid pathway, was increased in both incompatible and compatible interaction between plants and pathogens (Harllen *et al.*, 2004). (Chen *et al.*, 2000) demonstrated the existence in cucumbers of phenolic phytoalexine, which might be produced through a PAL pathway.

F. Genetics of plant defense

Attenuation of plant defenses however is not enough for a successful mutualistic interaction with AMF, upon natural recognition, a symbiotic program has to be activated (Reinhardt, 2007). This program should allow a redistribution of nutrients and active physical accommodation of the fungal symbiont inside root cell (Genera *et al.*, 2006). Both aspects may be regulated by dasmonates (Hause *et al.*, 2002). JA-responsive genes and genes involved in JA biosynthesis are expressed in arbuscule containing cells, and mycorrhizal roots are associated with increased level of endogenous JA. This increase occurs after the onset of mycorrhization, likely associated with fully established mycorrhizal. (Hause *et al.*, 2002). Elevated level of basal JA production could be related to increase resistance of mycorrhizal plants to pathogens and insects. Experimental evidences linking. Linking the JA pathway with primed deposition of cellose and enhanced resistance to oomycetces (Hamiduzzaman *et al.*, 2005) argue for a role of JA signaling in the primed papillae formation observed in mycorrhizal tomatoes upon infection with *Phytophthora* (Cordier *et al.*, 1998). Recently, a control role for JA in systematic immunity has been proposed in *Arabidopsis* (Turman *et al.*, 2007). It is tempting to speculate that JA also serves as endogenous signal in MIR, other hormones such as abcisic acid show altered levels in AM (Hause *et al.*, 2002) and may also affect plant resistance to challengers (Thaler *et al.*, 2004).

National Scene

In India, researches on AMF received attention in mid seventees, of 20th century. Over the years knowledge on various dimensions of its impact on host growth and development has expanded appreciably. The research efforts have been broadly directed towards ecological and taxonomical aspects, interaction with other microorganisms (particularly nitrogen fixing organisms as well as soil borne pathogens) side effects of pestisides on non-targeted myocrrhizal systems and other allied aspects.

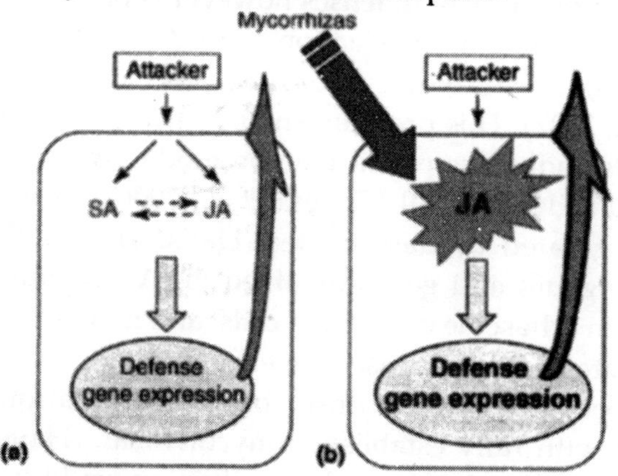

Fig. 24.3.: Model Illustrating Printing of JA-Dependent Responses in Mycorrhizal Plants. (a) Upon Attacker Recognition, the Plant Produces the Defense-Related Signals JA, ET, and SA in Different Proportions. Cross-Talk Among the Pathways they Co-ordinate Fine-Tunes the Appropriate Response. (b) Mycorrhiza Formation Primes the Tissues for a Quicker and More Effective Activation of JA-Dependent Defense Responses Upon Attack, Resulting in Enhanced Resistance. (Pozo, 2007)

Extensive studies on host pathogen and mycorrhizae interaction were intiated at Hisar. Studies clearly indicted that mycorrhization in legume crops resulted into a significant reduction *fusarium* wilt (Jalali and Thareja 1981). In (1999) Gupta *et al.* reported reduction in root rot of mungbean caused by *Macrophomina phaseolina.* Jalali and Jalali (1991) suggested that mycorrhization affected the activities of soil borne pathogens by competition, by formation of mechanical barrier against penetration, by favourable rhizosphere population which may compete with the pathogens, and libration of inhibitory compound. Singh (1990) reported suppression of *Meloidogyne javanica* in pre inoculated mycorrhizal roots. Bisht *et al.* (1985) obtained positive responses in terms of disease control ability in pigeonpea when simulataneous inoculation with blight pathogen *Phytopthora dreschleri f. sp. cajani* and mycorrhizal fungus *Gigaspora calospora* was carried out. Jalali *et al.* (1999) reported antioxidant metabolism was now known to facilitate the expression of resistance implants to pathogen attack. Attempts have been made to tailor genetically engineered mycorhizaal fungi for effective biological control to indicate that it is possible to transform ectomycorhizal fungi with value added genes such as insectidal proteins and other factors beneficial to the host plant (Hiremath and Podila, 2000). Recently, Singh *et al.*, 2007, Singh *et al.*, 2008 (a,b); Singh and Vyas 2007and Singh 2008 reported suppression of *Fusarium* wilt by inoculating in mycorrhized roots of tomato and *Cicer arietinum*.

CONCLUSION

Sustainability in agriculture has become the paradigm of present time. Mycorrhizal research persuits a relook on concept of synergy between soil, plant and fungal symbiont. One can see a closed chain of cause effect relationship as the resultant benefit of mycorrhizal symbiosis in the agrosystem. The fungal endophytes improve host growth, health and resistance to biotic and abiotic stresses; the host plant in turn becomes rich source of energy to the soil; the organisms accelerate soil aggregate formation; and life supporting soil structures thus formed facilitates better plant growth. In-depth more studies on non-nutrional influences of mycorrhizal endophytes on plant pathogen may yield more conclusive picture. This will facilitate to moniter appropriate use of pestisides under natural farming systems, and eventually make IPM technology more pragmatic and meaningful.

ACKNOWLEDGEMENT

Authors are thankful to Head, Department of Botany, Dr. H.S. Gour University, Sagar, MM and RKG thankfully acknowledge UGC for financially assistance.

REFERENCES

Allen, M .F. (1991). *The Ecology of mycorrhizae*. Cambridge University Press. Cambridge. Brown, V.K. and Gange, A.C. (1989) Differential effects of a above and below ground insect herbivory during early plant succession. *Oikos*. 54: 67–76.

Ames, R.N. and Linderman, R.G. (1978). The growth of Easter lily (*Lilium longiflorum*) as influenced by vesicular-arbuscular mycorrhizal fungi, *Fusarium oxysporum* and fertility level. *Canad. J. Bot*. 56:2773.

Andrade, G., Mihara K.L., Linderman, R.G., Bethlenfalvay G. (1997). Bacteria from rhizosphere and hydrosphere soils of different arbuscular mycorrhizal fungi. *Plant and Soil*.192 : 71–79.

Andrade, G., Mihara, K.L., Linderman, R.G and Bethlenfalavy, G.J. (1998). Soil Aggregation status and rhizobacteria in the mycorrhizosphere. *Plant Soil*. 202:89–90.

Atilano, R.A., Rich, J.R., Ferris, H and Menge, J.A. (1976). Effect of *Meloidogyne arenaria* on endomycorrhizal graphe (*Vitis vinifera*) rooting *J.Nematol*. 8:278.

Auge, R.M. and Stodola A.J.W. (1990). An apparent increase in symplastic water contribuites to greater turgor in mycorrhizal roots of droughted Rosa plants. *New phytol*.115:285–295

Avdiushko, S.A., Ye, X.S., Kue, J. (1993). Detection of several enzymatic activities in leaf prints of cucumber plant. *Physiological and Molecular Plant Pathology*, 42: 441–454.

Azcon-Aguilar, C. and Barea J.M. (1992). Interactions between mycorrhizal fungi and other rhizosphere micro-organisms.

Mycorrhizal functioning: an intergrative plant-fungal process, New York : Chapman and Hall, pp.163–198.

Azcon-Aguilar, C. and Barea, J.M. (1995). Saprophytic growth of arbuscular-mycorrhizal fungi. *Mycorrhiza structure, function, molecular biology* and *biotechnology*. Heidelberg, Germany: Springer-Verlag, pp. 391–407.

Azcon-Aguilar, C and Barea, J.M. (1996). Arbuscular mycorrhizas and biocontrol of soil borne plant pathogens, an overview of the biological mechanism involved. *Mycorrhiza* 6:457–464.

Baath, P. and Hayman, D.S. (1983). Plant growth responses to vesicular arbuscular mycorrhiza XIV. Intertraction with *Verticillium* wilt in tomato plants. *New Phytol.* 95:419.

Bagyaraj, D.J. (1984). Biological interactions with mycorrhizal fungi; In *VA mycorrhiza* (eds. Powell, C. and Bagyaraj, D.J.) NewYork: CRC Press, pp.131–153.

Bagyaraj, D.J., Manjunath, A. and Reddy, D.D.R. (1979b). Interaction of vesicular-arbuscular mycorrhiza with root knot nematode in tomato. *Plant & Soil* 51:397.

Bagyaraj, D.J., Manunath, A and Patil, R.B. (1979a). Interaction of vesicular-arbuscular mycorrhiza and *Rhizobium* and their effects on soybean in the field. *New Phytol.* 82:141.

Bali, M. and Mukerji, K.G. (1987). Interaction of vesicular-arbuscular mycorrhizae rhizosphere and rhizoplane fungi of *Gossypium hirsutum* and *Corchorus olitorius. In Mycorrhiza Round Table. Proc. Natl. Workshop,* Jawaharlal Nehru University, New Delhi, India, pp. 347–355.

Bali , M. and Mukerji, K.G. (1991). Effect of VAM fungi on *Fusarium* wilt of cotton and jute, *Conf. Mycorrhiza,* Madras, pp.233–234.

Bali, M. and Mukerji, K.G. (1991). Interaction between VA mycorrhizal fungi and root. *In Plant Roots and their Environment* (Eds., Michael, M.C. and Pearson, Elsevier, Amsterdam.

Baltruschat, H. and Schonbeck, F. (1972). The influence of endotrophic mycorhiza on the infestation of tobacco by *Thielaviopsis basicola* : Phytopatho. 84:172–188.

Barea, J.M., Andrade, G., Biancciotto, V., Dowling, D., Lohrke, S. Bonfante, P. O'Gara F., Azcon-Aguliar, C. (1998). Impact on mycorrhiza formation of *Pseudomonas* strain used as a inoculants for the biocontrol of soil borne plant fungal pathogens. *Applied and environmental microbiology*. 64:2304–2307.

Barea, J.M., Azcon R., Azcon–Aguilar. (2004). Mycorrhizal fungi and plant growth promoting rhizobacteria. *In*: Varma, A., Abott, L., Werner, D., Hampp, R. eds. *Plant surface microbiology*, Heidelberg, Germany : Springer-Verlag. Pp. 351–371.

Barea, J.M., Azcon, R., Azcon-Aguilar, C. (2002a). Mycorrhizosphere interactions to improve plant fitness and soil quality. Antonie van Leeuwenhoek International *Journal of General* and *Molecular Microbiology*. 81:343–351.

Barea, J.M., Toro, M., Orozco, M.O., Campos. E. and Azcon, R. (2002c). The application of isotopic 32p and ^{15}N-dilution techniques to evaluate the interactive effect of phosphate-salubilizing rhizobacteria, mycorrhizal fungi and *Rhizobium* to improve the agronomic efficiency of rock phosphate for legume crops. *Nutrient cycling in Agrocecosystems*. 63:35–42.

Barnard, E.L. (1977). *The Mycorrhizal Biology of Liriodendron tulipifera L. and its Relatiohship to Cylindrocladium Root Rot*. Ph.D. thesis, Duke Univ., Durham N.C. Pp. 147.

Bartschi, H., Gianinazzi-Pearson V and Veigh I. (1981). Vesicular-arbuscular mycorrhiza formation and root rot disease (*Phytophthora cinnamomi*) development in *Chmaecyparis lawsonian*. *Phytopathol. Z*. 102:213

Bathlenfalvay, G.J. and Barea, J.M. (1994). Mycorrhizae in sustainable agriculture. I Effects on seed yield and soil aggregation. *Am. J. Alturn, Agric*., 9:157–161.

Becker, W.M. (1976). *Quantification of Onion Vesicular-arbuscular Mycorrhiza and Their Resistance to Pyrenochaeta terrestris*. Ph.D. thesis, Univ. Illinios, Urbana, pp. 72.

Becker, W.M. (1976). *Quantification of Onion Vesicular-arbuscular Mycorrhiza and Their Resistance to Pyrenochaeta terrestris*. Ph.D. thesis, Univ. Illinios, Urbana. Pp. 72.

Bianciotto, V., Bonfante, P. (2002). Arbuscular mycorrhizal fungi: a specialized niche for rhizospheric and endocellular bacteria.81:365–371.

Bodker, L., Kjoller, R., Kristensen, K. and Rosendahl, S. (2002). Interactions between indigenous arbuscular mycorrhizal fungi and *Aphanomyces euteiches* in field grown pea. *Mycorrhiza*.12:7–12.

Borden, S. Higgins, V.J. (2002). Hydrogen peroxide plays a critical role in the defense response of tomato to *Cladosporecem fulvum*. *Physiological and Molecular Plant Pathology* 61:227–236.

Borowicz,V.A. (1997) A fungal root symbiont modifies plant resistance to an insect herbivore. *Oceologia*. 112:534–542.

Borowicz, V.A. (2001). Do arbuscular mycorrhizal fungi alter plant pathogen relations? *Ecology*. 82:3057–3068.

Breusegem, F.V., Vranova, E., Dat, J.F. and Inze, D. (2001). The role of active oxygen species in plant signal transduction. *Plant Science*. 161:405–414.

Brundrett, M.C. (1991) Mycorrhizas in natural ecosystems. Adv. *Ecol.Res*.21:171–313

Calvet, C. Pera, J., Barea, J.M. (1993). Growth response of marigold (Tagetes erecta) tio inoculums with *Glomus mosseae, Trichoderma aureoviridr* and *Pythium ultimum* in a peat -pertile mixture. *Plant ant soil*.148:1–6.

Caron, M., Fortin, J.A. and Richard, C. (1986b). Effect of phosphorus concentration and *Glomus intraradices* of *Fusarium* crown and root rot of tomatoes. *Phytopathology*. 76:942–946.

Caron, M., Richard, C. and Fortin, J.A. (1986a). Effect of pre-infestation of the soil by a vesicular arbuscular mycorrhizal fungus, *Glomus intraradices*, on *Fusarium* crown and root rot of tomatoes. *Phytoprotection*. 67:15–19.

Chakravarty, P. and Umestam, T. (1987). Mycorrhizal fungi prevent disease in stressed pine seedlings. *J. Phytopathol*. 118:335–340.

Chen, C.Q., Belanger, R.R., Benhamou, N. Paulitz, T.C. (2000). Defense enzymes induced in cucumber roots by treatment with plant growth-promoting rhizobacteria (PGPR) and *Pythium ophanidermatum*. *Physiological and Molecular Plant Pathology*. 56:13–23.

Chou, L.G. and Schmitthenner, A.F. (1974). Effect of *rhizobium japoncum* and *Endogone mosseae* on soybean root rot caused by Pythium ultimum and *Phytophtora megasperma var. Sojae, Plant Dis. Reptr*. 58:221.

Christie, P., Newman, E. and Campbell, R. (1978). The influence of neighbouring grassland plants on each others endomycorrhizas and root surface micro organisms. *Soil boil. Biochem*, 10:52.

Copper, K.M. and Grandison, G.S. (1986). Interaction of vesicular arbuscular mycorrhizal fungi of tomato and white clover susceptible to *Meloidogyne*.

Cordier, C., Gianinazzi, S. and Gianinazzi-Pearson, V. (1996). Colonisation patterns of root tissue by *Phytophthora nicotinae var. parasitica* related to reduced disease in mycorrhizal plants. *Plant soil*. 185:223–232

Cordier, C., Trouvelot, A., Gianinazzi, S., Gianinazzi-Pearson, V. (1996). Arbuscular mycorrhiza technology applied to micropropagated *Prunas avium* and to protection against *Phytopthora cinnamomi' Agronomie*. 16:679–688.

Crees, W.A., Johnson, G.V. and Barton, L.L. (1986). The role of endomycorrhizal fungi in iron uptake by *Hilaria Jamesii, Journal of plant Nutrition*. 9:547–556.

Daft, M.J. and Okusanya, B.O. (1973). *New Phytol*. 72:975–983.

Dar, G.H, Zargar, M.Y. Beigh, G.M., (1997). Biocontrol of *Fusarium* root rot in the common bean (*Phaseolus vulgaris L.*) by using symbiotic *Glomus mosseae* and *Rhizobium leguminosarum*. *Microb. Ecol*. 34:74–80.

Dar, G.H., Zargar, M.Y. Beigh, G.M. (1997). Biocontrol of *Fusarium* root rot in the common bean (*Phaseolus vulgaris L.*) by using symbiotic *Glomus mosseae* and *Rhizobium leguminosarum*. *Microb. Ecol*. 34:74–80.

Davis, R.M. and Menge, J.A. (1980). Influence of *Glomus faciculatus* on Phytphthora root rot of citrus. *Phytopathology* 70:447–452.

Davis, R.M. and Menge, J.A (1981). *Phytophthora parasitica* inoculation and intensity of vesicular-arbuscular mycorrhiza in Citrus. *New Phytol.* 87:705–715.

Davis, R.M., Menge, J.A. and Erwin, D.C. (1979). Influence of *Glomus- fasciculatum* and soil phosphorus on *Verticillium* wilt of cotton. *Phytopathology* 68:453–456.

Dehne, H.W. (1982). Interaction between vesicular-arbuscular mycorrhizal fungi and plant pathogens. *Phytopathology.* 725:1115–1119.

Dehne, H.W. and Schoenbeck, F. (1979). The influence of endotrophic mycorrhizae on the fusarium wilt of tomatoes. *Z. Pflanzenkr. Pflanzenschutz. Z.* 82:630–639.

Dehne, H.W. and Schonbeck, F. (1979). Untersuchungen zum Einfluss der endotrophen Mycorrhiza auf Pflanzenkrankheiten I. Ausbreitung von Fusarium oxysporum f.sp.lycopersici in Tomaten. *P hytopathology.* Z.95:105–110.

Dehne, H.W., Schoenbeck. and Baltruschat, H. (1978). The influence of endotrophic mycorrhiza on plant disease. 3. Chitinase activity and the ornithine-cycle. *Z. Pfllerankr. Pflanzenschutz.* 85:666–678.

Diedhiou, P.M., Hallmann, J. Oerke, E.C. and Dehne, H.W. (2003) Effects of arbuscular mycorrhizal fungi and a non pathogenic *Fusarium oxysporum* on *Meloidogyne incognita* infestation of tomato .*Mycorrhiza.* 13:199–204

Ducheshe, L.C. (1994). Role of ectomycorrhizal fungi in biocontrol. In *Mycorrhizae and Plant Health* (Eds. Pfleger, F.L. and Linderman, R.G.). Pp 27–45 APS Press Inc., St. Paul, Hung, L.L. and Graham, J.H.). Pp 193. Gainesville, USA.

Dumas-Gaudot, E. (2000). Modulation of host defence systems. In Arbuscular Mycorrhizas : Physiology and Fun Kluwer Academic Publishers. Pp. 173–200.

Elmer,W.H. (2002). Influence of formononetin and NaCl on mycorrhizal colonization and *Fusarium* crown and root rot of Asparagus. *Plant Diseases.* 86: 1318–1324.

Elsen, A., Declerck, S. and De Waele, D. (2001). Effects of *Glomus intrararadices* on the reproduction of the burrowing nematode (*Radopholus similis*) in dixenic culture. *Mycorrhiza.* 11:49–51.

Filion, M., St-Arnaud. M., and Fortin, J.A. (1999). Direct interaction between the arbuscular mycorrhizal fungus *Glomus intraradices* and different rhizosphere micro-organisms. *New phytol.*141:525–533

Founoune, H., Duponnois, R., Ba, A.M., Sall. S., Branget, I., Lorau, Neyra M., Chotte, J.L. (2002). Mycorrhiza Helper Bacteria stimulate ecotomycorrhizal symbiosis of *Acacia holosericea* with *Pisolithus albus. New Phytologists.* 153:81–89.

Fox, J.A. and Sposoff, L. (1972). Intrioduction of *Heterodera solanacearum and Endogone gigantea* on tobacco. *J. Nematol.* 4:224.

Frank, A.B. (1885). Uber die auf wurzelsymbiose bruhende Erharung gewisser Baume Durch underirdische Pilze. *Berchte der Deutsche Botanische Gesellschaff* .3:128–145

Frey-Klett, P., Chavatte, M., Calawse, M.L., Courrier S., Le Rouxc., Raaijmakers, J., Martirnotti, M.G., Pierrat, J.C, Garbaye, J. (2005). Ectomycorrhizal symbiosis affects functional diversity of rhizosphere fluorescent pseudomonads. *New phytologists.* 165:317–328.

Friese and Allen (1991). The spread of VA mycorrhizal hyphae in the soil: inoculum types and external hyphae architectaure. *Mycologia.* 83:409–418.

Fusconi, A., Gnavi, E., Trotta.and Berta, G. (1999). Apical meristom of tomato roots and their modifications induced by arbuscular mycorrhizal and soil borne pathogenic fungi. *New Phytopathology.* 142:505–516.

Gang, A.C., Brown, V.K., Alpin, D.M. (2003). Multitrophic link between arbuscular mycorrhizal fungi and insect parasitoids. *Ecol. Lett.* 6:1051–1055.

Gange, A.C., Brown , V.K. and Farmer, L.M. (1990). A test of mycorrhizal benefit in an early successional plants community. *New Phytologist*.115:85–91.

Gange, A.C. (2006). Insect mycorrhizal interactions: Patterns processes, and consequences. In indirect interaction webs:

Nontropic Linkages: Through induced plant Traits. Pp. 124–144.

Gange, A.C. and West, H.M. (1994). Interaction between arbuscular mycorrhizal fungi foliar-feeding insects in *Plantago lanceolata* L. *New Phytol.* 128:79–87.

Gange, A.C., Bower, E., Brown, V.K. (2002). Differential effects of insect herbivory on arbuscular mycorrhizal colonization. *Oecologia*. 131:103–112.

Gange,A.C., Brown, V.K., Alpin, D.M. (2005). Ecological specificity of arbuscular mycorrhizae: evidence from foliar and seed-feeding insects. *Ecology*. 86:603–611.

Garbaye, J. (1994). Helper bacteria: a new dimension to the mycorrhizal symbiosis. *New phytologist*. 128:197–210.

Garcia-Garrido, J.M. and Ocampo, J.A. (2002). Regulation of the plant defence response in arbuscular mycorrhizal symbiosis. *J. Exp.Bot.*53:1377–1386.

Garcia-Garrido, J.M. (2002). Regulation of the plant defence response in arbuscular mycorrhizal simbiosis. *J. Exp. Bot.* 53:1377–1386.

Genera, A., Chabaud, M., Timmers, T., Bonfante, P., Baker, D.G. (2005). Arbuscular mycorrhizal fungi elicit a novel intracellular apparatus in *Medicago truncatula* root epidermal cells before infection. *Plant Cell*. 17:3489–3499.

George, E., Marschner, H. and Jakobson, I. (1995). Role of arbuscular mycorrhizal fungi in uptake of phosphorous and nitrogen from soil. *Crit. Rev. Biotech.* 15:257–270.

Gerdemann, J.W. (1968). Vesicular-arbuscular mycorrhiza and plant growth. *Ann. Rev. Phytopathol.* 6:397.

Gianinazzi-Pearson, V., Dumas-Gaudot, E., Gollotte, A., Tahiri -Alaoui. and Gianinazzi, S. (1996). Cellular and molecular defence-related root responses to invasion by mycorrhizal fungi. *New Phytol.*133:45–57.

Gosling, P., Hodge, A.,Goodlars, G., Bending, G.D. (2005). Arbuscular mycorrhizal fungiand organic farming. *Agric.Eco.and Environ.* 113:17–35.

Gossling, P., Hodge, A., Goodlass, G., Bending, G.D. (2006). Arbuscular mycorrhizal fungi and organic farming. *Agriculture Ecosystem and environment*. 113:17–35.

Graham, J.H. and Menge, J.A. (1982). Influence of vesicular-arbuscular mycorrhizae and soild phosphours on take-all disease of wheat. *Phytopathology*. 72:95–98.

Grandison, G.S. and Cooper, K.M. (1986). Interaction of vesicular arbuscular mycorrhizae and cultivars of lucerne susceptible and resistant to *Meloidogyne hepla*. *J. Nematol.* 18:141.

Gryndler, M. (2000). Interactions of arbuscular mycorrhizal fungi with other soil organisms., In:Kapulnik. Y., Douds J. R. eds. *Arbuscular mycorrhizas: Physiology and function*. Dordrecht. The Netherlands : Kluwer Academic Publishers pp. 239–262.

Habte, M., Zhang, Y.C. and Schmitt, D.P. (1999). Effectiveness of *Glomus* species in protecting white clover against nematode damage.*Can. J. Bot.*77:135–139.

Hamiduzzaman, M.M., Jakab, G., Barnavon, L., Neuhaus, J.M., Mauch-Mani, B. (2005). Aminobutyric acid induced resistance against downy mildew in grapevine acts through the potentation of callose formation and jasmonic acid signaling. *Mol plant interact*.1075–1080.

Harley J.L. and Wilson J.M. (1959). The absorption of potassium by beech mycorrhizas. *New Phytol.* 56:281.

Harllen, S.A.S., Reginaldo S.R., Dirceu, M., Bernardo, A.H., Maria, C.B.P., Ann, M. (2004) Rhizobacterial induction of systemic resistance in tomato, plants : non-specificprotection and increases in enzyme activities. *Biological control*. 29:288–295

Harrier, L.A. and Watson, C.A. (2004). The potential role of arbuscular mycorrhizal (AM) fungi in the bioprotection of plants against soil-borne pathogens in organic and/or other sustainable farming system. *Pest Mangae. Sci.* 60:149–157.

Harrison, M.J. and Van Buuren, M.L. (1995). Phosphate transporter from the mycorrhizal fungi *Glomus versiforme*. *Nature*.378:626– 629.

Hause, B., Morsk, C., Isayenkov, S., Strack, D. (2007). Jasmonates in arbuscular mycorrhizal interations. *Phytochemistry*. 68:101–110.

Hause, B., Maier, W., Miersch, O., Kramell, R. and Strack, D. (2002). Induction of jasmonate biosynthesis in arbuscular mycorrhizal barley root. *Plant physiol.*130:1213–1220

Hock, B. and Varma, A. (1995). Mycorrhiza structure, function, molecular biology and biotechnology. Springer, Berlin, Germany.

Hodge, A. and Campbell, C.D.F.A.H. (2001).An arbuscular mycorrhizal fungus accelerates decomposition and accquires nitrogen directly from organic material. *Nature.* 413:297–299.

Hooker, J.E., Jaizme-Vega, M. and Aztikson, D. (1994). Biocontrol of plant pathogens using mycorrhizal fungi. *In: Impact of mycorrhizas on sustainable agriculture and natural ecosystems.* 191–200.

Hussey, R.S. and Ronacadori, R.W. (1982). Vesicular-arbuscular mycorrhizae my limit nematode activity and improve plant growth. *Plant Dis.* 66:9.

Hussey, R.S. and Roncadori, R.W. (1977). Interaction of *Pratylenchus brachyurus* and an endomycorrhizal fungus on cotton. *J. Nematol.* 9:270.

Johansson, J.F. and Paul, L.R., Finaly, R.D. (2004). Microbial interaction in the mycorrhizosphere and their significance for sustainable agriculture. *Microbiology Ecology.* 48:1–13.

Jung, W.J., Jin, Y.L., Kim, Y.C., Kim, K.Y., Park, R.D. Kim, T.H. (2004). Inoculation of *Paenibacillus illinoisensis* alleviates root mortality, activates of lignification-related enzymes, and induction of the isozymes in pepper plants infected by *Phytophthora capsici. Biological Control.*30:645–652.

Karagiannidis, N., Bletsos, F. and Stavropoulos, N. (2002). Effect of *Verticillium* wilt (*Verticillium dahliae Kleb.*) and mycorrhiza (*Glomus mosseae*) on root colonization, growth and nutrient uptake in tomato and egg plant seedlings. *Sci. Hortic.* 94:145–156.

Kasimdari, R.S., Smith, S.E., Smith, F.A. and Scott, E.S. (2002). Infulence of the mycorrhizal fungus *Glomus cornatum*, and soil phosphorous on infection and disease caused by binucleate *Rhizoctonia* and *Rhizoctonia solani* on mungbean (Vigna *radiata)*. Plant soil. 238:235–244

Kaye, J.W. Pfleger, F.L. and Stewart, E.L. (1984). Interaction of *Glomus fasciculatum* and *Pythium ultimum* in greenhouse growth poinsettia. *Canad. J.Bot* 62:1575.

Kellam, M.K. and Schenck, N.C. (1980). Interaction between a vesicular arbuscular mycorrhizal fungus and root-knot nematode on soybean. *Phytopathology* 70:293.

Kjoller, R. Rosendahl, S. (1996). The presence of the arbuscular mycorrhizal fungus *Glomus intraradices* influence enzymatic activities of the root pathogen *Aphanomyces euteiches* in pea roots. 6:487–491.

Koide, R.T. (1991). Transley Review No. 29: Nutrient supply, nutrient demand and plant responses to mycorrhizal infection. *New Phytol.*117.

Koide, R.T., Huenneke, L.F., Hamburg, S.P. and Mooney, H.A. (1988). Effect of applications of fungicide, phosphorus and nitrogen on the structure and productivity of an annual serpentine plants community. *Functional Ecology*. 2:335–344.

Koike, N., Hyakumachi, M.Kageyama, K., Tsuymu, S. and Doke, N. (2001). Induction of systemic resistance in cucumber against several diseases by plant growth-promoting fungus: Lignification and superoxide generation. *European Journal of Plant Pathology*. 107: 523–533.

Krishna, K.R. and Bagyaraj, D.J. (1983). Interaction between *Glomus faciculatum* and *Sclerotium rolfsii* in peanut; *Can. J. Bot.* 61: 2349–2351.

Lin, C.C and Kao, C.H. (2001). Cell wall peroxidase activity, hydrogen peroxide level and Nacl-inhibited root growth of rice seedlings. *Plant and soil*. 230:135–143

Linderman, R.G. (1988). Mycorrhizal interactions with the rhizosphere microflora. The mycorrhizosphere effect. *Phytopathology*. 78:366.

Linderman, R.G. (1994). Role of VAM in biocontrol mycorrhizae and plant health. In : Pfleger L, Linderman RG,eds.*Mycorrhizae and Plant health*.APS Press, St. Paul, NM, USA. Pp.1–26.

Linderman, R.G. (1998). Mycorrhizal interactions with the rhizosphere microflora: The mycorrhizospher effect.*Phytopathology*. 78:366–371.

Linderman, R.G. and Paulitz, T.C. (1990). Mycorrhizal and uninfected roots of *Eucalyptus fastigata* Deane and Maiden; *New Phytol*. 78 : 313–328.

Linderman, R.G. (1992) Vesicular-arbuscular mycorrhizae and soil microbial interactions. In: Mycorrhizae and Plant health. St. Paul, Minnesota , USA, 1–25.

Liu, J. (2007). Arbuscular mycorrhizal symbiosis is accompanied by local and systemic alterations in gene expression and an increase in disease resistance in the shoot. *Plant J* . 50:524–544.

Marschner, H. (1998) Mineral nutrition of higher plants. Academic Press limited, London, Great Britain.

Mathre, D.E. (1968). Photosynthetic activities of cotton plants infected with *Verticillium alboatrum Phytopathology*. 58:137.

Matsubara, Y., Kayukawa, Y.,Yano, M. and Fukui, H. (2001). Effect of arbulant disease. *In Soil-borne Plant Pathogens* (Eds., Schippers B and Gams). Academic Press, London. Pp.271

Matsubara, Y., Kayukawa, Y., Yano, M. and Fukui, H. (2000). Tolerance of asparagus seedlings infected with arbuscular mycorrhizal fungus to violet root rot caused by *Helicobasidium mompa*. *Journal of Japanese society of Horticultural sciences*.69: 552–556

McGraw, A.C. and Schenck, N.C. (1981). Effect of two species of vesicular-arbuscular mycorrhizal fungi on the development of *Fusarium* wilt of tomato. *Phytopathology*. 71:894.

Menge, J.A., Nemec, S., Davis, R.M. and Minassian, V. (1977). Mycorrhizal fungi associated with citrus and their possible interaction with pathogens. Proc. Intt. Soc. Citricult.3:872.

Mishra, M., Singh, P. and Vyas, D. (2007). Does AM fungi control *the Fusarium* wilt of tomato (Abstract in 48th annual conference association of microbiologist of India) AG-16:pp–149.

Mittal, N., Sharma, M., Saxena, G. and Mukerji, K.G. (1991). Effect of VA mycorrhiza on gall formation in tomato. *Root compatility news Letter*. 23:39–43.

Mohammad, M. Kazemi, H. (2002). Changes in peroxidase and polyphenol activity in susceptible and resistant wheat heads inoculated with *Fusarium graminearum* and induced resistance. *Plant Science*. 162: 491–498.

Morandi, D.,Bailey, J.A. and Gianinazzi-Pearson. (1984). Isoflavonoid accumulation in soybean roots infected with vesicular roots infected with vesicular-arbuscular mycorrhizal fungi; *Physiol. Plant Pathol.* 24: 357–364.

Mosse, B. (1973). Advances in the study of vesicular-arbusculr mycorrhiza; *Annu. Rev. Phytopathol.* 11:171–196.

Mukerji, K.G., Chamola, B., Pand, Sharma, Mamta. (1997). Mycorrhiza in control of plant pathogens. In, Management of threating plant diseases of National importance (Eds. Agnihotri VP, Sarbohy AK, Singh DV) Malhotra publishing house ,New delhi. pp-297–314

Norman, J.R., Atikson, D. and Hooker, J.E. (1996). Arbuscular mycorrhizal fungal induced alteration to root architecture in strawberry and induced resistance to the root pathogen. *Phytophthora fragariae. Plant and soil.*185:191–198.

Norman, J.R. and Hooker, J.E. (2000). Sporulation of *phytopthora fragariae* shows greater stimulation by exudates of non-mycorrhizal than by mycorrhizal strawberry roots, *Mycol. Res.* 104:1069–1073.

O'Bannon, J.H., Inserra, R.N.,Nemec, S. and Volvas, N. (1979). The influences of *Glomus mosseae* on *tylenchlus semipentrans* infected and uninfected citrus lemon seedlings. J.Nematol. 11:247.

O'Bannon, J.H. and Nemec, S. (1979). The response of citrus lemon seedlings to symbiont, *Glomus etunicatus,* and a pathogen *Radopholus similis, J. Namatol.* 11:270.

Oliveria ,A.A.R. and Zambolin, L. (1988). Influence of inoculum levels on the interaction between *Glomus etunicatum and Meloidogyne javaniace* on bean. *Pesquisa Agropecuari Brassilier.* 24:371.

Paget, D.K. (1975). The effect of *Cylindrocarpon* on plant growth responses to vesicular -arbuscular mycorrhiza. In endomycorrhiza (Eds. Sanders F E, Mosse, B and Tinker PB). pp 593–606. Academic Press London.

Paulitz, T.C. and Linderman, R.G. (1989). Interactions between fluorescent *pseudomonads* and VA mycorrhizal fungi. *New Phytol.* 113:37–45.

Paulitz, T.C. and Linderman, R.G. (1991). Mycorrhizal interaction with soil organisms, *Handbook of Applied Mycology, Dekker New York pp.* 77–129.

Pegg, G.F. and Jouglaekha, N. (1981). Assessment of colonization in *Chrysanthemum* growth under different photoperiods and infected with *Verticillium dabliae, Trans. Brit. Mycol. Soc.* 76:353.

Penninsi, E. (2004). The secret life of fungi. *Science* 304:1620–1622.

Pozo, M.J., Azcon-Aguliar., C.Dumas–Gaudot,E. and Bared, J.M. (1999). β 1,3 glucanase activities in tomato roots inoculated with arbuscular mycorrhizal and/or *Phytophthora parasitica* and their possible involvement in bioprotection .*Plant Sci.* 141:149–157.

Pozo, M.J., Slezack-Deschaumes, S., Dumas-Gaudot, E., Gianinazzi, S. Azcon-Aguilar (2002). Plant defense responses induced by arbuscular Mycorrhizal fungi. *In* : Gianinazzi, S., Schiiepp, H., Barea, J.M., Haselwandter K. *Mycorrhiza Technology from gene to bioproduction*.103–111.

Pozo, J.M. and Azcon-Aguliar, C. (2007). Unravelling mycorrhiza induced resistance. *Current Opnion in Plant Biology*.10:393–398.

Priestel, G. (1980). Wechsel benziehung zwischen der endotrophen mycorrhiza and. dem. Wurzel gallennematoden *Meloidogyne incognita* (Kofoid and White, 1919) chitwood, 1940 and Gurke. Diss. Univ. Hannover, W.Germany. Pp.103.

Raman, N. (1996). Role of myconhizae in biocontrol of plant pathogens. In *Current Trends in life Science. XXI recent Developments in Biocontrol of Plant Pathogens* (Eds. Mani Bhusanrao K and Mahadevan A) pp. 11–22 Today and Tomorrows Printers and Publishers, New Delhi.

Rambelli, A. (1973). The rhizosphere of mycorrhizae. In: Marks, GL and Koslowski TT (eds) :Ectomycorrhizae. Newyork, USA, 299–343.

Ramirez, B.N. (1974). *Influence of Endomycorrhizae on the Relationship of Inoculum Density of Phytophthora palmivora in Soil to Infection of Papaya Roots.* M.Sc. thesis. Florid., Gainville, pp. 45.

Rancadori, R.W. and Hussey, R.S. (1977). Interaction of the endomycorrhizal fungus *Gigaspora margarita* and root-knot nematode on cotton. *Phytopathology* 7:1507.

Ray, H., Douches, D.S. Hammerschmidt, R. (1998). Transformation of potato with cucumber peroxidase: Expression and disease response. *Physiological and Molecular Plant Pathology* .53:93–103.

Reinhardt, D. (2007). Programming good relations-development of the arbuscular mycorrhizal symbiosis. *Curr. Opin. Plant Bio.* 10:98–105.

Remy, W., Taylor, T.N., Hass, H. and Keep, H. (1994). Four hundred million-year old vesicular arbuscular mycorrhizae. Proc. Natl. ACad. Sci. USA 91 :11841–11843.

Rhodes, L.H. (1980). The use of mycorrhizae in crop production Systems. *Outlook Agriculture.* 10:275.

Ross, J.P. (1977). Influence of endogone mycorrhiza or *Phytophtora Phytopathology* 62:896.

Safir, G. (1968). The Influence of Vesicular-arbuscular Mycorrhiza on the *Pyrenochaeta terrestris.* M.Sc., thesis. Univ. Illinois. Urbana. USA.

Saleh, H.M. and Sokora, R.A. (1984). Relationship between *Glomus fascicculatum* root colonization of cotton and its effect on *Meloidogyne incognita. Nematologica.*30:230–237

Sanders, F.E. (1993). Modeling plant growth response to vesicular arbuscular mycorrhizal infection. *Adv. Plant Pathol.*9:135–166

Schenck, N.C. and Nicolson, T.H. (1977). A zoosporic fungus occurring on species of *Gigaspora margarita* and other vesicular-arbuscular mycorrhizal fungi. *Mycologia* 69:1049–1051

Schenck, N.C., Kinloch, R.A. and Dickson, D.W. (1975). Interaction of endomycorrhizal and root-knot nematode on soyabean. In Ectomycorrhiza (Eds. Sanders FE, Mosse, Tinker PB). Pp-605–617. Academic press. London.

Schoenbeck, F. (1979). Endomycorrhiza in relation to plant diseas. *In Soil-borne Plant Pathogens* (Eds., Schippers, B. and Gams). Academic Press, London.Pp. 271–292.

Schoenbeck, F. and Dehne, H.W. (1977). Damage to mycorrrhizal and non mycorrhizal cotton seedlings by *Thielaviopsis basicola*. *Plant Dis. Reptr.* 61:266–267.

Schoenbeck, F. and Dehne, H.W. (1979). Unterschungenzum Pilzliche spores parasiten. *Olpidium brassicae TMV*. *Z Pflanzenkrakh Pflanzenschutz* 86:103.

Schoenbeck, F. and Dehne, H.W. (1981). Mycorrhiza and Plant health. *Gesunde Pflazen* 33 : 86.

Schoenbeck, F. and Schinzer, U. (1972). Investigations on the Influence of endotrophic mycorrhiza on TMV lesion formation in *Nicotiana tobacum* L. var. *xanthi, Phytopathol*. Z. 73:78–82.

Shaul, O. Galili., Ginzberg., I., Chet, I., kapulnik, Y. (1999). Mycorrhiza -induced changes in disease severity and PRprotein expression in tobacco leaves. *Mol plant microbe interact* .12:1000–1007.

Sikora, R.A. and Schoenbeck, F. (1975). Effect of vesicular-arbuscular mycorrhiza (*Endogone mosseae*) on the population dynamics of the root-knot nematodes (*Meloidogyne incognita* and *M.hapla*). 8th Intern. *Plant Protection Congress*. Pp. 158.

Singh, P., Mishra, M. and Vyas, D. (2008) Phytoprotection role of AM fungi in the rhizosphere of tomato plant (Abstract in National seminar on Vision 2008: Current challenges in biological sciences) A76:pp–51.

Singh, P.K. (2008). To study microbial interaction of below ground organisms with special refernce to *Fusarium* wilt of chick pea. Ph.D. thesis Dr. Hari Singh University Sagar, M.P. India.

Singh, P.K., Mishra, M. and Vyas, D. (2008). Phytorotection role of AM fungi in the rhizosphere of tomato, In Proceedings of national seminar on Vision 2008: current challenges in Biological Sciences. (eds. Atri, D.C., Kandya, A.K., Khare, P.K., Vyas, D., Chourasia, B.K. and Singh, P.K.). Pp. 51.

Sitaramaah, K. and Sikora, R.A. (1982). Effect of an endomycorrhizal fungus. *Glomus macrocarpum*, on *Rotylenchulus reniformis* penetration and development on cotton. *Nematologia*. 24:69.

Siverding, E., (1991). Vesicular arbuscular Mycorrhiza management in tropical agrosystems. Schriftenreihe der GTZ NO. 224. Deutsche Gesellschaft fur Technische Zusammenarbeit GmbH, Eschborn, Germany.

Slezack, S., Dumas-Gaudot, E., Rosendahl, S., Kjoller, R., Paynot, M., Negrel, J., Gianinazzi, S. (1999). Endoproteolytic activities in pea roots inoculated with the arbuscular mycorrhizal fungus *Glomus mosseae* and/or *Aphanomyces eutuiches* in relation to bioprotection. *New Phytol.* 142:517-529.

Smith, G.S., Hussey, R.S. and Roncadori, R.W. (1986). Penetration and post-infection development of *Meloidogyne incognita* as affected by *Glomus intraradices* and *phosphorus*. *J.Nematol.* 18:29.

St.-Arnaud, M.C., Hamel, C.,Vimard ,B., Caron, Mand, Fortin, J.A. (1997). Inhibition of *Fusarium oxysporum f. sp. Dianthi* in the non -VAM species *Dianthus caryophyllus* by the coculture with *Tagetes patula* companion plants colonized by Glomus intraradices. *Can. J. Bot.*75:998-1005.

Stewar, F.E. and Pfleger, F.L. (1977). Development of poinsettia as influenced by endomycorrhizae, fertilizer and root rot pathogens *Pythium ultimum* and *Rhizoctonia solani*. *Florists Rev.* 159:7.

Strobel, N.E., Hussey, R.S. and Roncadori, R.W. (1982). Interactions of vesicular-arbuscular mycorrhizal fungi, *Meloidogyne* fungi, *Meloidogyne incognita*, and soil fertility on peach. *Phytopathology*. 72:690-694.

Sundaresan, P. (1989). The factors influencing the symbiosis between host plant and *Vigna unguiculata* and the mycorrhizal fungus *Glomus facicculatum*. Ph.D Thesis, Madurai Kamraj University, Madurai

Sundaresan, P., Ubalthoose, N., Raja. and Gunasekaran, P. (1993). Induction and accumulation of phytolaexins in cowpea roots infected with a mycorrhizal fungus *Glomus fasciculatum* and their resistance to *Fusarium* wilt. J. *Biosci.*18:291-301.

Syliva, D.M. and Chellemi, D.O. (2001). Interaction among root-inhabiting fungi and their implications for biological control of root pathogens. *Adv. Agron.* 73 :1–33.

Talavera, M. Itou, K., Mizukubo, T. (2001) Reduction of nematode damage by root colonization with arbuscular mycorrhiza (*Glomus spp.)* in tomato- *Meloidogyne incognita* (*Tylenchida: Melidogynidae)* and carrot -*Praylenchus penetrans* (*Tylenchida:Melidogynidae*) and carrot-*Pratylenchus penetrans(Tyienchida: Pratylenchidae*) pathosynthates. *Appl. Entomol. Zool.*36:387–392.

Thaler, J.S., Owen, B., Higgins, V.J. (2004). The role of the Jasmonate response in plant susceptibility to diverse pathogen with a range of life styles. *Plant Physio.* 135 :530–538.

Toro, M., Azcon, R., and Barea, J.M. (1997). Improvement of arbuscular mycorrhizal development by inoculation with phosphate solubilizing rhizobacteria to improve phosphate bioavailbility and nutrient cycling.63:4408–4412.

Torres-Barragan, A., Zavale-Tamejia E, Gonzalez-Chavez, C., Cepivorum. (1996). The use of arbuscular mycorrhizae to control onios. White rot (*Sclerotium cepivorum*) under filed conditions.*Mycorrhizae.* 6:253–257.

Traquiar, J.A. (1995). Fungal biocontrol of root disease: Endomycorrrhizal suppression of *Cylinderocarpon* root rot. *Canad. J.Bot.* 73 : 89–95.

Trotta, A.,Varese, G.C., Gnavi, E., Fusconi, A.,Sampo, S. and Berta, G. (1996). Interactions between the soilborne root pathogen *Phytophthora nicotinae var. parasitica* and the arbuscular mycorrhizal fungus *Glomus mosseae* in tomato plants. *Plant Soil.* 185:199–209.

Turman, W., Bennrtt, M.H., Kubigsetelting, I., Turnbell, C. and Grant, M. (2007). *Arabidopsis* systemic immunity uses conserved defense. Signalling pathways and is mediated by Jasmonates. *Proc. Nati. Acad. Sci.* 104:1075–1080.

Vaast, P., Caswell-Chen, E.P., Zasoski, R.J. (1998). Influences of a root lesion nematode, *Pratylenchus coffeae and two arbuscular* mycorrhizal fungi *Aculospora* and *Glomus clarum* on *Coffea (Ceffea Arabica L.). Biol. Fertil. Soil,* 26:130–135.

Van Bruggen, A.H.C., Grogan, R.G., Bogdanoff, C.P. and Waters, C.M. (1988). Corky root of lettuce in California caused by gram negative bacterium. *Phytopathology.*78:1139–1145.

Vigo, C., Norman, J.R. and Hooker, R. (2000). Biocontrol of *Pytopthora parasitica* by mycorrhizal fungi a onsequence of effects on infection loci. *Plant Pathol.* 49:509–514.

Waceke, J.W.,Waudo, S.W. and Sikora, R.A. (2001) Response of *Meloidogyne hapla* to mycorrhiza fungi inoculation.*Afric. J. Tec.* 2:63–70.

Whipps, J.M. (2004). Prospects and limitation for mycorrhizas in biocontrol of root pathogens, *Can. J. Bot.* 82:1198–1227.

Yano, K., Yamauchi, A. and Kono, Y. (1996). Localized alteration in lateral root development in roots colonized by an arbuscular mycorrhizal fungus.*Mycorrhiza.* 6:409–415.

Yao, M.K., Tweddell, R.J. and Desilets, H. (2002). Effect of two vesicular -arbuscular mycorrhizal fungi on the growth of micropropagated potato plantlets and on the extent of disease caused by *Rhizoctonia solani*. *Mycorrhiza.* 12:235–242.

Zaidi, R. and Mukerji, K.G. (1983). Incidence of vesicular-arbuscular mycorrhiza (VAM) in disease in healthy plants. *Indian J.Pl. Pathol.* 1:24–28.

Zambolim, L. and Schenck, N.C. (1983). Reduction of the effect of pathogenic root-infecting fungi on soybean by the mycorrhizal fungus *Glomus mosseae*. *Phytopathology.* 73:402–405.

Zheng, H.Z., Kim, Y.W., Lee, H.J., Park, R.D., Jung, W.J., Kim, Y.C., Lee, S.H. and Kim.,K.Y. (2004). Quantitative changes of PR proteins and antioxidant enzymes in response to *Glomus intraradices* and *Phytophthora capsici* in pepper (*Capsicum annum L.*) plants. *Journal of Microbiology and Biotechnology.* 14:53–5.

25

Population Dynamics and Diversity of Saline Tolerant AM Fungi and its Impact in the Restoration of Salt Affected Soils

A Case Study

S G Hiremath and H C Lakshman

Symbiotic relation between plants and microbes represents the climax of intricate relation, which refers to an association of living organisms that benefits both partners, enabling them to survive, grow and reproduce more effectively (Smith and Read, 1997). In the middle of the nineteenth century there has been great increase of interest in symbiosis, where both partners live in a state of physical and physiological equilibrium. The term symbiosis was soon extended to other cases of more or less evident mutualism in both plant and animal kingdom (Morton, 1999). The existence of such mutual relations has already been recognized in lichens, between algae and fungi, in corals, between algae and coelenterates, in root nodules of leguminous plants between roots and *Rhizobium* bacteria. In the regular fashion such

Post Graduate Department of Studies in Botany (Microbiology Lab.)
Karnatak University, Dharwad-580 003, India.

association of fungal mycelium with the roots of vascular plants is recognized as 'Mycorrhiza' which is subject of research interest in recent years (Manoharachary, 2009).

The term Mycorrhizal was coined by German Botanist A.B. Frank (1885), that created to reflect this reality came to us, moreover, from the combination of two words, one Greek *mikes* (fungus) and other Latin *rhiza* (roots), basically designates the symbiotic association between fungi and plant roots. Mycorrhiza associations have evolved through adaptive co-operation of two dissimilar organisms.

Mycorrhizae are usually divided into three morphologically distinct group; ectomycorrhiza, endomycorrhiza and ectendomycorrhiza (Peyronel *et al.*, 1969). There structural and functional diversity is determined by the plants and the fungi involved. Ectomycorrhiza associations are found in most of the tree species and completely cover the root surface forming mantle, leading to change in morphology, colour etc. Endomycorrhizal hyphae are external, intra and intercellular, penetrate the cortical cells (Harley, 1965). As far as endomycorrhizas are concerned, it is general term for all mycorrhizal type where fungus grows within the cortical cells. Endomycorrhizas are confined to order Glomales of Zygomycotina. The endomycorrhizas are characterized by inter and intracellular fungal growth in the root cortex, forming specific fungal structures referred to as vesicles and arbuscules. This characteristics growth gives endomycorrhizae alternate name Vesicular-arbuscular mycorrhiza (Dangeard, 1900). Endomycorrhiza invade the roots and penetrate in to the host cells, with minimal tissue damage. This relationship is so well balanced that, though number of host cells are colonized by fungal endophytes, no tissue damage is visible, showing that they have good survival mechanisms. On the basis of septation and the way of infection, endomycorrhiza are classified into Ericoid, Orchidoid and Monotropoid mycorrhiza where septate fungi associated with the roots of the members of Ericaceae, Orchidaceae and monotropaceae respectively (Bagyaraj, 1991). Ectendomycorrhiza are intermediate forms between ecto and endomycorrhiza which develop a well organized fungal sheath and hyphal penetration.

Vesicular-arbuscular mycorrhizal fungi are beneficial symbiotic microorganisms that colonize the root and increase the growth and yield of most crop plants. The improved plant growth is attributed to increased inorganic mineral nutrient uptake, organic mineral nutrition uptake, production of growth promoting substances, tolerance to drought, salinity and transplantation shock, resistance to plant pathogens and synergistic interactions with beneficial soil microorganisms like nitrogen fixers, phosphate solubilizers (Lakshman and Patil, 2003). AM Fungi bind soil in to semi stable aggregates, thus improving the structure of soil. Plant growth on industrial waste could be improved by incorporating AM Fungi.

Land degradation is currently recognized as one of the most important environmental problems worldwide due to the unprecedented level of human pressure on land resources over last several decades. This issue is particularly serious when it occurs in arid and semi-arid regions. Revegetation is one of the most important practices that can help in the reclamation of degraded lands (Lakshman, 2005). Use of mycorrhizas could form the basis of a sustainable method to support the survival of transplanted plants in revegetation by increasing the ability of plant root systems to take up nutrients (especially phosphorus) and water.

Different biotic and abiotic stresses such as drought, salinity, low temperature, high irradiance and nutritional factors have great impact on various physiological and biochemical processes of plants and they show a rapid molecular response to changing environmental conditions (Linderman, 1988; Lichenthaler, 1996). These environmental stress conditions are major factors limiting plant productivity and plant distribution. As plants have only limited mechanisms of stress avoidance, they require flexible means of adaptations to changing environmental conditions. (Polle and Rennenberg, 1993).

One of the best-known responses of plants to drought and other stresses is the accumulation of soluble, low molecular-mass solutes such as proline (i.e. proline concentration is more in stressed plants) (Paleg *et al.*, 1984). Photosynthesis is also greatly affected by salinity because of stomatal closure (Taleisnik, 1987). Dawson and Gibson (1987) reported a salinity-induced decrease in the photosynthetic rates of wheat and barley, while the water use efficiency (WUE) was marginally affected. The transpiration per unit leaf area in several crops has been shown to decrease in response to increasing salinity (Hoffman and Phene, 1971). Osmotic adjustment, as well as increased photosynthesis, may enable salinized plants to maintain better turgor and growth during periods of stress.

The introduction of arbuscular-mycorrhizal (AM) fungi to the sites with saline soils may improve early plant tolerance and growth (Jain *et al.*, 1989). Although improved salt tolerance of mycorrhizal plants can be related to enhanced mineral nutrition, particularly phosphorus (Graham, 1986), the effect of these fungi on salt tolerance may not only be limited to this mechanism. Hence, in a study of the interaction between AM fungi and salinity it is essential to determine if the effect of the fungus on P uptake is the primary mechanism by which it is able to increase salt tolerance or whether this effect is due to additional or alternative mechanisms. Arbuscular-mycorrhizal fungi may influence plant hormones (Danneberg *et al.*, 1992) or improve water uptake (Ruiz-Lozano and Azcon, 1995). Other mechanisms may include osmotic adjustment, which assists in the maintenance of leaf turgor and effects on physiological processes such as photosynthesis, transpiration, conductance and water use efficiency.

Out of various abiotic stresses, salinity (both of soil and water) is a serious problem in the arid and semi-arid regions of the world where the precipitation is not enough to leach the excess soluble salts from root zone. It is estimated that out of about 0.34 X 10^9 m ha (23% of cultivated land) world over (Szabolcs, 1989) and approximately 7 m ha of land in India (Sehgal and Abrol, 1994) is affected by salts. This area is gradually increasing year after year because of salt accumulation. The reclamation of 7 m ha of salt affected soils can lead to additional production of 35–50 million tones every year.

For the classification of soils under different category many authors have considered various criteria. The criteria like total dissolved salts and pH is considered most convenientfor grouping specific type of soil under saline or non-saline category. Coastal sand dunes (CSD's) are heterogeneous due to frequent changes in physicochemical conditions. Shoreline regression, salt spray, low nutrient availability, high temperatures and anthropogenic disturbances cause instability of CSD ecosystem (Maun, 1994). The most common environmental perturbations countered by the CSD plant community includes: high temperature, desiccation, low moisture retention of the soil, erosion, accretion and high salinity (Maun, 1994). Inadequate nitrogen, phosphorus, potassium and organic matter adversely affect the flora as well as founa of CSD's (Maun, 1994). Several studies on vegetation, restoration, and stabilization of CSD's are confined to temperate regions (Koske and Gemma, 1997). Coastal sand dunes of south west coast of India are reported to be moderately saline soil. Kumta is taluk place in north Canara district of Karnataka (India). This area has one riverine estuary (Aghnashini) and three coastal sand dunes (Dhareshawr, Gudeangadi and Vanalli). The whole locality is studied and considered to be moderately saline in nature. These moderately saline soils may be used as recreation site and also can be used as fields for saline tolerant crop varieties. For revegetating this area two major steps can be put forth as follows.

1. Screening of saline inhabitaning plants species.
2. Screening of saline tolerant AM fungi.
3. Mass multiplication of AM Fungi for the pot experiments.
4. Inoculation of saline tolerant AM fungi to coastal and non- coastal plants.

MATERIALS AND METHODS

(A) Screening of saline tolerant AM fungi on coastal sand dunes of Kumta in Karnataka

Kumta is a town in uttara kannada district in the Indian state of Karnataka. It is a sea-coast town and is about 40 miles (60 km) south of the city of Karwar, Geographically, it is lying in between (17′ to 43′ to 0° North latitude and 74′ to 10′ to 60° E. longitude). City has the altitude of 03 meters above the sea level.

Root and Rhizospheric soil samples are collected from plants growing on coastal sand dunes of Kumta in three major seasons of the year 2006. For each species 6 plants are sampled. The rhizospheric soil samples of individual plants within a species are mixed and complex sample is prepared.

Samples were collected along four parallel equidistant transects parallel to the ocean. In each transect, a total of 9–14 complex samples were collected. Sites in each transect were spaced 20 m apart, depending upon both plant species diversity and soil salinity levels. Samples were returned to the laboratory and the fine roots in each sample were removed, rinsed with tap water and fixed in formalin acetic alcohol

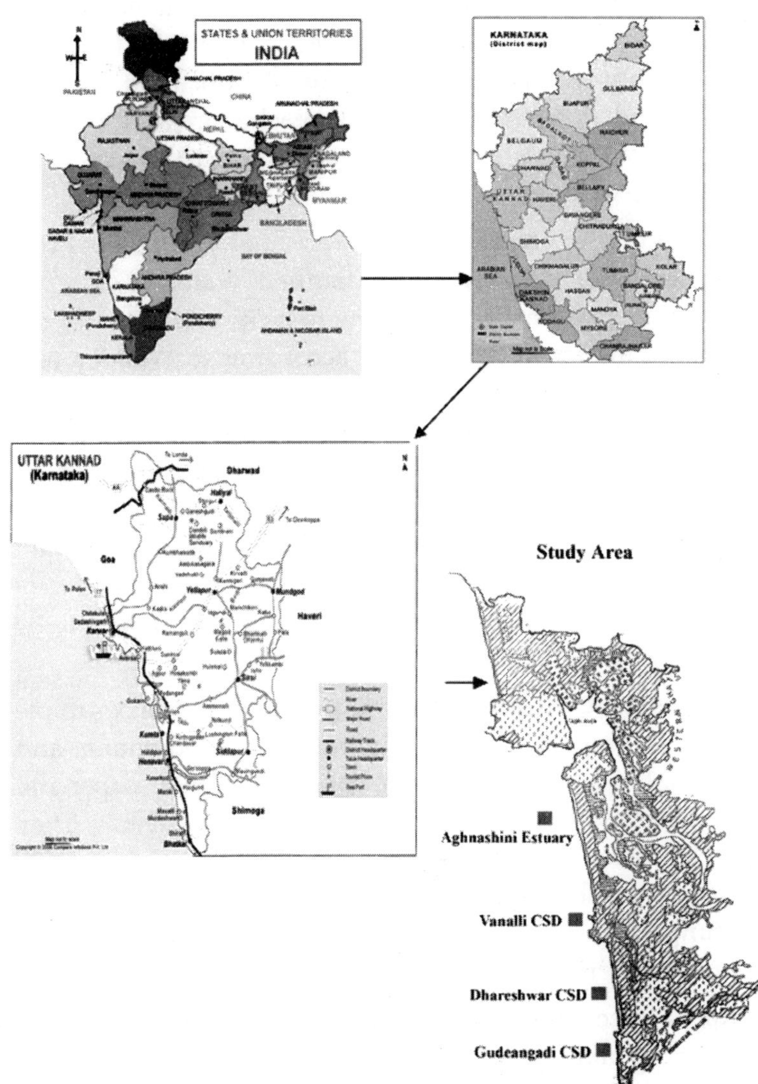

Fig. 25.1.: Showing the Location of Study Area (One Riverine Estuary and Three Coastal Sand Dunes of Kumta in Karnataka)

(FAA), for determination of root colonization. The soil samples were then air dried in the shade at laboratory temperature for spore counting. To get enough AM fungi for the green house experiments and for identification, the AM fungal populations from the coastal beaches are multiplied in green house using a mixture of plant species (Onion, Corn etc.) i.e. trap cultures.

Altogether 10 variables of soil were analyzed, soil pH was measured after dilution with distilled water (1/1 v/v). Nutrients were estimated according to Jackson (1963). Percent of organic matter is determined according to Walkley and Black's rapid method as described by Jackson (1963). Electric conductivity is measured using Bridge meter. Total exchangeable calcium, sodium, potassium was detected by flame emission spectrophotometry. The chloride content was detected by titration against silver nitrate with potassium chromate indicator. Total nitrogen was determined by Macro-Kjeldahl method. Available phosphorus was detected by Olsens *et al.* (1954) method.

Spores were extracted from 50 g of each complex sample by wet sieving and decanting method (Gedemann and Nicolson, 1963). The spores were collected on filter paper and washed with distilled water to spread the spores evenly. They were counted using dissecting microscope at X30 magnification. For character observation and identification, AM fungal spores were mounted in Polyvinyl alcohol lacto phenol and identified using Schenck, Perez's manual (1990).

Roots were rinsed with distilled water, cleared with 10% KOH for 15 min. at 90°C acidified in 2% HCL and stained using 0.05% Tryphan blue in lacto phenol according (Phillips and Hayman, 1970). Percentage of root length colonization was estimated by magnified intersection method (Mcgoniel *et al.*, 1990).

(B) Pot experiments to verify the effect of most frequently occurring saline tolerant AM fungi

Most of the previous research has used conditions in which plants were established in to the soil and salt concentrations were increased gradually or suddenly, followed by assessment of mycorrhizal development and function, but in this study pre-existing saline conditions were used to stimulate natural conditions, and pre-inoculated mycorrhizal transplants were used, so that effects of salinity on initial AM colonization were minimized. This method can be of practical importance in field conditions for saline land revegetation. Seedling could be grown in nursery conditions under non-saline conditions and sufficient time allowed to elapse between inoculation and imposition of salinity stress (or transplantation) to allow AM colonization in roots.

Earthen pots measuring 15 cm diameter containing 3 kg potting mixture were used in the present experiment to understand the growth responses of *Capsicum annuum* L. and *Solanum nigram* L. In case of *Erythrina indica* Lam. and *Jatropha curcas* L. 25 cm diameter earthen pots containing 8 kg potting mix were used. In either case pots lined with polythene sheet were used to avoid the leaching of the salts and are amended with 3 parts of garden soil and 1 part of sand.

Treatments

Ten treatments were set in triplicate on four selected experimental plants in polyhouse conditions is as follows.

A. Uninoculated control (UIC)

B. Only mycorrhiza (*Scutellospora erythropa* (Koske & Walker) Walker & Sanders) inoculated

C. Treated with (3 ds m^{-1}) salinity level without mycorrhiza

D. Treated with (3 ds m^{-1}) salinity level + Mycorrhiza (*Scutellospora erythropa* (Koske & Walker) Walker & Sanders)

E. Treated with (6 ds m^{-1}) salinity level without mycorrhiza

F. Treated with (6 ds m^{-1}) salinity level + Mycorrhiza (*Scutellospora erythropa* (Koske & Walker) Walker & Sanders)

G. Treated with (9 ds m^{-1}) salinity level without mycorrhiza

H. Treated with (9 ds m^{-1}) salinity level + Mycorrhiza (*Scutellospora erythropa* (Koske & Walker) Walker & Sanders).

I. Treated with (12 ds m^{-1}) salinity level without mycorrhiza

J. Treated with (12 ds m^{-1}) salinity level + Mycorrhiza (*Scutellospora erythropa* (Koske & Walker) Walker & Sanders).

Experimental pots are kept free of weeds, pests and irrigated thrice in a week in order to maintain moisture. Plants which were germinated in mycorrhiza inoculum itself were transplanted to pots after 15 day after their germination and were harvested after 30, 60 and 90 days interval after transplanting. Observation such as Shoot length (SL), Fresh weight of shoot (FWS), Dry weight of shoot (DWS), Root length (RL), Fresh weight of root (FWR), Dry weight of root (DWR), Number of leaves (NL), Number of floral buds (NB), Number of flowers (NFL), Number of fruits (NFR) are recorded. The spore count was carried out by wet sieving and decanting method (Gerdemann and Nicolson, 1963). Percent of root length colonization is estimated according to Phillips and Hayman (1970). After the harvest, experimental plants (shoot and root) are oven dried at 70°C until a constant weight is obtained to determine the dry weight. The uptake of 'P' was estimated according to Jackson (1963).

Table 25.1. : Showing the Physico Chemical Properties of Soil of One Estuary and Three Coastal Sand Dunes of Kumta in Karnataka

	Moisture %	pH	Electric Conductivity (M mhos/cm at 25° C(	Org. C %	Ca (µg /g)	Cl (µg /g)	Na (µg /g)	N (µg /g)	P (µg /g)	K (µg /g)
Aghinashini	0.46±0.01 d	4.90±0.10 c	3.23±0.08 c	0.02±0.001 b	75.25±0.03 d	73.34±0.01 d	42.38±0.04 d	10.90±0.04 a	90.33±0.02 b	7.28±0.07c
Dhareshwer	0.64±0.01 a	5.53±0.15 ab	3.84±0.03 a	0.07±0.01 a	142.13±0.02 c	91.46±0.01 c	68.50±0.06 c	10.43±0.01 b	84.55±0.02 d	8.64±0.04 a
Gudeangadi	0.57±0.02 b	5.73±0.05 a	3.50±0.04 b	0.08±0.01 a	152.36±0.02 b	97.61±0.03 a	75.38±0.03 a	10.37±0.11 b	86.33±0.02 c	7.46±0.15 b
Vanalli	0.51±0.01 c	5.40±0.10 b	3.91±0.03 a	0.03±0.003 b	167.44±0.02 a	92.50±0.15 b	70.31±0.14 b	9.46±0.01 c	100.23±0.15a	4.23±0.02 d

n=3 ± SD

Table 25.2. : Showing the Details of Plant Families on Aghanashini Estuary of Kumta in Karnataka in Three Major Seasons in the Year 2005-2006

Sl. No.	Family	Mat forming creepers	Prostrate or erect herbs and sedges	Climbers	Shrubs	Trees	Mangroves	Total plants in a family
1	Acanthaceae	-	02	-	-	-	01	**03**
2	Amaranthaceae	-	03	-	-	-	-	**03**
3	Anacordiaceae	-	-	-	-	02	-	**02**
4	Apocynaceae	-	-	-	01	-	-	**01**
5	Asclepediaceae	-	-	-	01	-	-	**01**
6	Casurinaceae	-	-	-	-	01	-	**01**
7	Chenopodiaceae	-	01	-	-	-	-	**01**
8	Convolvulaceae	01	-	-	-	-	-	**01**
9	Cyperaceae	-	01	-	-	-	-	**01**
10	Euphorbiaceae	-	02	-	02	-	-	**04**
11	Flagellariaceae	-	-	01	-	-	-	**01**
12	Geraniaceae	-	01	-	-	-	-	**01**
13	Gramineae	04	01	-	-	-	-	**05**
14	Labiatae	-	01	-	-	-	-	**01**
15	Leguminosae	-	02	01	-	02	-	**05**
16	Liliaceae	-	01	-	-	-	-	**01**
17	Malvaceae	-	01	-	-	-	-	**01**
18	Nyctaginaceae	-	01	-	-	-	-	**01**
19	Papaveraceae	-	01	-	-	-	-	**01**
20	Plumbaginaceae	-	01	-	-	-	-	**01**
21	Rhamnaceae	-	-	-	-	01	-	**01**
22	Solanaceae	-	01	-	-	-	-	**01**
23	Tamariaceae	-	-	-	01	-	-	**01**
24	Verbenaceae	-	-	-	02	-	01	**03**
Total No. of plants in a category		05	20	02	07	06	02	**42**

Each value is mean of three readings i.e. n=3

Table 25.3. :Showing Distribution of Plants Belonging to Different Families on Coastal sand Dunes of Dhareshwar of Kumta in Karnatak.

Sl. No.	Family	Mat forming creepers	Prostrate or erect herbs and sedges	Climbers	Shrubs	Trees	Mangroves	Total plants in a family
1	Acanthaceae	-	3	-	2	-	1	06
2	Aizoaceae	1	-	-	1	-	-	02
3	Amaranthaceae	-	5	-	-	-	-	05
4	Anacordiaceae	-	-	-	-	2	-	02
5	Apocynaceae	-	-	-	1	1	-	02
6	Asclepediaceae	-	-	-	1	-	-	01
7	Caryophyllaceae	-	2	-	-	-	-	02
8	Casurinaceae	-	-	-	-	1	-	01
9	Chenopodiaceae	-	2	-	-	-	-	02
10	Compositae	-	2	-	-	-	-	02
11	Convolvulaceae	1	-	-	-	-	-	01
12	Cyperaceae	-	1	-	-	-	-	01
13	Euphorbiaceae	-	3	-	3	1	-	07
14	Flagellariaceae	-	-	1	-	-	-	01
15	Gentinaceae	-	1	-	-	-	-	01
16	Geraniaceae	-	1	-	-	-	-	01
17	Goodeniaceae	-	1	-	1	-	-	02
18	Gramineae	7	3	-	-	-	-	10
19	Guttiferae	-	-	-	-	1	-	01
20	Labiatae	-	3	-	-	-	-	03
21	Leguminosae	1	4	2	1	4	-	12
22	Liliaceae	-	1	-	-	-	-	01
23	Lythraceae	-	-	-	-	1	-	01
24	Malvaceae	-	2	-	-	-	-	02
25	Menispermaceae	-	-	1	-	-	-	01
26	Myristicaceae	-	-	-	-	1	-	01
27	Myrsinaceae	-	-	-	-	1	-	01
28	Nyctaginaceae	-	1	-	-	-	-	01
29	Pandanaceae	-	-	-	-	1	-	01
30	Papaveraceae	-	1	-	-	-	-	01
31	Plumbaginaceae	-	1	-	-	-	-	01
32	Rhamnaceae	-	-	-	-	1	-	01
33	Rubiaceae	-	4	-	1	1	-	06
34	Salvadoraceae	-	-	-	-	1	-	01
35	Solanaceae	-	2	-	-	-	-	02
36	Tamariaceae	-	-	-	1	-	-	01
37	Verbenaceae	-	-	-	2	1	1	04
Total No. of plants in a category		10	43	04	14	18	02	91

Each value is mean of three readings i.e. n=3

Table 25.4. : Showing the details of plant families present on the coastal sand dunes of Gudeangadi of Kumta in Karnatakin the year 2005-06

Sl. No.	Family	Mat forming creepers	Prostrate or erect herbs and sedges	Climbers	Shrubs	Trees	Mangroves	Total plants in a family
1	Acanthaceae	-	01	-	02	-	01	04
2	Aizoaceae	-	-	-	01	-	-	01
3	Amaranthaceae	-	03	-	-	-	-	03
4	Anacordiaceae	-	-	-	-	01	-	01
5	Apocynaceae	-	-	-	01	01	-	02
6	Asclepediaceae	-	-	-	01	-	-	01
7	Caryophyllaceae	-	01	-	-	-	-	01
8	Casurinaceae	-	-	-	-	01	-	01
9	Chenopodiaceae	-	01	-	-	-	-	01
10	Compositae	-	01	-	-	-	-	01
11	Convolvulaceae	01	-	-	-	-	-	01
12	Euphorbiaceae	-	01	-	02	-	-	03
13	Geraniaceae	-	01	-	-	-	-	01
14	Goodeniaceae	-	01	-	01	-	-	02
15	Gramineae	06	02	-	-	-	-	08
16	Guttiferae	-	-	-	-	01	-	01
17	Labiatae	-	02	-	-	-	-	02
18	Leguminosae	01	03	02	-	03	-	09
19	Liliaceae	-	01	-	-	-	-	01
20	Malvaceae	-	01	-	-	-	-	01
21	Menispermaceae	-	-	01	-	-	-	01
22	Myristicaceae	-	-	-	-	01	-	01
23	Nyctaginaceae	-	01	-	-	-	-	01
24	Plumbaginaceae	-	01	-	-	-	-	01
25	Rhamnaceae	-	-	-	-	01	-	01
26	Rubiaceae	-	01	-	-	-	-	01
27	Salvadoraceae	-	-	-	-	01	-	01
28	Solanaceae	-	02	-	-	-	-	02
29	Tamariaceae	-	-	-	01	-	-	01
30	Verbenaceae	-	-	-	02	-	-	02
Total No. of plants in a category		08	24	03	11	10	01	57

Each value is mean of three readings i.e. n=3

Table 25.5. ; Showing the details of plant families present on coastal sand dunes of Vanalli of Kumta in Karnatak for the year 2005-06.

Sl. No.	Family	Mat forming creepers	Prostrate or erect herbs and sedges	Climbers	Shrubs	Trees	Mangroves	Total plants in a family
1	Acanthaceae	-	01	-	01	-	01	03
2	Amaranthaceae	-	04	-	-	-	-	04
3	Anacordiaceae	-	-	-	-	02	-	02
4	Caryophyllaceae	-	01	-	-	-	-	01
5	Casurinaceae	-	-	-	-	01	-	01
6	Compositae	-	01	-	-	-	-	01
7	Convolvulaceae	01	-	-	-	-	-	01
8	Cyperaceae	-	01	-	-	-	-	01
9	Euphorbiaceae	-	02	-	04	01	-	07
10	Flagellariaceae	-	-	01	-	-	-	01
11	Gentinaceae	-	01	-	-	-	-	01
12	Geraniaceae	-	01	-	-	-	-	01
13	Gramineae	06	02	-	-	-	-	08
14	Guttiferae	-	-	-	-	01	-	01
15	Labiatae	-	02	-	-	-	-	02
16	Leguminosae	01	04	02	-	03	-	10
17	Liliaceae	-	01	-	-	-	-	01
18	Malvaceae	-	01	-	-	-	-	01
19	Menispermaceae	-	-	01	-	-	-	01
20	Myristicaceae	-	-	-	-	01	-	01
21	Myrsinaceae	-	-	-	-	01	-	01
22	Nyctaginaceae	-	01	-	-	-	-	01
23	Plumbaginaceae	-	01	-	-	-	-	01
24	Rhamnaceae	-	-	-	-	01	-	01
25	Rubiaceae	-	01	-	-	-	-	01
26	Solanaceae	-	02	-	-	-	-	02
27	Tamariaceae	-	-	-	01	-	-	01
28	Verbenaceae	-	-	-	02	-	-	02
Total No. of plants in a category		**08**	**27**	**04**	**08**	**11**	**01**	**59**

Each value is mean of three readings i.e. n=3

Table 25.6. : Showing AMF Spores Recovered from Three Coastal Sand Dunes and one Estuary of Kumta in Karnataka in Three Major Seasons in the Year 2005-2006

Name of the AMF spore recovered	Aghinashini			Dhareshwer			Gudeangadi			Vanalli		
	S	M	W	S	M	W	S	M	W	S	M	W
Acaulospora scrobiculata Trappe.	-	-	+	+	+	-	-	-	+	+	+	-
Acaulospora undulata Sieverding.	+	+	+	-	+	-	+	+	-	-	-	+
Entrophospora schenckii Sieverding &Toro.	+	-	+	+	+	+	-	-	+	-	-	+
Gigaspora albida Schenck & Smith.	-	-	+	-	-	+	+	+	-	+	-	+
Gigaspora decipiens Hall & Abbott.	+	+	-	-	-	+	-	-	+	+	-	+
Gigaspora gigantea (Nicol. & Gerd.) Gerdermann & Trappe.	-	-	+	-	+	-	+	+	-	-	-	+
Gigaspora ramisporophora Spain, Sieverding & Schenck.	-	+	-	+	+	-	+	+	-	-	-	+
Glomus ambisporum Smith & Schenck.	-	-	+	+	-	+	-	-	+	+	+	-
Glomus botryoides Rothwell & Victor.	-	-	+	+	+	+	-	-	+	+	-	+
Glomus citricolum Tang & Zang.	+	+	-	-	-	+	-	+	-	+	-	+
Glomus corantum Schenck.	+	+	-	+	+	+	-	-	+	+	-	+
Glomus fasciculatum (Thaxter) Gerd. & Trappe emend.Walker & Koske.	-	-	+	+	-	+	+	+	-	+	+	-
Glomus gerdemannii Rose, Daniels & Trappe.	-	+	-	+	+	-	-	-	+	+	+	-
Glomus macrocarpum Tul.&Tul.	+	+	-	+	+	-	+	+	-	+	+	-
Glomus maculosum Miller & Walker.	+	+	+	+	+	+	-	+	-	-	-	+
Glomus magnicaule Hall.	-	-	+	-	-	+	+	+	-	+	+	-
Sclerocystis pachycaulis Wu & Chen.	+	-	+	+	-	+	-	-	+	+	-	+
Sclerocystis sinuosa Gerdermann & Bakshi.	+	-	-	+	+	+	+	+	-	+	+	-
Scutellospora arenicola Koske & Halvorson.	-	+	-	+	-	+	-	-	+	+	+	-
Scutellospora biornata Spain, Sieverding &Toro.	+	+	-	+	+	-	-	+	-	-	-	+
Scutellospora calospora (Nicol.& Gerd.) Walker & Sanders.	+	+	-	-	+	-	-	+	-	+	+	-
Scutellospora coralloidea (Trappe, Gerd & Ho) Walker & Sanders.	+	+	+	+	+	+	-	-	+	+	+	-
Scutellospora dipapillosa (Walker & Koske) Walker & Sanders.	-	-	+	-	-	+	+	+	-	+	+	-
Scutellospora dipurpurescens Mortan & Koske.	-	+	-	+	-	+	-	-	+	+	+	+
Scutellospora erythropa (Koske &Walker) Walker & Sanders.	+	+	+	+	+	+	+	+	-	-	-	+
Scutellospora gregaria (Schenck & Nicol.) Walker & Sanders.	-	-	+	+	+	-	+	+	-	-	-	+
Scutellospora pellucida (Nicol. & Schenck) Walker & Sanders.	+	+	-	+	+	-	+	+	-	-	-	+
Scutellospora persica (Koske & Walker) Walker & Sanders.	+	-	+	+	+	+	-	+	+	-	+	-
Scutellospora scutata Walker & Diederichs.	+	+	+	-	-	+	+	-	+	+	+	-
Scutellospora verrucosa (Koske &Walker) Walker & Sanders.	+	-	+	-	-	+	-	-	+	+	+	-

(+) = Present, (-) = Absent. Each value is mean of three readings i.e. n=3

Table 25.8. : Showing the Genus distribution of AM fungi on four coastal sand dunes of Kumta in Karnataka in three major seasons of the year 2005-2006

Name of the Place		*Acaulospora*	*Enterophosphora*	*Gigaspora*	*Glomus*	*Sclerocystis*	*Scutellospora*	Total No. of AM species
Aghanashini	S	1	1	1	4	2	8	17
	M	1	-	2	5	-	8	16
	W	2	1	2	5	1	7	17
Dhareshwer	S	1	1	1	7	2	8	20
	M	2	1	2	5	1	7	18
	W	-	1	2	7	2	8	20
Gudeangadi	S	1	-	3	3	1	5	13
	M	1	-	3	5	1	7	17
	W	1	1	1	4	1	6	14
Vanalli	S	1	-	2	8	2	7	20
	M	1	-	-	5	1	8	15
	W	1	1	4	4	1	5	16

S=summer, M=Monsoon, W=winter

Each value is mean of three readings i.e. n=3

Table 25.9. : Showing the effect of (*Scutellospora erythropa* (Koske &Walker) Walker & Sanders.) and Four Salinity Levels on Growth of *Erythrina indica* Lam. for 90 days

Treatments	Shoot length (cm)	Fresh weight of shoot (g)	Dry weight of shoot (g)	Root length (cm)	Fresh weight of root (g)	Dry weight of root (g)	Root colonization (%)	Number of leaves	Number of floral buds	Number of flowers	Number of fruits	'P' uptake in shoot (%)
						30 days						
Control	15.73 b	3.22 f	0.77 f	8.50 d	0.50 e	0.21 c	00	4.33 cd	00	00	00	0.08 c
AMF	16.20 a	5.11 a	1.80 a	9.80 a	0.65 a	0.28 a	60.54 a	6.33 a	00	00	00	0.14 a
Control (3ds/m)	15.26 c	3.00 g	0.67 g	8.30 d	0.44 g	0.18 e	00	3.66 de	00	00	00	0.07 d
AMF+ (3ds/m)	15.80 b	5.01 b	1.60 b	9.50 b	0.61 b	0.25 b	60.32 b	5.66 ab	00	00	00	0.11 b
Control (6ds/m)	14.40 d	2.86 h	0.55 h	7.76 e	0.42 h	0.14 g	00	3.33 e	00	00	00	0.06 e
AMF+ (6ds/m)	15.30 c	4.75 c	1.41 c	8.90 c	0.57 c	0.20 d	59.66 c	5.33 b	00	00	00	0.11 b
Control (9ds/m)	13.83 e	2.46 i	0.42 i	7.20 f	0.38 i	0.11 h	00	3.00 e	00	00	00	0.04 f
AMF+ (9ds/m)	14.40 d	4.21 d	1.24 d	8.40 d	0.51 d	0.18 e	59.12 d	5.00 bc	00	00	00	0.09 c
Control (12ds/m)	12.40 f	2.09 j	0.31 j	6.40 g	0.35 j	0.09 i	00	3.00 e	00	00	00	0.04 f
AMF+ (12ds/m)	13.70 e	4.00 e	1.11 e	7.63 e	0.47 f	0.15 f	59.10 d	3.33 e	00	00	00	0.08 cd
						60 days						
Control	61.50 e	47.963	10.61 f	18.46 e	5.62 f	1.99 f	00	15.66 cd	00	00	00	0.13 f
AMF	97.40 a	155.66 a	38.74 a	27.30 a	17.71 a	7.24 a	69.14 a	27.66 a	00	00	00	0.28 a
Control (3ds/m)	55.50 f	45.11 g	9.91 g	15.30 f	4.81 g	1.79 g	00	13.33 de	00	00	00	0.11 g
AMF+ (3ds/m)	95.36 b	133.24 b	29.70 b	23.63 b	16.79 b	5.87 b	68.45 b	18.00 c	00	00	00	0.25 b
Control (6ds/m)	50.50 g	37.39 h	7.94 h	13.36 g	3.99 h	1.64 h	00	12.00 e	00	00	00	0.09 h
AMF+ (6ds/m)	86.40 c	125.61 c	29.56 c	21.33 c	13.64 c	5.21 c	67.47 c	23.00 b	00	00	00	0.22 c
Control (9ds/m)	48.26 h	32.71 i	7.71 i	11.50 h	3.52 i	1.39 i	00	13.33 de	00	00	00	0.08 h
AMF+ (9ds/m)	71.50 d	107.37 d	25.45 d	20.43 d	12.62 d	5.09 d	66.23 d	15.00 d	00	00	00	0.18 d
Control (12ds/m)	45.40 i	32.31 j	7.44 j	9.20 i	2.95 j	1.33 j	00	14.00 de	00	00	00	0.07 i
AMF+ (12ds/m)	55.40 f	87.31 e	20.55 e	18.36 e	11.51 e	3.83 e	63.55 e	12.00 e	00	00	00	0.16 e
						90 days						
Control	83.33 f	86.69 f	17.74 f	18.43 f	6.93 f	3.01 f	00	22.33 b	00	00	00	0.16 e
AMF	113.43 a	281.15 a	53.86 a	27.50 a	33.01 a	10.45 a	75.16 a	36.66 a	00	00	00	0.36 a
Control (3ds/m)	76.53 g	70.61 g	14.31 g	17.50 g	6.45 g	2.78 g	00	14.00 de	00	00	00	0.16 e
AMF+ (3ds/m)	103.50 b	250.52 b	48.41 b	24.43 b	27.86 b	10.20 b	71.12 b	22.33 b	00	00	00	0.32 b
Control (6ds/m)	67.36 h	43.10 h	9.95 h	16.56 h	4.51 h	1.69 h	00	12.33 ef	00	00	00	0.13 f
AMF+ (6ds/m)	102.40 c	220.11 c	46.10 c	23.40 c	25.86 c	9.61 c	68.45 c	23.00 b	00	00	00	0.30 b
Control (9ds/m)	60.40 i	41.71 i	8.14 i	15.66 i	3.74 i	1.52 i	00	13.33 e	00	00	00	0.10 g
AMF+ (9ds/m)	97.36 d	204.34 d	41.86 d	22.40 d	25.81 d	8.61 d	66.35 d	16.66 c	00	00	00	0.27 c
Control (12ds/m)	59.46 j	38.105 j	8.13 i	15.46 i	3.51 j	1.51 j	00	10.33 f	00	00	00	0.09 g
AMF+ (12ds/m)	95.56 e	189.74 e	37.78 e	21.70 e	19.24 e	7.51 e	65.42 e	16.33 cd	00	00	00	0.25 d

Control = uninoculated control, AMF= Arbuscular mycorrhizal fungi (*Scutellospora erythropa*), AMF+3ds/m= Arbuscular mycorrhizal fungi (*Scutellospora erythropa*) + 3ds/m saline level, AMF+6ds/m = Arbuscular mycorrhizal fungi (*Scutellospora erythropa*) + 6ds/m saline level, AMF + 9ds/m= Arbuscular mycorrhizal fungi (*Scutellospora erythropa*) + 9ds/m, AMF + 12ds/m = Arbuscular mycorrhizal fungi (*Scutellospora erythropa*) + 12ds/m

*Mean values followed by the same letter within a column do not differ significantly at P=0.05 according to DMRT

Table 25.10. : Showing the effect of (*Scutellospora erythropa* (Koske & Walker) Walker & Sanders.) and Four Salinity Levels on Growth of *Solanum nigram* L. for 90 days

Treatments	Shoot length (cm)	Fresh weight of shoot (g)	Dry weight of shoot (g)	Root length (cm)	Fresh weight of root (g)	Dry weight of root (g)	Root colonization (%)	Number of leaves	Number of floral buds	Number of flowers	Number of fruits	'P' uptake in shoot (%)
						30 days						
Control	3.80 d	0.19 cd	0.03 f	3.03 e	0.02 f	0.01 c	00	6.66 b	00	00	00	0.06 e
AMF	4.73 a	0.29 a	0.09 a	4.46 a	0.07 a	0.01 a	38.43 a	8.66 a	00	00	00	0.11 a
Control (3ds/m)	3.43 e	0.18 d	0.03 f	2.70 f	0.01 g	0.00	00	6.00 bc	00	00	00	0.05 e
AMF+ (3ds/m)	4.60 a	0.25 b	0.08 b	4.20 b	0.07 b	0.01 b	37.52 b	6.00 bc	00	00	00	0.10 ab
Control (6ds/m)	3.10 f	0.14 ef	0.02 g	2.20 g	0.01 h	0.00	00	5.00 bc	00	00	00	0.05 ef
AMF+ (6ds/m)	4.30 b	0.21 c	0.07 c	3.80 c	0.06 c	0.01 c	35.85 c	6.33 b	00	00	00	0.09 bc
Control (9ds/m)	2.80 g	0.14 ef	0.02 h	1.80 h	0.01 i	0.00	00	4.33 c	00	00	00	0.04 f
AMF+ (9ds/m)	4.03 c	0.19 cd	0.06 d	3.50 d	0.05 d	0.01 c	32.33 d	5.00 bc	00	00	00	0.08 cd
Control (12ds/m)	2.36 h	0.12 f	0.01 i	1.40 i	0.01 i	0.00	00	5.00 bc	00	00	00	0.04 f
AMF+ (12ds/m)	3.83 d	0.17 de	0.05 e	3.00 e	0.05 e	0.00	30.28 e	5.33 bc	00	00	00	0.07 d
						60 days						
Control	50.30 d	8.33 f	3.07 f	22.60 b	1.59 f	1.25 e	00	66.66 d	00	00	00	0.08 e
AMF	59.70 a	14.15 a	6.31 a	26.10 a	2.41 a	2.08 a	46.28 a	135.66 a	00	00	00	0.14 a
Control (3ds/m)	45.60 g	5.06 g	2.59 g	13.30 g	1.51 g	1.05 f	00	61.00 e	00	00	00	0.07 f
AMF+ (3ds/m)	59.10 b	13.63 b	6.25 b	17.46 c	2.23 b	2.06 b	42.57 b	106.33 b	00	00	00	0.13 b
Control (6ds/m)	42.20 h	3.62 h	2.05 h	11.30 h	1.47 h	0.93 g	00	57.00 f	00	00	00	0.06 fg
AMF+ (6ds/m)	54.26 c	10.94 c	5.42 c	15.63 d	2.22 c	2.06 b	40.94 c	95.00 c	00	00	00	0.11 c
Control (9ds/m)	35.26 i	1.42 i	1.07 i	9.90 i	0.35 i	0.33 h	00	24.00 g	00	00	00	0.05 g
AMF+ (9ds/m)	49.66 e	9.13 d	4.42 d	14.33 e	2.09 d	2.02 c	37.63 d	93.00 c	00	00	00	0.11 c
Control (12ds/m)	31.66 j	1.27 j	0.82 j	9.40 j	0.31 j	0.23 i	00	22.00 g	00	00	00	0.04 g
AMF+ (12ds/m)	48.46 f	8.61 e	4.29 e	13.63 f	1.95 e	1.34 d	35.34 e	67.66 d	00	00	00	0.09 d
						90 days						
Control	59.50 f	10.17 f	4.46 f	26.33 b	1.96 d	1.49 f	00	109.00 d	7.66 d	00	48.33 b	0.12 d
AMF	68.43 a	15.55 a	6.88 a	29.53 a	3.32 a	3.24 a	53.43 a	164.66 a	16.66 a	17.00 a	55.33 a	0.18 a
Control (3ds/m)	51.63 g	9.64 g	3.64 g	22.20 e	1.83 e	1.34 g	00	95.00 e	6.66 d	5.00 cd	27.33 e	0.11 de
AMF+ (3ds/m)	67.43 b	15.39 b	6.57 b	24.43 c	2.90 b	2.74 b	50.34 b	144.66 b	4.66 e	6.33 bc	32.00 d	0.17 ab
Control (6ds/m)	42.73 h	7.91 h	3.21 h	20.50 f	1.58 f	1.24 h	00	92.00 f	7.00 d	5.66 bcd	42.33 c	0.09 e
AMF+ (6ds/m)	63.33 c	14.47 c	6.47 c	23.40 d	2.44 c	2.21 c	49.51 c	117.66 c	8.00 d	7.00 b	15.66 g	0.16 b
Control (9ds/m)	40.36 i	7.12 i	2.72 i	13.86 h	1.25 g	1.07 i	00	65.66 h	11.00 c	4.66 cde	15.00 g	0.08 f
AMF+ (9ds/m)	62.40 d	10.65 d	4.83 d	22.30 e	2.39 c	2.11 d	47.75 d	107.00 d	14.00 b	4.00 de	32.00 d	0.14 c
Control (12ds/m)	35.33 j	6.31 j	2.31 j	13.30 i	1.23 g	1.00 j	00	81.33 g	3.33 e	3.00 e	14.33 g	0.05 g
AMF+ (12ds/m)	61.33 e	10.43 e	4.52 e	18.70 g	2.38 c	2.05 e	45.57 e	94.66 e	11.00 c	4.33 de	24.00 e	0.12 d

Control = uninoculated control, AMF= Arbuscular mycorrhizal fungi (*Scutellospora erythropa*), AMF+3ds/m= Arbuscular mycorrhizal fungi (*Scutellospora erythropa*) + 3ds/m saline level, AMF+6ds/m = Arbuscular mycorrhizal fungi (*Scutellospora erythropa*) + 6ds/m saline level, AMF + 9ds/m= Arbuscular mycorrhizal fungi (*Scutellospora erythropa*) + 9ds/m, AMF + 12ds/m = Arbuscular mycorrhizal fungi (*Scutellospora erythropa*) + 12ds/m

*Mean values followed by the same letter within a column do not differ significantly at P=0.05 according to DMRT

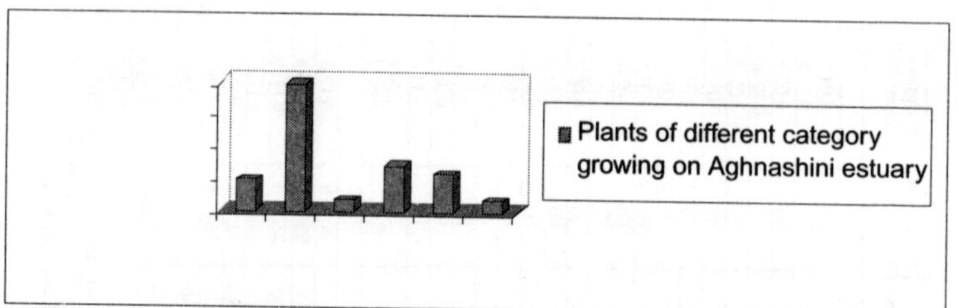

Fig. 25.2.:Showing the Distribution of Plants of Different Category on Aghanashini Reverine Estuary of Kumta

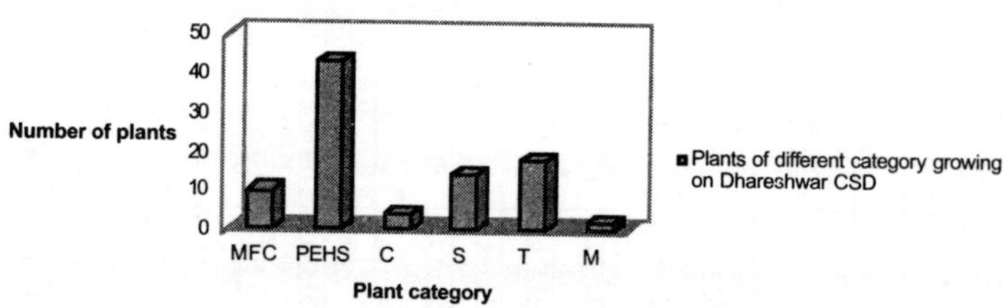

Fig. 25.3.: Showing the Distributation of Plants of Different Category on Dhareshwar Coastal Sand Dune of Kumta

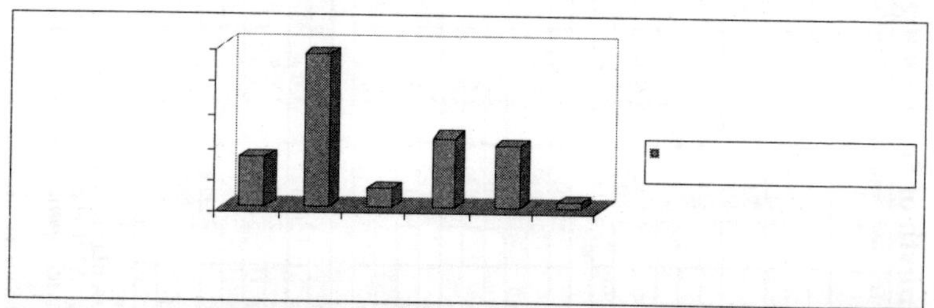

Fig 25.4. : Showing the Distributation of Plants of D ifferent Category on Gudeangadi Coastal Sand Dune of Kumta

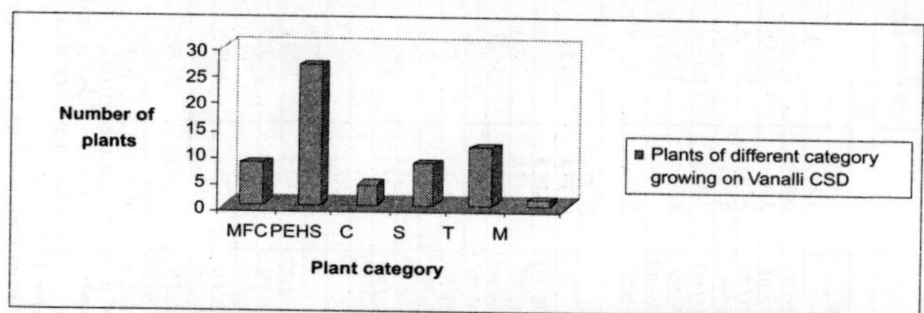

Fig 25.5.:Showing the Distribution of Plants of Different Category on Vanalli Coastal Sand Dune of Kumta

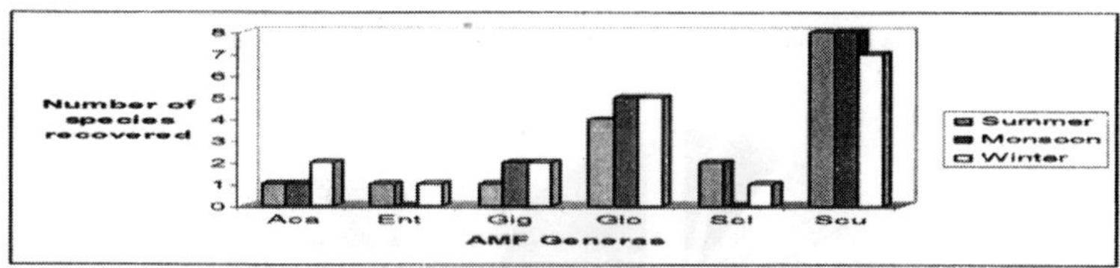

Fig. 25.6. : Showing the Number of AMF Species Recovered from Each of Six Genera in Three Major Seasons of the Year 2005-06 from Aghanashini Riverine Estuary of Kumta

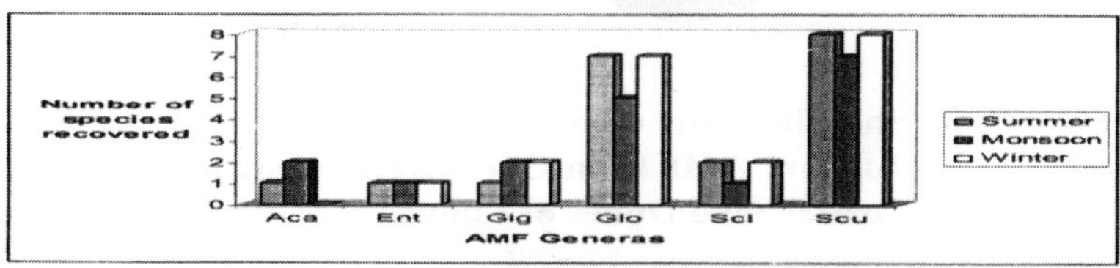

Fig. 25.7. : Showing the Number of AMF Species Recovered from each of Six Genera in Three Major Seasons of the Year 2005-06 from Dhareshwar Coastal Sand Dune of Kumta

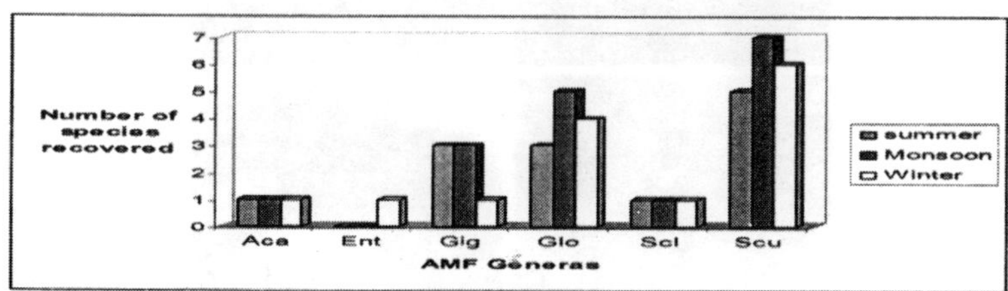

Fig. 25.8. : Showing the Number of AMF Species Recovered from each of Six Genera in Three Major Seasons of the Year 2005-06 from Gudeangadi Coastal Sand Dune of Kumta

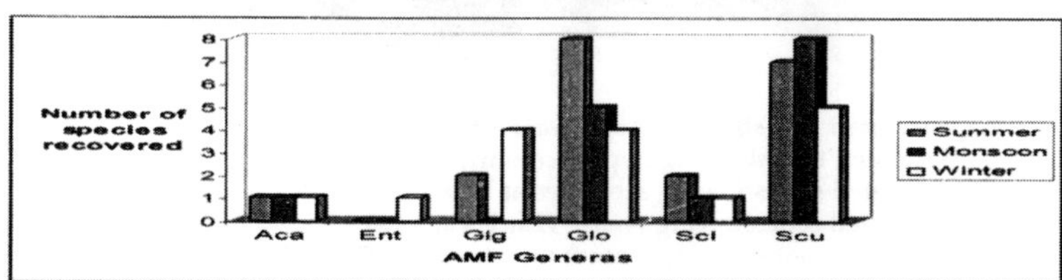

Fig. 25.9. : Showing the Number of AMF Species Recovered from Each of Six Genera in Three Major Seasons of the Year 2005-06 from Vanalli Coastal Sand Dune of Kumta

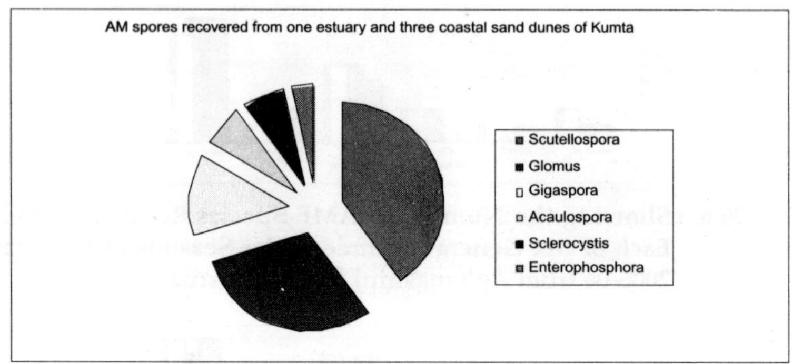

Fig. 25.10. Showing the Generawise Distribution of Saline Tolerant AMF from One Riverine Estuary and Three Coastal Sand Dunes of Kumta

PLATE-I

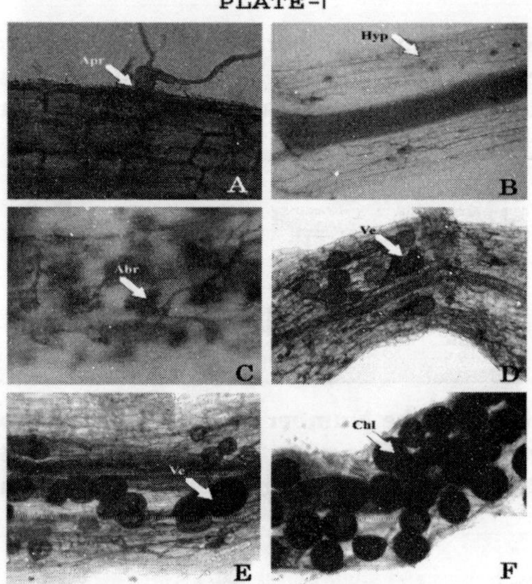

A) Macerated root showing Aspersoria at X400.
B) Macerated root showing hyphal colonization at X200.
C) Macerated root showing arbuscles at X400.
D) Macerated root showing young vesicles at X400.
E) Macerated root showing mature vesicles at X400.
F) Macerated root showing chlamydospores at X400.
Apr=Aspersoria.
Hyp= Hyphae
Abr=Arbuscles
Ve=Vesicles
Chl =Clamydospores

PLATE-II

Plate II Showing the effect of AM Fungi (*Scutellospora erythropa* (Koske & Walker) Walker & Sanders) and varied salinity level on the growth of *Erythrina indica* Lam for 90 days.

Picture 1 Represents all uninoculated control plants treated with various levels of salinity.

A. Uninoculated control (UIC).
B. Treated with (3ds m^{-1}) salinity level without mycorrhiza.
C. Treated with (6ds m^{-1}) salinity level without mycorrhiza.
D. Treated with (9ds m^{-1}) salinity level without mycorrhiza.
E. Treated with (12ds m^{-1}) salinity level without mycorrhiza.

Picture 2 Represents *Erythrina indica* Lam plants treated with AM fungi *Scutellospora erythropa* (Koske &Walker) Walker & Sanders. and varied levels of salinity.

A. Inoculated with AM fungi (*Scutellospora erythropa* (Koske &Walker) Walker & Sanders.) without salinity.
B. Treated with (3ds m^{-1}) salinity level + Mycorrhiza (*Scutellospora erythropa* (Koske &Walker) Walker & Sanders.)
C. Treated with (6ds m^{-1}) salinity level + Mycorrhiza (*Scutellospora erythropa* (Koske &Walker) Walker & Sanders).
D. Treated with (9ds m^{-1}) salinity level + Mycorrhiza (*Scutellospora erythropa* (Koske &Walker) Walker & Sanders).
E. Treated with (12ds m^{-1}) salinity level + Mycorrhiza (*Scutellospora erythropa* (Koske &Walker) Walker & Sanders).

PLATE - III

Plate III

Showing the effect of AM Fungi (*Scutellospora erythropa* (Koske &Walker) Walker & Sanders). and varied salinity level on the growth of *Solanum nigram* L. for 90 days.

Picture. 1 Represents all uninoculated control plants treated with various levels of salinity.

A. Uninoculated control (UIC).
B. Treated with (3ds m^{-1}) salinity level without mycorrhiza.
C. Treated with (6ds m^{-1}) salinity level without mycorrhiza.
D. Treated with (9ds m^{-1}) salinity level without mycorrhiza.
E. Treated with (12ds m^{-1}) salinity level without mycorrhiza.

Picture 2. Represents *Solanum nigram* L. plants treated with AM fungi *Scutellospora erythropa* (Koske &Walker) Walker & Sanders and various levels of salinity.

A. Inoculated with AM fungi (*Scutellospora erythropa* (Koske &Walker) Walker & Sanders.) without salinity.
B. Treated with (3ds m^{-1}) salinity level + Mycorrhiza (*Scutellospora erythropa* (Koske &Walker) Walker & Sanders.)
C. Treated with (6ds m^{-1}) salinity level + Mycorrhiza (*Scutellospora erythropa* (Koske & Walker) Walker & Sanders).
D. Treated with (9ds m^{-1}) salinity level + Mycorrhiza (*Scutellospora erythropa* (Koske &Walker) Walker & Sanders.)
E. Treated with (12ds m^{-1}) salinity level + Mycorrhiza (*Scutellospora erythropa* (Koske &Walker) Walker & Sanders).

RESULTS AND DISCUSSION

The present survey was undertaken on one estuary and three coastal sand dunes of Kumta in Karnataka. All plants growing in a natural conditions possessed AM as a regular component of the soil microflora. Genus *Scutellospora* was most predominant with its 12 species followed by *Glomus* (9 species), *Gigaspora* (4 species), *Acaulospora* (2 species), *Sclerocystis* (2 species) and *Enterophosphora* (1 species) respectively. *Scutellospora erythropa* (Koske & Walker) Walker & Sanders and *Scutellospora coralloidea* (Trappe, Gerd & Ho) Walker & Sanders are the most frequently occurring species. *Glomus corantum* Schenck., *Glomus macrocarpum* Tul. & Tul., *Glomus maculosum* Miller & Walker., *Sclerocystes sinusa* Gerdermann & Bakshi., *Scutellospora persica* (Koske & Walker) Walker & Sanders. and *Scutellospora scutata* Walker & Diederichs are the other species which were occurred very frequently.

Only 42 plants belonging to 24 families were recorded on Aghinashini estuary. The probable reason for this is may be human interference and edaphic factors, mainly it is due to removal of accumulated woody debris for fuel and fertilizer reduces the soil organic matter considerably. These interferences indirectly affect the AM Fungal activity that alters the plant community structure. Among the edaphic factors 'P' is the major limiting factor (Beena *et al.*, 2000). In general, highest percent of root colonization was recorded in winter moderate in monsoon and least in summer.

Highest of 91 plants belonging to 37 families and 57 plants representing 30 families have been screened for their AM association on coastal sand dunes of Dhareshwar and Gudeangadi respectively. Vanalli coastal sand dune had 59 plant species representing 28 families. These dunes are very beautiful and fit for recreation purposes. Moderate disturbance can be seen on all of these dunes because of their recreation value, fishing, boat lodge and shell removal. It was evident that moderately disturbed dunes possess high AM fungal diversity as well as plant diversity and vegetation cover (Rao and Meher-Homji, 1985; Kulkarni *et al.*, 1997 and Beena *et al.*, 2000). Moderate burial confers several advantages such as escape from surface predators, higher nutrient content, greater access to soil moisture, more space for root growth and insulation of seeds/ seedling roots against high temperature/desiccation (Maun, 1994) thus better floral and AM fungal diversity may be observed.

Varied percent of root colonization with respect to all major seasons of the year 2005-06 was observed on coastal sand dunes of Dhareshwar and Gudeangadi. In case of Vanalli, constant and proportionate increase or decrease in percent of root colonization was recorded with respect to three seasons. It was reported that, AM fungi preferentially associate with particles of decomposing organic matter (John *et al.*, 1983). Increased hyphal growth was observed with increase in organic matter (Joner and Jokobson, 1995). Dhareshwar and Gudeangadi coastal sand dunes showed varied degree of organic matter and hence varied percent of root colonization was recorded. Also, optimal colonization of plant roots, particularly in disturbed habitats, depends not only on the presence of extraradical hyphae or mycorrhizal root debris but, mainly, on the survival and well timed germination of AM fungal spores in the soil. This process can be altered by various biotic and abiotic factors, in particular by the association with soil microorganisms (Xavier and Germida, 2003). Indeed some bacterial populations called mycorrhizal helper bacteria, have beneficial effect on AM fungal growth. Helper bacteria not only improves mycorrhizal root colonization by stimulating extraradical hyphal growth but also by facilitating AMF spore germination (Garbaye, 1994; Gryndler *et al.*, 2000).

Bacteria associated with AMF spores colonize mainly the outer wall layer and rarely penetrate in to the inner layers (Walley and Germida, 1996; Maia and Kimbrough, 1998;

Bonfante-Fasolo and Schubert, 1987). Neverthless, some bacteria have been found in the cytoplasm of the AMF spores (MacDonald and Chandler, 1981; Bianciotto *et al.*, 2000). The role of bacteria associated with AMF is not clear. They may stimulate spore germination by eroding spore walls (Filippi *et al.*, 1998; Maia and Kimbrough, 1998) by producing stimulatory compounds such as CO_2 and other volatiles (Carpenter-Boggs, 1995), or by influencing AMF phosphorus acquisition.

In all coastal sand dunes and on an estuary spore number was recorded high in summer season, least in monsoon and moderate in winter season. Scanning electron microscopic studies of *Glomus geosporum* spore revealed that bacteria are possibly feeding on the outer hyaline spore layer (David *et al.*, 2005). It is known that, moisture always supports microbial and nematodal population. In monsoon season CSDs are enriched with moisture, which ultimately supports microflora and nematodal population which feeds on AM fungi, because of this population of AM spore will be lessened in monsoon season and ultimately spore count will be less. Degree of distraction is less in winter because of low moisture ultimately minimizing microflora and nematodes. In summer season microfloral and nematodal population is less active and hence degree of distruction is also less resulting huge number of spores. One more probable reason for increased AM spore count in summer season is most of the plant species like annuals completes their life cycle in winter or in summer season. During their death they produce less photosynthates which boosts AM fungi to produce chlamydospores for the continuation of their life cycle as a result of this spore count increases in summer season.

In case of *Erythrina indica* Lam. optimum results was obtained in plants after third harvest. After 90 days significantly increased (103.50 cm) shoot length, (250.52g) fresh weight of shoot, (48.41g) dry weight of shoot, (24.43 cm) root length, (27.86 g) fresh weight of root, (10.20g) dry weight of root, (71.12%) of root colonization, (22.33) number of leaves/plant and (0.32%) of 'P'uptake in shoot was recorded in plants treated with AMF + 3ds/m saline level.

NM and M plants treated with varied level of salinity showed significant difference in their dry weight of shoot and root. Uninoculated control plants have not shcwed root colonization. However, in mycorrhizal inoculated plants good percent of root colonization was recorded. Percentage of root colonization was recorded affected less in different levels of salinity in early stages. But in later stage percentage of root colonization was affected much significantly. Percent 'P' reacted in similar way as *Erythrina indica* Lam. though responded negatively to saline treatments, when compared with NM plants, M plants performed well with respect to their growth and showed more tolerance to different levels of salinity. During all three harvests no reproductive structures were recorded.

In case of *Solanum nigram* L. At 90 days interval, plants treated with AMF + 3ds/m salinity level showed increased (67.43cm) shoot length, (15.39g) fresh weight of shoot , (6.57g) dry weight of shoot, (144.66) leaves/plant, (24.43cm) root length, (2.90g) fresh weight of root, (2.74g) dry weight of root, (50.34%) root colonization and (0.17%) of 'P' in shoots. Reproductive structures which were remained absent for first two harvests can be seen in third harvest and highest of (16.66) buds/plant, (17.00) flowers/plant and (55.33) fruits/plant were recorded in plants treated with zero level of saline treated M plants. In case of saline treated plants, plants treated with AMF + 9ds/m salinity level showed highest of (14.00) buds/plant, plants treated with AMF + 6ds/m salinity level (Plate XI) showed highest of (7.00) flowers/plant and plants treated with 6ds/m salinity level without mycorrhiza showed highest of (42.33) fruits. Optimum of 0.17% 'P' was recorded in plants treated with AMF + 3ds/m salinity level amongst saline treated plants.

Solanum nigram L. plants have negatively responded to different levels of salinity. Response was more prevalent in NM plants. It was observed during all three harvests that percentage of root colonization was adversely affected by increasing soil salinity level though this effect was less significant in early stages. Plants showed significant decrease in all growth parameters with increased salinity level.

In case of pot experiments, pre-inoculation with AM fungi had significant role in promoting seedling growth and establishment of experimental plants in saline conditions. Improved nutrient uptake compared with non-mycorrhizal is probably the major mechanism involved in increasing seedling salinity tolerance; these findings are in consistent with early workers like (Marschner and Dell, 1994; Al-Karaki and Al-Raddad, 1997; Al-Karaki, 2000). These observations may be taken as highlight the potential of using AM pre-inoculated seedling of mycorrhizal responsive plants to revegetate saline lands so that AM fungi *Scutellospora erythropa* (Koske & Walker) Walker & Sanders) may be used as biological fertilizer in saline soil, is of great interest in agrobiology as AM fungi are involved in overcoming the limitations in the host plants imposed by the particular soil nutritive conditions. Coastal saline areas accumulated with sand are defiecient in organic matter and their fertility is very low. Presence of AM fungi is very critical for the regeneration of natural ecosystems in arid lands.

It is important to consider the physical and chemical status of the soil, besides nutrient concentration in the growing medium because both affect AM fungal growth in mycorrhiza inoculated plants. In the present study, it seems like AM fungi have a role in improving water relation of plants growing in sandy soils. There was decrease in plant growth response was recorded with advancement in their age. In all the three harvests, it was observed that *Erythrina indica* Lam, *Jatropha curcas* L. *Solanum nigram* L and *Capsicum annuum* L. plants exhibited similar trend. In case of mycorrhizal plants, plants treated with AMF + Zero level of salinity dominated other salinity treated plants and was followed by 3ds/m, 6ds/m, 9ds/m and 12ds/m saline treatments respectively. Decrease in plant growth was recorded with increase in salinity level and this is being clearly demonstrated in case of both non-mycorrhizal and mycorrhizal plants. No root colonization was recorded in uninoculated control plants. But the percentage of root colonization was recorded decreasing with increase in salinity level in mycorrhizal plants. Similarly with increase in salinity level percentage of 'P' was also recorded lowering both in NM and M plants.

In saline treated non-mycorrhizal and mycorrhizal plants, with increasing level of salinity there was decrease in growth parameters like shoot length, fresh weight of shoot, dry weight of shoot, root length, fresh weight of root, dry weight of root, per cent root colonization, number of leaves and percent of 'P' content in shoot was recorded. Decrease in growth parameters is more pronounced in non- mycorrhizal plants than mycorrhizal plants.

The exact mechanism by which mycorrhizal system tolerate unbalanced ionic conditions in salty soils is not known. The non- mycorrhizal plants under high salinity, have greater concentrations of potassium, sodium and phosphorus (Ojala *et al.*, 1983) where as mycorrhizal plants maintained a steady Na / K balance than non-mycorrhizal plants. Mycorrhizal advantage occurs through the increased concentration of phosphate and decreased concentration of sodium in shoot tissue in *Parthenium aggentatum* plants inoculated with *Glomus intraradices* compared to non-inoculated plants (Pfeiffer and Bloss, 1988).

Over all results showed growth and establishment in most of the experimental plants with [*Scutellospora erythropa* (Koske & Walker) Walker & Sanders] inoculation in saline conditions, compared with non inoculated conditions. Pre-inoculation increased mycorrhizal effectiveness and ensured that roots were well colonized under salt stress. Initial mycorrhizal colonization depends on germination of spores or other fungal propagules in the soil growth of hyphae through the soil and finally hypha enters the roots (Bowen, 1987; Bellgard, 1994). Previous studies in alfalfa and tomato showed that mycorrhizal colonization was negatively affected by increased soil salinity, possibly via the effects of salinity on initial colonization (Azcon and El-Atrash, 1997; Al-Karaki, 2000; Al-Karaki *et al.,* 2001). In this investigation seedling preinoculation allowed the fungi established before salinity stress was imposed. Consequently mycorrhizal colonization was significantly affected by increasing salinity level (Asghari, 2008).

Smith *et al.* (1986) and Jakobsen *et al.* (1992) reported increased 'P' inflow at early stage of mycorhizal plant growth compared to NM plants and declined with increasing plant age in *Allium cepa* and *T. subterraneum* respectively. Present investigation strongly supports the early workers contribution. 'P' content in mycorrhizal plants decreased with increasing age. This may be due to the biomass of the fungal tissues within the roots has been estimated to range from 3–16% of the total mycorrhizal root dry mass (Harries and Elder, 1987; Azcon, 1994).

CONCLUSION

Coastal sand dunes constitute major biomes of considerable ecological and economic importance. Below ground diversity, mycorrhiza is one of the major factors governing the plant distribution, diversity and functioning of coastal sand dune ecosystem. Elimination of mycorrhizal fungi decreases the plant equilibrium as well as productivity, which may result in destabilization. Future study should concentrate on the importance of mycorrhizae on coastal sand dune stabilization and application of stress tolerant mycorrhizae in agriculture.

REFERENCES

Al-Karaki, G. N. (2000). Growth and mineral acquisition by mycorrhizal tomato grown under salt stress. *Mycorriza*.10: 51–54.

Al-Karaki, G. N., Al-Raddad, A (1997). Effects of arbuscular fungi and drought stress on growth and nutrient uptake of two wheat genotypes diffrering in their drought resistance. *Mycorrhiza*. 7:83–88.

Al-Karaki, G. N., Hammad, R. and Rusan, M. (2001). Response of two tomato cultivars differing in salt tolerance to inoculation with mycorrhizal fungi under salt stress. *Mycorriza*. 11: 43–47.

Asgari, H. R. (2008). VA-Mycorrhizae improves salinity tolerance in pre-inoculated subterranean clover (*Trifolium subterraneum*) seedlings. *International J. of plant production*. 2(3):243–256.

Azcon, R. (1994). The role of AM fungi in nitrogen nutrition and metabolism by plants. *Mycorrhiza news*. 6(3): 1–5.

Azcon, R. and El-Atrash (1997). Influence of arbuscular mycorrhizae and phosphorus fertilization on growth, nodulation and N_2 fixation (15N) in *Medicago sativa* at four salinity levels. *Biol Fertil Soils*. 24: 81–86.

Bagyaraj, D.J. (1991). Ecology of vesicular-arbuscular mycorrhiza. In: Handbook of applied mycology. (eds. Arora, D.K., Rai, B., Mukerji, K.G. and Knudsen, G.R.) Vol.I *soil and plants. Marcel dekker, Inc., New York. Basel*. 3–34.

Beena, K. R., Raviraja, N. S. and Sridhar, K. R. (2000). Seasonal variations of arbuscular mycorrhizal fungal association with Ipomea pes-capre of coastal sand dunes, Southern India. *J. Environ. Biol*. 21: 341–347.

Bellgard, S. E. (1994). The bifunctional nature of the mycelial network associated with plants colonized AM fungi. *Mycorrhizal news*. 6(1)1–5.

Bianciotto, V., Lumini, E., Lanfranco, L., Minerdi, D., Bonfante, P. and Perotto, S. (2000). Detection and identification of bacterial endosymbionts in arbuscular mycorrhizal fungi belonging to the family *Gigasporaceae*. *Appl.Environ. Microbiol*. 66: 4503–4509.

Bonfante-Fasolo, P. and Schubert, A. (1987). Spore wall architecture of *Glomus* spp. *Can. J. Bot.* 65:539–546.

Bowen, G. D. (1987). The biology and physiology of infection and its development. In : Safir, G.R. Ecophysiology of VA mycorrhizal plants. *Boca Raton* Florida.: CRC press. 27–70.

Carpenter-Boggs, L., Loynachan, T. E. and Stahl, P. D. (1995). Spore germination of *Gigaspora margarita* stimulated by volatiles of soil-isolated *Actinomycetes. Soil Biol. Biochem.* 27:1445–1451.

Dangered, P. A. (1900). Le Rhizophagus populinus. *Botaniste.* 7: 285–287.

Danneberg, G., Latus, C., Zimmer, W., Hundeshagen, B., Schneider-Poetsch, H. J. and Bothe, H. (1992). Influence of vesicular-arbuscular mycorrhiza on phytohormone balances in maize (*Zea mays* L.). *J. Plant Physiol.* 141:33–39.

Dawson, J. O. and Gibson, A. M. (1987). Sensitivity of selected isolates from *Casuarina, AUocasuarina* and North American host plants to sodium chloride. *Physiol. Plant.* 70:111–211.

Filippi, C., Bagnoli, G., Citernesi, A. S. and Giovannetti, M. (1998). Ultra structural spatial distribution of bacteria associated with sporocarps of *Glomus mosseae. Symbiosis.* 24:1–12.

Frank, A. B. (1885). Veber die auf Wurzelsymbiose beruhende Ernahrung gewisser Baume durch Uniterirdische Pilze. *Ber.Dent. Bot. Gesell.* 3: 128–145.

Garbaye, J. (1994). Helper bacteria—a new dimension to the mycorrhizal symbiosis. *New Phytol.* 128:197–210.

Gerdemann, J. W. and Nicolson, T. H. (1963). Spores of Mycorrhizal Endogone species extracted from soil wet sieving and decanting. *Trans. Brit. Mycol. Soc.* 46: 235–244.

Graham, J. H. (1986). Citrus mycorrhizae: Potential benefits and interactions with pathogens. *Hort. Science.* 21: 1302–1306.

Gryndler, M. H., Hrselova and Striteska, D. (2000). Effect of soil bacteria on hyphal growth of the arbuscular mycorrhizal fungus *Glomus claroideum. Folia Microbiol.* 45:545–551.

Harley, J. L. (1965). Ecology of soil borne pathogens (Eds) K.F. Baker and W.C. Synder John.

Harriis, D. and Elder, P. (1987). Carbon requirements of VA mycorrhizae. *In:* Ecophysiology of VA mycorrhizal plants. Safir, R.G. (ed.) Florida, pp 93–105.

Hoffman, G. J. and Phene, C. J (1971). Effect of constant salinity levels on water-use efficiency of bean and cotton. *Trans. Am. Soc. Agric, Eng*. 117: 1103–1106.

Jackson, M. L. (1963). Soil chemical analysis Printice Hall Inc. New Delhi pp: 10-330.

Jacobsen, I., Abbott, L. K. and Robson, A. (1992). External hyphae of vesicular-arbuscular mycorrhizal fungi associated with *Trofoluim subterraneum* L. I. Spread of hyphae and phosphorus inflow into roots. *New Phytol*. 120:371–380.

Jain, R. K., Paliwal, K., Dixon. R. K. and Gjerstad, D. H. (1989). Improving productivity of multipurpose trees on substandard soils in India. - *J. For.* 87: 38–42.

John, T. V. S., Coleman, D. C. and Reid, C. P. P. (1983). *Ecology* 64: 957–959.

Joner, E. J. and Jokobson, I. (1995). 172: 221–227.

Koske, R. E. and Gemma, J. N. (1997). Mycorrhizae and succession in planting of beach grass in sand dune. *Amer. J. Bot.* 84:118–130.

Kulkarni, S. S., Raviraja, N. S. and Sridhar, K. R. (1997). Arbuscular mycorrhizal fungi of tropical sand dunes of west coast of India. *J.Coastal Res.* 13:931–936.

Lakshman, H.C. and Patil Geeta B. (2003) Effect of Glomus fasciculatum and rhizobium with phosphate application on Acacia auriculifolia. Microbiol world. 6(1):37–40.

Lakshman, H. C. (2005). Response of *Guizotia abyssinica* Cass (Niger) to AM fungi inoculation under different concentration of domestic sewage water. *In proceedings* of state level seminar on "Biodiversity and its conservation" at G.H. College, Haveri on 28th and 29th july 2005, pp 111–114.

Lichenthaler, H. K. (1996). Vegetation stress: an introduction to stress concepts in plants. *J.plant physiol*. 148: 4–14.

Linderman, R. G. (1988). Mycorrhizal interaction with with rhizosphere microflora. The mycorrhizosphere effect. *Phytopathology*. 78: 366–371.

MacDonald, R. M. and Chandler, M. R. (1981). Bacterium-like organelles in the vesicular-arbuscular mycorrhizal fungus *Glomus caledonius*. *New Phytol.* 89:241–246.

Maia, L. C. and Kimbrough, J. W. (1998). Ultra structural studies of spores and hypha of a *Glomus* species. *Int. J. Plant Sci.* 159: 581–589.

Manoharachary, C., Adholeya Alok and Kunwar, I.K. (2009). Mycorrhize some glimpses. *Mycorrhizal news*. 20(4):1–6.

Marschner, H. and Dell, B. (1994). Nutrient uptake in mycorrhizal symbiosis. *Plant Soil*. 159: 89–102.

Maun, M. A. (1994). Adaptations enhancing survival and establishment of seedling on coastal dune systems. *Vegetatio.* 111: 59–70.

Maun, M.A. (1998). *Can J. Bot*. 76: 713–738.

McGoniel, T.P., Miller, M. H., Evans D.G., Fairchild, G.L. and Swan, J.A. (1990). A new methode which gives an objective measure of colonization of roots by vesicular arbuscular mycorrhizal fungi. *New Phytol*. 115:495–501.

Morton, J. B. (1999). Evolutionary relationship among arbuscular mycorrhizal fungi in the endogonaceae. *Mycologia*. 82: 192–207.

Ojala, J. C., Jarrell, W. M., Menge, J. A. and Johnson, E. L. V. (1983) Influence of mycorrhizal fungi on the mineral nutrition and yield of onion in saline soil. *Agronomy Journal*. 75: 255–259.

Olsen, S. R., Cole, C. V., Watanabe, F. S. and Dean, L. A. (1954). Estimation of available phosphorus in soils with extraction with sodium bicarbonate. *U.S.Dept. Agric.Circ. 939, USA.*

Paieg, L. G., Stewart, G. R. and Bradbeer, J. W. (1984). Proline and giycine betaine influence protein solvation. - *Plant Physiol.* 75: 974–978.

Peyronel, B., Fassi, B., Fontana, A. and Trappe, J.M. (1969). *Mycologia,* 61: 410–411.

Pfeiffer, C. M. and Bloss, H. E. (1988). Growth and nutrition of guayule (*Parthenium argentatum*) in a saline soil as influenced by vesicular-arbuscular mycorrhiza and phosphorus fertilization. *New Phytol*. 108: 315–321.

Polle, A. and Rennenberg, H. (1993). Significants of antioxidants in plant adaptations to environmental stress. In: Fowden L., Mansfield, T. and Stoddart, T. (eds). *Plant adaptation to environmental stress*. Champman and Hall, pp 263-273.

Rao, T. A. and Mcher-Homji, V. M. (1985). Proc. Indian academic science (*Plant science*) 94: 505–523.

Ruiz-Lozano, J. M. and Azcon, R. (1995). Hyphal contribution to water uptake in myconfiizal plants as affected by the fungal species and water status. *Physiol, Plant*. 95: 472–478,

Schenck, N. C. and Perez, Y. (1990). Manual for the identification of VA-mycorrhizal fungi, *Synergistic publications*, USA.

Sehgal, J. and Abrol, I. P. (1994). Soil degradation in India: status and impact.Oxford and *IBH publication*. Co. Pvt Ltd. New Delhi, India. Pp 80.

Smith, S. E. and Read, D. J. (1997). Mycorrhizal symbiosis. London : *Academic press*. Pp 605.

Smith, S. E., St-John, B. J., Smith, F. A and Bromley, J. L. (1986). Effect of mycorrhizal infection on plant growth, nitrogen and phosphorus nutrition of glass house grown *Allium cepa* L. *New Phytologists*. 103: 359–373.

Taleisnik, E. L. (1987). Salinity effects on growth and carbon balance in *Lycopersicon esculentum* and *L. pennelti*. – *Physiol Plant*. 71: 213–218.

Walley, F. L. and Germida, J. J. (1996). Failure to decontaminate *Glomus clarum* NT4 spores is due to spore wall-associated bacteria. *Mycorrhiza*. 6: 43–49.

Xavier, L. J. C. and Germida, J. J. (2003). Bacteria associated with *Glomus clarum* spores influence mycorrhizal activity. *Soil Biol. Biochem*. 35: 471–478.

26

Prospects of Arbuscular Mycorrhizal Fungi in Bioremediation of Polluted Soils

H C Lakshman C Maya** and G Ravi Prasad***

INTRODUCTION

Soil is one of the most precious natural resources of a country, aptly described as 'the bridge between the inanimate and the living'. Both natural forces and human activity are degrading this resource. Nearly 57% of the land area is degarded due to many reasons. One of the reasons for this is rapid industrialization and the disturbance of land soil by the discharge of effluents, and the pollutants cause undesirable change in the soil ecosystem thus disturbing its physico-chemical setup and biological spectrum. Sewage effluents are a rich source of organic wastes, while industrial effluents contain many chemicals, including many heavy metals. It is estimated that the end of 2010 A.D. approximately 2400 million later sewagewill be released each day. Epidemic

* Post Graduate Department of Studies in Botany, Microbiology Laboratory, Karnatak University, Dharwad- 580003, India

** Department of Botany Bangalore University, Gynnana Bharati Campus, Bangalore - 560 056, India

diseases like Brucellosis, cholera, hepatitis and malaria are related with sewage. Besides heavy metals present in the sewage water pollute the environment. United Nations World Health Organization report that more than 2.5 crores of people affected by sewage related diseases every year (Pradhan *et al.*, 2000).

Bioremediation is a technology that has gained great popularity and is attracting attention as a remedial technology, in which arbuscular mycorrhizal fungi (AMF) are used to detoxify sewage water (Gaur and Adhloleya, 2004). The problems of sewage water can be minimal only by effective use of this water with beneficial microorganisms. In India municipalities, use this sewage water for cultivation of vegetables in farm fields. However, there are some heavy metals such as manganese, zinc, copper, cobalt, nickel, lead and cadmium etc., tend to accumulate in the environment and living systems. These metals directly affect the soil as well as accumulate the vegetables. It has been suggested that mycorrhizal colonization could also be of importance in plants colonizing industrial wastes (Nicolson, 1967) where plant cover in difficult to establish. Arbuscular mycorrhizal association is considered crucial for the survival and growth of the majority of plant species in natural ecosystems (Harley and Smith, 1983). Bioremediation is a technology that has gained great popularity and is attracting attention as a remedial technology in which microorganism are used to detoxify wastes.

Plants colonized with VA-mycorrhizal have been reported from the industrial waste, coal wastes and coal mine spoils (Daft and Nicolson, 1974; Aldon, 1975). VAM plants can grow on arid and eroded soils (Powell, 1977). These investigations have stressed the importance of arbuscular-mycorrhizal fungal relationship in allowing the successful recolonization and growth of haerbaceous plants on distrubed soils. Such disturbances may leave little viable inoculum capable of intiating mycorrhizal infection in plants colonizing the affected areas (Allen and Allen, 1980). These plants have tolerance to metal taxicants (Brandley *et al.*, 1981).

In recent days many laboratory and field experiments were conducted by using transplantation of VA-mycorrhizal saplings in top miled soil storage during the revegetation by using wheatgrass (Jak and Parkinson, 1982; Loree and Williams, 1987). It has already been established for these mine sites that Acacia spp. growing stock piled top soils suffer from delayed formation of arbuscular-mycorrhiza (Jasper *et al.*, 1989). In these studies, it was shown that losses in ineffectivity of the mycorrhizal fungi may occur during both the intial disturbance and subsequent storage. The re-establishment of mycorrhizal relation on mined soil thus comes to the reorganized as an important component of an overall reclamation strategy. The objective of the present study was to assess the percentage of arbuscular mycorrhiza fungal colonization stock piled mined spoils of Bisgod in Karnataka (Tables 26.1 – 26.3).

The endophytic fungi seem to be associated with plants producing large number of rootlets (Mosse, 1963). These findings signified the importance of VAM fungi in stock piled mined soil with higher manganese evels, with deficiency of P nutrition. These findings are supported to earlier work by (Lakshman, 1994 and 1995; Saralabai *et al.* (1995) reported that microbial activities are likely to be affected by the chemical and physical properties of the soil. But it is evident from the present findings that the stock piled mined soil was studied fora period of six months to one year. Even though percent of colonization in Graminae and Compositae did not altered much compared to early assessment when the saplings was two months old. This would depend on the inoculum required to satisfactorily intiate symbiosis and on the degree of host dependency during the sapling stage. It may be desirable to identify useful revegetation species with low mycorrhizal dependency for the use on disturbed sites with

Table 26.1 : VA-Mycorrhizal Status in Plants Growing on Stock Piled Mined Spoils of Bisgod in Karnataka India. (Lakshman, 2001)

Family / Scientific name	pH	Percent colonization	Spores/50 g Soils
Aslepiadaceae			
Hemidesmus indica R. Br.	6.8	13.4± 4.1	24.3 + 5.1
Compositae			
Ageratum conyzoides L.	6.4	46.3 + 4.2	34.9 + 2.4
Eupatorium adoratum Vahl.	7.1	4.1+ 2.2	17.5 + 1.3
S.F. Caesalpiniaceae			
Cassia tora linn.	7.2	18.2+ 5.2	21.3 +3.2
Commelinaceae			
Commellina communis L.	6.7	23.4+ 3.1	12.4 + 4.2
S.F. Mimosiaceae			
Acacia melanoxylon R.Bi.	7.3	16.4+ 5.3	31.7 +6.2
Mimosa pudica L.	6.3	15.5+3.2	19.4 + 5.1
Cyperaceae			
Cyperus rigidifolius stend	7.5	5.1+ 1.3	8.3 + 4.4
Euphorbiaceae			
Euphorbia sparcifolia L.	8.2	32.1+ 2.5	30.5 + 5.3
Graminae			
Cynodan dactylon Pers.	6.4	42.5+ 5.2	32.4 + 5.2
Malvaceae			
Sida veronifolia Lam.	6.3	17.4+ 4.5	18.5 + 4.2
Rhamnaceae			
Ziziphus trinervia Roxb.	7.3	9.3+ 5.0	16.4 + 3.1
Tiliaceae			
Corchorus acutangulus Lam.	7.2	27.4+ 1.2	14.3 + 2.2
C. mimosoidea L.	7.4	23.4 ± 4.1	39.5 ± 6.1
Crotalaria junceae L.	7.2	29.2 ± 4.3	42.4 ± 8.1
Phyllantus niruri L.	6.8	29.2 ± 4.3	33.1 ± 7.2
Eclipta alba Hassk	6.5	37.6 ± 5.2	39.3 ± 6.1
Elephantopus scaber Linn.	7.5	32.5 ± 6.1	41.1 ± 5.2
Euphorbia prostrata L.	7.3	21.0±6.1	37.2 ± 12.0
Cyperus diffbrnis Linn.	6.7	11.5± 3.3	35.4 ± 6.2
Chloris gayana Kunth	6.4	44.2 ± 4.2	78.2 ± 8.1

Table 26.2.VA-Mycorrhizal Status in Plants Growing in Undisturbed Sites Bisgod Forest Area Adjacent to Mined Area in Karnataka, India (Lakshman, 2001)

Family / Scientific name	pH	Percent colonization M ± S.E.	VAM Spore/509.soil M ± S.E.
Menisper maceae			
Coculus Villosus Dc.	7.6	53.3 ± 7.1	103.7+ 8.3
Gutiferaceae			
Caloplıyllum uightianum Anders.	6.9	34.2+ 4.4	57.4 + 2.2
Malvaceae			
Thespesia macrophylla Blume.	6.2	51.4+ 3.1	64.0 + 5.H
Elaeocarpaceae			
Elaeocarpus aristatus Roxb.	5.9	33.1+ 5.4	37.1 + 306
Rutaceae			
Toddalia aculeata pers.	6.1	72.5+ 2.2	110.01 + 3.1
Meliaceae			
Toona ciliata Roem.	7.5	32.2+ 1.1	107.0 + 4.3
Cipadessa fnriticcsa Blume	6.8	40.4+ 5.2	94.3 + 2.5
Rhamnaceae			
Zizyplıus mauritiana Lauck.	5.7	47.7+ 5.4	114.1 + 5.5
S.F. Mimosoiaceae			
Acacia catecha Willd.	8.2	53.4+ 6.3	102.3 + 3.4
Papilionaceae (S.F.)			
Dalbergia emarginata Roxb.	8.4	46.5+ 3.1	72.4 + 5.1
Plerocarpus marsupium Roxb.	7.5	47.4+ 4.4	89.1 + 4.4

Family / Scientific name	pH	Percent colonization	VAM Spore/509.soil
		M ± S.E.	M ± S.E.
Xylia xylocarpa			
Roxb.	6.5	39.3+ 6.1	93.3 + 5.6
Combretaceae			
T. bellerica Roxb.	7.1	72.4+ 3.5	98.1 + 2.3
T. chebula Retz.	6.3	79.5+ 7.2	124.4 + 3.4
T. alata Heyne.	6.8	84.2+ 5.1	120.1 + 5.5
Rubiaceae			
Stephegyne parxifolia Korth.	7.2	66.4+5.3	93.4 + 7.3
Compositae	7.3	94.2+ 5.2	78 4 + !;3
Elephantopus seaber Linn.	6.2	95.4+ 8.1	82 1 + 4,2
Eupalorium triplinerve Vahl.			
Apocynaceae	6.6	63.3+ 6.0	78.3 + 9.4
Wrightia tinctoria R. Br.			
Borginaceae	6.5	24.3+ 5.2	84.3 + 7.1
Cordia macleodii Hook.			
Verbenaceae	7.5	49.3+ 7.7	97.3 + 8.4
Tectona grandis Linn.			
Santalaceae	8.1	14.5+ 7.3	56.2 + 5.5
Santalum album Linn.			
Euphorbiaceae	7.4	77.6+ 6.3	97.8 + 5.4
Phyllanthus emblica L.			
Cyperaceae	6.3	13.2+ 6.5	48.3 + 8.4
Scirpus isolepis boeck.			
Graminae			
Brachiaria erneiformis.	5.4	100.0	117.3 + 3.2
Gricob.			

S.F. = Sub family, M=Mean value, S.E.=Standard Error. Each value is mean of 12 plants.

Table 26.3. : VAM Fungal Species Isolated from Plants Grown on Stock Mined Spoil of Bisgod in Karnataka, India (Lakshman, 2001)

VAM fungal species	Abbreva-tion	Gradation		Position of AMF species	
	1	2	3	4	5
Acaulospora					
Acaulospora foveta Trappe & Janons.	a	c	j	h	g
A. Apinosa wal. & Trappe.	b	j	g	1	d
Gigaspora					
Gigaspora decipiens Hall. & Abbott.	d	g	k	a	e
Glomus					
Glomus citricolum Tung & Zang.	f	g	d	e	a
G. etinicatum Becker & Gerd.	g	h	j	k	e
G. fasciculatum Walk. & Koske.	h	g	b	h	b
G. mosseae Gerd & Trappe.	k	f	a	n	c
Scutellospora					
Scutellospora fulgida Koske & Walker.	1	g	e	d	k
S. weresubiae Koske & Walker.	n	e	i	f	g

* Each value is mean of 12 plants.

depleted endophyte populations. To ensure the success of field inoculation, the most effective fungi must be used. The effectiveness of fungus in increased plant growth appeared to be correlated with its rate and extent of colonization (Abott and Robson, 1982; Jasper *et al.*, 1989).

The colonization levels of AM fungi of two families may indicate the degree to which biological activity has been restored in stock piled soils (Miller, 1985). This would lead to restoration of an ecosystem and combating environmental pollution on stock piled soils by introducing mycorrhizal plants. Colonized arbuscular mycorrhizal sappling transplantation to degraded mined soils from pot experiment is practically required. Detailed work is required concerning the dependence of pioneering plants on mycorrhizal colonization. Particularly root hair development, uptake of nutrients and the relative efficiency endemic endophytes in the uptake of Phosphorus. It will be important to investigate in field experiments. Whether the demonstrated early response to inoculation is maintained or whether the surviving arbuscular mycorrhizal fungi multiply to form an effective inoculum. Any changes in the soil handling process at either site that shortens or eliminates the stockpile stage is likely to improve the infectivity of respread soil and so reduce the requirement for inoculation.

Mining wastelands are those lands where large scale mining operations bring about degradation of the land and result in mine dumps. The importance of AM fungi in regeneration of different types of mining sites has also been emphasized. Extensive infection of AM fungi in plants colonizing the coal wastes was recorded by (Daft and Hacskalyo, 1976) who hypothesized that application of AM fungi may ensure successful reclamation of coal mine areas. Improvement in growth of *Atroplex canescens* (Williams *et al.*, 1974) of maize, alfalfa and other plants and of forage species (Lambert and Cole Jr, 1988) was obtained on-mine spoils when AM fungi were introduced. Similarly, use of AM fungi in reclamation of iron and chromate mine wastes has given positive responses (Mishra, 1992).

In mine wastes, AM fungi have been reported to bring an improvement in the growth of plants through an increased P uptake, a higher rate of photosynthesis and a reduction in stomatal and mesophyll resistance resulting from an altered hormonal balance and increased water flow (Allen, 1984). Altering levels of phytohormones by AM fungal formation (Allen *et al.*, 1981) and production of cytokinins and gibberellins due to *G. mosseae* (Barea Azcon-Aguilar, 1982) have also been reported.

HEAVY METAL POLLUTION AND IMPORTANCE OF AM FUNGI

Ecosytems have been contaminated with heavy metals due to various human and natural activities. The sources of metals in the soil are diverse, including burning of fossil fuels, mining and smelting of metalliferous ores, municipal wastes, fertilizers, pesticides, sewage sludge sediments, the use of pigments and batteries. These metals are commonly called heavy metals, although this term strictly refers to metallic elements with a specific mass higher than 5 g cm^{-3}, able to form sulphide (Adriano, 1984). Some of these metals are micronutrients necessary for Plant growth, such as Zn, Cu, Mn, Ni and Co (Marschner and Romheld, 1995) have not known biological function, such as Cd, Pb and Hg. Migration of these contaminants into non-contaminated areas as dust or leachates through the soil and spreading of heavy metal containing sewage sludge are examples of events that contribute contamination of our ecosystems.

It is well known that heavy metals cannot be chemically degraded and need to be physically removed or be immobilized. Traditoinonally, remediation of heavy metal contaminated soils involves either on-site management or excavation, and subsequent disposal to a landfill site (Parker, 1984). However, this method of disposal merely shifts the contamination problem elsewhere along with the hazards associated with transportation of contaminated soil and migration of contaminants from landfill into an adjacent

(Williams, 1988). Soil washing for removing contaminated soil is an alternative to excavation and disposal to landfill (Elliot *et al.*, 1989). This method is, however, costly and produces a residue in heavy metals, which will require further treatment or burial. Moreover, these physio-chemical technologies used for soil remediation render the land useless as a medium for plant growth, as they remove all biological activities. Therefore, sustainable on-site techniques for remediation of heavy metal contaminated sites need to be developed. Phytoremediation, the use of plants to remediate or clean-up contaminated soils can be used as a promising method to remove and/or stabilize soils contaminated with heavy metals. Moreover, it is essential to create a stabilizing plant cover composed of herbaceous species which are active in phytoremediation processes and produce high levels of root and shoot biomass (Leyval *et al.*, 1997) to prevent intensive wind erosion. Plants which appear spontaneously in such places are frequently devoid of mycorrhizal symbiosis and are mostly characterized by poorly developed shoot biomass when heavy metals are present (Pawlowska *et al.*, 1996). The lack of mycorrhiza can hamper the revegetation the metal-contaminated mine spoil or other degraded sites. The introduction of an AM fungal inoculum into these areas could be a strategy for enhancing the establishment of mycorrhizal herbaceous species. AM fungal isolates differ in their effect on heavy metal uptake by plants. Some reports indicate higher concentrations of heavy metals in plants due to AM fungi whereas, others have found a reduced plant concentration; for example, Zn and Cu in mycorrhizal plants (Heggo *et al.*, 1990). Thus, selection of appropriate isolates could be of importance for a given phytoremediation strategy. AM fungal species can be isolated from areas which are either naturally enriched by heavy metals or are old mine/industry waste sites in origin. In this context, AM fungi constitute an important functional component of the soil-plant system that is critical for sustainable productivity in degraded soils.

AM fungi are of importance as they play a vital role in metal tolerance and accumulation. External mycelium of AM fungi provides a wider exploration of by spreading beyond the root exploration zone. (Malcova *et al.*, 2003) thus providing access to greater volume of heavy metals present in the rhizosphere. A greater volume of metals is also stored in the mycorrhizal structures in the root and in spores. For example, concentrations of over 1200 mg kg^{-1} of Zn have been reported in fungal tissues of *G. mosseae* and over 600 $mgkg^{-1}$ in *G. versiforme* (Chen *et al.*, 2001). Another important feature of this symbiosis is that AM fungi can increase plant establishment and growth despite his levels of soil heavy metals, due to better nutrition water availability and soil aggregation properties associated with this symbiosis (Table 26.4). AM fungus is significant in the ecological improvement of rhizosphere (Azcon-Aguilar *et al.*, 2003).

Low level and sporadic nature of arbuscular mycorrhizae on mined spoils were less, comparatively those undisturbed sites. There are some questions as the means by which arbuscular fungal inoculum is dispersed (Rabtin, 1979) and the distance over which effective dispersal can occur (German and Trappe, 1974; Lakshman, 1992; Loree *et al.*, 1984; Singh, 1996) showed that worms, ants and birds could be agents for dispersal of spores. It is doubt whether this type of dispersal is a major countributor to long distance transport. It would likely be more important for within habitat movement of inoculum. The distance from mine sites to potentital AM fungal inoculum pools and the rates of dispersal and subsequent recolonization would likely be important factors that need to be considered in the planning of reclamation programmes. Success or failure of a revegetation strategy may in part be determined by the immigration rates of arbuscular mycorrhizal fungi in to these distrubed habitats. (Sambandan *et al.*, 1992) suggested that the slow rate of revegetation of disturbed lands in the semi arid and west costal region was due to the elmination or reduction in the number of AM fungal propagules.

In mine spoils where arbuscular mycorrhiza fungal inoculum levels are low or arbuscular mycorrhizal fungi present, as a result of reclamation practices, non-mycorrhizal plants speceis that colonize the sites could have competitive advantage over mycorrhizal dependent species, especially if nutrient become limiting (Loree *et al.*, 1984; Singh 1996). Arbuscular fungal mycorrhizal colonization of the grass highest in the plot with severest pollution damage and lowest in the least damaged plot (Sambandan *et al.*, 1992). If the rapid establishment of arbuscular mycorrhizae with plants grown on mined spoils is to be a major concern of a reclamation strategy, attention will have to be given to determining intial arbuscular mycorrhiza fungal inoculum densities in the mined spoils and the application rates of inorganic and organic amendments which donot suppress arbuscular mycorrhiza fungi development.

Table 26.4. : Summary of AM Fungi on Heavy Metal Interactions and Application in Phytoremedication [1a]

Heavy metal	Tolerant AM fungi	Host used	Reference
Cd	*Glomus mosseae*	*Trifolium repens*	[44]
	G. mosseae	*Trifolium subterraneum*	[45, 46]
Ni, Zn	Mixed AM inocula	*T. repens*	[47]
	Scutellospora dipurpurescens	*Agrostis capillaries*	[48]
	Glomus intraradices		[49]
	Glomus caledonium		[50]
	G. mosseae, Glomus macrocarpum	*Lygeum spartum*	[51]
	G. mosseae	*T. sbuterraneum*	[46]
Cd, Zn, Cu	*G. mosseae*	*Phaseolus vulgaris*	[52]
Zu, Cd, Cu, Ni, Pb	*G. caledonium*	*A. porrum, Sorghum bicolor*	[53]

AM FUNGI ON DROUGHT STRESS

Many reports available as literature show that AM fungal infections improve the drought tolerance of plants (Sieverding, 1981; Allen and Boosalis, 1983). Smith and Read (1997) studied the effects of mycorrhizal root colonization in onion plants exposed to water stress and found that the mycorrhizal plants were four times larger than the non-mycorrhizal plants. However, the mechanisms through which the mycorrhizal infections increase the drought resistance are still uncertain.

AM fungal infections are reported to cause a reduction in the resistance to water transport, leaf water potential, and stomatal resistance under water stress conditions (Allen *et al.*, 1981). These effects of mycorrhizal infection may be attributed to improved water uptake, increased photosynthesis or elevated cytokinin levels which stimulate stomatal opening (Allen *et al.*, 1981; Barea and Azcon-Aguilar, 1982). Hyphae ramifying into the soil are likely to increase the absorbing areas for water uptake and may be able to bypass the dry zones that often surround slow growing roots during periods of drought (Knee and Tinkar, 1977).

Rapid accumulation of proline is a metabolic consequence of water deficit in plants. Mycorrhizal plants have been shown to have a lower concentration of pro-line in both the root and the shoot implying a lowered physiological water stress (Nemec and Meredith, 1981; Tal *et al.*, 1979).

AM FUNGI ON SALINE AND ALKALINE SOILS

Nearly seven per cent of the land at the global level is saline due to the presence of excess soluble salts in the soil (Dudal and Purnell, 1986). There are conclusive evidences to show the importance of AM fungi in reclamation of saline soils (Pfeiffer and Bloss, 1988; Rozma *et al.*, 1986) mycorrhizal advantage occures through increased concentration of phosphate and a decreased concentration of sodium in shoot tissues in *Parihenium aggentatum* plants inoculated with *Glomus intraradices* compared to non-inoculnted plants (Pfeiffer and Bloss, 1988), due to diurnal changes in the leaf thickness as in *Aster tripolium* (Rozma *et al.*, 1986) increased tolerance to fluctuations in temperature and pH of the soil (Ross *et al.*, 1985).

The exact mechanism by which mycorrhizal systems tolerate unbalanced ionic conditions as in salty soils is not known. The mycorrhizal plants under high salinity, however, have greater concentrations of potassium, sodium and phosphorus (Ojala *et al.*, 1983; Allen and Cunningham, 1983) whereas AM plants maintained a steady Na/K balance than the non-mycorrhizal plants.

AM FUNGI ON SAND DUNES

Sandy areas accumulated with sand are deficient in organic matter and their fertility is very low. Presence of AMF is critical for the regeneration of natural ecosystem in arid lands. A number of workers have reported a high AMF status in desert plants and have emphasized the role of AMF fungi in establishment of plant cover and improving the fertility of the desert on the basis of their studies (Barea and Jeffries, 1995; Ernst *et al.*, 1984; Puppi and Riess, 1987). In addition to enhanced absorption of plant nutrients, AMF improves the physical structure of the soil too (Logan *et al.*, 1989).

It seems likely that AM-endophytes have a role in, improving water relations of plants in sandy areas. However, it is not clear whether this effect is a direct result of fungal invasion and can be attributed to improved waterflow through hyphae or a secondary response due to improved nutrition or physiological alterations of the host (Allen *et al.*, 1981; Barea and Azcon-Aguilar, 1982).

AM FUNGI ON INDUSTRIAL POLLUTANTS

Through their effluent spread, the industries create large tracts of unusable land. Presences of acidic or alkaline waste, and dyes or heavy metals are the main constraints in using such lands (Table 26.5). Quantitative and qualitative assessment of AM fungi in soils affected by industrial effluents has been made (Jagpal, 1988). It has been suggested that infection with AM fungi could be of importance in raising plants in land polluted by industrial waste. AM fungal infection is also found to increase the tolerance of AMF plants to zinc and cause an enhancement in the production of biomass (Dueck *et al.*, 1986). The mycorrhizal associations in some vegetable crops raised in soil with high concentration of some metal pollutants also cause a reduction in the amount of metals in their shoots (Schuepp and Bodmer, 1991). The translocation of potentially toxic heavy metals to the shoot seemed to be reduced in mycorrhizal plants possibly by retention of the metals in the hyphae of fungus within the root cells (Table 26.6–26.7). Many fungi accumulate heavy metals in their mycelium and it is suggested that fungal cell walls have a larger number of complexing sites for metallic cation binding, possible by special protein fractions. Therefore, in highly infected mycorrhizal roots, the fungus may provide the plant with additional metal binding sites by an increased surface area for metal absorption.

Extensive experiments have considered the influence of Al and heavy metals Fe, Mn, Zn, Cu, Pb, and Cd on development and efficacy of vesicular-arbuscular mycorrhiza in several tropical and subtropical plants (Fabig, 1982) AM fungal inoculation allowed better growth of plants irrespective of Al, Fe, Mn, Zn, Cu, Pb, and Cd up to a specific level (Dehn and

Schupp, 1988). The tolerant concentration levels were so high that even the highest amount of Al, Fe, Mn, Zn, and Cu did not cause even microscopically visible injuries. Effectiveness of mycorrhizal systems may become understood through more such studies with heavy metals and other toxicants.

Table 26.5. : Physio-chemical Factors of Industrial Effluents and Sewage Polluted Soils of Hubli-Dharwad in Karnataka, India (Lakshman, 1995)

Factors	C_1	C_2	C_3
	Sandy loam	Sandy silt loam	Sandy clay loam
PH	5.00 ±0.00	7.5 ±00.00	8.20 ±10.24
Moisture (%)	5.50 ±30.00	14.70 ±21.50	11.23 ±8.00
Temperature	36° C	32° C	29.7° C
Chloride	10.70 ±11.20	24.0 ±17.40	0.17182
Nitrogen (ppm)	68.359	124.34	245.23
Phosphorus (ppm)	27.55	33.12	126.20
Potassium (ppm)	109.8	126.30	201.51
Organic carbon	1.3 ±1.00	3.0 ±4.20	4.02 ±5.01
Manganese (ppm)	8.75	6.72	158.70
Zinc (ppm)	2.024	6.23	2.99
Copper (ppm)	1.214	2.98	4.09
Cobalt (ppm)	7.085	11.14	12.66
Nickel (ppm)	0.527	32.40	2.135
Lead (ppm)	2.25	2.863	2.750
Cadmium (ppm)	0.103	0.116	0.154

Mean value of 4 samples; C_1-Non-polluted soil; C_2-Polluted soil with industrial effluents; C_3 - Polluted soil with sewage.

Table 26.6. Arbuscular Mycorrhizal Fungal Colonization and Spore Population in Rhizosphere Plant Species Present in Soil Polluted with Industrial and Sewages of Hubli-Dharwad, Karnataka. (Lakshman, 1995)

Family / Plant species	% of AMF colonization	Spore population 50 g polluted soil
Acanthaceace		
Andrographis echoides Nees	48.10±0.51	105.1±9.3
Justicia simplex Linn.	39.12±1.22	89.14±5.4
Amaranthaceae		
Amaranthus Spinosous Linn.	10.00±00.00	21.14±5.32
A. viridis Linn.	02.00±00.00	17.10±9.14
Asciepiadaceae		
Calotropis gigantea R. Br.	42.16±14.02	77.12±6.30
Commelinaceae		
Commelina aibescens Hassk.	15.00±00.00	29.18±5.30
C; benghalensis Linn.	31.00±00.00	37.10±9.25
Compositae		
Agerantum conyzoides Linn	79.10±14.20	102.5±8.10
Blumea *oblique* (L) Druce	53.05±30.00	86.3±14.12
Bidens pilosa Linn.	58.40±41.23	79.0±7.33
Tridax Procuubens Linn.	54.15±50.00	64.4±4.40
Sonchus aspera Hill	36.22±19.12	121.5±3.20
S. oleraus Linn.	45.00±50.00	59.04±8.00
Viocaindoca L.	51.43±42.11	62.32±10.00
Vernonia Cinerea Schreb.	62.21±34.15	71.15±4.00
Cucurbitaceae		
Cucurmis trigonus Roxb.	54.14±81.21	83.14±7.64
Cyperaceae		
Cyperus bulbosus Linn.	05.00±00.00	23.12±8.14
*C. ferruginea(L)*vahl	09.05±00.00	19.30±5.10
C. *rotundus* Linn.	13.11±00.00	32.14±7.12
Euphorbiaceae		
Crotan balpindiana L.	48.02±26.01	56.24±5.12
Euphorbia hirta Linn.	56.40±14.20	81.10±2.01
Geraniaceae		
Oxalis Cornculata Linn.	24.11±00.00	31.00±00.00

Graminae		
Aritida hirtignma Stend.		
Brachiaria raptan (L.) Gard.	76.50±21.00	83.14±6.52
Chloris gayana Roxb.	51.01±14.12	107.21±9.40
C. roxburghiana Schult.	92.50±25.00	121.3±5.11
Cynodon dactylon (L.) pers.	88.40±12.10	104.5±4.50
Digtaria. Ciliaris (Retz.) Koeler.	56.05±11.00	96.32±7.15
Setaria FerticiV/ato (L.) Beanuv.	82.21±50.12	91.40±4.30
Malvaceae	67.14±32.10	72.05±14.00
Althaea ludwigii Linn.	44.05±12.10	93.42±9.61
Hibiscus ovalifoius vahl.	56.00±24.00	64.52±6.42
Sida ovata Frest	69.14±32.05	55.30±4.50
Minosiaceae		
Mimosa Pudica Linn.	43.20±41.20	38.15±10.00
Papilinaceae		
Desmodium trifolium DC.	31.18±52.31	105.10±4.00
Indigofera astragalina DC.	44.11±50.20	78.32±9.15
Ludwigia Perennis Linn.	53.34±42.31	74.50±6.02
Tephorsia lanceolata Linn.	53.34±42.31	83.14±8.15
Trigonella Corriculata Linn.	52.40±48.04	83.14±8.15
Zorria gibbosa Span	57.21±54.11	56.00±5.30
Portulacaceae		
Portulaca oleraceaeUnn.	61.14±61.02	112.10±2.00
P. quadrifides Linn.	00.00±02.00	19.12±15.22
Polygonaceae	00.00±02.00	19.12±15.22
Polygala erioptera Blum.	00.00±00.00	14.10±10.00
Rubiaceae		
Argostemma Verticillatum Wall.	7.01±00.00	16.05±6.12
Scropulariaceae	58.40±41.02	54.31±5.00
Bacopa monnierih.	64.03±33.10	81.22±11.01
Solanaceae		
Solanium nigrum Linn.	05.20±00.00	13.02±4.15
S. xanthocarpum Schrad.	72.25±10.00	63.14±7.04
Umbelliferae		
Centella asiatica Linn.	55.15±14.00	47.25±14.04
	27.10±00.00	29.40±8.15

Table 26.7. : AM Fungal Spore Species Recovered in Rhizospore Plants Growing in Non-Polluted Soils and Polluted with Industrial Effluent and Sewages of Hubli-Dharwad in Karnataka, India (Lakshman, 1995)

AMF Species	S1	S2	S3
Acaulospora			
Acaulospora foveata Trape & Janos	+	-	-
A. delata Morton	+	-	-
A. mella Spain & scheneck	+	-	-
A. spinosa walker & Trape	+	-	+
A. tuberculata Janos & Trappe.	+	+	-
Gigaspora			
Gigaspora albida Schenck & Smith.	+	-	-
G. *decipiens* Hall, b Abott.	+	-	-
G. *margrita* Beaker & Hall.	+	-	-
G. *rosea* Nicol. & Schenck.	+	-	-
Glomus			
Glomus aggregatum (Scheck & Smith), koske.	+	-	-
G. *boreale* (Thax.) Trappe. & Schenck.	+	-	+
G. *fasciculatum* (Thax.) Gerd. & Trappe.	+	-	+
G. *magnicaule* Hall.	+	-	-
G. *moaaeae* (Nicol & Gerd) Gerdd. & Trappe.	+	-	+
G. *tenue* (Green & Hall) Hal!.	+	+	+
G. *tortuosum* Schench & Smith	+	+	+
Sclerocystis			
Sclerocyslis clari spora Trappe.	+	-	-
S. sinuosa Gerd. & Bakshi.	+	-	-
Scutellospora			
Scutelospora (Hall.) walker. & Sanders.	+	-	-
S. *gilmorei* (Trappe Gerd.) Walker. & Sanders.	+	-	-
S. rubiformis Gerd. & Trappe.	+	-	-
S. tricalypta wal. & sanders.	+	-	-

Each Mean value is 8 samples; S, = Non-polluted soil, S2 = Polluted soil with industrial effluents; S3 = Polluted soils with Sewage + : present; - : absent.

CONCLUSIONS

Revegetation of oil-polluted areas with arbuscular mycorrhizal plants is challanging since it involves reconstruction of the above ground flora and the below ground microbiota. With this technology, the number of microbial communities, including bacteria, actinomycetes, saprotrophic and biotrophic fungi, is increased (Lakshman, 2009). These microorganisms contribute to the biological activity of the rhizosphere soil, where they improve toxic chemical degradation thus reducing the time factor in the remediation process (Khan, 1978; Cabello, 1997). Re-establishment of mycorrhizal fungi is necessary for successful plant succession, with improved nutrient uptake and the elimination of soil stress being key points for the plant survival. By introducing AMF associated with plant roots, there is an inoculum of mycorrhizal propagules, which will increase in the restored soil with time. In this way, new mycotrophic plants might be encouraged to colonize the area and the invasion of the soil by weed plants, which do not usually form mycorrhiza, will be limited. Increasing the bio-diversity of both microorganisms and plants is of crucial importance for high production in a restored ecosystem.

FUTURE PROSPECTS

Studies here conducted indicate that AM fungal technology may have a great deal to contribute in improving the productivity and profitability of stress lands. There will, however, be a need to select promising AMF species and/or strains suitable to each category of stress land. Possibilities of using a combination of symbiotic/non-symbiotic nitrogen fixers, phosphorus solubilizing microbes and AMF to give maximum benefit to the plants growing under stress conditions will have to be explored. Other possible measures such as leaching of excess salts from saline soils with water, drainage of water in water logged soils, lowering of soil pH by chemicals, etc. will also have to be integrated with the use of AMF and other useful soil microflora. A schedule will also have to be evolved for prescribing minimal dosages of fertilizers in combination with microbial supplements to ameleorate stress soils and to have optimum productivity from such soils.

REFERENCES

Abbott, L. K. and Robson, A. D. (1982). Infectivity of vesicular-arbuscular mycorrhizal fungi in agricultural soils. *Aust. J. Agri. Res.* 33:1–11.

Adriano, D. C. (1984). Trace Elements in Terrestrial Environments, Springer, Berlin, pp. 259.

Aldon, F. F. (1975). Endomycorrhizae enhance survival and growth of fourwing saltbush of coal mine spoils. USDA. For service research note RM vol. 294:2.

Allen, M. F., Smith W. K., Moore, T. S. (Jr), Christensen, M. (1981). Comparative water relations and photosynthesis of mycorrhizal and non-mycorrhizal *Bouteloua gracilis* H B K Lag ex steud. New *Phytologist.* 88: 683–693.

Allen, M. F. (1984). Physiology of Mycorrhizae: a key to understanding successful plant establishment. In: *VA-mycorrhizae and reclamation of arid lands.* S. E. Williams and M. F. Allen (eds.), Wyoming, USA, 69–80.

Allen, E. B. and Allen, M. F. (1980). Natural reestablishment of VA mycorrhizal fungi following strip mine reclamation in wyoming. *J. App. Ecol.* 17:139–147.

Allen, E. B. and Cunningham, G.L. (1983). Effects of VA mycorrhizae on spicta under three salinity levels. *New phytol.* 93:227–236.

Allen, M.F. and Boosalis, M.G. (1983). Comperative water relation of winter wheat by two VA mycorrhizal fungi. *New phytol.* 93:61–67.

Atmanav, Gaur. and Adhloleya Alok, (2004). Prospects of arbuscular mycorrhizal fungi in phytoremediation of heavy metal contaminated soils. Curreent Science. 86(4) : 528–534.

Azcon-Aguilar, C., Palenzuela, J., Roldan, A., Baulisla. S., Vallejo, R. and Barea, J. M. (2003). Analysis of the mycorrhizal potential in the rhizosphere of representative plant species from desertification-threatened Mediterranean shrublands. Appl. Soil Ecol. 22:29–37.

Barea, J. M. and Azcon-Aguilar, C. (1982). Production of plant growth-regulating substances by the vesicular-arbuscular mycorrhizal fungus *Glomus mosseae*. *Applied and Environmental Microbiology*. 43: 810–813

Barera, J. M. and Jeffries, P. (1995). Arbussculars mycorrhizas in sustainable soil plant systems. In: mycorrhiza structure, function, molecular biology and Biotechnology, A. Varma and B. Hock (Eds) Springer Verlag. Germany, pp 560.

Brandley, R., Burt, A. J. and Read, D. J. (1981). Mycorrhizal infection and resistance and heavy meta toxicity in Calluna valgaris. *Nature*. London. 292: 335–337.

Cabello, M. N. 1997. Hydrocarbon pollution: its effect on native arbuscular mycorrhizal fungi (AMF). *FEMS Microbiology Ecology*. 22, 233–236.

Chen, B., Christie, P. and Li, L. (2001). A modified glass bead compartment cultivation system for studies on nutrient afld trace metal uptake by arbuscular mycorrhiza. *Chemosphere*. 42:185–192.

Daft, M. J. and Hacskaylo, E. (1976). Arbuscular mycorrhizae in the anthracite and bituminous coal wastes of Pennsylvania. *Journal of Applied Ecology*. 13: 523–531.

Daft, M. J. and Nicolson, T. H. (1974). Arbuscular mycorrhizas in plants colonizing coal waster in Scotland. *New phytol*. 73:1129–1137.

Dehn, B. and Schupp, H. (1988). Influence of VA-mycorrhiza on the uptake and distribution of heavy metals in plants. *Proceedings of the 2nd European Symposium on Mycorrhizae*, Czechoslovakia, pp. 25.

Delval, C., Barea, J. M. and Azcon-Aguilar, C. (1999). Assessing the tolerance to heavy metals of arbuscular mycorrhizal fungi isolated from sewage sludge-contaminated soils. Appl. Soil Ecol., II, 261–269.

Diaz, G., Azcdn-Aguilar, C. and Honrubia, M. (1996). Influence of arbuscular mycorrhiza on heavy metal (Zn and Pb) uptake and growth of *Lygedum spartum* and *Anthyllis cytisoides. Plant Soil.* 180: 241–249.

Dodd, J.C. and Thompson, B. D. (1994). The screening and selection of inoculant arbuscular mycorrhizal and ectomycorrhizal fungi. *Plant soil,* 159: 149–158.

Dudal, R and Purnell, M. F. (1986). *Reclamation and Revegetation Research* 5: 169, Ecodevelopment of *Usar* land at Banthra, T. N. Khoshoo (ed). In: *Proceedings of the Symposium on Inputs of Science and Technology for the Development of Wastelands,* M. Sharma, T. N. Khoshoo and U S Srivastava (eds). *Proceedings of National Academy of Sciences,* Allahabad, India, 17–32.

Dueck, T. A., Visser, P. and Ernst, W. H. O. (1986). Vesicular-arbuscular mycorrhizae decrease zinc toxicity to grasses growing in zinc-polluted soil. *Soil Biol. Biochem.* 18:331–333.

Elliot, H. A., Linn, J. H. and Shields, G. A. (1989). Haz. Waste Haz Mater., 6: 223–228.

Ernst, W. H. O., Dium, W. E. and Van Oolbekking, G .T. (1984). Vesicular-arbuscular mycorrhiza in dune vegetation. *Acta Botanica Neerlandica.* 33(2): 151–160.

Fabig, B. (1982). *Einfluss von Al und den Schwermetallen* Fe Mn Zn Cu, Pb, *and* Cd *auf die Effizienz. Dissertation,* Institut fur Tropischen und subtropischen Pflanzenbau, Gottingen, pp. 181.

Fitter, A. H. (1986). Functioning of VA mycorrhiza on growth and water relation of red clover, *New phytol.* 89: 599–608.

Gerdemann, J.W. and Trappe, J. M. (1974). The Endogonaceaes in the pacific northwest, mycologia Memoir No. 5: pp. 65.

Gildon, A. and Tinker, P. B. (1981). A heavy metal-tolerant strain of a mycorrhizal fungus. *New Phytol.* 95: 263–268.

Guo, Y., George, E. and Marschner, H. (1996). Contribution of an arbuscular mycorrhizal fungus to the uptake of cadmium and nickel in bean and maize plants. *Plant Soil.* 184:195–205.

Harley, J.L. and Smith, S.E. (1983). Mycorrhizal symbiosis. Academic press London, 295 pp..

Heggo, A., Angle, J. S. and Chaney, R. L. (1990). Effects of vesicular arbuscular mycorrhizal fungi on heavy-metal uptake by soybeans. *Soil Biol. Biochem.* 22: 865–869.

Jagpal, R., Sharma, P. D. and Mukerji, K. G. (1988). Distribution of VAM in relation to pollution, *Proceedings of the 1st Asian Conference on Mycorrhizae,* Madras, India, 109–215.

Jasper, D. A., Abbott, L. K. and Robson, A. D. (1989). Acacia respond to additioons of phosphorus and to inoculation with VA-mycorriza fungi in stock piled during mineral sand mining. *Plant and soil.* 115:99–208.

Joner, E. J. and Leyval, C. (2001). Time-course of heavy metal uptake in maize and clover as affected by root density and different mycorrhizal inoculation regimes. *Biol. Fertil. Soils.* 33: 351-357.

Joner, E. J., Leyval, C. and Briones, R. (2000). Metal binding capacity of arbuscular mycorrhizal mycelium. *Biol. Fertil. Soils.* 226: 227–234.

Khan, A. G. (1978). VA mycorrhiza in plants colonizing black wastes from bituminous coal mining in the Illainsara region of New Southwales. *New phytol.* 81:53–63.

Lakshman, H. C. (1992). Spore dispersal by earthworms Dungrollers and Indian Fowls Con. *Nature.* 105:79–84.

Lakshman, H. C. (1994). Occurance of VA-mycorrhizae in some tropical hydrolytic and xerophytic plants of Dharwad district. *Karnataka J. Nat. Con.* 6(1): 29–35.

Lakshman, H. C. (1995). Isolation of spores and quantification of AM fungi in plants growing in sewage polluted soil and industrial effluents. *Eco. Plan. & Env.* 113 : 71–78.

Lakshman, H. C. (1999). VA-mycorrhizal survey of plants species colonizing beach ponds of Karwar in south India. *Asian. J. Microbiol & Env. Sc.* 11(1–2): 15–21.

Lakshman, H. C. (2001). Screening of AMF fungi in mined spoils of Bisgod in Karnataka, India. - Possible way of revegetation. Nature Con. 79(1):104–109.

Lambert, D. H. and Cole, J.R. H. (1988). Effect of mycorrhizae on establishment and performance of forage species in mine spoil, *Agron. J.* 72:257–260.

Leyval, C., Turnnu, K. and Haselwandter, K. (1997). Effect of heavy metal pollution on mycorrhizal colonization and function: physiological, ecological and applied aspects. *Mycorrhiza*. 33 : 139–153.

Liao, J. P., Lin, X. G., Cao, Z. H., Shi, Y. Q. and Wong, M. H. (2003). Interactions between arbuscular mycorrhizae and heavy metals under sand culture experiment. *Chemosphere*. 50: 847–853.

Logan, V. S., Clark, P. J., Allaway, W. G. (1989). Mycorrhizae and root attributes of plants of coastal sand dunes of New South Wales. *Australian Journal of Plant Physiology*. 16: 141–146.

Loree, M. A. and Williams, S. E. (1987). Colonization of western wheat grass Agropyron Smithi Rydb byh vesicular-arbuscular mycorrhizal fungi during the revegetation of surface mine. *New phytol*. 106:735–744.

Loree, M.A.J. and Williman, S.S.E. (1984). Vesicular-arbuscular mycorrhizae and severe land disturbance In : VA-mycorrhial and Redamation of Arid and Semiarid land (Ed. By S.E. Willian and H.F. Allen) pp. 1–14 University of Wyoming Agriculture Experimental Station; SA 261.

Malcova, R., Vosatka, M. and Gryndler, M. (2003). Effects of inoculation /with Glomus intraradices on lead uptake by Zea mays L. and Agrostis capillaris L. *Appl. Soil Ecol*. 23:55–67.

Marschner, H. and Romheld, V. (1995). Strategies of plants for acquisition of iron. *Plant Soil*. 262–274.

Miller, R. M. (1985). Some occurrence of vesicular-arbuscular mycorrhiza in natural and disturbed ecosystems of the Red Desert. *Can. J. Bot*. 57: 610–623 (1979).

Mishra, A.K. (1992). Strategies for reclamation of mine waste areas through help of selected microbes: problems and prospects, In: *Proceedings of the symposium on Strategies for Ecosystem Conservation*, ISCA, Indore, pp. 64.

Mosse, B. (1963). Vesicular-arbuscular mycorrhizal "An extreme form of fungal adaptation (in) Symbiotic associations" (Eds.) (Natman, P. S. and Mosse, B.) Cambridge University Press, Cambridge. Pp. 146–170.

Nemec, S. and Meredith, F. J. (1981). Amino Acid content of leaves in mycorrhizal and non-mycorrhizal citrus root stocks. Annals Botany, (London). 47: 351–355.

Nicolson, T. H. (1967). VA mycorrhiza a universal symbiosis Sci. Prog. Oxf. 55:561–569.

Nye. and Tinker, P. B. (1977). Solute movement in the soil root system, university of California Press, Berkeley, California. pp. 258.

Ojala, J. C., Jarrell, W .M., Menge, J. A., Johnson, E. L. V. (1983). Influence of mycorrhizal fungi on the mineral nutrition and yield of onion in saline soil. *Agronomy Journal.* 75: 255–259.

Parker, R. (1994). Environmental Restoration Technologies. EMIAA Yearbook, pp. 169–171.

Pawlowska, T. B., Blaszkowski, J. and Ruhling, A. (1996). The mycorrhizal status of plants colonizing a calamine spoil mound in southern Poland. *Mycorrhiza.* 6: 499–505.

Pfeiffer, C. M. and Bloss, M. E. (1988). Growth and nutrition of guayule (*Parthenium aggentatum*) in saline soil as influenced by vesicular-arbuscular mycorrhiza and phosphorus fertilization. *New Phytologist.* 108(3): 351–321.

Powell, C. L. (1977). Mycorrhizal in hill country soils. I. Spore bearing mycorrhizal fungi in thirty-seven soils. *N.Z.J. Agric. Res.* 20:53–57.

Pradhan, S. K., Sarkar, S. K. and Prakash, S. (2000). Effect of domestic sewage water on germination and growth of sun hemp *Ctotalaria juncea* L. *India J. Env. & Eco. Plan.* 3(1): 107–110.

Puppi, G. and Riess, S. (1987). Role of ecology of VA-mycorrhizae in sand dunes. *Angewante Botanik.* 67: 115–126.

Rabtin, S. C. (1979). Season and edaphic variation in AMF infection of grasses by *Glomus tenuis*. *New phytol.* 83:95–102.

Read, D. J. (1982). The biology mycorrhiza in the Ericales Proc. 5th NACOM.

Ross, J.A., Pond, E., Menge, J. A., and Jarrell, W. M. (1985). Effect of salinity on mycorrhizal onion and tomato in soil with and without additional phosphate. *Plant and soil.* 88:307–319.

Rozma, J., Arp, W.,Diggelen, J.,Van., Esbroek, M. Van Borekman, R. Punte, H. (1986). Occurrence and ecological significance of VAM in the soil marsh environment. Acfa *Botanica Neerlandica.* 35(4): 457–467.

Sambandan, K., Kannan, K. and Raman, N. (1992). Distribution of vesicular-arbusculur mycorrhizal fungi in heavy metal polluted soils of Tamil Nadu. J. Environ. Biol., 13: 159–167. 52.

Saralabai, V. C. and Vivekannandan, M. (1995). Ability of endomycorrhizae to whist and extreme stress. Proceedings of the 3rd National Conference of Mycorrhizal. New Delhi., 13–15 March. Pp 522–525.

Schuepp, H. and Bodmer, M. (1991). Complex response of VA-mycorrhizae to xenobiotic substances. *Toxicological and Environmental Chemistry.* 30, 193–199.

Sieverding, E. (1981). Influence of soil water regimes on VA-mycorrihzae Z. Ackerund planzenban. 150:400–405.

Singh,S.(1996). Effect of atmospheric pollution on mycorrhizal. Mycorrhiza News 8(1) :1–7.

Smith, S. E. & Read, D. J. (1997). *Mycorrhizal Symbiosis,* 2nd edn. London: Academic Press. Pp. 484

Tal, M. Katz, A. Heikin,H. Dehan, K. (1979). Salt tolerance in the wild relatives of the cultivated tomato: proline accumulation in *Lycopersicon esculentum* Mill, L, Peruvianum Mill, and *Solanum pennelli* Cor. treated with NaCI and polyethylene glycol. *New Phytologist.* 82: 349–352.

Vivas, A., Marulanda, A., Gomez, M., Barea, J. M. and Azcon, R. (2003). Physiological characteristics (SDH and ALP activties) of arbuscular mycorrhizal colonization as affected by *Bacillus thurigniensis* inoculation under two phosphorous levels. *Soil Boi. Biochem.* 35: 987–996.

Williams, S. E., Wollum, A .G. and Aldon, E .F. (1974). Growth of *Atriplex canescens* (Pursh) Nutt. improved by formation of vesicular-arbuscular mycorrhizae. Proceedings of Soil Science America. 38: 962–965.

Williams, G. M. (1988). Integrated studies into ground water pollution by hazardous wastes. In land Disposal of Hazardous Waste. Engineering and Environmental Issues (eds Gronow, J. R., Schofield, A. N. and Jain, R. K.), Horwood Ltd, Chichester, UK. Pp. 684.

Zak, J. C. and Parkinson, D. (1982). Initial vesicular-arbuscular mycorrhizal development of slender wheat grass on two amended mine soils. *Can. J. Bot.* 60:224–248.

27

Status of AM Fungi in Coastal-Non Coastal Area and its Role in Reclamation Strategies A Case Study on Two Saline Tolerant Plants

T C Taranath, S G Hiremath and H C Lakshman

INTRODUCTION

Arbuscular mycorrhiza (AM) fungi are a major component of the soil microbial community, forming symbiotic association with roots of over 90% of the terrestrial plants. The presence of AM fungi has been reported from different habitats and from wide variety of plants (Smith and Read, 1997; Lakshman *et al.*, 2001). Occurrence of the AM fungi, associated with different host plants grown in various types of soils is well documented (Koske and Gemma, 1997; Gupta *et al.*, 2002). They are known to colonize plant species of maritime sand dunes (Koske, 1975, and 1987; Lakshman *et al.*, 2000). Many studies have gained significance since AM fungi aid in the establishment of plant species in coastal sand dune by improving the uptake of water and nutrients especially

Post Graduate Department Studies in Botany, Karnatak University, DHARWAD- 580003 INDIA phosphorus (Nicolson, 1960; Moun, 1994). They also enhance dune stability by increasing the aggregation of sand particles (Sutton and Sheppard, 1976; Koske and Potson, 1984; Titus and Moral, 1998). Although research on AM fungi of coastal regions of India is very scanty except that (Kulkarni *et al.*, 1997; Mohan Kumar *et al.*, 1988), who have reported the occurrence of AM fungi in saline soils. The objectives of the present study were to screen for the VA-mycorrhizal colonization and isolation of indigenous AM fungal spores, in the rhizosphere of plant species growing, in coastal sand dune system of west coast (Karwar) of India. Again, preliminary growth response studies were conducted and the results of laboratory experiments demonstrated that AM plants can be used to restore natural vegetation on coastal sandy soils for reclamation programmes.

MATERIALS AND METHODS

Study area

The present study was undertaken in coastal sand dunes located at Karwar North-South coastal beaches of Karnataka (Fig. 1). Geologically study area lies in between 14° 42′ - 15° 02′ North longitude and 74° 08′ - 74° 34′ East latitude during October 2006 (Post mansoon season) and the vegetation exist as patches in a streach on sand dunes. The vegetated dune area selected for the study is about 10 km^2 which comprises five important beaches i.e., Sunkeri beach, Binaga beach, Arga beach, Talamatti beach and Baitkol beach. It supports several herbaceous species. Cultivated *Lantana camera, Erythrina indica* and *Casturina equiselifolia* in the fence parallel to the stretch of dune vegetation. Occasionally, the oceanic waves reach. The dune vegetation facilitates the deposition of Detritus. The climate is tropical, atmospheric temperature range between 26–32°C. The average annual rain fall is 980 mm, of which 86% of rainfall occurs during June–September due to south west mansoon.

The present investigation was carried out to,

(*i*) Examine the AM fungal status of the pioneering plants growing in coastal area.

(*ii*) Quantitative and qualitative estimation of AM fungal spores.

(*iii*) Measurement of physical and chemical characteristics of coastal sand dunes.

(*iv*) Role of AM fungi in and cooling revegetation of strategies.

EDAPHIC FEATURES

Soils of this coast are highly saline, during the dry months, as observed in electrical conductivity values. This has resulted in the accumulation of available potassium and some of the micro-nutrients such as boron and manganese in the soils. The other aspect of beach soil is its relationship with sub-coastal soils, especially the occurrence of mineral deposits adjacent to beach areas. This situation is indeed harsh and responsible for periodic fluctuations in its physico-chemical parameters. Despite this harshness, however, there are plants growing on them successfully. Their adaptive features are the results of the micro environment. The sand occurs as a thin layer, one to three feet in thickness, below occurs fine grained brownish sand. About 12,000 tons of silica sand is produced annually from these areas.

Particle size distribution of the samples was determined by sieving the soil samples through sieves, immediately after the collection, soil moisture percentage was estimated by drying known amount of samples at 85°C to a constant weight. Sand pH was detected after dilution with distilled water (1:1 w/v) using glass electrode. Organic carbon was determined by Walkey and Black's rapid titration method (1934), where organic carbon was oxidized to carbon dioxide by potassium dichromate left over after oxidation was titrated against Ferrous ammonium sulphate with diphenylamine indicator. Sand samples were dissolved in neutral ammonium acetate, Calcium, Sodium, Potassium Chloride content were determined by following the procedure of (Jackson, 1973). To estimate the available Phosphorus of sand, Olsen's method was employed. Available Nitrogen was estimated titrimetically. Sand samples were mixed with alkaline potassium permanganate and distilled. The generated ammonia was absorbed by 2% boric acid, excess of which was titrated against 2.5% sodium hydroxide with methyl red indicator (Jackson¸1973).

COLONIZATION OF AM FUNGI

Forty-two plant species growing on dune system were selected for the study (Table 27.2). Each plant species were excavated carefully from 5–25 cm depth of rhizosphere, feeder roots were collected in sterile polythene bags and brought to the laboratory and stored at 4°C at room temperature. Roots of individual plant species were washed in water, cleared with 10% KOH, acidified with 2N HCl, and stained in 0.05% tryphan blue in lactophenol (Phillips and Hayman, 1970). Quantification of root colonization of the AM fungi was carried out using the slide method (Giovannetti and Mosse, 1980).

SPORE DENSITY OF AM FUNGI

AMF spores were recovered by modified floatation-adhesion technique (Sutton and Barrow, 1971). Air dried 20 g of each sand sample was placed in 100 ml of distilled water in a cylinder, shaken vigorously for 30 sec and then allowed to stand for 10 sec. The supernatant was collected in a separating funnel. This procedure was repeated four times and all the supernatant in a separating funnel was filtered through whatman No. 4 filter paper. Settled material on filter paper was scanned under a dissecting microscope (20X). Individual spores and clumps of spores were transferred from filter paper to microscope slides and mounted with polyvinyl alcohol lacto-glycerol (PVLG) mountant (Polyvinyl alcohol, 8.33g; Distilled water, 50 ml; lactic acid, 50 ml, glycerine, 5 ml). Identification of AMF spores was made according to (Mortan, 1988; Mortan, 1990; and Schenck and Perez; 1990). The analysis of rhizosphere sand samples (60 samples: five sub samples per host plant species) yielded a total of 902 AM fungal spores. The relative abundance was calculated as follows:

$$\frac{\text{No. of sand samples possessing AM fungal spores of particular species} \times 100}{\text{Total No. of sand samples}}$$

The frequency of occurrences was calculated as follows:

$$\frac{\text{No. of AM fungal spores of particular species recorded out of 60 sand samples} \times 100}{\text{Total No. of all AM fungal spores recorded out of 60 sand samples}}$$

Similarly sand samples from non-vegetated dune adjacent to vegetated dune were also analysed.

RESULTS AND DISCUSSION

The soil was predominantly composed of coarse to medium sand and low in available nutrients (Table 27.1). Analyses of physicochemical characteristics of samples of vegetated and non-vegetated dunes reveal that there is significant difference in the levels of phosphate (p: 0.05). All the forty two plants species screened for arbascular mycorrhizal colonization contained either arbuscules or vesicles or both (Table 27.2). The level of colonization ranged from 0.0 to 48.1% with an over all average of 26 per cent. The lower AMF colonization was recorded in Amaranthaceae, cyperaceae and plumbaginaceae. Higher percent colonization was from 18 to 48% in Rubiaceae, Anacardiaceae, Verbenaceae, Casuarinaceae, Fabaceae, Malvaceae, Convolulaceae, Compositceae and mycorrhizal colonizal frequency (41%) in Rubiaceae, Compositceae and Fabaceae, where as Amaranthaceae and Cyperaceae completely lacked AM fungal colonization throughout root system, however vesicular colonization (4.1 to 9.0%) was observed in these two plants. The least arbuscular colonization was recorded in Labiataceae and Cyperaceae. A significantly higher number of AM spores were recovered which range from 24 to 113 per 25 g dried soil. On the other hand, optimum number of spores from 71 to 122 AMF spores was recovered in Fabaceae, Anacardiaceae, Compositae, Laminaceae and Rubiaceae. The edaphic features of vegetative and non-vegetative dune (Table 27.3), showed slightly higher concentration of nutrients. The lower number of spores from 24 to 53 was recovered in *Ipomaea pescarre, Zizyphus jujuba, Jatropha grandulifera* and *Vitex nigunda*.

With expectation of *I.pes-capre* (11/25g sand) spores in rhizosphere of all 39 plant species contained AM fungal spores (Table 27.2). Spore abundance range from 33-104/25g sand and an overall average of 73%/25g sand. Maximum No. of spores encountered in *Andropogon annulatus* (104/25g sand.) Followed by *Plumbago zeylanica and Leucas aspera* (101-103/25g sand). The spore was least in *Jatropa curcas* and *Oxalis corniculata* (36-37/25g sand). Spores of 33 AM fungi were found in association with plants of non-vegetated sand dunes (Table 27.4). Members of family Glomaccac (33%) were dominated followed by acaulosporaceae (21%) Gigasporaceae and Sckrocystaceae (15%). The genus *Glomus* was recovered most frequently followed by *Acaulospora, Gigaspora, Sclerocystis, Scutellospora* and *Enterophosphora*. Spores of *Glomus microaggregalum, Glomus mosseae, Glomus fasciculatum, Glomus intraradices* and *Gigaspora aurigloba* were dominant among the spore communities. Spore densities and frequencies were very low in *Sclerocystis coccogena. Sclerocystis clavispora, Sclerocystis pakistanica* and *Acaulospora dilatata* (0.1-2.1%).

Spores of 30 AM fungi were found associated with plants of vegetated coastal sand dune area (Table 27.5). Members of the family scutellisporaceae (36%) were dominated followed by the family Glomaceae (27%). Gigasporaceae (12%), Acauiosporaceae and Sclerocystaceae (6%). The genus

Scutellospora recorded most frequently followed by the *Glomus, Gigaspora, Sclerocystis* and *Enterophosphora*. Spores of *Scutellospora erythropa, Scutellospora coralloida, Scutellospora pellucida, Glomus citricolum, Glomus gigantea* and *Acaulospora scrobiculata* were dominant among the spore communities. Spore densities and frequencies were very low in *Sclerocystis sinuosa, Sclerocystis pachycaulis* and *Enterophosphora species* (0.1–0.2%).

Table 27.1. Soil Texture of Sand Dunes of Karwar Coastal Beaches of South West India (n=6; sd; Range of Parentheses)

Size class	Grain size (mm)	Vegetated dune	Non-Vegetated dune
Gravel	> 2.1	1.08 + 0.22 (0.75 – 1.29)	0.57 + 0.24 (0.35 – 0.86)
Very Coarse sand	1.0 – 2.1	5.98 ± 0.14 (6.55 – 6.57)	1.13 ± 0.10 (0.96 – 1.21)
Coarse sand	0.5 – 1.0	32.36 ± 5.35 (25.82 – 10.39)	36.98 ± 1.48 (35.51 – 38.33)
Medium sand	0.25 – 0.5	54.73 ± 2.88 (99.31 – 56.47)	51.04 ± 5.62 (43.01 – 56.17)
Fine sand	0.13 – 0.25	0.48 ± 2.46 (2.03 – 9.72)	80.1 ± 1.22 (6.52 – 10.20)
Very fine sand	0.06 – 0.14	0.35 ± 0.05 (0.27 – 0.43)	0.38 ± 0.30 (0.10 – 0.76)

Table 27.2. AM Fungal Colonization and Spore Number in Plants Growing in Coastal Sand Dune Beaches of South Western India

Sr. No.	Plant species	Habit	Root colonization			Spore No./25g air dried soil
			HC	VC	AC	
1.	*Acacia Arabica* Willd	T	48.1	18.4	14.3	82
2.	*Achyranthus aspera* L.	H	0.0	2.3	12.5	61
3.	*Acanthus ilicifolius* L.	S	20.7	11.6	18.0	53
4.	*Aleuropus lagopoides* Trin.	M.F.C.	22.4	15.3	31.5	81
5.	*Anacardium occidentale* L.	T	21.2	19.4	31.7	72
6.	*Andrographis paniculata* Nees	H	30.3	15.1	19.2	94

7.	*Andropogon annulatus* Forsk.	H	24.6	13.2	20.2	104
8.	*Amarantus oleraceus* Roxb.	H	0.00	1.4	7.3	89
9.	*Avicinnia officinalis* L.	T	18.0	0.0	11.8	79
10.	*Boerhaavia diffusa* L.	H	14.3	12.2	23.5	98
11.	*Calotropis gigaantea* R.Br.	S	13.1	12.0	26.4	74
12.	*Crotalaria juncea* L.	H.	15.2	15.1	20.6	48
13.	*Casurina equisetifolia* Forst	T	12.3	22.1	14.7	86
14.	*Clerodendron inerme* Gacrtn.	S	26.2	20.0	11.2	92
15.	*Clitorea ternatea* L.	C	24.5	14.2	12.1	87
16.	*Cynodon dactylon* Pers.	H	13.2	11.8	20.5	77
17.	*Cyperus bulbosus* Vahl.	H	0.0	0.0	11.2	83
18.	*Datua fastuosa* Linn.	H	12.2	11.3	14.7	68
19.	*Derris paniculata* Bak.	T	14.2	10.3	14.2	58
20.	*Erythrina indica* Lam.	T	14.1	11.2	10.4	71
21.	*Euphorbia rosea* Retz.	H	14.2	9.6	16.0	49
22.	*Hydrophylax maritima* L.f.	H	10.8	16.2	14.5	62
23.	*Indigofera aspalathoides* Vahl	M.F.C.	11.6	14.9	11.2	52
24.	*Ipomaea pes-capre* Sweet.	M.F.C.	34.5	21.6	13.7	24
25.	*Ixora elongata Hyene.*	S	26.3	12.5	18.2	69
26.	*Jatropha curcas* L.	S	13.8	11.3	14.1	36.2
27.	*Jatropha Glandulifera* Roxb.	S	19.3	20.7	15.4	42.0
28.	*Leucas aspera* Spr.	H	28.2	21.3	11.3	113.2
29.	*Mangiferas indica* L.	T	11.7	14.5	12.2	52.3
30.	*Mimosa pudica* L.	H	31.3	20.5	14.2	101.0
31.	*Myristica malabarica* Lam.	T	24.9	8.6	16.1	82.0
32.	*Oxalis corniculata* L.	H	23.1	11.5	11.8	37
33.	*Oldenlandia aspera* (Retz) DC.	H	26.2	21.2	16.3	79
34.	*Paspalum vaginatum* Sw.	M.F.C.	14.6	19.9	15.4	92.2
35.	*Perotis indica* O. ktz.	M.F.C.	15.2	24.5	16.6	74.3
36.	*Plumbago zeylanica* Linn	H	11.3	7.8	16.5	109.0
37.	*Polycarpaea coryymbosa* Lam	H	18.2	22.5	16.8	122
38.	*Sida acuta* Burm.	H	21.5	24.2	18.1	96.5
39.	*Spinifex squarrosus* L.	M.F.C.	30.5	26.3	22.8	91
40.	*Sporobolus virginicus* Kunth.	M.F.C.	11.0	11.1	21.4	53
41.	*Vitex trifolia* L.f.	S	40.2	15.1	30.0	49
42.	*Zizyphus jujube* Lam.	T	11.6	14.3	22.1	33

T=tree, S=Shrub, H=herb, C=climber and M.F.C. = Mat forming climber.

Table 27.3. : Edaphic Characteristics of the Sand Dune of Karwar Coastal Beaches of South Western India (n–6; sd; Range in Parentheses)

Characteristics	Vegetated dune	Non vegetated dune
Moisture***	1.72 ± 0.16 (1.51 – 1.92)	1.36 ± 0.17 (1.18 – 1.55)
pH **	6.71 ± 0.02 (6.66 – 6.68)	7.14 ± 0.02 (7.11 – 7.16)
Organic carbon %	0.04 ± 0.01 (0.03 – 0.04)	0.02 ± 0.01 (0.01 – 0.03)
Nitrogen µg/g***	6.56 ± 1.05 (6.45 – 6.73)	1.47 ± 0.20 (1.28 – 1.81)
Phosphate µg/g**	71 ± 8.92 (61 – 80)	57 ± 5.70 (50 – 70)
Potassium µg/g***	28.11 ± 0.10 (27.00 – 27.46)	8.75 ± 0.43 (8.56 – 9.48)
Calcium µg/g*	14.68 ± 1.55 (12.23 – 15.30)	16.40 ± 1.00 (14.29 – 16.31)
Sodium µg/g***	8.75 ± 0.14 (8.59 – 9.47)	28.14 ± 0.22 (27.01 – 27.46)
Chloride µg/g*	802.44 ± 36.30 (707 – 891)	810.70 ± 36.61 (791 – 892)

Table 27.4. AM Fungal Spores Recorded from Sand Dune Vegetation at Karwar Coastal Beaches of South West India

Sl. No	AMF spore species	Relative abundance (%)	Frequency
1.	*Acaulospora scrobiculata* Trappe	36.2	13.7
2.	*Acaulospora undulata* Sieverding	34.1	18.3
3.	*Entrophospora schenckii* Sieverding & Toro	1.2	0.2
4.	*Gigaspora albida* Schenck & Smith	23.3	18.5
5.	*Gigaspora decipiens* Hall & Abbott	32.5	21.4
6.	*Gigaspora gigantean* (Nicol. & Gerd.) Gerdermann & Trappe.	37.4	26.3
7.	*Gigaspora ramisporophora* Spain, Sieverding & Schenck	31.7	30.2

8.	*Glomus ambisporum* Smith & Schenck	23.1	16.5
9.	*Glomus botryoides* Rothwell & Victor	32.7	23.7
10.	*Glomus citricolum* Tang & Zang.	24.8	14.6
11.	*Glomus convolatum* Gerdermann.	41.3	11.3
12.	*Glomus fasciculaturm* (Thaxter) Gerd.&Trappe emend, Walk & Koske	31.9	21.2
13.	*Glomus gerdemannii* Rose, Daniels & Trappe	32.9	12.2
14.	*Glomus macrocarpum* Tul & Tul.	41.5	14.5
15.	*Glomus maculosum* Miller & Walker.	26.3	16.6
16.	*Glomus magnicaule* Hall.	18.2	20.4
17.	*Sclerocystis pachycaulis* Wu & Chen.	2.1	0.1
18.	*Sclerocystis sinuosa* Gerdemann & Bakshi	2.3	0.1
19.	*Scutellispora arenicola* Koske & Halvorson	36.8	22.4
20.	*Scutellispora biornata* Spain, Sieverding & Toro	29.9	19.3
21.	*Scutellispora biornata* Spain, Sieverding & Toro	26.1	14.8
22.	*Scutellispora calospora* (Nicol. & Gerd.) Walker & Sanders	40.2	12.5
23.	*Scutellispora coralloides* (Trappe, Gerd & Ho) Walk & Sanders	22.5	28.3
24.	*Scutellispora dipapillosa* (Walker & Koske) Walker & Sanders	27.3	22.0
25.	*Scutellispora dipurpurescens* Mortan & Koske	29.4	19.0
26.	*Scutellispora erythropa* (Koske & Walker) Walker & Sanders.	42.6	36.1
27.	*Scutellispora gregaria* (Schenck & Nicol.) Walker & Sanders	31.7	29.4
28.	*Scutellispora pellucida* (Nicol. & Schenck) Walker & Sanders	18.3	27.2
29.	*Scutellispora persica* (Koske & Walker) Walk & Sanders	25.4	26.5
30.	*Scutellispora scutata* Walker & Dicdcrichs.	41.5	28.1
31.	*Scutellispora verrucosa* (Koske & Walker) Walker & Sanders.		

Table 27.5. AM Fungal Spores Population Associated with Coastal Sand Dune Non-Vegetated at Karwar Coastal Beaches of South Western India

Sl. No	AMF spore species	Relative abundance (%)	Frequency
1.	*Acaulospora bireticulata* Rothwell & Trappe	18.3	8.9
2.	*Acaulospora delicata* Walker & Pfeiffer & Bloss	28.4	15.2
3.	*Acaulospora dilatata* Mortan.	16.1	5.4
4.	*Acaulospora laevis* Gerd. & Trappe.	15.4	7.3
5.	*Acaulospora myriocarpa* Spain Sieverding & Schenck	13.2	6.3
6.	*Acaulospora rehmii* Sieverding & Toro.	11.5	8.2
7.	*Acaulospora scrobiculata* Trappe.	41.5	18.2
8.	*Entrophospora colombiana* Spain & Schenck.	5.3	1.9
9.	*Entrophospora schenckii* Sieverding & Toro.	7.2	2.1
10.	*Gigaspora aurigloba*	11.8	6.2
11.	*Gigaspora decipiens* Hall & Abbott.	14.6	5.5
12.	*Gigaspora gigantean* (Nicol. & Gerd.) Gerdemann & Trappe	17.3	10.2
13.	*Gigaspora margarita* Beacker & Hall	19.4	7.2
14.	*Gigaspora sensulata*	13.8	9.2
15.	*Glomus bakshiensis*	32.2	8.7
16.	*Glomus constrictum* Trappe.	48.3	23.8
17.	*Glomus fasciculatum* (Thaxter) Gerd. & Trappe emend. Walker & Koske.	32.3	13.8
18.	*Glomus fragile* Berk & Brome.	8.5	13.5
19.	*Glomus globiferum* Koske & Walker.	35.5	11.2
20.	*Glomus intraradices* Schenck & Smith.	38.4	19.2
21.	*Glomus maculosum* Miller & Walker.	32.8	14.3
22.	*Glomus microagggregatum* Koske, Gemma & Olexia	35.6	21.2
23.	*Glomus microcrapum* Tul & Tul	31.5	19.3
24.	*Glomus mosseae* Gerd & Trappe	42.5	18.5

25.	*Glomus multicaule* Gerdemann & Bakshi	36.6	17.4
26.	*Sclerocystis clavispora* Trappe	4.2	0.1
27.	*Sclerocystis coccogena* (Pat.) V. Hohn.	3.1	0.1
28.	*Sclerocystis pakistanica* Iqbal & Bushra	3.1	0.1
29.	*Sclerocystis rubiformis* Gerdemann & Trappe	2.2	0.2
30.	*Sclerocystis sinuosa* Gerdemann & Bakshi	3.1	0.2

However, it has been emphasized in this paper, much more needs to be known about the micro-morphological, Physiological and genetic variability in VAM at the different levels of their organizations (Intra and inter specific population ecosystem). Conservation and efficient utilization of their bio-diversity are of crucial importance for sustainable production system for the equilibrium of natural plant community structure (Walker, 1995).

It is widely accepted that AM fungi contribute significantly to plant growth and survival (Koske, 1975; Allen, 1991; Lakshman, 1994). Any stress imposed on the ecosystem affect the plant and inevitably influences the symbiotic relationship. Kulkarni *et al.* (1997) AM fungi in 12 plant species along south west coast (Mangalore of India). All plant species had AM fungal spores in the rhizosphere and spores of nine species of AM fungi were recovered. Of this *Glomus clarum* and *Glomus pustulatum* were found common. But in the present study, *Glomus citricola, Glomus microcarpum* and *Glomus fasciculatum* were common. And there were more predominated *Scutellispora pellucida, Scutellispora, coralloidea, Scutellospora verrucosa* and *Scutellospora dipapillosa* were isolated, in addition to *Gigaspora ramisporophora* and *Gigaspora gigardia.* Despite of repeated examination, *Oldenlandia aspera* (Roth.) DC. (Rubiaccae) have not shown AM fungal root colonization (Mohan kumar *et al.*, 1988; Sridhar, 2006), however in the present study 53% of the roots of *Oldenlandia aspera* (Roth.) DC. had been colonized. Normally the members of Caryophyllaecae, Cyperaccae are not colonized by AM fungi (Gerdemann, 1968: Louis. 1990; Beena *et al.*, 2001a), but in the present study *Potycarpea corymbosa* (Caryophyllaceae) had an AM fungal root colonization of 55% and a spore number in rhizosphere of 122/50g. soil, however the root AM fungal colonization of *Achyranthus aspera* (Amarantaceae), *Cyperus hulbosus* had least colonization of arbuscules and hyphae. These findings are in consistent with early workers. Generally, production of AM fungi is maximum at the end of growing season in many plants (Koske and Potson, 1984). Though the roots of *Ipomoea pes-capre* Sweet. had 53% of colonization of AM fungi, rhizosphere contain 24 spores/50 g soil. This in contrast to Koske and Potson (1984) work did not recover any AM fungal spores. This plant had maximum relative cover (60%) at flowering stage in September 2004. During subsequent months in a seasonal study the plant cover decreased drastically with seed setting and several AM fungal spores were recovered (Lakshman, H.C. and Hiremath, S.G., Unpublished observation). The number of arbuscules (21.6) and vesicles (34.5) in *Ipomoea pes-capre* in September 2004 was almost equal (18–22 cm^{-1} root). The vesicles were maximum when compared to other plant species on the coastal sand dune. It is interesting to study the extent of arbuscules, hyphae and vesicles. When *Ipomoea pes-capre* produce AM fungal spores in its rhizosphere.

Presence of AM fungal spores in non-vegetated coastal sand dune suggests that they are dispersed by biotic or abiotic factors (Lakshman *et al.*, 2000). Many plants sets seeds in September along with flowers, their dispersal usually take place by wind and water waves. The dispersal of AM fungal spores along with seeds may have definite advantage in establishment of plants as well as AM fungi. Besides these factors, dispersal of spores might be dependent on distribution of roots. *Ipomoea pes-capre* produces roots at nodal regions, *Aeluropus lagopoides, Indigofera aspalatnoides, Paspalum vaginatum, Protis indica, Spinifex squarrosus* and *Sporobohis*

Virginicus are mat forming climbers has deep roots in addition to horizontal spreading. Members of Poaceae were also common in different coastal sand dunes of study area and they possess horizontal rhizomes and smaller biomass of tibrous roots. Such plants might function as spreading zones of AM fungi in these sand dunes.

Koske and Gemma (1997) found no correlation between frequency of occurrence and relative abundance of AM fungal spores on sand dunes of Hawaii. They concluded that observed difference in frequency of occurrence and spore abundance of AM Fungal species was due to unequal spatial distribution. In this study a strong correlation (r=0.89; < 0.005) was found between overall frequency of occurrence and relative abundance of AM fungal spores suggesting possible uniform distribution of spores in vegetated coastal sand dune. These results are in consistent with the findings of (Shridhar, 2006) on coastal sand dune legumes of South West coast of India.

Louis (1990) studied the distribution of AM Fungi along the sand dunes of the coastal area reclaimed in Singapore and did not find significant correlation between soil nutrients (N, P and K) or other soil properties and abundance of AM spores. In the present study significant difference of physico-chemical features of vegetated and non-vegetated dune might be due to the influence of plants as well as AM fungi. The mycorrhizal roots of thirty eight plants (Table 27.1) were however, heavily colonized by internal mycelium (Hyphae, Vesicles and Arbuscles) and it appears that the close association between mycorrhizal roots of the various plant species in coastal area ensures a rapid establishment of mycorrhizal species and elimination of non-mycorrhizal areas. There are evidences for mycorrhizal fungal hyphae connection between roots of separate plants that allow for nutrient transfer and rapid and extensive infection and colonization (John *et al.*, 1983). In this nutrient limited ecosystem, mycorrhizai roots were colonized and the interconnecting mycorrhizal fungal network may be most effective method for nutrient transfer and spreading of rnycorrhizal infection. The lower available phosphorus status in the current study m vegetated dune ($P = 0.05$) might had been influenced by increased plant uptake. Similarly significant high level of sodium and potassium ($P<0.001$) in the vegetated coastal dune might be influenced by accumulation by plants and fungi in the rhizosphere. The higher nitrogen level ($P<0.001$) in vegetated coastal dune is attributed to increased synthesis of nitrogenous compounds and decomposition of detritus in the rhizosphere. The diversity of glomalean population has mainly been studied in coastal sand dunes (Giovannetti 1985, Lakshman *et al.*, 2000). The activity of glomalean AM fungi under experimental conditions is affected by many adaphic factors, of which pH plays a very important role (Beena *et al.*, 2001a). The relationship between AM Fungi, host plants and adaphic factors in coastal sand dunes needs further study.

Soil particles by hyphae; b) roots with hyphae create conditions suitable to form microaggregates; c) roots and hyphae enmesh and bind the microaggregates to large macroaggregates (Beena *et al.*, 2001b). Besides fungal hyphae, polysaccharides produced by the hyphae firmly bind the microaggregates (Tisdall, 1991). One-gram stable rnacroaggregate is known to have about 50 m hyphae (Tisdall, 1991). *Phcseolus vulgaris* showed 10 g/kg aggregates without AM fungal colonization, while colonization *of Glomus* resulted in 54 g/kg (Sutton and Sheppard, 1976). Aggregates of sand dunes are known to remain intact even after the death of roots or hyphae (Koske and Poison, 1984). The AM fungi (Acaulosporaceae and Glomaceae) are known to produce extracellular polysaccharide. □ 1–3 glucan, which serves the soil aggregation (Koske and Gemma, 1990). A stable glycoprotein, glomalin produced by *Gigaspora gigantea* is responsible for soil aggregation (Wright and Upadhyaya, 1998).

In summary, coastal sand dunes (CSDs) constitute major biomes of considerable ecological and economic importance. Members of plant family Poaceae are dominant plant species in temperate dunes, while Asteraceae, Convolvulaceae, Fabaceae and Poaceae in tropical dunes. Stress-tolerant plant species and associated microbes of CSDs need special attention for stabilization. Below ground diversity of mycorrhiza is one of the major factors governing the plant distribution, diversity and functioning of CSD ecosystem. Elimination of mycorrhizal fungi decreases the plant equilibrium as well as productivity, which result in destabilization. Future studies should concentrate on the importance of mycorrhizae on CSD stabilization and application of stress-tolerant mycorrhizae in agriculture.

MATERIAL AND METHODS FOR TWO SALINE TOLERANT PLANTS

Earthen pots of 15 cm diameter containing 3 kg potting mixture are used to study the growth responses in *Erythrina indica* Lam. and *Jatropha curcas* 25 cm diameter earthen pots containing 9 kg potting mixture were used. In either case, pots were amended with 2 parts of carrier material, 1 part of garden soil and one part of sand. Soil used for the experiment was analysed for its physico-chemical properties by the method outlined by Jackson (1973).

Collection of seeds:

Seeds of *Solarium nigram* L. and *Erythrina indica* Lam. were collected from plants growing in botanical garden, Karnatak University, Dharwad. *Jatropha curcas* L. seeds were collected from forest department branch office, Dharwad. Before sowing, seeds were surface sterilized with 0.02% sodium hypochlorite and their after washed 4–5 times with sterile distilled water.

MYCORRHIZAL INOCULUM PRODUCTION

The AM species [*Scutellospora erythropa* (Koske & Walker) Walker & Sanders], was mass multiplied following the procedures of Sreenivasa and Bagyaraj (1988) using Foxtail millet, onion or *Sorghum vulgare* as host plant (Plate VI). Hoagland plant nutrient solution minus phosphorus was given to host plants between intervals of 15 days. Sterile distilled water was used to iriigate host plants every alternate day in order to maintain moisture.

INOCULATION OF AM FUNGI

15 g of air dried AM inoculum was given to each pot as thin layer, 2 cm below the soil surface (except uninoculated control). The inoculum consisted of 8 g rhizospheric soil (100 chlamydospores/ 50 g soil approximately) and 2 g of root bits of host plant with hyphae and sporocarps.

Treatments

Six treatments ware set in triplicate on four selected experimental plants in polyhouse conditions are as follows :

(A) Uninoculated control (UIC)

(B) Mycorrhiza (*Scutellospora erythropa*) inoculated

(C) Mycorrhiza (*Scutellospora erythropa*) + Leaf compost

(D) Mycorrhiza (*Scutellospora erythropa*) + Charcoal

(E) Mycorrhiza (*Scutellospora erythropa*) + Perlite

(F) Mycorrhiza (*Scutellospora erythropa*) + Peat

Experimental pots were kept free of weeds, pests and irrigated every alternate day in order to maintain moisture. The plants were harvested after 30, 60 and 90 days interval after sowing. Observations such as Shoot length (SL), Fresh weight of shoot (FWS), Dry weight of shoot (DWS), Root length (RL), Fresh weight of root (FWR), Dry weight of root (DWR), Number of leaves (NL), Number of floral buds (NB), Number of flowers (NFL), Number of fruits (NFR) were recorded. The spore count was carried out by wet sieving and decanting method (Gerdemann and Nicolson, 1963). Percent of root length colonization is estimated according to Phillips and Hayman (1970). Percent of root colonization is calculated by the following formula:

Percent of root colonization=Number of root bits showing the colonization X 100/Total number of root bits observed. After the harvest, experimental plants (Shoot and root) are oven dried at 70°C until a constant weight is obtained to determine the dry weight. The uptake of P is estimated according to Jackson (1973).

Table 27.1: Showing the Phyisco-Chemical Properties of the Soil Used for the Pot Experiment

Soil parameters	Potting mixture Garden soil: sand (1:1) Parameter values n=2±SD
Soil type	Sandy loamy
Soil moisture (%)	28.25 ± 0.01
pH	6.64 ± 0.03
Electric conductivity	0.523 ± 0.01
Organic carbon (%)	0.480 ± 0.01
Zinc (%)	2.17 ± 0.01
Copper(%)	1.10 ± 0.01
Magnesium (%)	1.24 ± 0.04
Nitrogen (%)	1.26 ± 0.02
Phosphate (%)	0.25 ± 0.02
Potassium (%)	2.46 ± 0.01

Table 27.2. Showing the Effect of *Scutellospora Erythropa* (Koske & Walker) Walker & Sanders and Four Carrier Materials on Growth of *Erythrina indica* Lam. for 90 Days

Treatments	Shoot length (cm)	Fresh weight of shoot (g)	Dry weight of shoot (g)	Root length (cm)	Fresh weight of root (g)	Dry weight of root (g)	Root colonization (%)	Number of leaves	Number of floral buds	Number of flowers	Number of fruits	Spores/ 50g soil	'P' uptake in shoot (%)
							30 days						
Control	15.70 f	3.21 f	0.77 f	8.36 f	0.48 f	0.19 f	00	4.00 c	0	0	0	00	0.08 e
AMF	16.00 e	5.10 e	1.73 e	9.50 e	0.64 e	0.27 e	15.85 e	5.66 ab	0	0	0	16.33 d	0.11 d
AMF+LC	25.30 a	12.22 a	2.03 a	16.30 a	1.24 a	0.38 a	25.24 d	7.00 a	0	0	0	18.66 c	0.12 c
AMF+Ch	17.63 d	8.41 d	1.76 d	10.03 d	0.65 d	0.29 d	65.43 a	5.66 ab	0	0	0	30.00 a	0.16 a
AMF+Per.	21.40 b	11.96 b	2.00 b	14.23 b	0.99 b	0.33 b	42.16 b	7.00 a	0	0	0	24.33 b	0.14 b
AMF+Peat	20.20 c	10.21 c	1.87 c	11.40 c	0.96 c	0.31 c	32.32 c	5.33 b	0	0	0	20.00 c	0.14 b
							60 days						
Control	15.06 c	6.86 f	0.89 f	11.30 f	0.80 f	0.38 f	00	7.66 d	0	0	0	00	0.10 e
AMF	23.40 b	11.61 e	3.68 e	13.20 e	0.85 e	0.44 e	20.42 e	8.66 d	0	0	0	24.00 e	0.12 d
AMF+LC	40.20 a	65.41 a	10.11 a	36.46 a	7.18 a	2.11 a	32.43 d	14.00 a	0	0	0	28.33 d	0.15 c
AMF+Ch	32.40 a	22.41 d	3.89 d	15.70 d	1.75 d	0.91 d	69.55 a	10.33 d	0	0	0	53.00 a	0.21 a
AMF+Per.	38.50 a	30.41 b	5.96 b	23.66 b	5.60 b	1.80 b	48.26 b	15.00 a	0	0	0	46.33 b	0.17 b
AMF+Peat	35.43 a	23.48 c	5.41 c	18.50 c	5.43 c	1.10 c	35.93 c	12.00 b	0	0	0	32.66 c	0.16 bc
							90 days						
Control	30.33 f	13.51 f	1.10 f	12.10 f	0.99 f	0.49 f	00	11.00 c	0	0	0	00	0.12 e
AMF	34.60 e	55.65 e	6.65 e	12.66 e	1.04 e	0.68 e	25.45 e	13.33 b	0	0	0	31.00 e	0.18 d
AMF+LC	69.13 a	101.10 a	24.41 a	47.46 a	14.42 a	3.61 a	50.94 d	18.00 a	0	0	0	36.33 d	0.19 d
AMF+Ch	48.80 d	69.77 d	12.10 d	21.66 d	2.36 d	2.22 d	99.82 a	13.00 b	0	0	0	66.33 a	0.25 a
AMF+Per.	63.30 b	95.20 b	22.84 b	40.30 b	10.11 b	2.91 b	78.74 h	14.00 b	0	0	0	57.66 b	0.23 b
AMF+Peat	50.40 c	90.56 c	18.41 c	23.20 c	8.45 c	2.63 c	60.15 c	13.00 b	0	0	0	43.00 c	0.21 c

Control=uninoculated control, AMF=Arbuscular mycorrhizal fungi (*Scutellospora erythropa*), AMF+LC=Arbuscular mycorrhizal fungi (*Scutellospora erythropa*)+leaf compost, AMF+Ch=Arbuscular mycorrhizal fungi (*Scutellospora erythropa*)+Charcoal, AMF+Per.=Arbuscular mycorrhizal fungi (*Scutellospora erythropa*)+Perlite, AMF+Peat=Arbuscular mycorrizal fungi (*Scutellospora erythropa*)+Peat.

* Mean values followed by the same letter within a column do not differ significantly at P=0.05 according to DMRT

Table 27.3.: Showing the Effect of *Scutellospora erythropa* (Koske & Walker) Walker & Sanders and Four Carrier Materials on Growth of *Jatropha curcas* L. for 90 Days.

Treatments	Shoot length (cm)	Fresh weight of shoot (g)	Dry weight of shoot (g)	Root length (cm)	Fresh weight of root (g)	Dry weight of root (g)	Root colonization (%)	Number of leaves	Number of floral buds	Number of flowers	Number of fruits	Spores/ 50g soil	'P' uptake in shoot (%)
							30 days						
Control	14.33 f	1.30 f	0.36 f	2.80 f	0.36 f	0.04 f	00	4.00 b	0	0	0	00	0.05 e
AMF	23.36 c	18.63 c	2.96 c	15.50 b	1.18 c	0.26 c	38.76 a	5.33 a	0	0	0	27.00 a	0.16 a
AMF+LC	26.36 a	26.51 a	4.04 a	17.33 a	1.28 a	0.29 a	25.15 c	5.00 ab	0	0	0	19.66 c	0.14 b
AMF+Ch	20.20 d	17.46 d	2.69 d	11.83 c	0.94 d	0.23 d	15.23 e	5.00 ab	0	0	0	14.66 d	0.13 b
AMF+Per.	25.20 b	21.30 b	3.33 b	17.50 a	1.26 b	0.27 b	15.85 d	5.33 a	0	0	0	19.66 e	0.14 b
AMF+Peat	19.56 e	11.33 e	2.17 e	11.50 d	0.76 e	0.15 e	29.43 b	4.33 ab	0	0	0	23.33 b	0.15 b
							60 days						
Control	14.60 e	2.50 f	0.42 f	4.76 f	0.85 f	0.18 f	00	5.00 ab	0	0	0	00	0.07 d
AMF	24.50 c	28.46 c	6.18 c	17.30 c	3.34 c	0.49 c	58.45 a	4.33 b	0	0	0	36.33 a	0.18 a
AMF+LC	26.50 a	42.18 a	8.77 a	22.50 a	3.96 a	0.67 a	31.44 c	5.66 a	0	0	0	30.33 d	0.16 b
AMF+Ch	20.50 d	26.41 d	3.67 d	15.60 d	1.60 d	0.41 d	20.78 d	4.33 b	0	0	0	24.00 a	0.15 b
AMF+Per.	25.50 a	40.14 b	7.33 b	17.83 b	3.47 b	0.58 b	20.74 e	5.00 ab	0	0	0	22.00 c	0.13 c
AMF+Peat	20.50 d	11.86 e	2.36 e	15.40 d	1.30 e	0.36 e	38.67 b	5.00 ab	0	0	0	31.00 b	0.18 a
							90 days						
Control	14.86 e	2.81 f	0.51 f	10.30 e	1.46 f	0.32 f	00	4.00 b	0	0	0	00	0.09 e
AMF	25.23 c	35.56 c	7.38 c	19.50 c	4.70 c	0.80 c	97.45 a	5.00 b	0	0	0	37.33 a	0.24 a
AMF+LC	26.80 a	75.40 a	10.40 a	29.53 a	7.44 a	0.98 a	50.15 c	8.00 a	0	0	0	32.00 b	0.18 c
AMF+Ch	21.53 d	28.67 d	5.55 d	18.73 d	2.41 d	0.68 d	25.18 e	4.00 b	0	0	0	27.33 a	0.16 d
AMF+Per.	26.40 b	75.25 b	9.06 b	24.46 b	5.91 b	0.81 b	25.26 d	7.30 a	0	0	0	25.00 c	0.17 c
AMF+Peat	21.30 d	12.11 e	2.85 e	18.63 d	2.40 e	0.61 e	59.65 b	4.00 b	0	0	0	34.33 b	0.20 b

Control=uninoculated control, AMF=Arbuscular mycorrhizal fungi (*Scutellospora erythropa*), AMF+LC=Arbuscular mycorrhizal fungi (*Scutellospora erythropa*)+leaf compost, AMF+Ch=Arbuscular mycorrhizal fungi (*Scutellospora erythropa*)+Charcoal, AMF+Per.=Arbuscular mycorrhizal fungi (*Scutellospora erythropa*)+Perlite, AMF+Peat=Arbuscular mycorrizal fungi (*Scutellospora erythropa*)+Peat.

* Mean values followed by the same letter within a column do not differ significantly at P=0.05 according to DMRT

RESULTS AND DISCUSSION

Growth response studies in *Erythrina indica* Lam

In general there was a significant increased growth in plants treated with mycorrhiza (*Scutellospora erythropa*) + leaf compost (shown in table 33 and plate VII). After 30 days of sowing plant showed increased (25.30cm) Shoot length, (12.22g) Fresh weight of shoot, (2.03g) Dry weight of shoot, (16.30cm) Root length, (1.24g) Fresh weight of root, (0.38g) Dry weight of root and (7.00) number of leaves. All these parameter values were found dominating over other treatments like Mycorrhiza + Perlite, Mycorrhiza + Peat, Mycorrhiza + Charcoal, only mycorrhiza and UIC respectively. Highest 65.43% root colonization, 30 AMF spores/50g soil and 0.16% of 'P' was recorded in plants treated with mycorrhiza + Charcoal which is followed by Mycorrhiza + Perlite, Mycorrhiza + Peat, Mycorrhiza + leaf compost and only Mycorrhiza. Percent of root colonization and spore number/50g soil was not recorded in UIC plants.

It was after 60 days, significant increased in growth was observed in plants treated with mycorrhiza + leaf compost and showed increased (40.20 cm) shoot length, (65.41 g) Fresh weight of shoot, (10.11 g) Dry weight of shoot, (36.46 cm) Root length, (7.18 g) Fresh weight of root, (2.11 g) Dry weight of root and (14.00) number of leaves. This treatment found dominating on other treatments like Mycorrhiza + Perlite, Mycorrhiza + Peat, Mycorrhiza + Charcoal, only mycorrhiza and UIC respectively. Higher of 69.55% root colonization, Highest of 53 AMF spores / 50 soil and 0.21% of 'P' was recorded in plants treated with mycorrhiza + Charcoal which are followed by Mycorrhiza + Perlite, Mycorrhiza + Peat, Mycorrhiza + leaf compost and only Mycorrhiza. No percent of root colonization and no spores per 50 g soil were recorded in UIC plants.

It was after 90 days, plants growth was optimum and continued to follow the same trend that was observed after 30 and 60 days. Plants treated with mycorrhiza + leaf compost showed increased (69.13cm) shoot length, (101. l0g) Fresh weight of shoot, (24-41g) Dry weight of shoot, (47.46cm) Root length, (14.42g) Fresh weight of root, (3.61g) Dry weight of root and (18.00) number of leaves. They were found dominating all other treatments with respect to parameters considered above. Highest of 99.82 % root colonization 66.33 AMF spores / 50g soil and 0.25% 'P' were recorded in plants treated with mycorrhiza + Charcoal. However, Either AM fungi nor carrier materials used did not boosted *Erythrina indica* Lam plants to produce floral buds, flowers and fruits (refer Fig. 27.8 for the effect of AM fungi + perlite on growth of *Erythrina indica* Lam).

Growth response studies in *Jatropha curcas* L

After 30 days significant increase in growth of plants treated with mycorrhiza + leaf compost was observed and results dominated other treatments like Mycorrhiza + Perlite, only mycorrhiza, Mycorrhiza + Charcoal, Mycorrhiza + Peat and UIC shown in (Table 34 and plate VII). Plants treated with mycorrhiza + leaf compost showed increased (26.36cm) shoot length, (26.51 g) fresh weight of shoot, (4.04g) dry weight of shoot, (1.28cm) fresh weight of root and (0.29g) dry weight of root. Highest root length of 17.50 cm and 17.33 cm is recorded in plants treated with Mycorrhiza + perlite and mycorrhiza + leaf compost respectively. Highest leaves (5.33) were recorded in plants treated with only mycorrhiza and mycorrhiza + perlite. Highest of 38.76% root colonization and highest of 0.16% 'P' was recorded in plants treated with only mycorrhiza, maximum of 27 AMF spores/50g soil were recovered in plants treated with only mycorrhiza.

It was after 60 days, plants treated with mycorrhiza + leaf compost showed increased (26.50cm) shoot length, (42.18g) fresh weight of shoot, (8.77g) dry weight of shoot, (22.50cm) root length, (3.96g) fresh weight of root, (0.67g) dry weight of root and (5.66) leaves and ruling over other treatments like mycorrhiza + perlite, only mycorrhiza, mycorrhiza + peat, mycorrhiza + charcoal and UIC respectively. Only mycorrhiza treated plants showed highest (58.45%) of root colonization also, highest of 36.33 AMF spores/ 50g soil and maximum of 0.18% uptake of 'P'.

Similarly, after 90 days, plants treated with mycorrhiza +leaf compost showed (Fig .9) increased (26.80cm) shoot length, (75.40g) fresh weight of shoot, (10.40g) dry weight of shoot, (29.53cm) Root length, (7.44g) fresh weight of root, (0.98g) dry weight of root and (8.00) leaves/plant. Highest of 97.45% of root colonization, 37.33 AMF spores per 50g soil and 0.24% uptake of 'P' was recorded in plants treatment with only mycorrhiza. Floral buds, flowers and fruits were not detected through out the experiment.

DISCUSSION

All experimental plants, treated with mycorrhiza + leaf compost performed well and showed increased Shoot length, Fresh weight of shoot, Dry weight of shoot, Root length, Fresh weight of root, Dry weight of root and number of leaves when compared with mycorrhiza and UIC. However, percent of root colonization, spore number per 50 g soil and percent 'P' uptake is found maximum in experimental plants when treated with specific carrier material. For example, percent of root colonization, spore number per 50 g soil and percent 'P' uptake is higher in *Erythrina indica* and *Solarium nigram* plants treated with mycorrhiza + charcoal. In case of *Jatropha curcas* which has reacted negatively to the presence of carrier material with respect to percent root colonization, spore number per 50 g soil and percent 'P' uptake in shoot showed highest in plants treated with only mycorrhiza. *Capsicum annuum* showed highest percent of root colonization, spore number per 50 g soil and percent 'P' uptake in plants treated with mycorrhiza + perlite.

Gour *et al.* (2000) and Gour and Adholeya (2002) have developed methods of on farm production of AM fungi, where beds of soil measuring 60cm × 60cm × 16cm were used after fumigation. Primary inoculation (pot culture) was inoculated and plants like *Zea mays, Sorgham sudanese* and *Daucucs carota* were grown for three years (each for 4 months). They could get 2.5×10^6 propagules per bed which was 10 fold increase in inoculum from year 1 to 3 year.

The same method has been modified by using leaf compost (2:1, Soil: leaf compost). Period was also reduced to one year and host plants like vegetables and forage crops were selected in order to get huge quantity of AM propogules. Surprisingly, they could get increased shoot weight (51–119% in vegetables and 10–70% in forage crops) but reduced inoculum propogules (up to 40 fold fewer inoculum propogule than 3 year methods). In the present study experimental plants treated with mycorrhiza + leaf compost showed increased fresh weight compared to all other treatments but had reduced percent of root colonization, spore number per 50g soil and 'P' uptake. Thus, present study is in accordance with Gour *et al.* (2000) and Gour and Adholeya (2002).

In present study, charcoal treated plants showed better response (except *Jatropha curcas* L.) with regard to per cent root colonization, spore number per 50 g soil and per cent of 'P' uptake. It is because; the application of charcoal to soil stimulates the colonization of crops by indigenous AM fungi. Similar findings were reported by Pai *et al.*, 1994. The effect of charcoal is ascribed to its physico-chemical properties. Charcoal is porous, weakly alkaline and does not serve as substrate for saprophytes. The AM fungi, sensitive to competition from saprophytes can easily extend their extra radical hyphae in to charcoal buried in soil and sporulate in the particles (Rajanisiriwong *et al.*, 1987). Charcoal particles act as micro-habitats for AM fungi to survive and later grow in to the soil, which makes charcoal suitable as a carrier of AM fungal inoculum (Rajanisiriwong *et al.*, 1987).

Perlite facilitates better aeration and does not serve as substrate for saprophytes. Therefore like charcoal it also can serve as carrier material. In the present study *C. annuum* treated with mycorrhiza + perlite showed highest percent root colonization, spore number per 50 g soil and percent of 'P' uptake in shoot compared to other treatments.

Peat is used for the improvement of productivity of soil (Powell *et al.*, 1995) and as soil less media for the AM culture. Literature regarding how peat acts for the better survival of AM fungi is very meager but, probably, peat has better capacity to hold the moisture which is further helpful for the growth of beneficial microbes which help AM fungal spore germination and to reproduce them.

It is observed that, in the presence of carrier material, all experimental plants reacted differently in percent of root colonization, spore number per 50 g soil and percent of 'P' uptake in shoot. In general, Irrespective of their growth, plants which showed highest percent of root colonization in the presence of one carrier material consequently showed highest percent of 'P' uptake in shoot in shoot and highest spores per 50 g soil in the presence of same carrier material. This situation may vary with plant species in the presence of carrier material.

It is concluded from present study that, all the experimental plants (except *Jatropha curcas* L.) reacted positively to all carrier materials and showed increased per cent of root colonization. The remaining three plant species reacted differently to different carrier material and their reaction is directly proportional to AM fungal propogule production. Only positively reacting plant species may be selected for the on farm production of AM propogule. Further research is warranted on study of the mechanism of interaction of carrier material and host plant, also how this interaction may be helpful in AMF propogule production.

REFERENCES

Allen, M. F. (1991). The Ecology of Mycorrhizae. Cambridge Univ. Press. New York. Pp. 184.

Beena, K. R., A. B. Arun, N. S. Raviraja, and K. R. Sridhar (2001a). Arbuscular mycorrhizal status of *polycarpaea corymbosa* (Carophyllaceae) on sand dunes of west coast of India. *Ecol. Env. Cons.* 7: 355–363.

Beena, K. R., A. B. Arun, N. S. Raviraja, and K. R. Sridhar (2001b). Association of arbuscular mycorrhizal fungi with plants of coastal sand dunes of west coast of India. *Tropical Ecology* 42: 213–222.

Gerdemann, J. W. (1968). Vesicular arbuscular mycorrhiza and plant growth. Ann. Rev. Phytopath. V. 105: pp. 397–418

Gerdemann, J. W. and Nicolson, T. H. (1963). Spores of Endogone species extracted from soils by wet sieving and decanting. Trans. Brit. Mycol. Soc. V. 46 : pp. 235–244.

Giovannetti, M. (1985). Seasonal variation of vesicular-arbuscular mycorrhizas and endogonaceous spores in a maritime sand dune. *Trans. Br. Mycol. Soc.* 84: 679–689.

Giovannetti, M. and Mosse, B. (1980). An evaluation of technologies for measuring vesicular arbuscular mycorrhizal infection inroots. *New Phytologist*, 84: 489–500.

Gour, A. and Adholeya, A. (2002). Arbuscular mycorrhizal inoculation of five tropical fodder crops and inoculum production in marginal soil amended with organic matter. *Biol. Ferti. Soil.* 35: 214–218.

Gour, A., Adholeya, A. and Mukerji, K. G. (2000). On-farm production of VAM inoculum and vegetable crops in marginal soil amended with organic matter. *Tropical Agriculture*, 77: 21–26.

Jackson, M. L. (1973). Soil chemical analysis. Prentice Hall, New Delhi.

John, T. V. S., Coleman, D. C. and Reid, C. P. P. (1983). Association of vesicular-arbuscular mycorrhizal hyphae with soil organic particles. *Ecology*, 64: 957–959.

Koske, R. E. (1975). *Endogone* spores in Australian sand dunes. *Can. J. Bot.* 53:668–672.

Koske, R. E. (1988). Vesiculars-arbuscular mycorrhizae of some Hawaiian dune plants. Pacific Science. V. 42 : pp. 217–229.

Koske, R. E. and Gemma, J. N. (1990). VA mycorrhizae in vegetation of Hawaiian Costal strand: evidence for codispersal of fungi and plants. *Amer. J. Bot*, 77: 466–474.

Koske, R. E. and J. N. Gemma (1990). VA mycorrhizae in vegetation of Hawaiian Coastal strand: evidence for codispersal of fungi and plants. *Amer. J. Bot*. 77:466–674.

Koske, R. E. and J. N. Gemma (1997). Mycorrhizae and succession in plantings of beach grass in sand dunes. *Amer. J. Bot*. 84:118–130.

Koske, R. E. and Potson, W. R. (1984). Are VA mycorrhizae required for sand dune stabilization ? Bio Science V. 34: pp. 420–424.

Kulkarni, S. S., Raviraj, N. S. and Sridhar, K. R. (1997). Arbuscular mycorrhizal fungi of tropical sand dunes of east coast of India. J. Coastal Res. V. 13(3): pp. 931–936.

Lakshman, H. C. (1994). Occurrence of VA mycorrhizae in some tropical hydrophytic and Xerophytic plants of Dharwad District, Karnataka. J. Nat. Con. V.6 (1): pp. 29–35.

Lakshman, H. C., Mulla, F. I. and Inchal, R. F. (2000). Some coastal potential plants with AM fungi and their use in land reclamation programme. In Proce. Coastal zone Management. S.D.M.C.E.T. Dharwad. Spl. pub. 2 : 189–196.

Leonard, R. I. and F. W. Judd (1999). The biological flora of coastal dunes and wetlands *impomoea imperati* (Vahl) Griseb.*J. Coastal Res*. 15: 645–652.

Louis, I. (1990). A mycorrhizal survey of plant species colonizing coastal reclaimed land in Singapore. *Mycologia* 82: 772–778.

Maun, M. A. (1994). Adaptations enhancing survival and establishment of seedling on coastal dune ecosystem. *CRC Crit. Rev. Aquat. Sci*. 1: 661–681.

Mohan Kumar, V., Ragupathy, S., Nirmala, C. B. and Mahadevan, A. (1988). Distribution of VAM in the sand beach soils of Madras, Current Sci. V. 57 : pp. 367–368.

Morton, J. B. (1990). Evolutionary relationships among arbuscular mycorrhizal fungi in the Endogonaceae. *Mycologia*, 82: 192–207.

Morton, J. B. (1998). Taxonomy of VA mycorrhizal fungi: classification, nomenclature and identification. *Mycotaxon*, 32: 267–324.

Nicolson, T. H. (1960). Vesicular arbuscular mycorrhizal a universal plant symbiosis. Sci. Prog. Oxford. 55: 561–581.

Pai, G., Bagyaraj, D. J., Padmavati Ravindra, T. and Prasad, T. G. (1994). The water relations of mycorrhizal and non-mycorrhizal cowpea (Vigna unguiculata) grown at three soil moisture regimes. *Indian J. Micro.* 34: 317–322.

Phillips, J. M. and Hayman, D. S. (1970). Improved procedures for clearing roots and staining parasitic and vesicular mycorrhizal fungi for rapid assessment of infection. Trans. Br. Mycol. Soc. V. 53: pp. 158–161.

Powell, C. L., Bagyaraj, D. J., Clark, G. E. and Cladwell, K. I. (1985). Inoculation with arbuscular mycorrhizal fungi in greenhouse production of Asparagus seedlings. *Newzealand J. Agri. Res.* 28: 293–298.

Rajanisiriwong, W., Nopamornbodi, O. and Thamsurrkal, S. (1987). *In: Mycorrhizae in next decades*. (Eds.) Sylvia, D. M., Hung, L. L. and Graham, J. H. University of Florida Gainesville, Florida, pp. 217.

Schenk, N. C. and Perez, Y. (1990). Manual for the identification of VA-mycorrhizal fungi, 3rd edition. Gaines Ville, Florida, University of Florida. Pp. 249.

Smith, S. E. and Read, D. J. (1997). Mycorrhizal symbiosis. San Diego: Academic Press. N. Y. California pp. 605.

Sreenivasa, M. N. and Bagyaraj, D. J. (1988). Selection of suitable substrate for mass multiplication of *Glomus fasciculatum*. *Plant soil*, 106: 289–290.

Sridhar, K. R. (2006). Arbuscular mycorrhizal fungi of coastal sand dunes. *Techniques in mycorrhizae*, 30–42.

Sutton, J. C. and B. R. Sheppard (1976). Aggregation of sand dune soil by endomycorrhizal fungi. *Can. J. Bot.* 54:326–333.

Sutton, J. C. and Barron, G. L. (1971). Population dynamics of Endogone spores in soil. *Can. J. Bot*, 50: 1909–1914.

Tisdall, J. M. (1991). Fungal hyphae and structural stability of soil. *Aust. J. Soil Res.* 29 :729–743.

Titus, J. H. and Del Moral R. (1998). Vesicular arbuscular mycorrhizae influence Mount St. Helens pioneer species in green house experiment Oikos 8193: pp. 495–510.

Walker, C. (1995). AM or VAM: What's in a word? In: Mycorrhizal structure, function, molecular biology and biotechnology (Eds. A. Varma and B. Hock). Springer Verlag. Pp. 25–26.

Wright, S. F. and A. Upadhyaya (1998). A survey of soils for aggregate stability and glomalin, a glycoprotein produced by hyphae of arbuscular mycorrhizal fungi. *Plant and Soil* 198 : 97–107.

28

Importance of Microorganisms in Solubilization of Phosphorus in Different Soils

[1]*Girish*

INTRODUCTION

Microorganisms, bacteria, fungi and algae and to some extent protozoans play important role in mineralization of soil. Mineralization is essential for the maintenance of life on our planet. Soil microorganisms serve as biochemical agents for the mineralization of organic carbon, nitrogen, sulphur, phosphorous and other micro elements. As a result of mineralisation, these elements are returned to the environments to be used over again in the form of cycle. The cyclic moments of chemical elements between the living organisms and the environment form biogeochemical cycles. The ability to soloubilize poorly soluble mineral phosphates has been shown invitro for many commonly occurring soil bacteria, although the invitro data have generally been equivocal the ecological paradox of growth limiting amount of soluble 'P' in the midst of large amounts of unavailable fixed 'P' provides a strong evolutionary basis for the development

1. Department of Microbiology, Basaveshwar Science College, Bagalkot – 587 101, India.

of mineral phosphate solubilizing trait. Thus, a shift to a bio-process based technology will greatly increase the available phosphate reverses (Goldstein *et al.*, 1993). The mechanise of MPS has been a subject of analysis for a long time and still a matter of curiosity. The different mechanisms so far hypothesized has been reviewed herein.

Organic acids have been implicated to chelate the cationic counterpart of 'P' ion and release soluble Pi in the medium (Katznelson and Bose, 1959). Analysis of culture filtrates of many MPS bacteria have shown the production of Mono, di and tricarboxylic acids. The amount of acids liberated by these microbes roughly is more than 5 percent of the carbohydrate consumed (Banik and Dey, 1982).

The Indian soils are mostly low to medium in available forms of phosphorus. The recent reports based on the analysis of 3.65 million soil samples showed that 42 percent of soil samples fall in low category, 38 percent in medium and 20 percent in high category (Motsara, 2002). It has been observed that vast alluvial tracts of north, south, east and west regions of the country either are low or medium in available P. Therefore, 80 percent of the soils in India need P application at recommended rates. In high P soils, considering the P removal—of crops, application of some quantity of P is essential to stop mining from soils so—as to sustain the yields in future (Wani, 1980; Lindermann, 1988).

The average P fertility index of soils in Haryana, Gujarat, Uttara Pradesh, Orissa, Arunachal Predesh, Pondichery and Andaman and Nicobar islands in low and those of Assam, Nagaland, Tripura, West Bengal, Himachal Pradesh, Goa, Madhya Pradesh, Punjab, and Karnataka is medium.

Phosphorus is one of the essential nutrients for plant growth. It is an integral part of the cellular activities of living organisms. It occurs in soils as inorganic phosphate, produced by weathering of parent rock or as organic phosphate derived from decayed plant, animal or microorganisms. The mineral phosphate solubilizing microbes, which acidify the substrate and possibly the rhizosphere, have been used in unison with rock phosphate so that the phosphorus in the rock phosphate can be made available in the soil for plant uptake. Development of superior MPS strains for use along with rock phosphate should be given immense importance. A package of inoculant containing materials that increase the release of Pi from rock phosphate should include the PSM bacteria (Gaur, 2002). The use of such approach would reduce the pollution caused by excess application of phosphate fertilizer. Additionally, the by-products of the reaction of phosphates.

IMPORTANCE OF PHOSPHATES IN AGRICULTURE

Phosphorus is a queen of element next to nitrogen. It is most essential element for plant growth and its productivity. Phosphate concentration is an important factor that limits primary productivity of many ecosystems. The precipitation of phosphorus greatly limits primary productivity in marine habitat. Several industrial wastes contain inorganic phosphate. Phosphates are also used in many detergents, which are passed are into streams, rivers and oceans where they settle down at the bottom as sediments. A lot of phosphates are lost through leaching (Shinde and Patil, 1985; Swaby and Sperber, 1988). The leached and sedimented phosphates go out of circulation. They become available only after long geological periods as fuel or fertilizer. In any ordinary soil the amount of exchangeable phosphate may be 1.6 to 2.1×-10^{-4} moles per 100 g with approximately 0.1 mg in the associated solution. The majority of soil phosphorus reaction products are only sparingly soluble, having solubilities in the micro molar range. These conditions combine to make P an important living factor for biological growth in many natural environments.

Because of phosphorus relative scarcity in available form, phosphate is often the most limiting factor in plant growth. In regions of high human population density and intensive soil cultivation in India, phosphate shortage is an agricultural problem.

Altering organic phosphate by microorganisms

In the soil the organic phosphate cannot be utilized directly by plants. The decomposers like bacteria, fungi and actinomycetes convert the organic phosphorus in to inorganic form. Minerals and raw materials are released during this process. The phenomenon is known as mineralization. This process is catalysed by phosphate enzymes. Microorganisms produce these enzymes. Some bacteria and fungi produce another enzyme called phytase. Phytase releases soluble inorganic phosphate from phytic acid (inositol hexophosphate).

$$\text{Insoitol hexophosphate} + 6H_2O \xrightarrow{\text{Phytase}} \text{Insoitol} + 6PO_4^{-3}$$

Thus the inorganic phosphate can be directly absorbed by plants for recycling.

Developing new techniques for reintroducing solubilizers lands for sustainable agriculture and forestry and for restoring natural areas. Therefore phosphorus occur in three forms and its cycles as

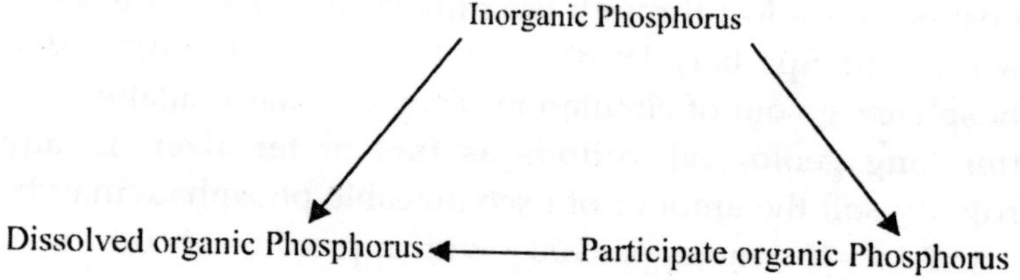

ALTERING SOLUBILITY PHOSPHATE BY BACTERIA

In soils bacteria may attack soil particles. There are differences in bacterial population in soil. Their distribution is not uniform because population increase where there is a source of organic substrate. For example, near the plant roots the population may be 100 times higher than in the rest of the soil. Most of the bacteria are heterotrophs and take part in the decomposition of plant residues. Microorganisms are present in all soils whether acidic or alkaline, water logged or well drained and in regions that are hot or cold or dry, here phosphorus occur in the form of inorganic state, then involve in the solubilization of inorganic phosphorus. In recent days, many laboratory experiments have been well documented that P strait species of *Pseudomonas, Mycobacterium, Micrococcus, Flavobacterium, Bacillus polymyxa* and *Enterohacter ashuriae* could solubilize soil phosphate. When they grow in the presence of glucose as carbon source and ammonia or nitrate as the Nitrogen source (Lajtha and Schilesinger, 1988; Ihner and Schinner, 1995).

Bacteria like *Thiobacillus ferroxidous, Bacillus thurengiensis* convert non-available inorganic phosphorus present in soil into an available form utilizable by crop plants. These bacteria also produce iron chelating substances i.e., pseudobactin called siderophores, which chelate the iron present in the root zone. This iron becomes non-available to harmful microorganisms and in this manner, crop plants are protected from them. Similarly many actinomycetes solubilise inorganic phosphates in soil. They restricted to an aerobic environment, and tend to be more dominant in warm soils than cold. Most important P solubilizing actinomycetes are as: *Streptomyces, Nacardia, Streptoverticillium, Thermoactinomycetes, Micromonospora, Strep to sporangium, Ample llariella* and *microbispora.*

About 50% of the cultivable lands in India are deficit in available form of soil P and only 20–25 % of applied phosphates fertilizers are utilized by crop forming insoluble complexes with metal cat ions and clay mineral complexes.

Added P is also readily fixed in most soils that its concentration in solution is always meager. Barber (1980) observed that the distance traversed by phosphate ions through soil in four days is between 0.80 to 0.006 cm. Thus, phosphate ions are practically immobile and hence its placement in root zone/moist layer is prerequisites for its efficient utilization. P in this root zone exists as an orthophosphate ion P in this root zone exists as orthophosphate ions (H_2PO_4) and its availability to plant is dependent on its replenishment.

ALKALINE AND ALKALI SOILS

An alkaline soil has a pH greater than 7.0 and is includes with calcareous soils (one containing calcium carbonate and a large amount of exchangeable calcium) because there are many characteristics common to both soils. In general these soils occur largely in arid regions.

An alkali soils has a high exchangeable sodium percentage and may not contain excess soluble salts.

Saline - alkali soils refer to soils that contain an excess of both soluble salts and exchangeable sodium. Conductivity of the saturation extract of this soil is > 4 mmhos/cm at 25°C and the exchangeable sodium percentage is >15 mmhos.

Non-saline-alkali soils have conductivity of the saturation extract < 4 mmhos/cm at 25°C and exchangeable sodium percentage > 15 and pH between 8.5 and 10.0. These soils frequently occur in semiarid and arid regions in small irregular areas. Biological activity of these soils is normally low which can be improved by amending gypsum, followed by cropping as well as by growing grasses and trees. This amendment is costly and practically difficult.

ALKALIZATION OF SOIL

Soil particles, as consequence of the electric charge at their surface, adsorb and train cat ion on their surfaces. These adsorbed cat ions may be other cat ions that occur in the soil solution. Sodium, calcium and magnesium cat ions are always readily exchangeable while other cat ions like potassium and ammonium are exchange with great difficulty, hence said to be fixed. Calcium and magnesium are the principal cat ions found in the soil solution and one the exchange complexes of normal soils in arid regions. When excess soluble salts accumulate in these soils, sodium frequently becomes the dominant action in the soil solution calcium and magnesium ions in the soil solution are more strangle adsorbed by the exchangeable complex than sodium.

INORGANIC P IN ALKALINE SOILS

Native P in alkaline and calcareous soils occur chiefly in the form of hydroxyapatite, fluorapatite, chlorapatite, Wagnerite, and wavelike and in organic compounds. Other minerals of Fe and A1 phosphates and other calcium and magnesium phosphates also occur.

Carbon dioxide produced from the decomposition of organic matter and respiration of roots and microorganisms, reacts with water to from carbonic acid that reacts with tricalcium phosphate (TCP) to form water-soluble monocalcium phosphate. This reacts with calcium ions and calcium carbonate to form insoluble TCP in strongly alkaline soils.

PHOSPHATE SOLUBILIZING MICROORGANISM (PSM) AS BIOFERTILIZERS

In vivo phosphate solubilization is more complex than *in vitro*, which is one of the reasons for not getting consistent crop response. The phosphate biofertilizers (phosphotacterin) was first used in USSR as soil seed inoculants in an area of 14 million ha and increased the crop yield 5 to 10 percent. In India, phosphobacteria was tested in 1963 on wheat under field trails were conducted and significant increase was noticed. Subsequently several field trails were conducted under different agro climatic conditions with cereals, pulse, oilseeds and vegetables. The field and pot trials with PSM with or without phosphates fertilizers, RPs, pyrites or hydroxyl patties showed increase in yield and P uptake from marginal to significant levels (5–20%). The response was more-in organic matter amended soils. But doubts were raised on the contributions of PSM under field conditions, as the results were not consistent. The PSM are involves not only in P solubilization but other factors like release of phothormones, nitrogen fixation, mineralization and mobilization of other nutrients, antagonism to plant pathogens, promotion of plant growth promoting rhizosphere microorganisms (PGPRMO) etc., are equally responsible for increased crop yield (Goldstein, 1986).

The PSM have shown synergistic effect with other rhizosphere microorganisms under laboratory and field conditions for P- solubilization and crop response. The PSM have been tested for P- solubilization *in vitro* with *Azotobacter, Azosprillium, Rhizobium,* and Bacillus etc. The various crops showing positive response to inoculation include wheat, cotton, lavender, paddy, soybean, chickpea, moonbeam, etc. The vesicular arbuscular mycorrhia (AM) invade plant roots and form symbiotic relationship. These organisms help the host in uptake of nutrients, water and are also antagonistic to pathogens.

INTERACTION WITH OTHER SOIL MICROORGANISMS

The PSM have shown synergistic effect with other rhizosphere microorganisms when tested under laboratory and field conditions for P-solubiliztion.

INTERACTION WITH DIAZOTROPHS

The PSM have been tested for P conditions with many diazotrophic bacteria like *Azotobacter. Azospirrillum, Rhizobium.* Bacillus and others. Synergistic effect was observed with these organisms as plants are benefited by availability of P as well as N (Ocampo *et al.,* 1975; Gaur, 1985; Alagawadi and Gaur,

1988). The various crops, showing positive response to inoculation of both types of organisms include wheat, cotton, lavender, paddy. soybean, chickpea and moonbeam.

INTERACTION WITH MYCORRHIZAL FUNGI

The arbuscular mycorrhiza (AM) invade plant roots (endomycorrhizal fungi) and form symbiotic relationship. These organisms help the host in absorption of P, micronutrients, water and show antagonism to plant pathogens particularly in P deficient soils. However, 32 P labeled studies has indicated that mycorrhizae does not help solubilization but facilitate P uptake by increasing absorptive surface area of the roots (Arrora and Gaur, 1978). The dual inoculation of PSM and AM have synergistic effect on plant growth and P uptake as former increase the P availability and the latter the root surface area (Subbarao, 1988; Lakshman, 1996, 2001). The increased yield and P uptake by dual inoculation of PSM and AM has been observed on several crops by many workers (Azcon *et al.*, 1976; Lee and Sattar and Gaur, 1989; Gaur, 1990).

ENRICHMENT OF COMPOST ROCK PHOSPHATE

In sustainable agriculture, the emphasis is laid on the use of organic matter and natural resources to fill the gap of nutrients depleted by crops. The enrichment of compost by adding low-grade rock phosphate offers a potential preposition for the effective utilization of insoluble P. The enriched organic manure can be prepared at the farmers field and offer a low cost alternate to chemical fertilizers. The enrichment of compost with 5% single super phosphate has been done but the amendment with insoluble rock phosphate during composting is a more rational approach as solubilization of insoluble P occurs during composting. The compost able materials (farm and animal residues) is mixed well with compost and soil (90:5:5) to which 25% of rock phosphate is added on dry weight basis. The mixture after moistening is put into the pit and allowed to decompose for 90 days with mixing at regular intervals (15 days). About 50% of the insoluble P is converted into in to soluble P (6–8% P_2O_5) in the phosphocompost. The field trials conducted with phosphoscompost on pearl millet, wheat, pigeon pea, green gram and cluster bean showed comparable results with recommended doses of super phosphate even in neutral to alkaline soils (Reganold *et al.*, 1990). The use of aspergillums amatory an efficient rock phosphate solubilizer has found beneficial in improving the quality of rock phosphate amended compost (Gaur, 2002). Table 28.1.

Table 28.1. : Showing the Response of the Crops to Phosphate Solubilizing Organisms

Crop	Inoculant	Observation	Reference
Honne	PSM, VAM	Increase in biomass	Lakshman, 1996
Coral Tree	Azotobacler and PSM	Increase in biomass and P uptake	Lakshman.2001
Black gram	P Solubilizing microorgnnisms	Yield Increase	Tomnr, 1998
Groundnut	Psoliibilizing bncleiiii	Meld increase	I'atcl and Thaktir. I99S
Greengram	P solubilizing fungi	Biomass increase	Gaur, 1990; Rao and Rao, 1994
Cotton	Azotobacter and PSM	Increase in yield	Kundu and Gaur, 1980
Mungbean	PSM, VAM	Yield increase	Singh and Kapoor, 1998
Barley	Phosphobacterium	Yield increase and P uptake	Kundu and Gaur, 1982,
			Krasilnckov et al., 1957
	Pseudomonas and Azotobacter seedling,	Increase in yield	Muskow, 1961
	Pseudomonas and Bacillus	Yield and P increase	Taha et al., 1964
Maize	Phosphobacterium	Yield increase	Dzhenaev, 1965

	Pseudomonas sp.	Biomass increase	EdiPremonoeA al. 1996
Potatoes	Pseudomonas and Bacillus	Yield increase	Kundu and Gaur, 1980; Shikina, 1961
Peas	Phosphobacterium	Yield increase	Gaur, 1990
Tomatoes	Phosphobacterium	Increase in yield and sugar content	Zavyalov, 1958
Cowpea	Bacillus, Pseudomonas and Azotobacter	Increase in Biomass and N., fixation	Gaur, 1990;
Paddy	Pseudomonas, Bacillus, Aspergillus	Increase in yield	Gaur, 1990, Chandra and Singh, 1996
Chickpeas	Bacillus, Pseudomonas, Azotobacter	Increase in yield and	P Gaur, 1990
Soyabean	Pseudomonas striata	Increase grain yield and P uptake	Gaur, 1979; Gaur, 1985

FUNGI

The cells of most fungi are joined to form filaments (hyphae), which are usually 5–20 mm in diameter and can be several centimeter in length. When they colonize leaf litter and compost, several hyphae often become aggregated as rhizomorphus, which aids in absorption of phosphorus (PO_4^{-3}) ions. Mycorrizal fungi belongs to phycomycetes a symbiotic relationship with higher plant roots. These fungi can tolerate to more acidic soil, arbuscular mycorrhizal fungi, produce running hyphae that extend large distance outward, infecting multiple roots from many plants. In-addition, to these have a finely branched absorbing hyphical network that intensively searches the soil surrounding individual roots for nutrients. These fungi appears to have mineral enzymatic capacity but their intense exploration increases the surface area for uptake of HPO_4^{-3} especially those that are not moved by mass flow. Ectomycorrhizae and related types of mycorrhizae such as arbutiod and ectoendo mycorrhizae form rhizomorphus, and they allow mycelium to extend long distances in search of resources and new plants, and highly branched hyphae (belong to both basideoumycetes and ascomycetes) that can take up elements (phosphorus) directly or break down organic compounds, and transporting organic resources, especially organic acids.

Some results in 32 p labeled, in very P deficient soils suggested another possible function of AM-mycorrhiza. In these soils plants took up virtually no phosphate unless they had mycorrhiza. Table 28.2 shows this negligible uptake of P in non-mycorrhizal plants compared to mycorrhizal plants.

Table 28.2. Showing Phosphorus from 32P Labeled Soils by Sonic Non-Mycorrhizal (NM) and Mycorrhizal (M) Plants.

Plant species	Method	Soil	NM ppm/ plant	M
Centrosemapubescens	T	Brizil	Nil	367
	T		36	1720
	S		24	693
Paspaluim notatum	T	Brazil (2)	Nil	1634
Fraxinus excelsior	T	Mixed deciduous forest	50	1590
Fragaria Sp.	T	Mixed deciduous forest	5	2152

T = inoculated and later transplanted into 32 P labeled soil,
S = sown and inoculated in 32 P labeled soil.

PHOSPHATE – INDUCED BY ALGAE

In most lakes, the growth of phyto plankton and aquatic plants is restricted by the limited availability of phosphate and a usable form of nitrogen. However, when these nutrients enter lakes in the effluent from sewage disposal plants and as farm land run off of animal wastes and excess fertilizer. Eventually, large amounts of algae die and decompose after which the aerobic microorganisms of decay and deplete the lake's dissolved oxygen, causing die-offs of fishes and other aerobic organisms (Naplekoka, 1967). However, there are evidences that soils of these lakes they liberate nutrients by feeding as bacteria, an indication of the complexity of nutrient mainly of limited phosphorus in soils. Importantly, human-made and catastrophic natural disturbances often have the same basic effects on beneficial microorganism in soil, and many practices for restoring phosphate solubilizing organisms can be based as successional processes that occur naturally. Developing new techniques for reintroducing P solubilizers leads to sustainable agriculture and forestry and for restoring natural areas.

Importantly, human-made and catastrophic natural disturbance often have the some basic effects on beneficial microorganism in soil, and may practices for restoring phosphate solubilizing organisms can be based as successional process that occur naturally. Developing new techniques for reintroducing p-solubilizers on lands for sustainable agriculture and forestry and for restoring natural areas.

PROSPECTS

The pressure on land and the application of imbalanced plant nutrients and indiscriminate use of pesticides has resulted in poor soil health and deterioration of soil fertility. If appropriate action is not taken, it may lead to unpreventable loss of soil fertility. Phosphate solubilizing microorganisms have shown the potential to solubilize mineral phosphate. Large unexploited reserves of low grade rock phosphate are available in many countries which are not suitable for manufacture of phosphates fertilizer but can be utilized along with PSM for direct application in different types of soil and via phosphor-compost. The results reported with PSM have shown interest in this in view of increased emphasis on sustainable agriculture and organic farming. The phosphate solubilizing microorganisms are being used as seed or soil inoculants for improving the phosphorus nutrition of crops. Their use in phosphate compost for improving the quality of organic manure and available P content has been demonstrated. Calcareous soils the various forms of calcium phosphate predominate. Calcium phosphate is present as the carbonate-appetite. In alkaline calcareous sails, the phosphate solubility increases above pH 8.5, is at minimum within the range 7.6 to 8.5, and increase again below pH 7.6.

The concentration (or activity) of P in the soil solution in alkaline or calcareous soils will be largely governed by three factors. 1. Calcium ion activity 2. amount and particle size of free calcium carbonate in the soil and 3. amount of clay present. The activity, a large amount of finely divided calcium carbonate and a large amount of calcium-saturated clay. In order to maintain a given level of phosphate activity in the soil solution, it is necessary to add larger quantity of phosphate fertilizers to such soils.

REFERENCE

Alagawadi, A.R. and Gaur, A.C. (1988). Associated effect of Rhizobium and phosphate solubilizing bacteria on the yield and nutrient uptake of Chickpea. Pl. Soil, 105:241–246.

Arrora, D. and Gaur. A. C. (1978). Periodic microbial solubilization of 32p labeled hydroxyapatite. *Indian J. Microb.* 18: 193–195.

Azcon, R. Barea, J. M. and Hayman, D. S. (1976). Utilization of Rock phosphate in alkaline soil by plants inoculated with mycorrhizal fungi and PSB soil biol Biochem., 8:135–138

Bagyaraj. D. J. and Verma. A. (1995). Interaction between arbuscular mycorrhizal fungi and their importance in sustainable agriculture in arid and semiarid tropics. *Ad. Micro* Ecol. England.

Banik, S. and Dey, B.K. (1982). Available phosphate content of an alluvial soil as influenced by inoculation of some isolated phosphate solubilizing microorganisms. Pl. Soil, 69:353.

Gaur, A. C. (1990). Phosphate solubilizing micro organisms as biofertilizer. Omega scientific Publisher, New Delhi, pp. 176.

Gaur, A. C. (2002). Microbial mineral phosphate solubilization in. *National Symposium on mineral phosphate soluhilization* U.A.S. Dharwad. Pp–5.

Gaur, A. C. (2002). *Microbial technology jor composing of Agricultural residues by improved methods* ICAR, New Delhi, pp–78.

Gaur, A. C. (1985) Phosphate solubiziling micro ogranisms and their role in plant growth and crop yield. Proc. Naat. Symp. on soil Biology, HAU, Hissar, pp 125–138.

Ghosh, A. B. and Hasan, R. (1979). Phosphorus fertility status of soils of India In: phosphorus in soils. Crops and Fertilizers *Bull Ind. Soc. Soil Sci.* 21:1–8.

Goldstein, A. H. (1986). Bacterial solubilization of mineral phosphates: historical perspective and future prospects *A. .1. Alt Agri.* 1:51–57.

Illmer, P. and Schinner, F. (1995). Solubilization of inorganic calcium phosphates Soil Biol. *Bio chem* 27:257–263.

Katznelson, H. and Bose, B. (1959). Metabolic activity and phosphate dissolving ability of bacterial isolates from wheat root rhizosphere and non-rhizosphere soil. Con. J. microbiol. 5:79–85.

Kucey, R.M.N. (1987). Increase in 'P' uptake by wheat and filed beans with P-solubilizing bilaji and VAM fungi. App.. Environ. microbiol, 53: 2699–2703.

Lajtha, K. and Schilesinger, W. H. (1988). The biochemistry of phosphorus cycling and phosphorus availability along a desert soil chronoscquence *Ecology.* 69: 24–39.

Lakshman. H. C. (1996). Interaction between VAM and beneficial micro organisms. *Em: Ecol. And Counser* 117:79–85.

Lakshman. H. C. (2001). Effect of indigenous VAM and PSM on Indian Coral tree. *I.-J. Ecopl, Envirn.* 101(1):27–32.

Lee, A. and Bagyaraj. D. J. (1986). Effect of inoculation VAM and either rock phosphate dissolving bacteria or thiobacilli on dry matter production and uptake of 'P' by tomato crop. N. Z. J. Agric. Res., 29: 521–531.

Lindermann, R. G. (1988). Mycorrhizal interactions with the rhizospheric microflora. *Phytopathology* 78: 366–371.

Mogal, H. F. and Desai, P. B. (1992). Microbes and bioleaching Everyman's Science 102: 124–125.

Motsara, M. R. (2002). Available nitrogen phosphorous and potassium: Status of Indian soils as depicted by soil fertility maps. Fertilizer news 47(8): 15–21.

Naplekoka, N.N. (1967). Release of relatively immobile phosphates by fungi and actinomyeetes growing as cellulose. *Sov. Soil Sci.* 11: 1495–1502.

Ocampo, J .A., Barea, J. M. and Montoya, E. (1975). Interaction between Azotobacter and Phosphobacteria and their establishment in the rhizosphere Can. J. Microbiol. 21:1160–1165.

Reganold, J. P., Paperdick, R. I. and Parr, J. E. (1990). Sustainable agriculture. *Scientific America* 262(2): 112–120.

Sattar, M.A. and Gaur, A.C. (1989). Effect of Vesicular arbuscular mycorrhiza and phosphate solubilizing microorganisms on the yield phosphate uptake up of lentil (*Lense esculenta* moench). Thai, J. Agri. Sci, 22:129–136.

Shinde, B. N. and Patil, A. J. (1985). Use of phosphate solubilizing cultures and rock phosphate as P–availability to wheat in black calcareous soil *Curr. Res.* I: 105–106.

Subba Rao, N. S. (1988). *Biofertilizers In Agriculture New Delhi:* Oxford and IBH publishing Co. Pvt. Ltd.

Swaby, R. J. and Sperber, J. I. (1958). Phosphate dissolving micro-organisms in rhizosphere of legumes in nutrition of legumes, Academic Press N. K. Pp. 289–294.

Wani, P. A. (1980). Studies as phosphate solubilizing microorganism a review *J. of Maha, Agri. Uni.* 5: 144–147.

29

Role of AM Fungi in Reclamation of Mined Soils

[1]*M G Nadagouda and* [2]*H C Lakshman*

INTRODUCTION

AM application an further be extended for improvement of tree species for plantation programmes and reclamation of wastelands. This study will include the analysis of stress factors existing in various wasteland situations and growth responses of introduction of specific legumes under inoculation of *Rhizobium* and AM fungi. The wastelands areas could be deforested midlands, coastal saline mangrooves areas of Orissia or coal mine overburdens. Mine areas reflect additional heavy metal toxicity problem and the mangroove areass depict society problems. Plants growth responses are poor in all types of wastelands because of poor soil quality due to low nutrient content, particularly N_2 and phosphorous and organic carbon along with implanted pH situation occurring as a common problem. Most of the soil of these

1. Department of botany, A.S.M. College for Women, Gandhinagar, Bellary-583 101, INDIA.
2. P.G. Studies in Botany (Microbilogy Lab) Karnatak University, Dharwad-580003, India

wastelands were drought prone with poor water holding capacity.

Arbuscular mycorrhizal colonization is extensive in root system of many plant communities of the semi-arid land (Davidson and Christensen, 1977; Reeves *et al.*, 1979; Loree and Williams, 1987; Allen and Allen, 1980). The ubiquitous of arbuscular mycorrhizas and their apparent role in nutrient cycling suggest that they have a powerful functional influence on many dry ecosystem (Trappe,1981). Plants colonizing on mined spoils and those introduced during land vegetation frequently become infected (Miller, 1979; Danienson, 1985). When mining begins, vegetation is removed and the top 25 to 40 cm of soil is stock piled. This stock piled spoil need pioneering species in revegetation on of denuded land. However, these plants depend on an adequate supply of P to achieve their potential for rapid growth (Jasper *et al.*, 1988). It is likely that P-deficient soils from stock piled during mineral sand mining need additional phosphorus and inoculation with mycorrhizal fungi (Jasper *et al.*, 1989; Lakshman, 1997). Mining activities have a deleterious effect on plant and soil microbial community health there must be ameliorated with the addition of amendment in order to effect successful revegetation. Amendments provide a nutrient source and an environment that enhanced the development of fully structured and functional soil microbial community Vissel (1985).

The primary object of this study was to investigate the colonization dynamics of the VA-mycorrhizal fungi during early recovery of herbs/shrubs growing on the mined spoils of Bisgod-Karnataka (South India). Mycorrhizal status and the frequency of Endogonaceae spores associated with plant species. From the data gathered in this study on disturbed and undisturbed communities, the rate of AM in the succession and subsequent stabilization with surface reclamation may then be elucidated.

MATERIALS AND METHODS

In the present study Bisgod mined area have been selected. And its geographical location laying in between 56'N latitude and 14°, 43' to 15°, 09' E longitude, in Karnataka (South India). Investigation of A-mycorrhizal assessment was carried out on both disturbed, undisturbed vegetation and in freshly dumped top mined spoils. The samples from plants occurring on each site were collected. They were cleared free of soil, and immediately placed them in Formalin, Acetic acid and Alcohol in the ratio of 5:5:90 (FAA) fixation. Returning to the laboratory for further analysis. Roots were cut into 1 cm segments, cleared with 10% KOH and stained 0.05% trypan blue in lactophenol using the method of Philips and Hayman (1970). Root segments were viewed under microscope at 150 X and scanned for AMF infection. Endogonaceous spores were extracted from 50 g. Soil samples surrounding the roots by wet sieving and decanting technique Gerdeman and Nicolsan (1963). AMF fungal spores were identified with the aid of manual written by Schenck and Perez, (1990). Green house experiments were conducted, by using (1:1) Sterilized forest soil : Sterilized mined spoil). Indigenous *G. fasciculatum* was used as an inoculum on *Cassia sirceae* Sw. for one year. All the data concerned to plant growth, biomass production, phosphorus in shoots was determined by adopting the technique of Olson *et al.*, (1954) and percent of AMF colonization, spore population was assessed following Read *et al.* (1976). And Gridline intersect method was used to estimate in proportion of root length colonization on the slides following the technique of (Giovannetti and Moss, 1980).

RESULTS AND DISCUSSION

Arbuscular mycorrhizal colonization was very meager among the plants grown on mined spoils shown (Table 29.1). There was lower colonizations 9–13 percent, in Cyperaceae, Rhamnaceae and Ascelpiadaceae. Increased colonization from

Table 29.1. : Arbuscular Mycorrhizal Status in Plants of Stock Piled Mined Spoils of Bisgod Mined Spoils in Karnataka, India

Family / Scientific name	pH	Percent colonization	Spores/50 g Soil
Aslepiadaceae			
Hemidesmus indica R. Br.	6.8	13.4± 4.1	24.3 + 5.1
Compositae			
Ageratum conyzoides L.	6.4	46.3+ 4.2	34.9 + 2.4
Eupatorium adoratum Vahl.	7.1	4.1+ 2.2	17.5 + 1.3
S.F. Caesalpiniaceae			
Cassia tora linn.	7.2	18.2+ 5.2	21.3 + 3.2
Commelinaceae			
Commellina communis L.	6.7	23.4+ 3.1	12.4 + 4.2
S.F. Mimoisaceae			
Acacia melanoxylon R.Br.	7.3	16.4+ 5.3	31.7 + 6.2
Mimosa pudica L.	6.3	15.5+ 3.2	19.4 + 5.1
Cyperaceae			
Cyperus rigidifolius stend	7.5	5.1+ 1.3	8.3 + 4.4
Euphorbiaceae			
Euphorbia sparcifolia L.	8.2	32.1+ 2.5	30.5 + 5.3 .
Graminae			
Cynodan dactylon Pers.	6.4	42.5+ 5.2	32.4 + 5.2
Malvaceae			
Sida veronifolia Lam.	6.3	17.4+ 4.5	18.5 + 4.2
Rhamnaceae			
Ziziphus trinervia Roxb.	7.3	9.3+ 5.0	16.4 + 3.1
Tiliaceae			
Corchorus acutangulus Lam.	7.2	27.4+ 1.2	14.3 + 2.2
C. mimosoidea L.	7.4	23.4 ± 4.1	39.5 ± 6.1
Crotalaria junceae L.	7.2	29.2 ± 4.3	42.4 ± 8.1
Phyllantus niruri L.	6.8	29.2 ± 4.3	33.1 ± 7.2
Eclipta alba Hassk	6.5	37.6 ± 5.2	39.3 ± 6.1
Elephantopus scaber Linn.	7.5	32.5 ± 6.1	41.1 ± 5.2
Euphorbia prostrata L.	7.3	21.0± 6.1	37.2 ± 12.0
Cyperus diffornis Linn.	6.7	11.5 ± 3.3	35.4 ± 6.2
Chloris gayana Kunth	6.4	44.2 ± 4.2	78.2 ± 8.1

32–45 percent was in Euphorbiaceae, Compositae and Graminae. Spore population was not correlated with percent of colonization, except the families of Cyperaceae, Commellinaceae and Graminae. The distributed forest area of 25 plants of their AMF colonization and spore population was assessed (Table 29.2). Here percentage of colonization and spore population more or less correlated with each other, except the families Cyperaceae, Elacocarpaceae, Santalaceae and Guttiferae, families had lower colonization (14–34%), and spores were 48–56 per 50g soil, and therefore isolated fungal spore species were limited (Table 29.3). Recovered spores were more in disturbed areas of adjacent mined soils. In the study, most of the examined rhizospheric soil sample revealed the *Glomus* species was predominantly seen. This indicated that, the development of A-mycorrhiza was more significant

Table 29.2. : Arbuscular Mycorrhizal Status of Undisturbed Sites Bisgod Forest Area Adjacent to Mined Area in Karnataka, India

Family / Scientific name	pH	Percent colonization M ± S.E.	VAM Spore/509.soil M ± S.E.
Menisper maceae			
Coculus Villosus Dc.	7.6	53.3 ± 7.1	103.7+ 8.3
Gutiferaceae			
Calophyllum uightianum Anders.	6.9	34.2+ 4.4	57.4 + 2.2
Malvaceae			
Thespesia macrophylla Blume	6.2	51.4+ 3.1	64.0 + 5.H
Elaeocarpaceae			
Raxb. *Elaeocarpus aristatus*	5.9	33.1+ 5.4	37.1 + 306
Rutaceae			
Toddalia aculeata pers.	6.1	72.5+ 2.2	110.01 + 3.1
Meliaceae			
Toona ciliata Roem.	7.5	32.2+ 1.1	107.0 + 4.3
Cipadessa fntiticcsa			

Blume	6.8	40.4+ 5.2	94.3 + 2.5
Rhamnaceae			
Zizyphus mauritiana Lauck.	5.7	47.7+ 5.4	114.1 +5.5
S.F. Mimosoiaceae			
Acacia catecha willd.	8.2	53.4+ 6.3	102.3 + 3.4
Papilionaceae (S.F.)			
Dalbergia emarginata Roxb	8.4	46.5 + 3.1	72.4 + 5.1
Pterocarpus marsupium Roxb.	7.5	47.4+ 4.4	89.1 + 4.4
Xylia xylocarpa Roxb.	6.5	39.3+ 6.1	93.3 + 5.6
Combretaceae			
T. bellerica Roxb.	7.1	72.4+ 3.5	98.1 +2.3
T. chebula Retz.	6.3	79.5+ 7.2	124.4 + 3.4
T. alata Heyne.	6.8	84.2+ 5.1	120.1 + 5.5
Rubiaceae			
Stephegyne parvifolia Korth	5.8	66.4+ 5.8	93.4 + 7.3
Compositae	7.3	94.2+ 5.2	78 4 + 1.3
Elephantopus seaber Linn.	6.2	95.4+ 8.1	82.1 + 4.2
Eupatorium triplinerve Vahl.			
Apocvnaceae	6.6	63.3+ 6.0	78.3 + 9.4
IVrightia tinctoria R. Br.			
Borginaceae	6.5	24.3+ 5.2	84.3 + 7.1
Cordia macleodii Hook.			
Vcrbenaceae	7.5	49.3+ 7.7	97.3 + 8.4
Tectona grandis Linn.			
Santalaceae	8.1	14.5+ 7.3	56.2 + 5.5
Santalum album Linn.			
Euphorbiaceae	7.4	77.6+ 6.3	97.8 + 5.4
Phyllanthus emblica L.			
Cyperaceae	6.3	13.2+ 6.5	48.3 + 8.4
Scirpus isolepis boeck.			
Graminae			
Brachiaria erneiformis Gricob.	5.4	100.0	117.3 + 3.2

S.F. = Subfamily
M = Mean value, Each value is mean of 12 plants
S.E. = Standard error.

Table 29.3. : AM fungal Species Isolated from Plants Grown on Stock Mined Spoil of Bisgod

AM fungal species	Abbreviation	Gradation		Position in	
Acaulospora					
Acaulospora foveta Trappe & Janons.	a	c	j	h	g
A. Apinosa wal. & Trappe.	b	j	g	1	d
Gigaspora					
Gigaspora decipiens Hall. & Abbott.	d	g	k	a	e
Glomus					
Glomus citricolum Tung & Zang.	f	g	d	e	a
G. etinicatum Becker & Gerd.	g	h	j	k	e
G. fasciculatum Walk. & Koske.	h	g	b	h	b
G. mosseae Gerd & Trappe.	k	f	a	n	c
Scutellospora					
Scutellospora fiilgida Koske & Walker.	1	g	e	d	k
S. weresubiae Koske & Walker	n	e	i	f	g

in natural vegetation (Table 29.4). Growth response studies showed the *Cassia sirceae* have shown significant growth, biomass production and spore population (Table 29.5). The percent of A-mycorrhizal levels at the end of final harvest was more significant in mycorrhizal plants, it is mainly, because mycorrhiza colonized plants which received indigenous AMF inoculum. Control plants did not exhibited mycorrhizal colonization nor spore population. There was favourable increased P level concentration in shoots.

Low level and sporadic nature of Arbuscular mycorrhizae on mined spoils were less, comparatively those undisturbed

Table 29.4. : AM Fungal Species Isolated from Rhizospheric Soils of C. *Siraceae* Amended by Mined Spoils and Forest Soil Plus VAM

AM fungal species	Abbre-viation	Position in gradation			
		1	2	3	4
Acaulospora					
Acaulospora dilatata Morton.	a	m	g	h	k
A. longula Spain & Schenck.	b	k	f	n	0
A. mellea Spain & Schenck	c	e	b	f	d
A. sporocarpia Berch.	d	i	k	h	n
Gigaspora					
Gigarpora decipiens Hall & Abbott.	e	h	1	c	f
G. fragile Trappe & Gerd.	f	e	k	0	g
Glomus					
Glomus collosum Sieverding.	g	d	a	1	h
G. dimorphicum Mortin & walker.	h	e	j	k	n
G. fasciculatum (Thaxter) Gerd. & Trappe. Wai. & Koske.	i	0	1	g	a
G. macrocarpum Tul. & Tul.	j	e	m	h	j
G. mosseae (Nicol. & Gerd.) Gerd. & Trappe.	k	d	k	n	e
G. multicaule Gerd. & Bakshi.	e	k	m	e	d
Scuteliospora					
Scutellospora nigra (Redhead) Wai. & Sanders.	m	e	d	b	k
S. rubiformis Gerd. & Trappe	n	0	a	i	m
S. Werrucosa (Kos. & Wai.) Wai. & Sanders.	0	g	n	k	a

Table 29.5 : Effect of *Glomus fasciculatum* on *cassia siraceae* sw. by Proportional Emendation of Mined Spoils and Forest Soils

Treatment in months	plant height (cm)	Shoot fresh (g)	Shoot dry (g)	Root fresh (g)	Root dry wt.	Root/ shoot ratio	% of VAM coloni-zation	No. of spores/ 50 g. soil	Total shoot dry wt. (g)	'P' (%) in
NM	6.9 + 4.3	4.1 + 2.1	1.5+0.3	1.3+0.1	0+0.52	0.346	2.02	0.05		
3 M : 1:0.25	10.4+8.2	5.4+4.3	2.1+2.0	1.8+2.1	0.0+0.66	0.314	19.3+1.2	19.1+2.1	2.76	0.17
6M : 1:0.50	17.4+4.4	6.7+5.1	2.4+3.2	2.1 + 1.4	0.0+0.97	0.404	24.7+4.4	56.3+56	3.37	0.19
9M : 1: 0.75	20.3+6.2	14.5+3.5	7.4+3.2	6.3+0.3	1.3+5.2	0.176	69.6+3.3	98.2 + 5.1	8.70	0.21
12M : 1:1	29.7+7.5	6.3+7.4	7.8+6.8	7.5+1.6	2.6+4.6	0.333	87.3+4.6	107.3 + 3.6	10.40	0.21

Each value is the means of 3 triplicates
M.S. : Mined spoil. F.S. : Forest soil
3 M to 12 M = 3 to 12 months
0.25, 0.50, 0.75, 1.0 = Mine spoil range, 1 = forest soils

sites. There are some questions as the means by which A-fungal inoculum is dispersed Trappe (1981) and the distance over which effective dispersal can occur (Drighton, 1991). Mukherjee and Sharma, 1992; Lakshman, 1992) showed that worms, ants and birds could be agents for dispersal of spores. It is doubt whether this type of dispersal is a major contributor to long distance transport. It would likely be more important for within habitat movement of inoculum. The distance from mine sites to potential Arbuscular fungal inoculum pools and the rates of dispersal and subsequent recolonization would likely be important factors, that need to be considered in the planning of a reclamation programmes. Success or failure of a revegetation strategy may in part be determined by the immigration rates of Arbuscular mycorrhizal fungi in to these disturbed habitats. Revees *et al.*, (1979) suggested that the slow rate of revegetation of distributed lands in the semi arid and west coastal region was due to the elimination or reduction in the number of Arbuscular fungal propagules.

In mine soils where Arbuscular mycorrhizal inoculum levels are low or Arbuscular mycorrhizal fungi present, as a result of reclamation practices, non-mycorrhizal plants species that colonize the sites could have competitive advantage over mycorrhizal dependent species, especially if nutrient become limiting (Loree *et al.*, 1984; Singh, 1996). Arbuscular mycorrhizal colonization of the grass highest in the plot with severest pollution damage and lowest in the least damaged plot (Sambandon *et al.*, 1992; Johnson, 1993). If the rapid establishment of Arbuscular mycorrhizae with plants grown on mined spoils is to be a major concern of a reclamation strategy, attention will have to be given to determining initial Arbuscular mycorrhizal inoculum densities in the mined spoils and the application rates of inorganic and organic amendments which do not suppress Arbuscular mycorrhizal development but allow for optimal plant growth and survival.

A great many experimental factors probably influence the colonization and need to be studied, some examples are, toxicity, soil compaction and local anaerobiosis, biological interactions. In addition to chemical analysis of soils of both mycorrhizal, and non- mycorrhizal plants are warranted. It may be necessary to add inoculum to the site to ensure the rapid development of Arbuscular mycorrhizae, the type of amendment applied to a mine spoil could determine such parameters as the recovery sequence, the rate of recovery and ultimately the level at which ecosystem functioning become stabilized. By introducing mycorrhizal plants in degraded areas reclamation and revegetation practices. However, the present investigation recommended detailed studies on physiochemical properties of mined, forest and soil used for pot experiments.

REFERENCES

Chapperton, M.J. and Parkinson. D. (1990). The effect of So on the vesicular-arbuscular mycorrhizae associated with a submontane mixed grass prairie in Albertay Canada *Can J. Bot.* 68 (8) : 1946–1950.

Danienson, R.M. (1985). Mycorrtriza and reclamation of stressed terrestrial environments. In : Soil reclamation process; Microbial analyses and applications (Ed. by R.L. Tata III & D.A. Klein), pp. 173–201. Marcel-Dekker, Inc., New York and Basel.

Davidson, D.E. and Christenson. M. (1977). Root micro fungal and mycorrhizal associations in a shortgrass prairie. In the Below ground ecosystem: A synthesis of plant associated processes (Ed. by J.K. Marshall), pp. 279–287. Colorado state University, Range Science Department Science Series 26. ft. collins.

Drighton, J. (1991). Acquisition of nutrients from organic resources by mycorrhizal autotrophic plants. Experientia, 47:362–369.

Gerdmann, J.W. and Nicolson. T.H. (1963). Spores of mycorrhizal Endogone species extracted from soil by wet sieving and decanting technique. *Trans. Brit. Mycel. Soc.* 46 : 234–235

Giovannetti, M. and Mosse, B. (1980). An evaluation for measuring VAM infection in roots. New phytologists. Vol. 84. Pp:489–500.

Jasper, D.A, Robson. A.D and Abbott. L.K. (1988). Revegetation of an iron-ore mine-nutrient requirements for plant growth and the potential role of vesicular -arbuscular (VA) mycorrhizal fungi. *Aust. J. Soil. Res.* 26 : 497–507.

Jasper, D.A., Abbott. L.K, and Robson. A.D. (1989). Acacia respond to additions of phosphorus and to inoculation with VA-mycorrhizal fungi in sails stock piled during mineral sand mining. *Plant and soil.* 115 : 99–208.

Johnson, N.C. (1993). Can fertilization of soil select less mutualistic mycorrhizae. *Ecol. appli.* 3 : 749–757

Lakshman, H. C. (1997). Screening of AM fungi in mined area and its utilization in reclamation programmes. Journal of current science. 109:1951–199.

Lakshman, H.C. (1992). Spore dispersal of Endogonaceae by worms, wasps and Indian Domestic fowls. *J. Natcon.* 42 : 105–109.

Loree, M.A.J. and Williman, S.S.E. (1984). Vesicular-arbuscular mycorrhizae and severe land disturbance In : VA-mycorrhizal and Redamation of Arid and Semiarid lands (Ed. by S.E. Willian and H.F. Allen) pp. 1–14 University of Wyoming Agriculture Experimental Station; SA 261.

Loree, M.A.J. and Williams, S.E. (1987). Colonization of western wheat grass *Agropyron smithii* Rydb. by vescular arbuscular mycorrhizal fungi during the revegetation of a surface mine. *New Phytol.* 105 : 735–744

Miller. R.M. (1979). Some occurrence of vesicular-arbuscular mycorrhiza in natural and disturbed ecosystems of the Red Desert. *Can. J. Bot.* 57 : 610–623.

Mukerji, K. G. and Sharma, M. (1992). Use of VAM fungi in afforestation in semi-arid zones of India. Proc II Asian Mycorrhizal Cong., Chaingmai, Thailand (in press).

Olsen. S.R., Ode, C.V., Watanabe F.S. and Dean L.A. (1954). Estimation of available phosphorus in soils by extraction with sodium bicarbonate, Circ. 936. USDA. Sap. of documents of US. Govt. printing office, Washington 25, D.D. pp. 19.

Read, D. J., Koucheki H. K. and Hodgson J. (1976). Veiscular arbuscular mycorrhiza in natural vegetation system I. The occurrence of infection, New phytol 77: 641–653.

Reeves, F. B., Wagner. D., Moorman. T. and Kiel. J (1979). The role of ensdomycorrhizae in revegtation practices in semi-arid west. I. a. comparison of incidence of mycorrhizae in severely disturbed – natural environment Amer. J. Bot. 66:6–13.

Sambandon, K., Kannan, K. and Raman, N. (1992). Distribution of Vesicular arbuscular mycorrhizal fungi in heavy metal polluted soils of Tamil Nadu. India *J. Env. Biol.* 13 (2) : 159–167.

Schenck. N.C. and Perez, Y. (1990). Manual for the identification of VA-mycorrhizal fungi, third edition, synergistic publications, Gainesville, U.S.A. Pp. 286

Singh, S. (1996). Effect of atmospheric pollution on Mycorrhizal. *Mycorrhiza News* 8 (1) : 1–7.

Skujins, J. and Allen M.F. (1996). Use of mycorrhizae for land rehabilitation. *Uircen J.* 2 : 161–176.

Trappe, J.M. (1981). Mycorrhizae and productivity. Advances in food processing systems for and semi-arid lands. Edited by J. Manassan.

Visser, S., Brishey, E.J. (1985). Academic press. Inc. New York. Pp. 581–599.

Zak, J.C. and Parkinson, D. : Effect of Surface amendation of two mine spoils in Alberta, Canada, on Vesioular-arbuscular mycorrhizal development of slender wheat grass : a 4 year study. *Can. J. Bot.* 61 : 798–803 (1983).

Index

I

J

K

L

M

N

O

P

R

S